Genetic Engineering: Emerging Concepts and Technologies

Genetic Engineering: Emerging Concepts and Technologies

Edited by Patrick Faraday

SYRAWOOD
PUBLISHING HOUSE
New York

Published by Syrawood Publishing House,
750 Third Avenue, 9th Floor,
New York, NY 10017, USA
www.syrawoodpublishinghouse.com

Genetic Engineering: Emerging Concepts and Technologies
Edited by Patrick Faraday

International Standard Book Number: 978-1-68286-516-3 (Hardback)

Cataloging-in-Publication Data

Genetic engineering : emerging concepts and technologies / edited by Patrick Faraday.
p. cm.
Includes bibliographical references and index.
ISBN 978-1-68286-516-3
1. Genetic engineering. 2. Genetics. 3. Biotechnology. I. Faraday, Patrick.
QH442 .G46 2018
660.65--dc23

Printed in China.

TABLE OF CONTENTS

PREFACE

Genetic engineering is the modification of an organism's genome by using biotechnology. It is used to modify genes and to change their genetic trait. There are a number of techniques that are used in genetic engineering, such as gene isolation, gene targeting and regeneration. These techniques are applied in a number of fields including industrial biotechnology and medicine. As this field is emerging at a rapid pace, the contents of this book will help the readers understand the modern concepts and applications of the subject. In this book, using case studies and examples; constant effort has been made to make the understanding of the difficult concepts of genetic engineering as easy and informative as possible, for the readers.

The information contained in this book is the result of intensive hard work done by researchers in this field. All due efforts have been made to make this book serve as a complete guiding source for students and researchers. The topics in this book have been comprehensively explained to help readers understand the growing trends in the field.

I would like to thank the entire group of writers who made sincere efforts in this book and my family who supported me in my efforts of working on this book. I take this opportunity to thank all those who have been a guiding force throughout my life.

Editor

1

Expression of Animal Anti-Apoptotic Gene Ced-9 Enhances Tolerance during *Glycine max* L.–*Bradyrhizobium japonicum* Interaction under Saline Stress but Reduces Nodule Formation

Germán Robert[1,2], Nacira Muñoz[1,2], Mariana Melchiorre[1], Federico Sánchez[3], Ramiro Lascano[1,2]*

1 Instituto de Fisiología y Recursos Genéticos Vegetales, Centro de Investigaciones Agropecuarias-INTA, Córdoba, Argentina, **2** Cátedra de Fisiología Vegetal, Facultad de Ciencias Exactas Físicas y Naturales, Universidad Nacional de Córdoba, Córdoba, Argentina, **3** Departamento de Biología Molecular de Plantas, Instituto de Biotecnología, Universidad Nacional Autónoma de México, Cuernavaca, Morelos, México

Abstract

The mechanisms by which the expression of animal cell death suppressors in economically important plants conferred enhanced stress tolerance are not fully understood. In the present work, the effect of expression of animal antiapoptotic gene Ced-9 in soybean hairy roots was evaluated under root hairs and hairy roots death-inducing stress conditions given by i) *Bradyrhizobium japonicum* inoculation in presence of 50 mM NaCl, and ii) severe salt stress (150 mM NaCl), for 30 min and 3 h, respectively. We have determined that root hairs death induced by inoculation in presence of 50 mM NaCl showed characteristics of ordered process, with increased ROS generation, MDA and ATP levels, whereas the cell death induced by 150 mM NaCl treatment showed non-ordered or necrotic-like characteristics. The expression of Ced-9 inhibited or at least delayed root hairs death under these treatments. Hairy roots expressing Ced-9 had better homeostasis maintenance, preventing potassium release; increasing the ATP levels and controlling the oxidative damage avoiding the increase of reactive oxygen species production. Even when our results demonstrate a positive effect of animal cell death suppressors in plant cell ionic and redox homeostasis under cell death-inducing conditions, its expression, contrary to expectations, drastically inhibited nodule formation even under control conditions.

Editor: Jin-Song Zhang, Institute of Genetics and Developmental Biology, Chinese Academy of Sciences, China

Funding: This work was supported by grants from Agencia de Promoción Científica y Tecnológica, Argentina; Instituto Nacional de Tecnología Agropecuaria (INTA); and Scientific and Technological Cooperation Program between Ministerio de Ciencia, Tecnología e Innovación Productiva de la República Argentina (MINCYT), and CONACYT de México. The funders had no role in study design, data collection and analysis, decision to publish, or preparation of the manuscript.

Competing Interests: The authors have declared that no competing interests exist.

* Email: lascano.ramiro@inta.gob.ar

Introduction

Programmed cell death (PCD) is a genetically regulated process of cellular suicide and is well known to play a fundamental role in a wide variety of developmental and physiological functions in animals, plants, and fungi [1–3]. A key feature of PCD is the requirement of energy to the control and execution of the death in an orderly manner [4,5]. In plants, PCD is essential for cell homeostasis and specialization, playing an important role in plant development and adaptive responses to different stress conditions such as salinity, cold stress, hypoxia and pathogen attack [6–8].

In metazoans, from humans to *Caenorhabditis elegans*, the central regulators of PCD are well characterized and conserved involving pro- and anti-apoptotic protein such as APAF-1/CED-4 and BCL-2/CED-9, and executing protein family caspasas/CED-3 [9,10]. Interestingly, although these regulators are absent in the genomes of plants and yeast, the effects of animal pro- and anti-apoptotic proteins has been studied in transgenic plants [11–18]. According to the localization of these heterologous proteins in plant cells, it is proposed that cell death suppressors contribute to maintain the organelles homeostasis preventing the generation/release of death signals, similar to what occurs in animals [11,13]. However, there are limited data regarding the mechanisms through which the animal cell death suppressors modulate the plant physiology.

Remarkably, the expression of PCD suppressors in plants result in agronomical beneficial features such as improved tolerance to a variety of biotic and abiotic stresses [11–16]. Increased the biological nitrogen fixation in legumes is a main objective for the agriculture, and different strategies had been explored towards this objective. During the natural or stress induced senescence, which involve cell death processes, the biological nitrogen fixation metabolism is impaired, affecting both quality and quantity of legume yields [19,20]. Therefore, the development of strategies to increase the tolerance to a variety of stresses is highly relevant. To the best of our knowledge, the effect of animal PCD suppressor has not been tested in legumes.

Soybean (*Glycine max* L.) culture is strongly affected by drought and salinity [21,22]. Likewise, the soybean-rhizobia symbiotic interaction process is also severely affected by stress conditions [19,20]. Our group has reported that salt stress, but not osmotic stress, negatively affect the early stages of the *Glycine max* L.-*Bradyrhizobium japonicum* interaction such as root hairs deformations and viability [23], and how these short-term treatments affect nodule formation [24]. In this context, two root hairs death-inducing conditions were identified: sub lethal salt stress treatments combined with *B. japonicum* inoculation (inoculated 50 mM NaCl) and severe salt stress (150 mM NaCl). Interestingly, in root hairs under sub lethal salt stress conditions (50 mM NaCl), the symbiotic interaction with *B. japonicum* induced a sustained increase of intracellular reactive oxygen species (ROS) levels in a similar pattern to that observed in response to pathogenic elicitors [23,25]. In contrast, under 150 mM NaCl, the intracellular ROS production decreased from the beginning of treatment independently of the presence of the symbiont [23].

The aim of the present work was to evaluate if the expression of Ced-9 from *Caenorhabditis elegans* could improve the stress tolerance of legume-rizobia symbiotic interaction and the biological nitrogen fixation process. Transgenic soybean-hairy roots expressing Ced-9, obtained with Agrobacterium rhizogenes [26,27], were subjected to the above described cell death-inducing conditions, in order to evaluate root cell viability, redox and ionic parameters associated and nodule development.

Results

Root hairs death-inducing stress conditions

Two-day soybean seedlings were subjected 30 min under previously reported root hairs death-inducing conditions. Moderate salt stress treatments (50 mM NaCl) combined with *B. japonicum* (inoculated 50 mM NaCl) and severe salt stress (150 mM NaCl) were the cell death conditions for root hairs [23]. However, under these root hairs death-inducing conditions the roots were kept alive. These results were observed by Evans Blue staining and DNA degradation analysis (Fig. 1A, 1B and 1C). Root hairs DNA degradation was observed while roots maintained the chromatin integrity (Fig. 1C).

Malondialdehyde content (MDA), which is an intermediary metabolite of lipid peroxidation used as oxidative stress marker, was measured in root hairs. The MDA level increased in inoculated and inoculated 50 mM NaCl treatments, whereas non significant differences were observed in salt stress alone (50 mM NaCl and 150 mM NaCl) respect to control (Fig. 2A). Likewise, with the purpose to discriminate ordered or non-ordered death processes, the levels of adenosine-5′-triphosphate (ATP) were quantified [4,28] in root hairs subjected to root hairs death-inducing conditions. No significant changes were observed at 50 mM NaCl (Fig. 2B). Under inoculated 50 mM NaCl treatment, root hairs had increased ATP levels as well as under inoculated control treatment (Fig. 2B). Conversely, under 150 mM NaCl treatment, root hairs showed a slight, but not significant decrease in ATP levels respect to the control (Fig. 2B).

CED-9 expression ameliorates root hairs death-inducing conditions effects

In order to evaluate the effect of the animal cell death suppressor, wild type and CED-9 transgenic hairy roots were obtained by infection with *A. rhizogenes* K599 strain. It should be pointed out that the differentiation and development of hairy roots were conducted without antibiotic selection, thus the resulting K599-CED9 composite plants contained both transgenic and wild type hairy roots (Fig. S1A). The expression level of the transgene in hairy roots was tested by qPCR using Ced-9 specific-derived primers (Fig. S1B) and the identity of qPCR product was verified by nucleotide sequencing. The K599-CED9 hairy roots were shorter than wild type hairy roots obtained by infection with untransformed *A. rhizogenes* (K599-empty) (Fig. S2A).

Roots hairs nuclear morphology was evaluated in K599-empty and K599-CED9 hairy roots incubated 30 min under control, or root hair death-inducing conditions (Fig. 3; Fig. S3). The nuclear morphology of root hairs was evaluated by acridine orange and ethidium bromide (AO/EB) staining and observed with confocal microscopy. Acridine orange is a dual-fluorescence dye that interacts with DNA and RNA, and it also serves as a pH indicator.

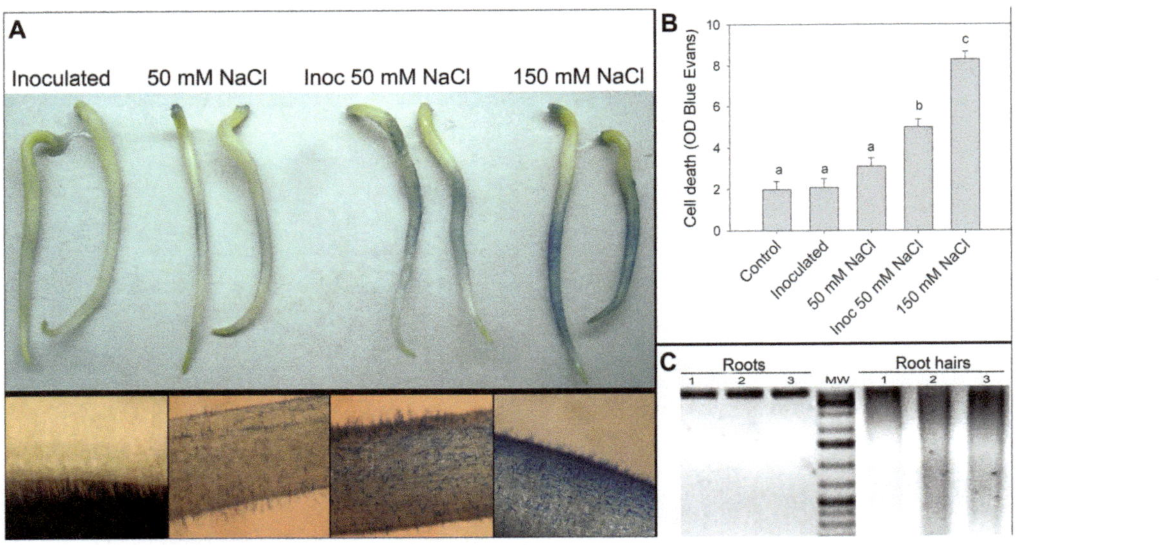

Figure 1. Root hairs death inducing stress conditions. Two-day soybean seedlings were subjected 30 min to control, inoculated with *B. japonicum* (inoculated), inoculated with *B. japonicum* in presence of 50 mM NaCl (inoc 50 mM NaCl), and 150 mM NaCl conditions. A) Evans blue staining of roots showing loss of membrane integrity (upper image) and detail of root hairs staining (image below). B) Evans Blue stain intensity was measured by the image analyzer program Optimas 6.1. C) DNA degradation in a representative sample pool of root and root hairs. 1: Control, 2: inoculated 50 mM NaCl, 3: 150 mM NaCl. 2 µg of DNA were loaded on a 2% TAE agarose gel and stained with ethidium bromide. Data are means ± SE of four independent experiments (two roots per experiment). Different letters indicate significant differences between treatments ($p<0.05$, DGC test).

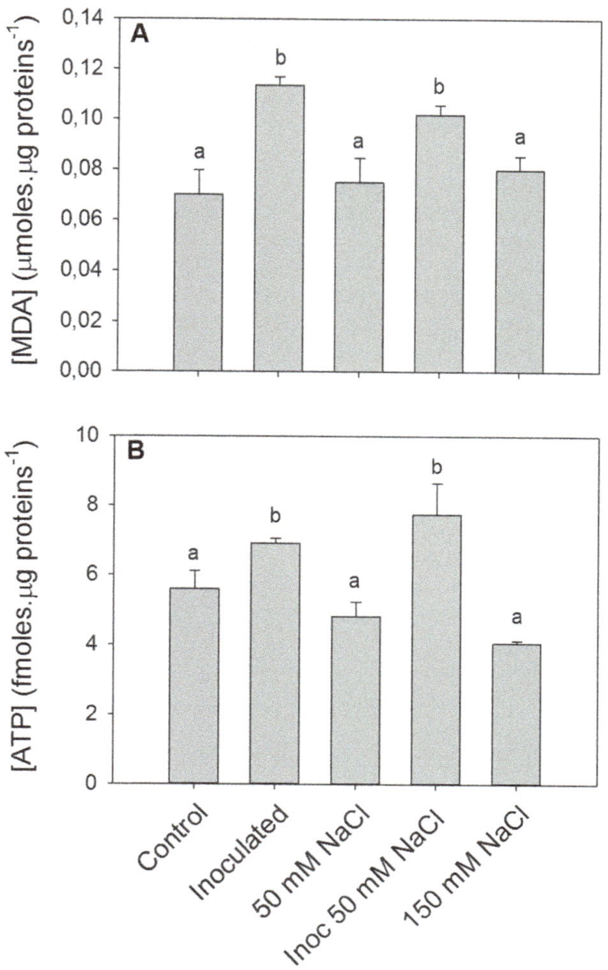

Figure 2. Evaluation of malondialdehyde (MDA) and Adenosine-5′-triphosphate (ATP) levels in root hairs. Two-day soybean seedlings were subjected 30 min to control, inoculated with *B. japonicum* (inoculated), 50 mM NaCl, inoculated with *B. japonicum* in presence of 50 mM NaCl (inoc 50 mM NaCl), and 150 mM NaCl conditions. Then, root hairs were extracted from roots and MDA and ATP content were evaluated. Data are means ± SE of four (MDA) and three (ATP) independent experiments (root hairs from 200 roots per experiment). Different letters indicate significant differences between treatments ($p<0.05$, DGC test).

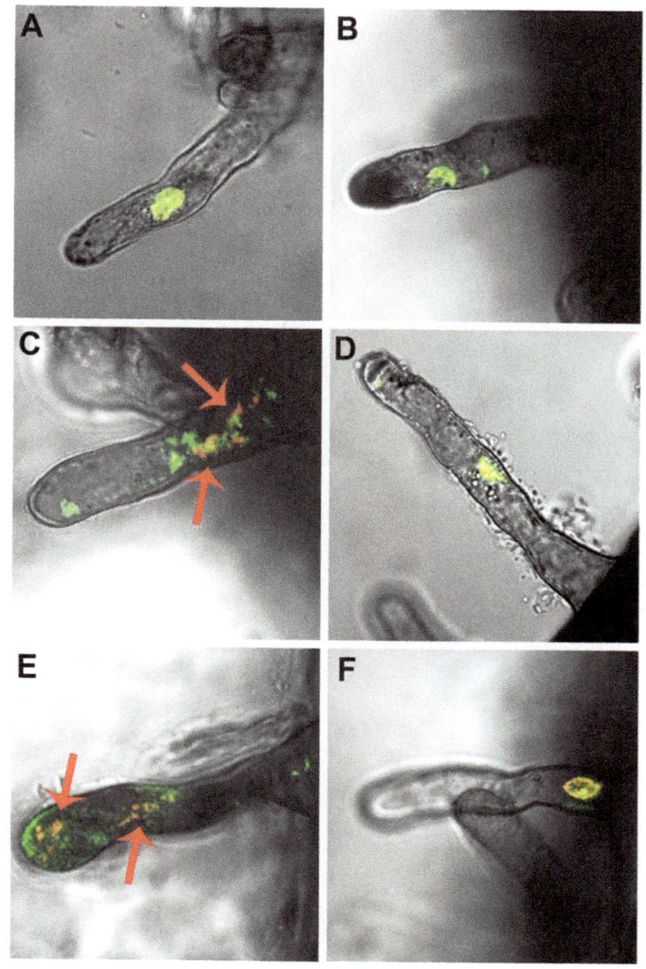

Figure 3. Ced-9 expression ameliorates root hairs death-inducing conditions effects. K599-empty (A, C, E) and K599-CED9 (B, D, F) hairy roots were subjected 30 min under control (A and B), inoculated with *B. japonicum* in presence of 50 mM NaCl (inoc 50 mM NaCl) (C and D), and 150 mM NaCl (E and F). Nuclear morphology of root hairs was evaluated. Arrows indicates nuclear fragmentation. Images were taken with a Zeiss confocal microscope. The excitation was performed simultaneously at 488 nm and emission filter BP 500–530 IR and BP 565–615 IR for AO and EtBr, respectively (image overlay).

Ethidium bromide binds to DNA by intercalating between the bases, but it is membrane impermeant so generally excluded from viable cells [28]. Hence, roots were incubated 30 min with AO/EtBr to allow entry of the probes. The nuclei of root hairs in the control treatments exhibited an orthodox conformation, and similar size and shape between K599-empty and K599-CED9 hairy roots (Fig. S3). Under root hair death-inducing conditions, root hairs of K599-empty hairy roots showed significantly higher nuclear fragmentation than root hairs of K599-CED9 hairy roots (Fig. 3). Furthermore, an increase in AO staining was observed in K599-empty hairy roots particularly under 150 mM NaCl conditions (Fig. 3).

CED-9 effects on root ion and redox homeostasis under hairy roots death-inducing conditions

Whole hairy roots were used to perform the biochemical determinations due to the low yield of root hairs in hairy roots. However, as previously observed, root hair death-inducing conditions did not induce roots death. Then, treatment time was adjusted to 3 h, when positive Evans Blue staining (Fig. S2A), but not DNA degradation (Fig. S2B) could be observed. This result indicates an early stage of root cell death. It also was noted that K599-CED9 hairy roots showed more membrane selectivity than K599-empty hairy roots (Fig. S2A).

K599-empty hairy roots showed a dramatic decrease in potassium levels under 150 mM NaCl treatment, whereas in K599-CED9 hairy roots this decrease was much less pronounced. Nevertheless, no significant differences were observed in potassium content between control and inoculated 50 mM NaCl treatments in any transgenic or wild type genotype (Fig. 4A). However, sodium content in hairy roots increased in a dose dependent manner in both K599-empty and K599-CED9 hairy roots (Fig. 4B). Moreover, no significant differences in the concentration of sodium were observed between K599-empty and K599-CED9 in any of the treatments performed (Fig. 4B). Likewise, calcium concentration decreased markedly in K599-empty hairy roots only under 150 mM NaCl treatment (Fig. 4C), whereas the calcium

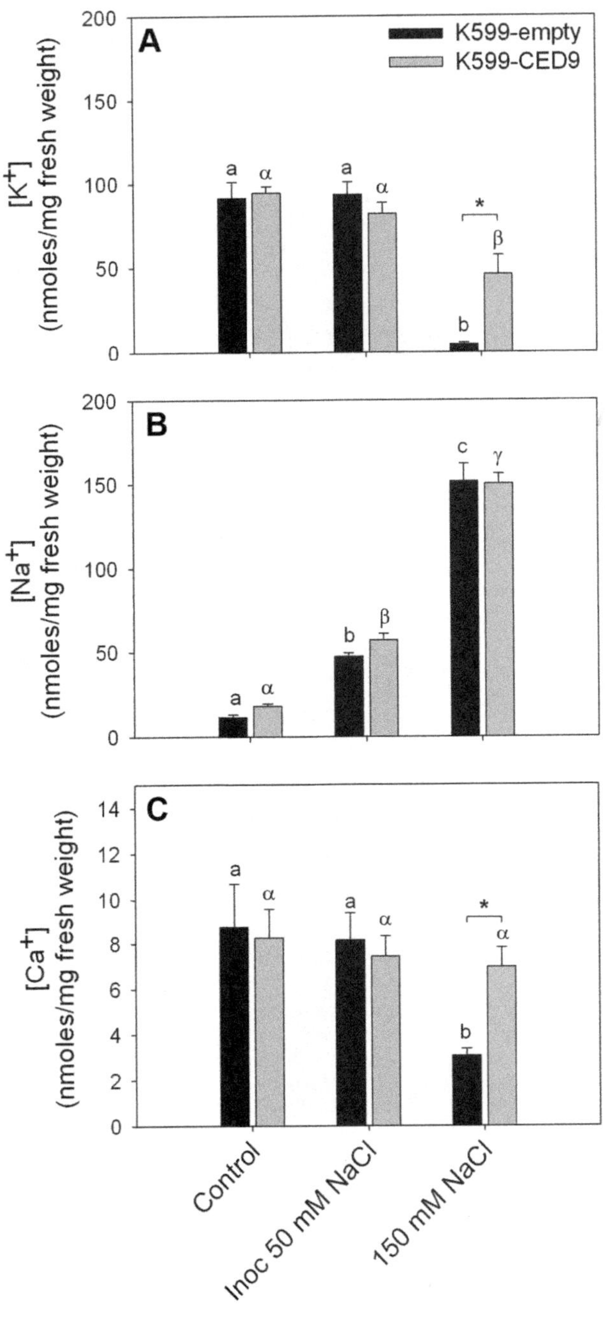

Figure 4. Ced-9 affects ion relationship during hairy roots death-inducing stress conditions. K599-empty (dark bars) and K599-CED9 (grey bars) hairy roots were subjected 3 h to control, inoculated with *B. japonicum* in presence of 50 mM NaCl (inoc 50 mM NaCl), and 150 mM NaCl conditions, and then potassium, sodium and calcium ions were quantified by high pressure liquid chromatography. Data are means ± SE of five independent hairy roots. Different Latin and Greek letters indicate significant differences between treatments in K599-empty and K599-CED9 hairy roots, respectively ($p<0.05$, DGC test). Asterisks indicate significant differences between hairy roots genotypes ($p<0.05$, DGC test).

content in K599-CED-9 hairy roots did not change in stress treatment respect to control (Fig. 4C).

In order to characterize changes in the redox state during hairy roots death-inducing conditions, the levels of MDA, hydrogen peroxide (H_2O_2), antioxidant capacity, ascorbic acid ratio (reduced form/total) and ATP in K599-empty and K599-CED-9 hairy root were quantified (Fig. 5 and Fig. 6). MDA content increased in K599-empty hairy roots under stress conditions (Fig. 5A). Interestingly, K599-CED9 hairy roots did not show significant differences under any stress treatments regarding control (Fig. 5A). In accordance, H_2O_2 levels increased in K599-empty hairy roots under inoculated 50 mM NaCl and 150 mM NaCl treatments, whereas K599-CED9 hairy roots did not show significant differences in H_2O_2 content between control and stress treatments (Fig. 5B). Moreover, differences in H_2O_2 content between K599-empty and K599-CED9 were only observed under 150 mM NaCl treatment (Fig 5B). The Ferric Reducing Ability of Plasma (FRAP) assay showed raises in antioxidant capacity in K599-empty hairy roots under 150 mM NaCl (Fig. 5C), although the ascorbic acid ratio was reduced (Fig. 5D). Under roots death-inducing conditions, K599-CED9 hairy roots did not show differences in both antioxidant capacity and ascorbic acid ratio respect to control (Fig. 5C and 5D).

ATP levels increased in K599-empty hairy roots inoculated in presence of 50 mM NaCl treatment while no significant difference were observed under 150 mM NaCl (Fig. 6), similarly to the results observed in root hairs (Fig. 2). In contrast, K599-CED9 hairy roots had increased ATP content under all stress conditions, including 150 mM NaCl treatment, showing significant differences respect to K599-empty hairy roots (Fig. 6).

Ced-9 expression inhibit nodule formation in hairy roots

Strikingly, K599-CED9 composite plants showed a reduction of 60% in the number of nodules compared to K599-empty composite plants under control conditions (Fig. 7A and 7B). As it was previously mentioned, each soybean K599-CED9 composited plants developed both transgenic K599-CED9 and non-transgenic hairy roots (Fig. S1). In the nodulation assay, K599-CED9 composite plants had both hairy roots with and without nodules. Hairy roots from K599-CED9 composite plants were separated according nodulated and non nodulated and the presence of Ced-9 transgene was examined by PCR. This experiment clearly showed that in K599-CED9 hairy roots nodulation was dramatically inhibited, while in the same composite plant, those non-transgenic hairy roots were nodulated (Fig. 7C).

Discussion

Soybean-rhizobia symbiotic interaction is severely affected by salt stress, showing a reduction on number and weight of nodules in plants salinized with 26 mM NaCl [19,20]. Our group had studied the effects of salt stress conditions on early events of *Glycine max* L.-*B. japonicum* symbiotic interaction, where undescribed root hairs death-inducing conditions were identified: sub lethal salt stress treatments combined with *B. japonicum* (inoculated 50 mM NaCl) and severe salt stress (150 mM NaCl) (Figure 1). During the early events of symbiotic interaction, a fast and transient increase of intracellular ROS generation take place in root hairs [23,25], whereas a sustained ROS production was reported when the symbiotic interaction occurred under 50 mM NaCl [23]. A similar root hair ROS kinetic was observed in response to pathogenic elicitors [25]. In contrast, under 150 mM NaCl conditions, intracellular ROS production diminished from the beginning of treatment [23]. Hence, the initial hypothesis was that the expression of anti-apoptotic proteins from animals, which have no homologues identified in plants, modulates redox homeostasis and delays senescence and death processes of the plant-symbiont system in legumes under stress conditions.

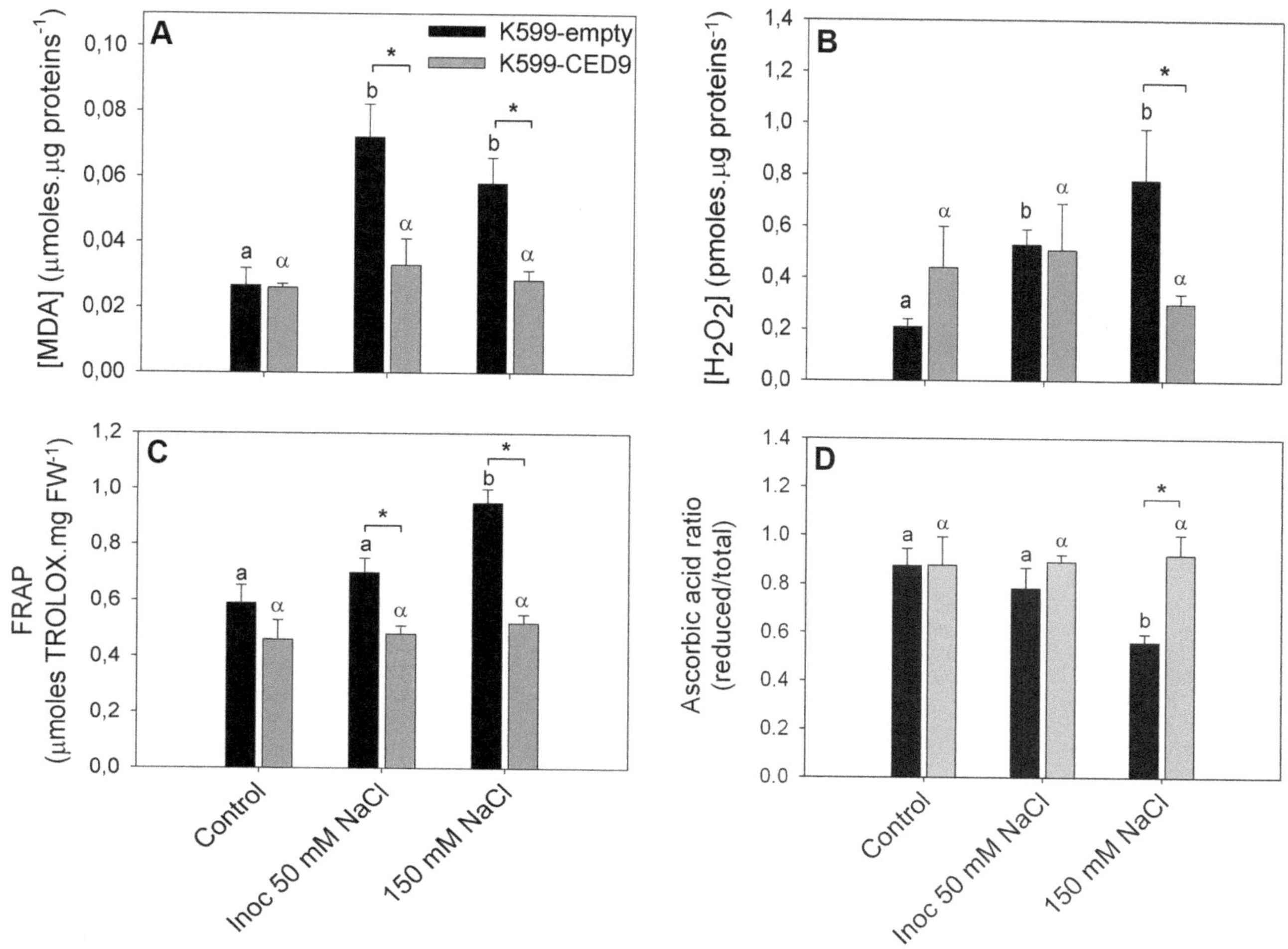

Figure 5. Ced-9 expression prevents ROS generation under hairy roots death-inducing stress conditions. K599-empty (dark bars) and K599-CED9 (grey bars) hairy roots were subjected 3 h under control, inoculated with *B. japonicum* in presence of 50 mM NaCl (inoc 50 mM NaCl), and 150 mM NaCl conditions, and redox parameters were evaluated. A) MDA content, B) H_2O_2 content, C) FRAP (Ferric Reducing Ability of Plasma) assay, D) Reduced/total ascorbic acid ratio. Data are means ± SE from five independent hairy roots. Different Latin and Greek letters indicate significant differences between treatments in K599-empty and K599-CED9 hairy roots, respectively ($p<0.05$, DGC test). Asterisks indicate significant differences between hairy roots genotypes ($p<0.05$, DGC test).

First at all, in this work we characterized these two root hairs death-inducing conditions. There are two main ways to execute cell death: ordered (programmed-like) and non-ordered (necrosis). The ATP level is a fine parameter to distinguish ordered from non-ordered cell death types in mammalian cells [4,5]. During apoptosis, ATP has to be maintained high to allow the formation of the apoptosome [29,30], while ATP depletion has been observed during necrosis [31–33]. However, this observation has not been clearly observed in plants [34–37]. Casolo et al [38] found that in soybean cell cultures, low H_2O_2 concentration induces PCD, which is accompanied by a slight decrease in ATP. In addition, ATP depletion after PCD induction in *A. thaliana* [39] and tobacco BY-2 cells [40] has also been reported. It has been reported that environmental stimuli can produce different types of cell death depending on the stimulus intensity and the ATP availability within the cell [41]. Here, we have determined that root hairs death induced by inoculation in presence of 50 mM NaCl showed characteristics of ordered-process, with increased ROS generation, MDA and ATP levels, whereas the cell death induced by 150 mM NaCl treatment showed non-ordered or necrotic-like characteristics, like decreases in ROS production and ATP levels (Figure 2). Furthermore, the differences observed in Evans Blue staining between these death-inducing treatments (Figure 1A) also indicates the differences in the stress intensity which would lead to the execution of ordered-death or necrosis-like processes. Moreover, the increased MDA and ATP levels observed in control inoculated root hairs (Figure 2) would be due to the increased metabolic activity during early responses of the symbiotic interaction [25,42,43].

The expression of cell death suppressor Ced-9 from *C. elegans* inhibited or at least delayed cell death under root hairs death-inducing conditions (Figure 3). Furthermore, an increase in AO staining was observed especially under 150 mM NaCl conditions, which would indicate cellular acidification. Interestingly, the expression of Ced-9 affected both ordered and necrotic-like death events in root hairs (Figure 3), and it has also been documented in animal systems [44–46], suggesting similar functionality level between the components of the mechanisms of cell death in plants and animals. However, few works have evaluated homeostatic and physiological parameters in transgenic plants in order to understand the effects of the expression of Ced9. Shabala and coworkers [16] have demonstrated that the expression of Ced-9 delays the onset of leaf senescence symptoms under salt and oxidative stress conditions altering the flow patterns of K^+ and H^+ across the

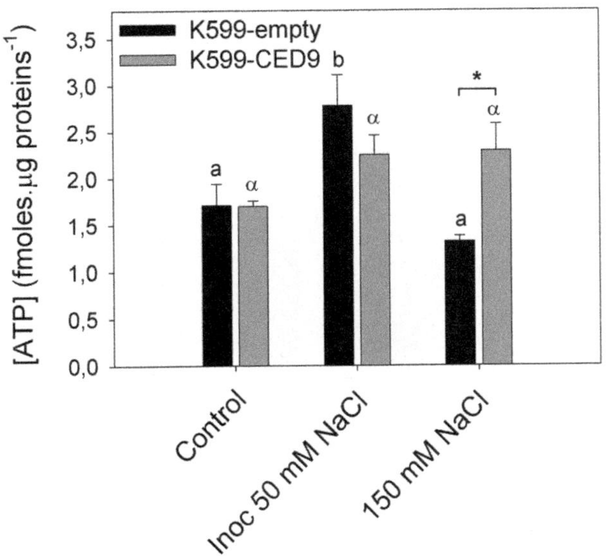

Figure 6. Effects of Ced-9 expression on adenosine-5′-triphosphate (ATP) levels in control and stressed hairy roots. K599-empty (dark bars) and K599-CED9 (grey bars) hairy roots were subjected 3 h to control, inoculated with *B. japonicum* in presence of 50 mM NaCl (inoc 50 mM NaCl), and 150 mM NaCl conditions and ATP content was evaluated. Data are means ± SE of four independent hairy roots. Different Latin and Greek letters indicate significant differences between treatments in K599-empty and K599-CED9 hairy roots, respectively ($p<0.05$, DGC test). Asterisks indicate significant differences between hairy roots genotypes ($p<0.05$, DGC test).

plasma membrane. Consistent with Shabala's results [16], K599-CED9 hairy roots showed altered potassium content respect K599-empty hairy roots only under 150 mM NaCl condition, whereas non-significant differences were observed under inoculated 50 mM NaCl between transgenic and wild type hairy roots. This result indicates different causes of death and therefore, different mechanisms of action of CED-9 under these stress conditions (Figure 4B). Moreover, CED-9 effects on K^+-efflux may be due to sustained levels of Ca^{2+} which subsequently may affect the opening and closing balance of non-selective cation channels (NSCC) [47,48] (Figure 4C), but Ca^{2+} subcellular localization approaches are required to verify this hypothesis. However, Ced9-expression had no effect on sodium influx which increased in dose dependent manner, similarly to that observed in K599-empty hairy roots (Figure 4B).

It has been shown that under saline stress ROS generation are induced [49–51] leading to oxidative damage [52,53]. In this regard, it has been suggested that anti-apoptotic genes from animals would suppress ROS generation or promote its removal in plants [7,14]. However, to best of our knowledge, there are no redox studies to support this hypothesis since these conclusions were based on visual observations such as a lack of decoloration in transgenic leaves under stress conditions [13,14] and chlorophyll content in salt stressed leaves [16]. In this work, we reported redox effects of Ced-9-expression in soybean hairy roots under stress conditions (Figure 5 and 6). Increases in antioxidant capacity in K599-empty hairy roots (Figure 5C) could indicate a response to oxidative stress induced by hairy root death-inducing conditions (Figure 5A and 5B); while no changes were observed between treatments in K599-CED9 hairy roots (Figure 5A, 5B, 5C and 5D). These results demonstrated that the expression of Ced9 prevents ROS generation in hairy roots under stress conditions.

On the other hand, the mammalian homologous of CED-9 may regulate metabolic efficiency in neurons through interaction with the mitochondrial F_1F_0 ATP synthase in the inner membrane [54]. Likewise, Qiao et al [13] suggested a possible contribution of Bcl-xL and Ced-9 to improved mitochondrial membrane potential when were expressed in plants. In this regard, this work demonstrated that K599-CED-9 hairy roots had improved metabolism assessed as ATP content (Figure 6), particularly in severe salt conditions.

Strikingly, despite of improved metabolism and tolerance to death-induced stress conditions, K599-CED9 hairy roots had a significant inhibition of its nodulation capacity (Figure 7). Moreover, given that cell death process is an early control of the number of nodules [55,56], we expected that the expression of Ced-9 could impact positively on the nodulation process. Taking into account that one of the main action of Ced-9 is the ionic flux control, it is possible that its expression in legume could adversely affect the ion flux signatures that occur during rhizobium perception [57,58]. Likewise, It has been reported in animals that CED9 interact with proteins involved in vesicular traffic and autophagy [59], which in turn have participation in organogenesis events [60–62]. In this regard, we have the hypothesis and also relevant unpublished data showing that CED9 expression, which have no homologues identified in plants, could affect nodule organogenesis by interacting with vesicular traffic and autophagy proteins conserved in plants.

In summary, in this work we characterized the effects of Ced-9-expression on soybean hairy root under different, ordered-like and necrosis-like root hair and root death-inducing conditions. In this respect, we demonstrated that part of improved tolerance given by Ced-9 expression is based on the maintenance of ionic and redox homeostasis capacity. However, contrary to expectations, Ced-9 expression drastically inhibited nodule formation, and consequently the expression of animal cell death suppressors seems not to be an adequate strategy to increase the nitrogen content derived from biological fixation.

Materials and Methods

Bacterial strain and plant material

Soybean seeds (*Glycine max* L. DM4800) were disinfected with sodium hypochlorite 5% (v/v) for 5 min and germinated in the dark for 48 h on filter paper moistened with distilled water. The seeds were incubated at 28 and 37 °C during the first and second 24 h periods, respectively, to promote the growth of roots and root hairs. *Bradyrhizobium japonicum* USDA 138 was cultured in yeast extract mannitol (YEM) medium [63] at 28 °C with constant agitation for 5 days (3×10^9 cells mL^{-1}). The bacteria were washed and resuspended in sterile water.

Binary vector and *A. rhizogenes* strains

The binary vector pBI2113-Ced-9 has an efficient promoter cassette overexpressing the Ced-9 gene (GenBank accession number L26545), a *Caenorhabditis elegans* homolog of Bcl-xL which was kindly provided by Dr. Yuko Ohashi [12,64]. Cucumopine-type *A. rhizogenes* strain K599 was used to infect cotyledon axes regions. *A. rhizogenes* K599 with pBI2113-Ced-9 was grown in Luria-Bertani (LB) medium containing kanamycin (Km) at 50 µg mL^{-1}. To get fresh cells, *A. rhizogenes* K599 was grown on LB plates containing Km and incubated 48 h at 28 °C. Cells were collected from these plates and diluted into 1 mL of sterile water. For control hairy roots (K599-empty), a fresh culture of *A. rhizogenes* K599 lacking the binary vector was grown in LB medium without antibiotics.

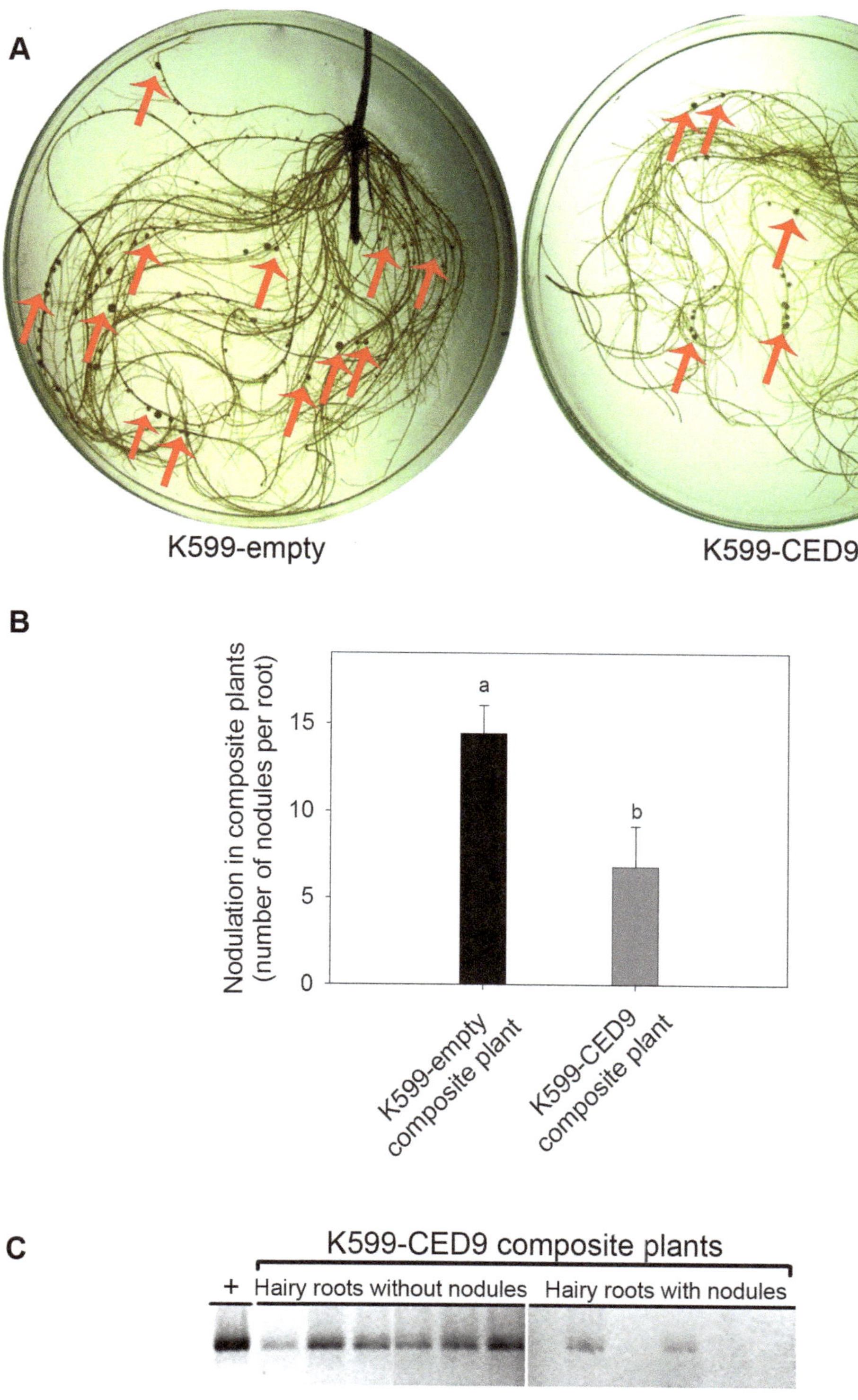

Figure 7. Ced-9 expression inhibits nodule formation in hairy roots. K599-empty and K599-CED9 composite plants were inoculated with *B. japonicum* USDA138 and after 14 days the nodulation was evaluated. A) Image of nodules in K599-empty and K599-CED9 hairy roots. B) Number of nodules in K599-empty (black bar) and K599-CED9 (grey bar) composite plants. Total nodules were counted and divided by the total number of hairy roots, including hairy roots without nodules (nodule number per hairy root). C) Nodulated and non-nodulated hairy roots of K599-CED9 composite plants were separated and Ced-9 transgenic hairy roots were identified by Ced-9 PCR (+ indicates Ced-9 positive control). Data are means ± SE of four composite plants. Different letters indicate significant differences between composite plant genotypes ($p<0.05$, DGC test).

A. rhizogenes-mediated root transformation

Induction of *A. rhizogenes*-mediated root transformation protocol was modified from Estrada-Navarrete [27]. Briefly, after germination, sprouts were inoculated by injection directly into the cotyledonary nodes with a syringe and transferred to a hydroponic double tube system and incubated in a growth chamber under 16 h photoperiod (350 µmol m^{-2} s^{-1}) at 26±2 °C. The smaller tube contained the sprout watered with B&D solution [65] supplemented with 8 mM KNO_3 and it was within a larger tube that serves as moist chamber. Typically, soybean plants infected by *A. rhizogenes* started to show tumors approximately 5 days after inoculation. Twelve days after *A. rhizogenes* infection, plantlets exhibited numerous induced hairy roots per wound site. Primary root was removed from the plant by cutting approximately 1 cm below the cotyledon nodes and the composite plants were placed in plastic trays with B&D solution with or without KNO_3 depending on the treatment to be performed.

Root hairs death-inducing conditions

After germination, sprouts were incubated 30 min in aerated tubes that contained sterile water (control), *B. japonicum* (inoculated), 50 mM NaCl, *B. japonicum* in presence of 50 mM NaCl and 150 mM NaCl. Root hairs from roots subjected to different stress treatments were extracted by peeling the root zone containing young root hairs, which were immediately frozen in liquid air. Peeling was performed under a magnifying glass by making an incision with a scalpel in root and pulling the epidermal tissue containing the root hairs using a fine-tipped clamp. Root hairs of approximately 200 roots generate sufficient material for a sample.

Hairy roots death-inducing conditions

Once primary root was removed from the plant by cutting below the cotyledon nodes, the composite plants were placed in aerated plastic trays with B&D solution supplemented with 8 mM KNO_3 and incubated in a growth chamber under 16 h photoperiod (350 µmol m^{-2} s^{-1}) at 26±2 °C during two weeks. Hairy roots were subjected to stress treatments for 3 h and then, they were immediately frozen in liquid air.

Cell death evaluations

Cell death evaluations were performed by Evans Blue staining and DNA degradation analysis. Evans Blue is a dye used in the determination of cell viability [13] due to its inability to permeate intact cell membranes. When cells lose the membrane potential, the dye diffuses within the cell and it may visualized by conventional microscopy. The roots were incubated 10 min with Evans Blue 0.05% (w/v) in water or each NaCl levels assayed.

Genomic DNA was isolated using CTAB [66]. In brief, the samples were homogenized to a fine powder using a mortar and pestle under liquid nitrogen and thawed in CTAB extraction buffer (2% w/v CTAB, 1.4 M NaCl, 20 mM EDTA, 100 mM TRIS-HCl pH 8.0). RNase A was added and the homogenate was incubated for 30 min at 37 °C. DNA was extracted twice with an equal volume of chloroform:isoamylalcohol (24:1 v/v) and precipitated with 0.6 vols. of isopropanol. For visualization of the DNA degradation, equal amounts of DNA (2 µg) were loaded on a 2% TAE agarose gel and stained with ethidium bromide.

The nuclear morphology of hairy roots was evaluated by acridine orange (AO, 50 µg/mL) and ethidium bromide (EtBr, 50 µg/mL) staining and observed with Zeiss confocal microscopy. AO and EtBr are dyes that intercalate DNA and fluoresce under UV light. The orange color and the presence of the dispersed chromatin in the cytoplasm indicate that the cells have lost integrity of the nuclear membrane and are in a very late stage of death. Both dyes are excited by 488 nm and emission was observed at 500–530 nm and 565–615 nm for AO and EtBr, respectively.

MDA and ATP quantification in root hairs and hairy roots

The samples were homogenized using a mortar and pestle under liquid nitrogen and thawed in 3% (v/v) trichloroacetic acid (TCA) then centrifugation was carried out at 13,000 g, 4 °C during 15 min.

MDA levels were quantified according Heath and Packer [67]. Briefly, 100 µL of sample were mixed with 100 µL of 20% TCA + 0.5% thiobarbituric acid (TBA), incubated at 90°C for 20 min and ice cold rapidly. The mix was centrifuged at 13,000 g for 10 min. The absorbance of the supernatant was read at 532 nm y 600 nm.

The concentration of ATP present in each of the samples was determined with a GloMax luminometer using a bioluminescent detection reagent (ENLITEN rLuciferase/Luciferin; Promega) according to manufacturer. The amount of ATP present in the sample was calculated from the measured relative light units using a standard curve spanning the relative light unit range obtained from the samples [68].

Na^+, K^+ and Ca^+ determination in hairy roots

Ion quantification was performed by high pressure liquid chromatography (Shimadzu LC2010) with Shim-pack IC–C3 column and non-suppressed system. Hairy roots segments (50 mg) were placed into 1 mL of 0.1 N nitric acid during 3 days. Samples were passed through 0.22 µm pore size MF Millipore cellulose membrane filters and diluted 6 times with MQ water. Then, 30 µL of samples were analyzed. The mobile phase was oxalic acid 2.5 mM and the time of chromatography was 20 min with a flow rate of 1.2 mL min^{-1}. Quantitative analysis was done with multicationic standards by software LCSolution.

Fluorometric H_2O_2 determination in hairy roots

The samples were homogenized in 3% (v/v) trichloroacetic acid (TCA) then centrifugation was carried out at 12,000 g, 4 °C during 15 min. 100 µL of supernatant was mixed with determination buffer (100 mM potassium phosphate pH 7.4, 4.5 U/mL horse radish peroxidase (HRP) and 1 mM p-hydroxyphenilacetic acid). Duplicates were incubated with catalase in order to diminish unspecific fluorescence. Hydrogen peroxide fluorescence was measured with spectrofluorometer Shimatzu at 371–414 nm excitation and emission respectively [69].

FRAP (Ferric Reducing Ability of Plasma) assay

The samples were homogenized using a mortar and pestle under liquid nitrogen and ethanol 80% was added. Then, centrifugation was carried out at 12,000 g, 4 °C during 10 min. 100 µL of supernatant was mixed with reaction buffer (5 mL of Acetate buffer 0.3 M pH 3.6; 0.5 mL of TPTZ 10 mM (2,4, 6 Tris (2 pyridyl) s-triazine) diluted in 40 mM HCl and 0.5 mL of $FeCl_3$ 200 mM) in a microplate on ice. Samples were removed from the ice and read at 600 nm after 20 min. A standard curve with TROLOX was done to calculate FRAP capacity in samples [70].

Ascorbic acid determination

Samples were prepared for ascorbate analyses by homogenizing material in 1 mL of 3% trichloroacetic acid. The homogenate was centrifuged at 10 000×g at 4°C for 15 minutes and the supernatant was collected for analyses of ascorbate. Total ascorbate content was determined according to Gillespie and Ainsworth [71] with modifications [72]. The reaction mixture for total ascorbate contained a 50 µL aliquot of the supernatant, 15 µL of 150 mM phosphate buffer (pH 7.4) and 15 µL of 10 mM DTT; samples were incubated at room temperature for 10 min.

After that, the mix was incubated for 60 min at 37 °C with 15 µL of 0.5% N-ethylmaleimide (NEM), 16.6% orthophosphoric acid (H_3PO_4), 1.33% α-α'bipyridyl, and 30 µL of 3% $FeCl_3$. The samples were measured at 525 nm in ELISA MRX II. Reduced ascorbic acid content was determined using the same protocol, except for the addition of NEM and DTT.

Protein content

Protein content was determined spectrophotometrically at 578 nm according to Bradford [73] with bovine serum albumin (BSA) as a standard.

Evaluation of nodulation in hairy roots

Primary soybean root was removed twelve days post-infection with *A. rhizogenes* by cutting approximately 1 cm below the cotyledon nodes and the composite plants were placed in aerated plastic trays with B&D solution with 8 mM KNO_3 in a growth chamber under 16 h photoperiod (350 µmol m^{-2} s^{-1}) at 26±2 °C. After two days, the nutrient solution was replaced by B&D without KNO_3 and the inoculation with *B. japonicum* USDA138 was performed. The nodule number was assessed 14 days post-inoculation.

RNA extraction

Samples were homogenized in a cold mortar with TRIzol Reagent (1:10 µg plant tissue:µL reagent), mixed for 1 min and incubated at room temperature for 5 min. Then, 0.2 mL chloroform per mL of TRIzol Reagent was added and incubated at room temperature for 3 min. After incubation, the samples were centrifuged at 12,000 g at 4°C for 15 min and the aqueous phases were transferred to clean tubes. RNA was precipitated by adding 1 vol. of isopropanol, incubated at room temperature for 10 min and centrifuged at 12,000 g at 4°C for 15 min. The precipitate was washed with 70% ethanol and the samples were centrifuged again at 12,000 g, 4°C for 15 min. The precipitate was dried and resuspended in DEPC water and its concentration was quantified using a NanoDrop 3300 spectrometer (Thermo Scientific). Purified RNA was treated with DNase I (Invitrogen) to remove genomic DNA, according to the manufacturer's instructions.

qPCR

RNA DNA-free (1 to 2.5 µg) was used with oligo(dT) for first strand cDNA synthesis using the Moloney Murine Leukemia Virus Reverse Transcriptase (M-MLV RT, Promega) according to the manufacturer's instructions. For each primer pair, the presence of a unique product of the expected size was checked on ethidium bromide–stained agarose gels after PCR reactions. Absence of contaminant genomic DNA was confirmed in reactions with DNase treated RNA as template. The qPCR reaction was performed using iQ Universal SYBR Green Supermix Bio Rad.

Amplification of actin (forward primer 5′-AACGACCT-TAATCTTCATGCTGC-3′ and reverse primer 5′- GGTAA-CATTGTGCTCAGTGGTGG-3′) and EF1α (forward primer 5′-GGTCATTGGTCATGTCGACTCTGG-3′ and reverse primer 5′-GCACCCAGGCATACTTGAATGACC-3′) was used to normalize the amount of template cDNA. The gene-specific primer pairs employed for the detection of transcripts of Ced-9 was: forward primer 5′-CTACGAACGAGCAGAAGCTGAA-3′ and reverse primer 5′- CAAGCTGAACATCATCCGCCCA-3′. qRT-PCR was performed in thermocycler iQ5 (BioRad) at 59°C with iQ SYBR Green Supermix (BioRad), according to the manufacturer's instruction.

Statistical analyses

Data were analyzed using analysis of variance (ANOVA) followed by the DGC test model with InfoStat software [74].

Supporting Information

Figure S1 K599-CED9 composite plants developed both transgenic and wild type hairy roots. Development of transgenic (K599-CED9) and non-transgenic (K599-empty) hairy roots in a soybean K599-CED9 composite plant. A) Identification of K599-CED9 hairy roots by Ced-9 PCR (+: positive control, numbers from 1 to 5: independent hairy roots of the same K599-CED9 composite plant). B) Ced-9-expression levels in K599-CED9 hairy roots measured by qRT-PCR and normalized to elongation factor 1α (EF1α).

Figure S2 Hairy roots death-inducing stress conditions. A) Evans blue staining of K599-empty and K599-CED9 hairy roots showing loss of membrane integrity after 3 h of treatment with 150 mM NaCl. B) DNA degradation in hairy roots incubated 3 h under control and stress conditions. 1: Control, 2: inoculated with *B. japonicum* in presence of 50 mM NaCl, 3: 150 mM NaCl. 2 µg of DNA were loaded on a 2% TAE agarose gel and stained with ethidium bromide.

Figure S3 Nuclear morphology in root hairs. K599-empty (A, C, E) and K599-CED9 (B, D, F) hairy roots were subjected 30 min under control (A and B), inoculated with *B. japonicum* (C and D), and 50 mM NaCl (E and F), and root hairs nuclear morphology was evaluated. Images were taken with a Zeiss confocal microscope. The excitation was performed simultaneously at 488 nm and emission filter BP 500–530 IR and BP 565–615 IR for AO and EtBr, respectively. AO: acridine orange channel, EtBr: ethidium bromide channel, AO/EtBr: image overlay.

Acknowledgments

Authors thank Dr. Yuko Ohashi for providing Ced-9, *Caenorhabditis elegans* homolog of Bcl-xL construct and Dr. Jesús Montiel for sharing the hydroponic double tube system to obtain hairy roots.

Author Contributions

Conceived and designed the experiments: GR RL. Performed the experiments: GR NM. Analyzed the data: GR NM MM FS RL. Contributed reagents/materials/analysis tools: FS MM RL. Wrote the paper: GR NM MM FS RL.

References

1. Bozhkov PV, Lam E (2011) Green death: revealing programmed cell death in plants. Cell Death Differ. 18: 1239–1240.
2. Fuchs Y, Steller H (2011) Programmed cell death in animal development and disease. Cell 147: 742–758.
3. Teng X, Cheng WC, Qi B, Yu TX, Ramachandran K, et al. (2011) Gene-dependent cell death in yeast. Cell Death and Disease. 2: 1–9.
4. Lemasters JJ, Qian T, Bradham CA, Brenner DA, Cascio WE, et al. (1999) Mitochondrial dysfunction in the pathogenesis of necrotic and apoptotic cell death. J. Bioenerg. Biomembr. 31: 305–319.
5. Atlante A, Giannattasio S, Bobba A, Gagliardi S, Calissano P, et al. (2005) An increase in the ATP levels occurs in cerebellar granule cells en route to apoptosis in which ATP derives from both oxidative phosphorylation and anaerobic glycolysis. Biochim. Biophys. Acta 1708: 50–62.

6. Katsuhara M, Kawasaki T (1996) Salt stress induced nuclear and DNA degradation in meristematic cells of barley roots. Plant Cell 37: 169–173.
7. Pennell R, Lamb C (1997) Programmed Cell Death in Plants. Plant Cell 9: 1157–1168.
8. Huh GH, Damsz B, Matsumoto TK, Reddy MP, Rus AM, et al. (2002) Salt causes ion disequilibrium-induced programmed cell death in yeast and plants. Plant J. 29: 649–659.
9. Jacobson MD, Weil MM, Raff MC (1997) Programmed cell death in animal development. Cell 88: 347–354.
10. Ameisen JC (2002) On the origin, evolution, and nature of programmed cell death: a timeline of four billion years. Cell Death Differ. 9: 367–393.
11. Chen S, Dickman MB (2004) Bcl-2 family members localize to tobacco chloroplasts and inhibit programmed cell death induced by chloroplast-targeted herbicides. J. Exp. Bot. 55: 2617–2623.
12. Mitsuhara I, Malik K, Miura M, Ohashi Y (1999) Animal cell-death suppressors Bcl-x(L) and Ced-9 inhibit cell death in tobacco plants. Curr. Biol. 9: 775–778.
13. Qiao J, Mitsuhara I, Yazaki Y, Sakano K, Gotoh Y, et al. (2002) Enhanced resistance to salt, cold and wound stresses by overproduction of animal cell death suppressors Bcl-xL and Ced-9 in tobacco cells - their possible contribution through improved function of organella. Plant Cell Physiol. 43: 992–1005.
14. Xu P, Rogers SJ, Roossinck MJ (2004) Expression of antiapoptotic genes bcl-xL and ced-9 in tomato enhances tolerance to viral-induced necrosis and abiotic stress. PNAS 101: 15805–15810.
15. Dickman MB, Park YK, Oltersdorf T, Li W, Clemente T, et al. (2001) Abrogation of disease development in plants expressing animal antiapoptotic genes PNAS 98: 6957–6962.
16. Shabala S, Cuin T, Prismall L, Nemchinov LG (2007) Expression of animal CED-9 anti-apoptotic gene in tobacco modifies plasma membrane ion fluxes in response to salinity and oxidative stress. Planta 227: 189–97.
17. Kawai-Yamada M, Jin L, Yoshinaga K, Hirata A, Uchimiya H (2001) Mammalian Bax-induced plant cell death can be down-regulated by overexpression of Arabidopsis Bax Inhibitor-1 (AtBI-1). PNAS 98: 12295–12300.
18. Lacomme C, Santa Cruz S (1999) Bax-induced cell death in tobacco is similar to the hypersensitive response. PNAS 96: 7956–7961.
19. Singleton P, Bohlool B (1984) Effect of salinity on nodule formation by soybean. Plant Physiol. 74: 72–76.
20. Delgado J, Ligero F, Liuch C (1994) Effects of salt stress on growth and nitrogen fixation by pea, faba-bean, common bean and soybean plant. Soil Biol. Biochem. 26: 371–376.
21. Ashraf M, Wu L (1994) Breeding for salinity tolerance in plants. Critical review of plant science 13: 17–42.
22. Sionit N, Kramer PJ (1977) Effect of water stress during different stages of growth of soybean. Agron J. 69: 274–278.
23. Muñoz N, Robert G, Melchiorre M, Racca R, Lascano R (2012) Saline and osmotic stress differentially affects apoplastic and intracellular reactive oxygen species production, curling and death of root hair during *Glycine max* L.–Bradyrhizobium japonicum interaction. Environ. Exp. Bot. 78: 76–83.
24. Muñoz N, Rodriguez M, Robert G, Lascano R (2013) Negative short-term salt effects on the soybean–Bradyrhizobium japonicum interaction and partial reversion by calcium addition. Func. Plant Biol. 41: 96–105.
25. Cárdenas L, Martinez A, Sanchez F, Quinto C (2008) Fast, transient and specific intracellular ROS changes in living root hair cells responding to Nod factors (NFs). Plant J. 56: 802–813.
26. Oldroyd G, Harrison M, Udvardi M (2005) Peace talks and trade deals. Keys to long-term harmony in legume-microbe symbioses. Plant Physiol. 137: 1205–1210.
27. Estrada-Navarrete G, Alvarado-Affantranger X, Olivares JE, Díaz-Camino C, Santana O, et al. (2006) Agrobacterium rhizogenes transformation of the Phaseolus spp: a tool for functional genomics. MPMI 19: 1385–1393.
28. Newell JM, Leigh RA, Hall JL (1998) Vacuole development in cultured evacuolated oat mesophyll protoplasts. J. Exp. Bot. 49: 817–827.
29. Desagher S, Martinou JC (2000) Mitochondria as the central control point of apoptosis. Trends Cell Biol. 10: 369–377.
30. Kroemer G, Reed JC (2000) Mitochondrial control of cell death. Nat Med. 6: 513–519.
31. Zong WX, Thompson CB (2006) Necrotic death as a cell fate. Genes Dev. 20: 1–15.
32. Galluzzi L, Kroemer G (2008) Necroptosis: a specialized pathway of programmed necrosis. Cellhttp://www.ncbi.nlm.nih.gov/pubmed/19109884 135: 1161–1163.
33. Vanlangenakker N, Vanden Berghe T, Krysko DV, Festjens N, Vandenabeele P (2008) Molecular mechanisms and pathophysiology of necrotic cell death. Curr Mol Med. 8: 207–220.
34. Vacca RA, Valenti D, Bobba A, Merafina RS, Passarella S, et al. (2006) Cytochrome c is released in a reactive oxygen species-dependent manner and is degraded via caspase-like proteases in Tobacco Bright-Yellow 2 cells en route to heat shock-induced cell death. Plant Physiol. 141: 208–219.
35. Valenti D, Anna R, Concetta M, Pinto D, De Gara L, et al. (2007) In the early phase of programmed cell death in Tobacco Bright Yellow 2 cells the mitochondrial adenine nucleotide translocator, adenylate kinase and nucleoside diphosphate kinase are impaired in a reactive oxygen species-dependent manner. Biochim. Biophys. Acta. 1767: 66–78.
36. Azad K, Ishikawa T, Ishikawa T, Sawa Y, Shibata H (2008) Intracellular energy depletion triggers programmed cell death during petal senescence in tulip. J. Exp. Bot. 59: 2085–2095.
37. Chen X, Wang Y, Li J, Jiang A, Cheng Y, et al. (2009) Mitochondrial proteome during salt stress-induced programmed cell death in rice. Plant Physiol. Biochem. 47: 407–415.
38. Casolo V, Petrussa E, Krajnakova J, Macri F, Vianello A (2005) Involvement of the mitochondrial K(+)ATP channel in H_2O_2 or NO-induced programmed death of soybean suspension cell cultures. J. Exp. Bot. 56: 997–1006.
39. Tiwari BS, Belenghi B, Levine A (2002) Oxidative stress increased respiration and generation of reactive oxygen species, resulting in ATP depletion, opening of mitochondrial permeability transition, and programmed cell death. Plant Physiol. 128: 1271–1281.
40. Mlejnek P, Dolezel P, Procházka S (2003) Intracellular phosphorylation of benzyladenosine is related to apoptosis induction in tobacco BY-2 cells. Plant Cell Environ. 26: 1723–1735.
41. Papucci L, Formigli L, Schiavone N, Tani A, Donnini M, et al. (2004) Apoptosis shifts to necrosis via intermediate types of cell death by a mechanism depending on c-myc and bcl-2 expression. Cell Tissue Res. 316: 197–209.
42. Wan J, Torres M, Ganapathy A, Thelen J, DaGue BB, et al. (2005) Proteomic analysis of soybean root hairs after infection by *Bradyrhizobium japonicum*. MPMI 18: 458–467.
43. Jones KM, Sharopova N, Lohar DP, Zhang JQ, VandenBosch K, et al. (2008) Differential response of the plant Medicago truncatula to its symbiont Sinorhizobium meliloti or an exopolysaccharide-deficient mutant. PNAS 105: 704–709.
44. Tsujimoto Y, Shimizu S, Eguchi Y, Kamiike W, Matsuda H (1997) Bcl-2 and Bcl-xL block apoptosis as well as necrosis: possible involvement of common mediators in apoptotic and necrotic signal transduction pathways. Leukemia 3: 380–382.
45. Saldeen J (2000) Cytokines induce both necrosis and apoptosis via a common Bcl-2-inhibitable pathway in rat insulin-producing cells. Endocrinology 141: 2003–2010.
46. Fukuda K, Yamamoto M (1999) Acquisition of resistance to apoptosis and necrosis by Bcl-xL over-expression in rat hepatoma McA-RH8994 cells. J Gastroenterol Hepatol. 14: 682–690.
47. Demidchik V, Maathuis F (2007) Physiological roles of nonselective cation channels in plants: from salt stress to signalling and development. New Phytol. 175: 387–404.
48. Demidchik V, Shabala S, Davies J (2007) Spatial variation in H2O2 response of Arabidopsis thaliana root epidermal Ca2+ flux and plasma membrane Ca2+ channels. Plant J. 49: 377–386.
49. Mazel A, Leshem Y, Tiwari BS, Levine A (2004) Induction of salt and osmotic stress tolerance by overexpression of an intracellular vesicle trafficking protein AtRab7 (AtRabG3e). Plant Physiol. 134: 118–128.
50. Leshem Y, Melamed-Book N, Cagnac O, Ronen G, Nishri Y, et al. (2006) Suppression of Arabidopsis vesicle-SNARE expression inhibited fusion of H_2O_2-containing vesicles with tonoplast and increased salt tolerance. PNAS 103: 18008–18013.
51. Leshem Y, Seri L, Levine A (2007) Induction of phosphatidylinositol 3-kinase-mediated endocytosis by salt stress leads to intracellular production of reactive oxygen species and salt tolerance. Plant J. 51: 185–197.
52. Borsani O, Valpuesta V, Botella M (2001) Evidence for a role of salicylic acid in the oxidative damage generated by NaCl and osmotic stress in Arabidopsis seedlings. Plant Physiol. 126: 1024–1030.
53. Slesak I, Miszalski Z, Karpinska B, Niewiadomska E, Ratajczak R, et al. (2002) Redox control of oxidative stress responses in the C-3-CAM intermediate plant *Mesembryanthemum crystallinum*. Plant Physiol. Biochem. 40: 669–677.
54. Alavian KN, Li H, Collins L, Bonanni L, Zeng L, et al. (2011) Bcl-xL regulates metabolic efficiency of neurons through interaction with the mitochondrial F_1F_O ATP synthase. Nat. Cell Biol. 13: 1224–1233.
55. Vasse J, de Billy F, Truchet G (1993) Abortion of infection during the Rhizobium meliloti-alfalfa sybiotic interaction is accompanied by a hypersensitive reaction. Plant J. 4: 555–566.
56. Baron C, Zambryski PC (1995) The plant response in pathogenesis, symbiosis, and wounding: variations on a common theme? Annu. Rev. Genet. 29: 107–129.
57. Cárdenas L, Holdaway-Clarke T, Sanchez F, Quinto C, Feijo J, et al. (2000) Ion changes in legume root hairs responding to Nod factors. Plant Physiol. 123: 443–452.
58. Felle HH, Kondorosi E, Kondorosi A, Schultze M (2000) How alfalfa root hairs discriminate between Nod factors and oligochitin elicitors. Plant physiol. 124: 1373–1380.
59. Takacs-Vellai K, Vellai T, Puoti A, Passannante M, Wicky C, et al. (2005) Inactivation of the autophagy gene bec-1 triggers apoptotic cell death in C. elegans. Current Biology, 15: 1513–1517.
60. Fujiki Y, Yoshimoto K, Ohsumi. (2007) An Arabidopsis homolog of yeast ATG6/VPS30 is essential for pollen germination. Plant Physiol. 143: 1132–1139.
61. Lee Y, Kim ES, Choi Y, Hwang I, Staiger CJ, et al. (2008) The Arabidopsis phosphatidylinositol 3-kinase is important for pollen development. Plant Physiol. 147: 1886–1897.

62. Blanco FA, Meschini EP, Zanetti ME, Aguilar OM (2009) A small GTPase of the Rab family is required for root hair formation and preinfection stages of the common bean-Rhizobium symbiotic association. Plant Cell 21: 2797–2810.
63. Vincent J (1970) A Manual for the practical study of the root-nodule bacteria. Oxford: Blackwell Scientific Publications. 164p
64. Mitsuhara I, Ugaki M, Hirochika H, Ohshima M, Murakami T, et al. (1996) Efficient promoter cassettes for enhanced expression of foreign genes in dicotyledonous and monocotyledonous plants. Plant Cell Physiol. 37: 49–59.
65. Broughton W, Dilworth M (1971) Control of leghaemoglobin synthesis in snake beans. Biochem. J. 125: 1075–1080.
66. Murray MG, Thompson WF (1980) Rapid isolation of high molecular weight plant DNA. Nucleic Acids Res. 8: 4321–4325.
67. Heath RL, Packer L (1968) Photoprotection in isolated chloroplasts I kinetics and stoichiometry of fatty acid peroxidation. Arch Biochem Biophys. 125: 189–198.
68. Rodríguez M, Taleisnik E, Lenardon S, Lascano R (2010) Are Sunflower chlorotic mottle virus infection symptoms modulated by early increases in leaf sugar concentration? J. Plant Physiol. 167: 1137–1144.
69. Guibault GG, Brignac PJ, Juneau M (1968) New substrates for the fluorometric determination of oxidative enzymes. Anal Chem 40: 1256–1263.
70. Benzie IFF, Strain JJ (1996) The Ferric Reducing Ability of Plasma (FRAP) as a Measure of "Antioxidant Power": The FRAP Assay. Anal. Biochem. 239: 70–76.
71. Gillespie K, Ainsworth E (2007) Measurement of reduced, oxidized and total ascorbate content in plants. Nat Protoc. 2: 871–874.
72. Rodríguez M, Muñoz N, Lenardon S, Lascano R (2013) Redox-related metabolites and gene expression modulated by sugar in sunflower leaves: similarities with Sunflower chlorotic mottle virus-induced symptom. Redox Rep. 18: 27–35.
73. Bradford MM (1976) A rapid and sensitive method for the quantitation of microgram quantities of proteins utilizing the principle of protein–dye binding. Anal Biochem. 72: 248–254.
74. Di Rienzo JA, Casanoves F, Balzarini MG, Gonzalez L, Tablada M, et al. (2012) Grupo InfoStat, FCA, Universidad Nacional de Córdoba, Argentina. InfoStat Software estadístico. Available: http://www.infostat.com.ar Accessed 2013 Oct 02.

2

Tubular Overexpression of Gremlin Induces Renal Damage Susceptibility in Mice

Alejandra Droguett[1], Paola Krall[1], M. Eugenia Burgos[1], Graciela Valderrama[1], Daniel Carpio[3], Leopoldo Ardiles[1], Raquel Rodriguez-Diez[4], Bredford Kerr[2], Katherina Walz[2], Marta Ruiz-Ortega[4], Jesus Egido[4], Sergio Mezzano[1]*

1 Division Nephrology, School of Medicine, Universidad Austral de Chile, Valdivia, Chile, **2** Centro de Estudios Científicos, Valdivia, Chile, **3** Hystopathology Division, School of Medicine, Universidad Austral de Chile, Valdivia, Chile, **4** Cellular Biology in Renal Diseases Laboratory, Universidad Autónoma Madrid, Madrid, Spain

Abstract

A growing number of patients are recognized worldwide to have chronic kidney disease. Glomerular and interstitial fibrosis are hallmarks of renal progression. However, fibrosis of the kidney remains an unresolved challenge, and its molecular mechanisms are still not fully understood. Gremlin is an embryogenic gene that has been shown to play a key role in nephrogenesis, and its expression is generally low in the normal adult kidney. However, gremlin expression is elevated in many human renal diseases, including diabetic nephropathy, pauci-immune glomerulonephritis and chronic allograft nephropathy. Several studies have proposed that gremlin may be involved in renal damage by acting as a downstream mediator of TGF-β. To examine the *in vivo* role of gremlin in kidney pathophysiology, we generated seven viable transgenic mouse lines expressing human gremlin (GREM1) specifically in renal proximal tubular epithelial cells under the control of an androgen-regulated promoter. These lines demonstrated 1.2- to 200-fold increased GREM1 expression. GREM1 transgenic mice presented a normal phenotype and were without proteinuria and renal function involvement. In response to the acute renal damage cause by folic acid nephrotoxicity, tubule-specific GREM1 transgenic mice developed increased proteinuria after 7 and 14 days compared with wild-type treated mice. At 14 days tubular lesions, such as dilatation, epithelium flattening and hyaline casts, with interstitial cell infiltration and mild fibrosis were significantly more prominent in transgenic mice than wild-type mice. Tubular GREM1 overexpression was correlated with the renal upregulation of profibrotic factors, such as TGF-β and αSMA, and with increased numbers of monocytes/macrophages and lymphocytes compared to wild-type mice. Taken together, our results suggest that GREM1-overexpressing mice have an increased susceptibility to renal damage, supporting the involvement of gremlin in renal damage progression. This transgenic mouse model could be used as a new tool for enhancing the knowledge of renal disease progression.

Editor: Jean-Claude Dussaule, INSERM, France

Funding: Supported by Ciberdem, Redinren, Fondecyt 1120480, Fondecyt 1080083, Fondecyt 1100821, Chile. The funders had no role in study design, data collection and analysis, decision to publish, or preparation of the manuscript.

Competing Interests: The authors have declared that no competing interests exist.

* Email: smezzano@uach.cl

Introduction

Kidney disease and renal failure are worldwide health problems and increasing research efforts are required to understand the molecular mechanisms underlying kidney injury to identify new therapeutic approaches.

Gremlin is a highly conserved secreted protein that is present both on the external cell surface and within the ER-Golgi compartment of different cell types [1,2], affecting diverse biological processes such as growth, differentiation and development [2]. At early stages of development, gremlin is expressed in the mesoderm and inhibits BMP (bone morphogenetic protein) signaling by binding to and blocking BMP activity [3]. Thus, gremlin is considered a BMP antagonist. The *Grem1*-null mouse was the first *in vivo* gremlin model. This model was neonatally lethal, and the embryos presented kidney and lung defects [4]. Gremlin is critical in nephrogenesis but is quiescent after birth and absent in the normal adult kidney [5]. The gremlin gene may be induced in human mesangial cell cultures that are exposed to high glucose levels and is expressed in the kidneys of diabetic rats [6–8]. In biopsies obtained from patients with diabetic nephropathy, we have observed gremlin expression in areas with tubule-interstitial fibrosis, and it colocalizes with transforming growth factor-β (TGF-β) [9]. Moreover, gremlin is also expressed in cellular glomerular crescents and in the tubular and infiltrating interstitial cells of human biopsies of pauci-immune glomerulonephritis and chronic allograft nephropathy, broadening the range of activity to a more global role for gremlin in renal diseases [10,11]. We have recently shown that recombinant gremlin directly regulates profibrotic events in cultured tubulo-interstitial cells and acts as a mediator of TGF-β responses [12]. Blockade of gremlin in experimental models of diabetes, using heterozygous *grem1* mice or gene silencing, has been shown to ameliorate renal damage, including proteinuria and fibrosis, suggesting that gremlin contributes to renal damage progression [13,14]. Furthermore, gremlin overexpression in rat lungs results in a partly reversible lung fibrosis through the activation of alveolar epithelial cell proliferation [15]. Interestingly, there are no data regarding the direct effect of gremlin in the kidney.

With the interest to investigate the potential role of gremlin in the kidney in physiological and pathological conditions *in vivo*, we generated viable transgenic (TG) mice expressing the human gremlin gene (GREM1) in the renal proximal tubular cells under the control of a specific kidney androgen-regulated promoter (KAP) that can be used as a molecular "on-off" switch.

We found that GREM1-expressing mice presented normal renal function, but they were more susceptible to developing renal damage induced by folic acid (FA) administration (a known experimental model of acute renal injury), suggesting that gremlin plays a pathogenic role in renal damage. These mice could be used as an *in vivo* experimental model to study the role of gremlin in renal diseases, such as diabetic nephropathy.

Materials and Methods

GREM1 cloning

Because human and murine mRNAs and proteins for gremlin exhibit high homology (89% and 98%, respectively), all of the experiments were performed using the human sequence. GREM1 cDNA was purchased from the Mammalian Gene NIH Collection (Bethesda, Maryland USA). To facilitate the detection of human gremlin (as opposed to the endogenous mouse protein), we added a c-myc tag to the 3′ portion of GREM1 using PCR with the forward primer 5′AGTGCGGCGGCTGAGGACCC GCCGC-ACTGACAT-3′ and the reverse primer 5′-ATAGCCGCCGCT-TACAGATCCTCTTCTGAGATGAGTTTTTGTTCATCCA-AATCGATGGATATGC-3′. To add another signal to the transgene to facilitate detection in transfected cells, we inserted an e-GFP sequence downstream of human gremlin as follows. The IRES-eGFP sequence was obtained by PCR using a pIRES2-EGFP plasmid (Clontech Mountain View, CA, USA) as the template with the following primers: IRES-eGFP-F (5′-TACAT-TAATGGGCCCGGGATCCGCCCCTC-3′) and IRES-eGFP-R (5′-GGCCATATGCGCCTTAAGATACATTGATG-3′). The GREM1-c-myc and IRES-eGFP fragments were independently cloned into a pGEMT-Easy vector and then sequenced (Macrogen, Seoul, Korea) to confirm the modifications and absence of additional mutations. Next, both the GREM1-c-myc and IRES-eGFP fragments were subcloned into a modified pCDNA3 vector using the *Eco*RI and *Not*I restriction sites, respectively.

Transfection of EBNA293 and HK-2 cells

To determine whether the pCDNA3-GREM1-myc-IRES-eGFP generated stable proteins, EBNA293 and human renal proximal tubulo-epithelial (HK2 cell line, ATCC CRL-2190, Virginia, USA) cells were both grown in RPMI with 10% fetal bovine serum (FBS), 1% non-essential amino acids, 100 U/ml penicillin, 100 µg/ml streptomycin, Insulin Transferrin Selenium (5 µg/ml) and hydrocortisone (36 ng/ml) in 5% CO_2 at 37°C and were then transfected. EBNA293 cells were transfected with 0.6 µg of plasmid using LF2000 (Invitrogen, Carlsbad, CA, USA). Immunofluorescence was performed using a mouse anti-c-myc antibody (M4439, clone 9E10, Sigma, St. Louis, USA) at a 1/500 dilution followed by a donkey anti-mouse IgG conjugated to Cy3 antibody (Jackson ImmunoResearch, West Grove, PA, USA) at a 1/800 dilution. Images were captured using a Zeiss Axiovert 100M epifluorescence microscope. HK2 cells were grown in the aforementioned conditions and at 60–70% of confluence; cells were growth-arrested in serum-free medium for 24 hours before experiments. HK2 cells were transiently transfected for 48 hours using FuGENE (Roche, Basel Switzerland) and the pCDNA3-GREM1-c-myc-IRES2-eGFP plasmid vector. To study the expression of GREM1 in renal-specific tubulo-epithelial cells, HK2 cells were transiently transfected with pKAP-GREM1-c-myc-IRES2-eGFP to confirm the promoter activity. Immunocytochemical studies were performed in cells grown on coverslips. The cells were then fixed in Merckofix (Merck, Darmstadt, Germany) and permeabilized with 0.2% Triton-X100 for 1 min. After blocking with 4% bovine serum albumin and 8% of the corresponding serum (secondary antibody) for 1 hour, the cells were incubated with primary antibodies overnight at 4°C, followed by an Alexa Fluor 633-conjugated antibody (Invitrogen) at a 1/300 dilution for 1 hour. Negative control samples were processed in the absence of primary antibody. The cells were mounted in Mowiol 40–88 (Sigma) and examined using a Leica DM-IRB confocal microscope.

Transgenesis, genotyping and colony expansion

The pKAP plasmid, which has been previously shown to be specific in males [16–20], was used as the backbone for transgene generation. The pKAP plasmid was modified by excising an exon of the human angiotensinogen gene and the poly (A) signal. First, the IRES-eGFP sequence was subcloned downstream of GREM1-c-myc into pGEMT-Easy. The resulting construct was digested with *Sph*I and *Nde*I, and the 2.3 kb fragment containing GREM1-myc IRES-eGFP was subcloned into the modified pKAP plasmid. The 4.1 kb-long transgene was isolated with *Ase*I and *Eco*RV and purified using the QIAEx II kit (QIAGEN, Valencia, CA, USA). Five hundred molecules were microinjected into hybrid C57BL/6J x CBA/J zygotes, which were then transplanted into 13 pseudopregnant mothers. The born mice were maintained in a specific pathogen-free mouse facility in a 12-hour light:dark cycle with access to food and water *ad libitum*. The sacrifice of animals was done with administration of anaesthesia and analgesia, following the protocols approved by the Committee on the Ethics of Animal Experiments of Universidad Austral de Chile (Permit Number:20.2011), and FONDECYT Ethics Committee, and according to the NIH Guidelines. The founders and pups were screened using PCR with the following primers: *GREM1* intron 1F (5′-GCCAGTA AGGAATTCTAATAGG-3′), *KAP* promoter F (5′-ATGAGGACTCTAA TGCGTACAT-3′) and *GREM1* exon 2R (5′-TCCAAATCGATGGATA TGCAAC-3′). The PCR reaction generated two differential products (820 bp endogenous; 1040 bp TG). Once the genotype of the founders was confirmed, the founders were mated with pure C57BL/6J mice to expand the colony. F1 mice were screened using PCR as previously described and used for further molecular and phenotypic characterization studies.

Mouse characterization

First, kidneys from 4 to 5 week-old mice were analyzed by indirect immunofluorescence using a rabbit anti-GFP antibody at a 1/100 dilution (Invitrogen) and immunohistochemistry (IHC) to detect c-myc. Following confirmation that the female transgenic mice did not express the transgene, these mice were used only for mating and to evaluate the off/on system by the administration of 2.5 mg of testosterone via an intraperitoneal (i.p.) injection over five consecutive days [17,18]. Male TG and wild-type (WT) mice of five selected lines were also analyzed using western blotting analyses. The kidneys were dissected from both WT and TG mice homogenized in lysis buffer (125 mM Tris pH 6.8 and 1% SDS) supplemented with 1X protease inhibitor cocktail (Sigma P8340). Twenty micrograms of protein was electrophoresed in a 4–12% SDS-PAGE gel and transferred onto a PVDF membrane (Bio-Rad, Dreieich, Germany). The membrane was incubated with

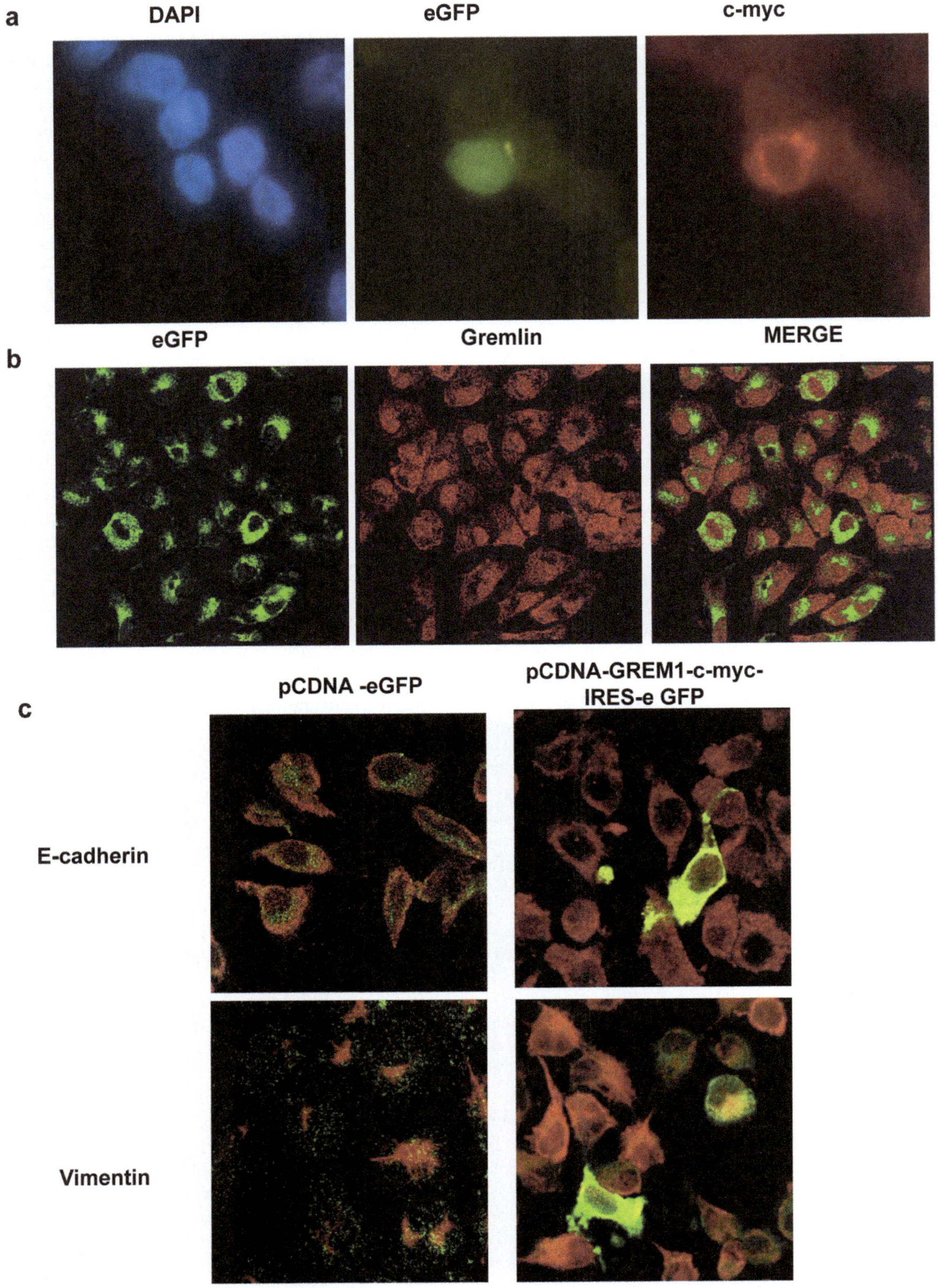

Figure 1. In vitro validation of the GREM1 plasmid. (a) EBNA293 cells were transfected with pCDNA3-GREM1-c-myc-IRES-eGFP, as described in the methods. Immunofluorescence shows that eGFP (green) and c-myc (red) are expressed in the same transfected EBNA293 cell. Nuclei were stained with DAPI (blue) (1000x). (b) In HK-2 cells transfected with pCDNA3-GREM1-c-myc-IRES-eGFP, GREM-1 expression was evaluated by immunocytochemistry using an antibody against GREM-1, followed by a secondary TRICT antibody (red staining). The figure shows eGFP and GREM-1 expressed in the same cell (800x). (c) Confocal immunofluorescence of HK-2 transfected cells, showing the loss of E-cadherin and induction of vimentin in eGFP-positive GREM-1-expressing cells (1600X). E-cadherin and vimentin immunostaining was detected with secondary anti-FITC antibodies (green).

anti-c-myc at 1/2500 or anti-actin at 1/5000 (Sigma). Densitometric analysis of the immunoreactive bands was performed using Quantity One software (Bio-Rad).

mRNA expression

Total RNA was extracted with TRIzol according to the manufacturers instructions and quantified using Qubit reagent (Invitrogen). RNA was treated with DNase I (Ambion, Austin TX, USA) to remove potential contamination and reverse transcribed using random primers and the ImProm-II kit (Promega) to synthesize double-stranded cDNA. qPCR was performed with the commercial reagent Maxima SYBR Green qPCR Master Mix (Promega, Madison WI, USA) to determine *GREM1, cyclophilin* and *GAPDH* mRNA expression levels using the following primers: human GREM1 F (5′-CCCGGGGAGGAGGTGCTGGAGT-3′); human GREM1 R (5′-CCGGATGTGCCTGGGGATGTA-GAA-3′); mouse cyclophilin1 F (5′-GCAGACATGGTCAACCC-CACCG-3′); mouse cyclophilin1 R (5′-GAAATTAGAGTTGT-CCACAGTCGG-3′); mouse GAPDH F(5′-TCCGCCCCTTCT-GCCGATG-3′); and mouse GAPDH R (5′-CACGGAAGGCC-ATGGCAGTGA-3′). PCR product specificity was verified by melting curve analysis, and all of the real-time PCR reactions were performed in triplicate. The $2^{-\Delta\Delta CT}$ method was used to analyze the relative changes in gene expression levels [21].

Induction of acute renal damage in TG mice

The adult TG lines A and D and wild-type (WT) male littermates aged 4 to 5 months were used. TG and WT mice were injected i.p. with 250 mg/kg body weight of FA (Sigma F7876), dissolved in the vehicle 0.3 M sodium bicarbonate (veh). Control animals, both WT and TG, received 0.3 ml of veh. Additional studies were done in transgenic homozygous mice from line A, injected with FA or vehicle (used as control, because there were not wild type littermates). Spot urine and serum were collected on days 0, 7 and 14 from all of the animals and analyzed for proteinuria and creatininuria using Bradford assay (Bio-Rad) and a Creatinina Wiener Lab Kit (Wiener Laboratorios, Rosario, Argentina), respectively. Seven or 14 days after the injection, the animals were anesthetized with 2% 2,2,2-tribromethanol (Sigma) dissolved in 2-methyl-buthanol (Sigma). The kidneys were removed, decapsulated and cut along the sagittal plane. The left kidney was fixed in 4% formaldehyde, while the right kidney was immediately frozen in liquid nitrogen and processed for RNA and protein extraction. The specimens were embedded in paraffin and cut into 4 μm tissue sections for further histological (PAS/Masson)

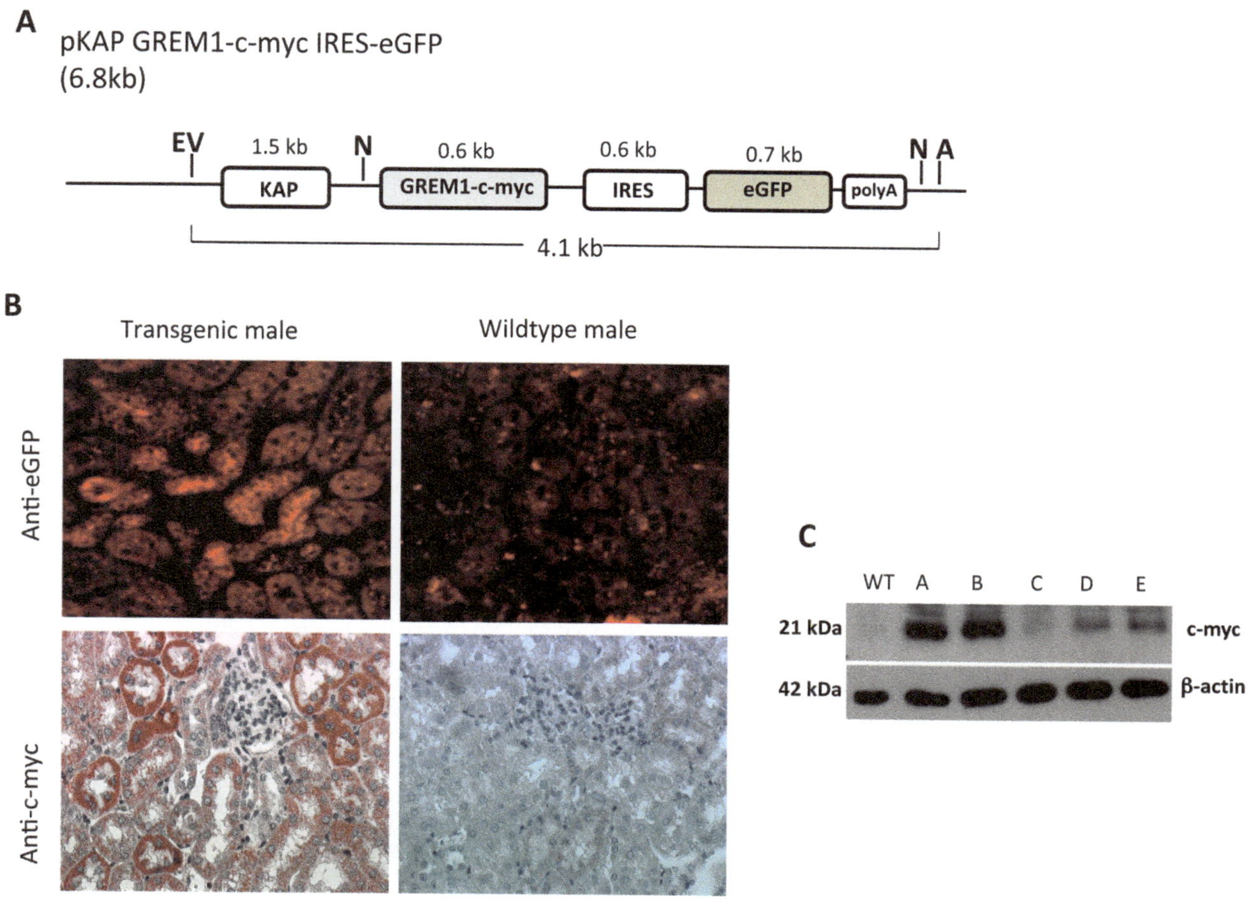

Figure 2. Generation and validation of transgenic mice with specific tubular GREM1 overexpression. (a) Illustration of the pKAP GREM1-c-myc-IRES-eGFP plasmid. Restriction sites used for transgene isolation are indicated with EV (*EcoRV*) and A (*AseI*). (b) **eGFP and c-myc detection in renal tubular epithelial cells of transgenic mice.** Immunofluorescence against eGFP and immunohistochemistry for c-myc (peroxidase immunostaining) to detect these proteins in the kidney tissue of transgenic males from line A and WT mice (400x). (c) Kidneys were dissected from WT and transgenic male mice of lines A, B, C, D and E, and isolated proteins were subjected to western blotting using an antibody against c-myc (1:1000); anti β-actin (1:2500) was used as a loading control. GREM1 expression was determined by densitometric analysis of the c-myc/β-actin ratio and normalized to transgenic line C expression.

Table 1. Molecular characterization of transgenic mice.

Line	Copy number	Transmittance F1–F2 (%)	eGFP signal (IIF)	GREM1 levels (mRNA)	GREM1 levels (c-myc)	Male/Female (%)	Ectopicity
A	44	54	+++	190 ± 92	7.54	55/45	2
B	2–3	53	+	3.9 ± 0.4	6.38	52/48	1
C	2	50	++	2.2 ± 0.7	1.00	45/55	2
D	8–18	81***	++	10.5 ± 11.1	2.07	53/47	2
E	24	49	++	9.7 ± 7.2	2.4	48/52	1
F	11	46	+	54.6 ± 29.9	n.d	48/52	1

Qualitative and quantitative data from the transgenic lines are shown. Copy number and GREM1 expression levels were determined using real-time PCR. The eGFP signal was qualitatively analyzed. c-myc levels were quantified by densitometric analysis in each line and normalized to line C. Ectopic expression is indicated as the number of the eight analyzed extrarenal organs that demonstrated increased for GREM1 expression compared with wild-type expression. M, male; F, female; n.d., no data. *** $p < 0.001$.

and IHC studies using antibodies against gremlin and αSMA, F4/80, CD3 and PCNA.

Histological analysis and IHC

Tubular and interstitial lesions were graded from 0 to 4 and analyzed as previously described [22]. IHC for different markers was performed following heat-induced epitope retrieval (microwaving for 10 min in citrate buffer), and sections were incubated overnight with rabbit anti-human gremlin 1/300 (AP6133a, Abgent, San Diego, CA, USA), followed by incubation with Impress anti-rabbit reagent (Vector, Burlingame, CA, USA); or mouse anti-αSMA 1/100 (DAKO, Carpinteria, CA, USA), monoclonal anti-c-myc clone 9E10 1/300 (Thermo, Rockford IL, USA) or anti-PCNA (PC10, DAKO) followed by incubation with the M.O.M. Immunodetection kit (PK 2200 Vector). All of the tissue sections were developed using AMEC red chromogen (SK 4285, Vector) or DAB, and counterstained with hematoxylin. Interstitial infiltrating cells were detected by mean of F4/80 (monocytes/macrophages) and CD3 (T lymphocytes) antibodies. F4/80 was detected by using the MA1-91124 antibody (dilution: 1/100, THERMO, Rockford, IL, USA) followed by Immpress Reagent Kit (MP 7444, Vector, USA) and CD3 was detected using Trilogy epitope retrieval (Cell Marque, Rocklin, USA) and A 0452 (181–195) antibody (dilution: 1/200, DAKO, USA) followed by horseradish peroxidase streptavidin (dilution 1:500, SA-5004 Vector, USA), reveled with DAB, and counterstained with hematoxylin.

Murine Gremlin was detected using anti-murine Gremlin antibody (dilution: 1/20 AF 956, R&D Systems, Minneapolis, MN, USA) overnight at 4°C, and the reaction was developed with Impact NovaRed SK-4805 (Vector, USA).

Image analysis and quantification of the IHC signals were performed using the KS300 imaging system, version 3.0 (Zeiss). For each sample, the mean staining area was obtained by an analysis of 20 fields (20x). The staining score is expressed as the mm^2/dens

Statistical analysis

The results were expressed as the means ± SEM. Two-tailed chi-square tests were performed to determine the statistical significance of the viability of the transgene and the proportion of female:male pups born in the TG lines. A factorial ANOVA followed by the Tukey test was performed to compare proteinuria and *Grem1* mRNA expression. The Mann-Whitney U-test was performed to compare the tubular/interstitial lesions and Grem1, αSMA, F4/80 and CD3 IHC signals in the TG and WT mice injected with FA. A Spearman rank correlation was performed to determine the correlation between GREM1 and TGF-β, and Kruskal-Wallis analysis was performed to evaluate endogenous murine gremlin and PCNA expression. Values of $p< 0.05$ were considered significant.

Results

Generation of TG GREM1 mice

Validation of the GREM1 expression vector. Expression of pCDNA3-GREM1-c-myc-IRES-eGFP in EBNA293 cells revealed that this construct produced stable proteins of c-myc-GREM1 and eGFP, and both were expressed simultaneously in these cells (Figure 1a). To distinguish the human TG protein from the murine endogenous gremlin, a c-myc tag was fused to the GREM1 protein, and we used eGFP as reporter for transgene expression. *In vitro* experiments in EBNA293 cells were performed to detect the c-myc signal and showed an expression

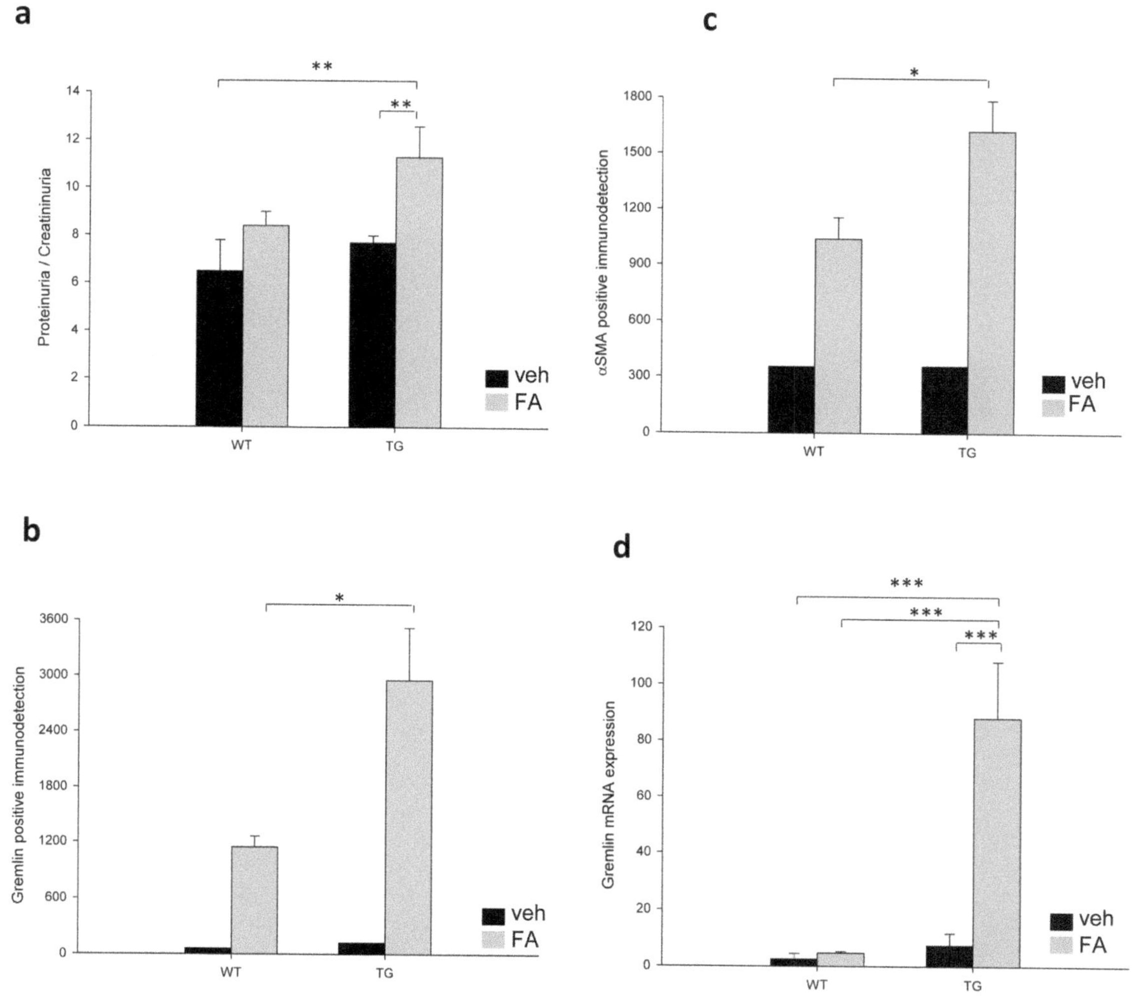

Figure 3. FA injection induces proteinuria and GREM1 and αSMA expression in transgenic line D. (a) The urinary protein to creatinine ratio (μg/mg) was examined in each experimental group. Positive IHC signals were quantified for (b) gremlin and (c) αSMA with KS300 image analyzer software. (d) Gremlin mRNA expression levels were determined using real-time PCR. The four parameters were significantly increased in GREM1-overexpressing transgenic mice. Data are shown as the mean ± SEM of 5-6 mice per group * $p < 0.05$; ** $p < 0.01$; *** $p < 0.001$. TG vs WT control

pattern consistent with GREM1 subcellular localization [1,2]; in a similar pattern to that found in human renal biopsies [9–11], where this protein is localized in cytoplasm and nucleus of the affected tubular epithelial cells. HK-2 cells transfected with pCDNA3-GREM1-c-myc-IRES-eGFP were positive for GREM1 (Figure 1b), which confirmed that eGFP could be used as a reporter for *in vivo* experiments and that c-myc could be used to differentiate human GREM1 in TG animals.

To validate our TG construction, exclude loss of function of the GREM1-c-myc-fused protein and determine whether GREM1 was functional, several experiments were performed in HK2 cells. We have previously demonstrated that stimulation with recombinant GREM1 protein in HK2 cells induces phenotypic changes related to epithelial to mesenchymal transition (EMT) [12]. Transient transfection of these cells with pCDNA3 GREM1-c-myc IRES-eGFP also induced characteristic EMT features, such as the downregulation of E-cadherin immunostaining and an induction of vimentin expression, as well as changes in the cell phenotype to a fibroblast-like morphology, confirming that our expression vector displayed similar effects to the GREM1 recombinant protein (Figure 1c).

Thus, to specifically induce the expression of GREM1 in proximal tubular renal cells, we used the promoter of kidney androgen-regulated protein (pKAP) to drive transgene expression because it is transcriptionally active only in these cells and its activity is testosterone dependent.

Transgene isolation was performed by digestion of the pKAP GREM1-c-myc IRES-eGFP plasmid (Figure 2a) with *Ase*I and *Eco*RV, which generated three fragments of 4.1, 1.5 and 1.2 kb. The 4.1 kb fragment was purified and quantified to obtain 500 molecules/picoliter and then microinjected into C57BL/6J x CBA/J hybrid zygotes. Ninety percent (232/259) of the microinjected zygotes were transferred to pseudopregnant mothers. Eight of the 57 (14%) pups born were confirmed to be TG using PCR. All of the founders reached the age of 3 months, and seven of the founders were subsequently mated with pure C57BL/6 to generate F1 mice.

Molecular and phenotypic characterization of GREM1 mice

TG lines (named in alphabetic order (A–G)) demonstrated normal fertility and litter sizes. Transgene transmission from all of the lines was verified using PCR analysis and appeared normal in

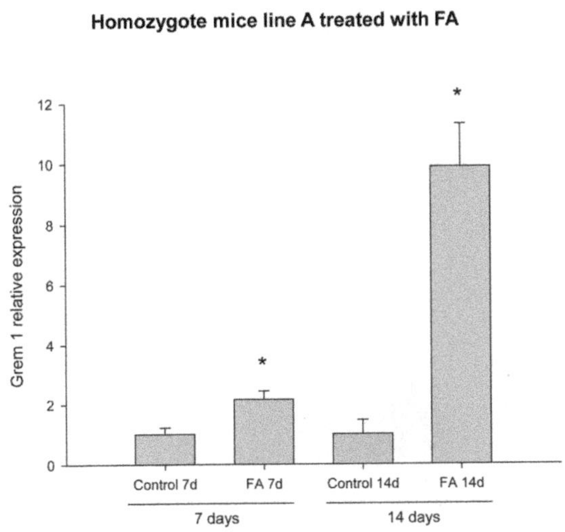

Figure 4. Effect of FA administration on GREM1 expression in transgenic line A homozygous mice. Gremlin expression in transgenic mice was examined 7 and 14 days after treatment with FA. The GREM1 relative expression in homozygous mice of transgenic line A was increased at 7 days after injection with FA and remained increased at 14 days. Data are shown as the mean ± SEM of 4-9 mice per group * p≤0.0049 TG-FA vs TG-Veh, used as control because no WT littermates of the transgenic homozygotes mice were available.

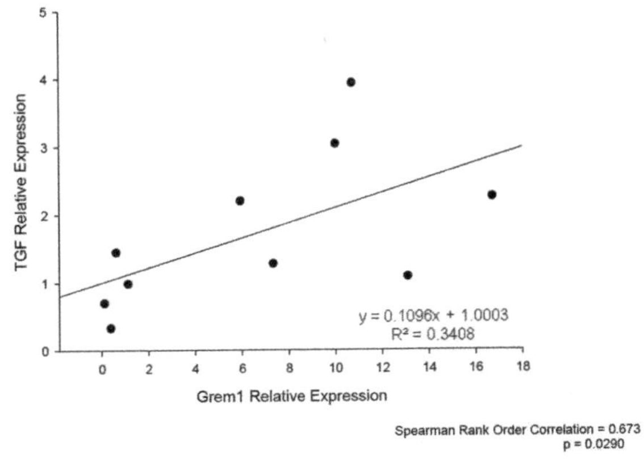

Figure 5. Correlation of TGF-β and GREM1 expression in transgenic line A homozygous mice. TGF-β gene expression was measured in FA-injected GREM1 transgenic mice. We observed a strongly positive correlation between TGF-β and GREM1 expression (p≤ 0.029, R = 0.67; 4-9 mice per group).

the first and second (F1 and F2) generations in all lines but was significantly increased in line D ($p \leq 0.001$) (Table 1). In all TG lines, Mendelian ratios were observed between TG/WT males and females at birth, discarding an effect of transgene expression on male survival during gestation (Table 1).

Males and females were subjected to immunofluorescence analysis for eGFP and immunohistochemistry for c-myc. Males showed specific tubular epithelial expression of eGFP that was qualitatively variable in each TG line (Table 1). The eGFP and c-myc signals were not detected in WT males (Figure 2b). Moreover, this signal was not detected in TG females but could be induced in testosterone-treated TG females (data not shown), verifying the functionality of the promoter as previously described [18,20].

To confirm that the TG mice expressed a stable TG protein, western blotting analysis for c-myc-tagged GREM1 was performed. The assay detected a protein with the expected molecular weight (~21 kDa) in the kidneys of the TG lines, and this c-myc-tagged protein was absent in WT mice (Figure 2c). Densitometric analysis revealed that the GREM1 expression levels were variable in each TG line and ranged between 1 (line C exhibited the lowest expression) and 7.5 (line A exhibited the highest expression) (Table 1). Line G was not used in further experiments because the RT-PCR analysis detected ectopic expression of GREM1 in seven extra-renal tissues. The other six TG lines showed low extra-renal expression (cerebellum, data not shown). The levels of GREM1 mRNA in the kidney ranged between a 2- and 200-fold increase compared with the WT mice and were related to the transgene copy number (2 to 44 transgene copy number) (Table 1). Proteinuria (urine protein to creatinine ratio) was monitored in the first generation of males every 2 weeks until the age of 6 months in the three groups (WT and TG lines B and D), and no abnormal increases were observed compared with the WT mice, indicating that GREM1 expression alone was not sufficient to develop renal injury. Moreover, no histological lesions were observed at 6 months of age in any group.

Renal injury induction in GREM1 TG mice

To determine the effect of GREM1 in renal damage *in vivo*, we selected mice from two lines and challenged these mice with a model of FA-mediated acute renal failure [23].

The line D, TG mice (8–18 transgene copies) were evaluated first. Seven days after FA injection, TG mice presented a significant increase in proteinuria compared with the treated WT mice (Figure 3a). However, the morphological lesions score found at 7 days, did not reach statistical significance between TG-FA compared with WT-FA (TG 9.0 ± 1.0 vs. WT 8.0 ± 2.1; $p = 0.328$, $n = 5$–6 mice per group). To further evaluate profibrotic markers, IHC for αSMA, as the first phenotypic marker of activated fibroblasts, was performed. In FA-TG mice, αSMA was markedly upregulated, showing a significant increase compared with FA-treated WT mice (Figure 3c).

Gremlin expression was also evaluated using real-time PCR. In FA-injected TG mice, (TG-FA) gremlin mRNA expression showed a nearly 10-fold increase compared with veh-injected mice and was significantly higher than FA-treated WT mice (WT-FA) (Figure 3d). Similar findings were observed by IHC, with a significant increase in gremlin staining after FA injection in TG mice compared with WT-FA mice (Figure 3b). Furthermore, we found a significant correlation between αSMA and gremlin IHC signals ($F_{1,10}$: 14.34; $p < 0.0356$; $r = 0.5892$), indicating that gremlin is associated with the variation in αSMA ($r = 0.5892$). Taken together, these results indicated that the effect observed *in vitro* [12] could be corroborated *in vivo*.

To further evaluate the effect of gremlin in FA-induced damage and determine if this was a transient or sustained effect, we performed additional studies using homozygous mice from TG line A and additionally evaluated its expression 14 days after treatment with FA. This line had the greatest number of copies of GREM1 (88 transgene copies) and gremlin mRNA. By real time PCR, we demonstrated that Grem1 was significantly increased at 7 and 14 days after FA injection (Figure 4) and by IHC gremlin renal staining was also increased at 7 and 14 days, with higher level at 14 days after treatment (score, expressed as mean of percentage/mm^2: TG-VEH; 14, TG-FA at 7 days; 343, TG-FA at 14 days; 862, $p \leq 0.0049$), indicating that GREM1 expression in TG mice induces higher renal synthesis of gremlin following FA administration. Additionally, gremlin has been proposed as a downstream mediator of TGF-β [6], and we have previously

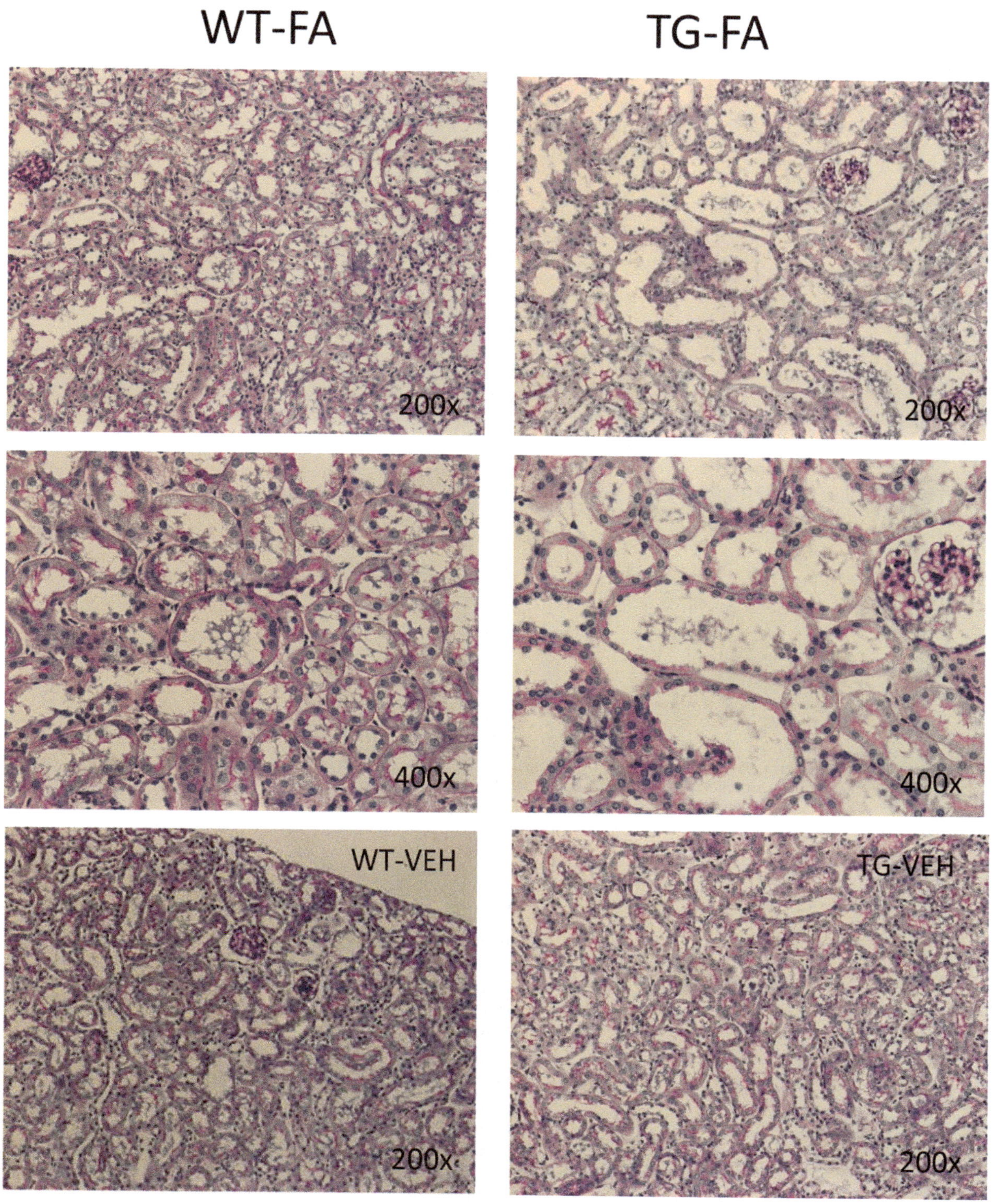

Figure 6. Histological analysis (PAS, Masson) of FA-injected mice in transgenic line A homozygous mice. TG-FA mice (right column) showed more severe morphological lesions (tubular dilatation, flattening of tubular epithelial cells, hyaline casts, interstitial infiltrating cells and mild interstitial fibrosis) compared with WT mice (left column). $p < 0.05$ (200x-400x). Control vehicle treated mice are shown at bottom. Figure shows representative mice of each group of 14–18 studied.

reported that TGF-β induced gremlin expression in renal tubular cells *in vitro* [12]. Therefore, to further evaluate profibrotic factors, TGF-β gene expression was measured. In FA-injected GREM1 TG mice, an increased relative renal expression of TGF-β at 7 and 14 days after treatment was observed compared with WT mice. Moreover, a strongly positive correlation was found between TGF-β and gremlin expression ($p \leq 0.029$, $R = 0.67$) (Figure 5). Renal histological analysis at 14 days showed more severe morphological lesions (tubular dilatation, epithelium flattening, hyaline casts, interstitial cell infiltration and mild interstitial fibrosis) in TG-FA mice compared with WT-FA mice ($p = 0.0037$ Fisher's test, and Mann-Whitney $p < 0.01$) (Figure 6).

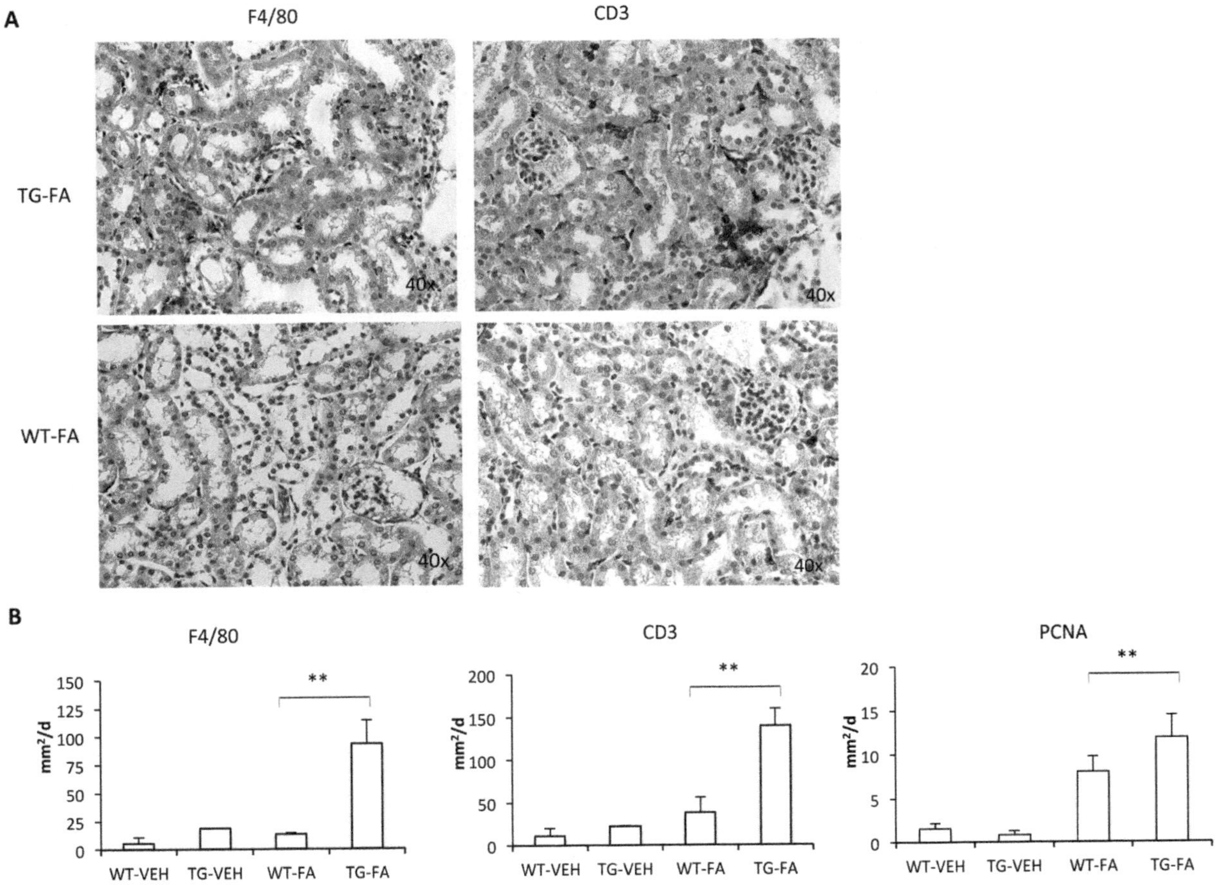

Figure 7. FA injection induces interstitial cell infiltration in transgenic line A homozygous mice. The inflammatory cell infiltration was characterized by immunohistochemistry with anti-F4/80 (monocytes/macrophages) and anti-CD3 (T cells) antibodies. (A and B). Representative immunostaining of one mouse from each group (x400 magnification). (C) Quantification of positive IHC signals were quantified for (a) F4/80, (b) CD3 and (c) PCNA using KS300 image analyzer software. All parameters were significantly increased in transgenic mice. Data are shown as the mean ± SEM of 14–18 mice per group * $p < 0.05$; ** $p < 0.01$ vs WT-FA.

To further evaluate inflammatory cell infiltration and proliferation, IHC for F4/80 (murine macrophages), CD3 (murine T cells), and PCNA was performed. In TG-FA mice, all markers were strongly upregulated, showing a significant increase compared with WT-FA mice (Figure 7). Also, a positive correlation was found between gremlin expression and PCNA at 14 days ($R = 0.88$, $p \leq 0.0019$; data not shown)

Discussion

We reported here, for the first time, the generation of TG mice expressing GREM1 in a sex- and renal tubular cell-specific manner with no evident lethal effects and normal renal function and morphology. This mouse model was designed as a molecular tool to analyze the effect of gremlin expression in renal damage. Our results suggest that under normal conditions, GREM1 in adult tubular cells is not sufficient to cause renal damage. However, in response to acute renal injury caused by FA injection, GREM1 TG mice presented exacerbated renal damage, suggesting that gremlin could participate in renal damage progression *in vivo*. Gremlin is a developmental gene involved in renal morphogenesis due to its role as a BMP antagonist, but its function in the adult kidney is unknown. Several *in vitro* studies have evaluated the effect of gremlin in renal cells; however, the *in vivo* function has not been investigated. In tubular epithelial cells *in vitro*, overexpression of this expression vector resembles the effect of recombinant GREM1, inducing phenotypic changes related to EMT [12].

These pKAP GREM1-overexpressing mice, which presented a specific GREM1 expression pattern in kidney tubular epithelial cells, demonstrated normal renal function and morphology. The specific pKAP promoter [20], motivated the specific cell type and hormone-regulated targeting of transgene expression. In addition, due to the developmental role of gremlin, expression driven by the pKAP promoter prevented any lethal effects at embryonic stages and generated a testosterone-dependent off/on switch in TG females.

Studies in TG mice overexpressing profibrotic factors, such as connective tissue growth factor (CTGF-CCN2) in different tissues have shown similar findings, as observed in our renal GREM1 TG mice. Although *in vitro* studies have shown that recombinant CCN2 increased extracellular matrix production [24], as observed with gremlin, several *in vivo* studies have shown that CCN2 alone is not sufficient to cause ongoing profibrotic changes. In the kidney, podocyte-specific CCN2-TG mice (in C57BL/6 background) exhibit no glomerular abnormalities, proteinuria or matrix accumulation [25]. In C57BL/6 mice, systemic CCN2 administration has been shown induce a transient overexpression of profibrotic genes at day 5, but it is not sufficient to induce progressive fibrosis [26], as observed following CCN2 overexpression in rat lungs [27]. In a mouse skin model, only coinjection of CCN2 and TGF-β1, not either cytokine alone, caused persistent fibrosis [28].

Several authors have suggested that gremlin could be considered as a mediator of renal injury in diabetic nephropathy, based on experimental studies showing a beneficial effect of gremlin inhibition [13,14]. In response to FA-induced acute renal damage, GREM1 TG mice developed higher proteinuria after 7 and 14 days than WT mice. Tubular GREM1 overexpression was associated with renal upregulation of profibrotic factors, such as TGF-β and αSMA, recruitment of F4/80 and CD3 positive cells, and increased cell proliferation in TG mice challenged with FA compared with WT mice. Furthermore, as hypothesized these GREM1 overexpressing mice developed more severe histological damage in response to FA injection at 14 days, particularly those mice with more transgenic copies and at the time of more gremlin expression.

In biopsies from patients with diabetic nephropathy, we have demonstrated that GREM1 is expressed in areas of tubular-interstitial fibrosis and that it colocalizes with profibrotic markers, including TGF-β, αSMA and vimentin. These changes also correlate directly with renal dysfunction, as shown by serum creatinine levels [9]. All these data suggest a role for gremlin in the pathogenesis of kidney damage.

Indeed, in the transgenic mice we found more acute tubular injury induced by folic acid than chronic damage progression. However it is important to note, than acute tubular injury is clearly associated with progression towards end stage renal disease [29], and on the other hand we found a significant interstitial cell infiltration that is commonly considered as the major initial mechanism leading to renal fibrosis.

Some evidence supports a potential interrelation between TGF-β1 and gremlin responses. We have previously demonstrated that *in vitro* blockade of endogenous gremlin by a specific siRNA inhibits TGF-β1-induced profibrotic gene overexpression and extracellular matrix production in renal fibroblasts. Moreover, gremlin blockade inhibits TGF-β1-mediated phenotypic-changes in tubular epithelial cells [12]. Even more, recently, it has been reported that gremlin likely induces endogenous TGF-β/Smad signaling, resulting in podocyte injury in mouse podocytes cultured in high glucose conditions [30]. Many data suggest that gremlin could be an important promoter of fibrosis in different pathologies, including liver fibrosis and lung diseases, particularly pulmonary hypertension, idiopathic pulmonary fibrosis and cancer invasion [31-35], as we have shown here in an experimental model of renal damage.

Our results suggest that GREM1-overexpressing mice have an increased susceptibility to renal damage, supporting the involvement of gremlin in renal damage progression.

Acknowledgments

The authors thank Dr. Juan Young for gifting the plasmid pMeCP2-flag IRES-eGFP (CECs, Chile) and Dr. Curt Sigmund from the University of Iowa, USA, for providing the pKAP2 plasmid. We thank Ms. Vanesa Marchant for her technical help.

Author Contributions

Conceived and designed the experiments: AD PK LA KW MRO JE SM. Performed the experiments: AD PK MEB GV RRD BK. Analyzed the data: AD PK MEB GV DC LA RRD BK KW MRO JE SM. Contributed reagents/materials/analysis tools: AD PK DC LA RRD BK KW MRO JE SM. Wrote the paper: AD PK BK MRO JE SM.

References

1. Topol LZ, Marx M, Laugier D, Bogdanova NN, Boubnov NV, et al. (1997) Identification of dmr, a novel gene whose expression is suppressed in transformed cells and which can inhibit growth of normal but not transformed cells in culture. Mol Cell Biol 17: 4801–4810.
2. Topol LZ, Bardot B, Zhang Q, Resau J, Huillard E, et al. (2000) Biosynthesis, post-translation modification, and functional characterization of Drm/Gremlin. J Biol Chem 275: 8785–8793.
3. Hsu DR, Economides AN, Wang X, Eimon PM, Harland RM (1998) The Xenopus dorsalizing factor Gremlin identifies a novel family of secreted proteins that antagonize BMP activities. Mol Cell1: 673–683.
4. Michos O, Panman L, Vintersten K, Beier K, Zeller R, et al. (2004) Gremlin-mediated BMP antagonism induces the epithelial-mesenchymal feedback signaling controlling metanephric kidney and limb organogenesis. Development 131: 3401–3410.
5. Roxburgh SA, Murphy M, Pollock CA, Brazil D (2006) Recapitulation of embryological programmes in renal fibrosis–the importance of epithelial cell plasticity and developmental genes. Nephron Physiol 103: 139–148.
6. McMahon R, Murphy M, Clarkson M, Taal M, Mackensie H, et al. (2000) IHG-2, a mesangial cell gene induced by high glucose, is human gremlin. Regulation by extracellular glucose concentration, cyclic mechanical strain, and transforming growth factor-beta1. J Biol Chem 275: 9901–9904.
7. Murphy M, Godson C, Cannon S, Kato S, Mackenzie HS, et al. (1999) Suppression subtractive hybridization identifies high glucose levels as a stimulus for expression of connective tissue growth factor and other genes in human mesangial cells. J Biol Chem 274: 5830–5834.
8. Lappin DW, McMahon R, Murphy M, Brady HR (2002) Gremlin: an example of the re-emergence of developmental programmes in diabetic nephropathy. Nephrol Dial Transplant 17s9: 65–67.
9. Dolan V, Murphy M, Sadlier D, Lappin D, Doran P, et al. (2005) Expression of gremlin, a bone morphogenetic protein antagonist, in human diabetic nephropathy. Am J Kidney Dis 45: 1034–1039.
10. Mezzano S, Droguett A, Burgos ME, Aros C, Ardiles L, et al. (2007) Expression of gremlin, a bone morphogenetic protein antagonist, in glomerular crescents of pauci-immune glomerulonephritis. Nephrol Dial Transplant 22: 1882–1890.
11. Carvajal G, Droguett A, Burgos ME, Aros C, Ardiles L, et al. (2008) Gremlin: a novel mediator of epithelial mesenchymal transition and fibrosis in chronic allograft nephropathy. Transplant Proc 40: 734–739.
12. Rodríguez Diez R, Lavoz C, Carvajal G, Rayego-Mateos S, Rodriguez Diez RR, et al. (2012) Gremlin is a downstream profibrotic mediator of transforming growth factor-beta in cultured renal cells. Nephron Exp Nephrol 122: 62–74.
13. Roxburgh SA, Kattla JJ, Curran SP, O'Meara YM, Pollock CA, et al. (2009) Allelic depletion of grem1 attenuates diabetic kidney disease. Diabetes 58: 1641–1650.
14. Zhang Q, Shi Y, Wada J, Malakauskas SM, Liu M, et al. (2010) In vivo delivery of Gremlin siRNA plasmid reveals therapeutic potential against diabetic nephropathy by recovering bone morphogenetic protein-7. PLoS One 5: e11709.
15. Farkas L, Farkas D, Gauldie J, Warburton D, Shi W, et al. (2011) Transient overexpression of Gremlin results in epithelial activation and reversible fibrosis in rat lungs. Am J Respir Cell Mol Biol 44: 870–878.
16. Ding Y, Davisson RL, Hardy DO, Zhu L, MerrilL DC, et al. (1997) The kidney androgen-regulated protein promoter confers renal proximal tubule cell-specific and highly androgen-responsive expression on the human angiotensinogen gene in transgenic mice. J Biol Chem 272: 28142–28148.
17. Ding Y, Sigmund C (2001) Androgen-dependent regulation of human angiotensinogen expression in KAP-hAGT transgenic mice. Am J Physiol Renal Physiol 280: F54–F60.
18. Lavoie JL, Lake-Bruse KD, Sigmund CD (2004) Increased blood pressure in transgenic mice expressing both human renin and angiotensinogen in the renal proximal tubule. Am J Physiol Renal Physiol 286: F965–F971.
19. Sachetelli S, Liu Q, Zhang SL, Liu F, Hsieh TJ, et al. (2006) RAS blockade decreases blood pressure and proteinuria in transgenic mice overexpressing rat angiotensinogen gene in the kidney. Kidney Int 69: 1016–1023.
20. Li H, Zhou X, Davis DR, Xu D, Sigmund CD (2008) An androgen-inducible proximal tubule-specific Cre recombinase transgenic model. Am J Physiol Renal Physiol 294: F1481–F1486.
21. Livak KJ, Schmittgen TD (2001) Analysis of relative gene expression data using real-time quantitative PCR and the 2(-Delta Delta C(T)) Method. Methods 25: 402–408.
22. Zoja C, Corna D, Camozzi D, Cattaneo D, Rottoli D, et al. (2002) How to fully protect the kidney in a severe model of progressive nephropathy: a multidrug approach. J Am Soc Nephrol 13: 2898–2908.
23. Ortega A, Rámila D, Ardura JA, Esteban V, Ruiz-Ortega M, et al. (2006) Role of parathyroid hormone-related protein in tubulointerstitial apoptosis and fibrosis after folic acid-induced nephrotoxicity. J Am Soc Nephrol 17: 1594–1603.
24. Phanish MK, Winn SK, Dockrell ME (2010) Connective tissue growth factor-(CTGF, CCN2)- a marker mediator and therapeutic target for renal fibrosis. Nephron Exp Nephrol 114: e83–92.

25. Yokoi H, Mukoyama M, Mori K, Kasahara M, Suganami T, et al. (2008) Overexpression of connective tissue growth factor in podocytes worsens diabetic nephropathy in mice. Kidney Int 73: 446–455.
26. Alfaro MP, Deskins DL, Wallus M, DasGupta J, Davidson JM, et al. (2013) A physiological role for connective tissue growth factor in early wound healing. Lab Invest 93: 81–95.
27. Bonniaud P, Margetts PJ, Kolb M, Haberberger T, Kelly M, et al. (2003) Adenoviral gene transfer of connective tissue growth factor in the lung induces transient fibrosis. Am J Respir Crit Care Med 168: 770–778.
28. Mori T, Kawara S, Shinozaki M, Hayashi N, Kakinuma T, et al. (1999) Role and interaction of connective tissue growth factor with transforming growth factor-beta in persistent fibrosis: a mouse fibrosis model. J Cell Physiol 181: 153–159.
29. Gentle ME, Shi S, Daehn I, Zhang T, Qi H, et al. (2013) Epithelial cell TGF-β signaling induces acute tubular injury and interstitial inflammation. J Am Soc Nephrol 24:787–799
30. Li G, Li Y, Liu S, Shi Y, Chi Y, et al. (2013) Gremlin aggravates hyperglycemia-induced podocyte injury by a TGFβ/Smad dependent signaling pathway. J Cell Biochem 114: 2101–2112.
31. Guimei M, Baddour N, Elkaffash D, Abdou L, Taher Y (2012) Gremlin in the pathogenesis of hepatocellular carcinoma complicating chronic hepatitis C: an immunohistochemical and PCR study of human liver biopsies. BMC Res Notes 5: 390.
32. Costello CM, Cahill E, Martin F, Gaine S, Mc Loughlin P (2010) Role of Gremlin in the lung: development and disease. Am J Respir Cell Mol Biol 42: 517–23.
33. Cahill E, Costello CM, Rowan SC, Harkin S, Howell K, et al. (2012) Gremlin plays a key role in the pathogenesis of pulmonary hypertension. Circulation 125: 920–930.
34. Koli K, Myllärniemi M, Vuorinen K, Salmenkivi K, Ryynänen MJ, et al. (2006) Bone morphogenetic protein-4 inhibitor gremlin is overexpressed in idiopathic pulmonary fibrosis. Am J Pathol 169: 61–71.
35. Karagiannis GS, Berk A, Dimitromanolakis A, Diamandis EP (2013) Enrichment map profiling of the cancer invasion front suggests regulation of colorectal cancer progression by the bone morphogenetic protein antagonist, gremlin-1. Mol Oncol7:826–839.

3

An Enzymatic Platform for the Synthesis of Isoprenoid Precursors

Sofia B. Rodriguez, Thomas S. Leyh*

The Department of Microbiology & Immunology, The Albert Einstein College of Medicine, Bronx, New York, United States of America

Abstract

The isoprenoid family of compounds is estimated to contain ~65,000 unique structures including medicines, fragrances, and biofuels. Due to their structural complexity, many isoprenoids can only be obtained by extraction from natural sources, an inherently risky and costly process. Consequently, the biotechnology industry is attempting to genetically engineer microorganisms that can produce isoprenoid-based drugs and fuels on a commercial scale. Isoprenoid backbones are constructed from two, five-carbon building blocks, isopentenyl 5-pyrophosphate and dimethylallyl 5-pyrophosphate, which are end-products of either the mevalonate or non-mevalonate pathways. By linking the HMG-CoA reductase pathway (which produces mevalonate) to the mevalonate pathway, these building block can be synthesized enzymatically from acetate, ATP, NAD(P)H and CoA. Here, the enzymes in these pathways are used to produce pathway intermediates and end-products in single-pot reactions and in remarkably high yield, ~85%. A strategy for the regio-specific incorporation of isotopes into isoprenoid backbones is developed and used to synthesize a series of isotopomers of diphosphomevalonate, the immediate end-product of the mevalonate pathway. The enzymatic system is shown to be robust and capable of producing quantities of product in aqueous solutions that meet or exceed the highest levels achieved using genetically engineered organisms in high-density fermentation.

Editor: Y-H Percival Zhang, Virginia Tech, United States of America

Funding: This study was supported by NIH GM54469, NIH AI068989, NIH GM106158. The funders had no role in study design, data collection and analysis, decision to publish, or preparation of the manuscript.

Competing Interests: The authors have declared that no competing interests exist.

* Email: tom.leyh@einstein.yu.edu

Introduction

Improved access to large numbers of pure proteins, and a rapidly increasing repertoire of well characterized enzymes, isoenzymes and mutants have substantially increased the potential to utilize *in situ* metabolic pathways, or concatenated enzymatic reactions, in the synthesis of complex natural and synthetic products. Enzymes have been honed over evolutionary time to accomplish specific catalytic tasks [1,2]. Many are extremely efficient, regio-selective catalysts, while others exhibit broad substrate specificities that can provide flexibility in synthetic schemes. Indeed, significant efforts are underway to develop enzymes whose catalytic properties have been altered to achieve specific synthetic goals [3–5]. Enzymatic synthesis has been used to produce numerous valuable compounds [6–16] and often provides significant enhancements in yield, purity, production time and cost when compared to traditional chemical synthetic methods [17,18]. Considerable effort is being expended to develop cell-free enzymatic systems for the production of biofuels, including dihydrogen [19] and butanol [20], biomass conversion to starch [21], and high-energy-density biobatteries [22]. While enzymatic synthesis will never replace traditional synthesis, it provides a valuable adjunct to traditional approaches particularly when the objective is to build complex natural products.

The medicinal values of isoprenoids have been documented as early as 168 BC [23,24]. Today, we are only beginning to understand the social and commercial potential of this enormous, diverse family of natural compounds, which is estimated to contain approximately 65,000 unique structures [25]. Biotechnology companies are attempting to synthesize isopreonoid-based medicines, cosmetics [26], flavors [27], fragrances [28] and biofuels [29–31] by genetically engineering plants and bacteria to produce desired isoprenoids in commercial quantities [29,32–34]. Recent efforts along these lines include attempts to genetically engineer organisms to produce artemesinin (an antimalarial) at costs that will significantly expand third-world access to this drug [35,36], and to produce isoprenoid-based fuels [30,31].

The carbon backbones of isoprenoids are assembled from two fundamental building blocks, isopentenyl 5-pyrophosphate and dimethylallyl 5-pyrophosphate [37–40]. By linking the HMG-CoA reductase pathway, which produces mevalonate, to the mevalonate pathway, these building blocks can be enzymatically assembled from acetate, ATP, NAD(P)H, and CoA (Fig 1). Alternatively, they can be synthesized using the so-called non-mevalonate pathway [41], which is mechanistically more complex and less well defined [42]. Here, ten enzymes, including those that comprise the HMG-CoA reductase and mevalonate pathways [43–48] are strategically employed to accomplish efficient, high-yielding (>85%) single-pot syntheses of the intermediates and endproducts of the mevalonate pathway. Labeling strategies that regio-specifically position carbon and hydrogen isotopes into the building-block backbone are developed and used to synthesize and purify isotopomers of the immediate endproduct of the mevalonate pathway, diphosphomevalonate (DPM, Fig 2) [44]. Finally, the enzymatic system is shown to be robust and capable of producing pathway end-products in simple, aqueous solutions at levels that match or exceed the highest reported levels, which are only achieved using high-density fermentation.

Materials and Methods

Materials

Lactate dehydrogenase (rabbit muscle), pyruvate kinase (rabbit muscle), and inorganic pyrophosphatase (Baker's yeast) were purchased from Roche Applied Science. (*R*, *S*)-[$^2H_3$]methyl-mevalonolactone, (*R*, *S*)-mevalonolactone, acetyl-CoA, glutamate dehydrogenase (bovine liver), acetyl-CoA synthetase (Baker's yeast), myokinase (rabbit muscle) and lysozyme (bovine) were purchased from Sigma. Sodium acetate (^{13}C, 99%), sodium acetate (^2H, 99%) and D_2O (99%) were purchased from Cambridge Isotope Laboratories, Inc. All other chemical reagents were of the highest grades available. Plasmids pET28efTR (encodes a bi-functional enzyme, *Enterococcus faecalis* acetoacetyl-CoA thiolase/HMG-CoA reductase), pET28efS2 A100G (encodes *Enterococcus faecalis* HMG-CoA synthase), and pET28-efR (encodes *Enterococcus faecalis* HMG-CoA reductase) were generous provided by Prof. V. W. Rodwell [43]. Mevalonate kinase (*Staphylococcus aureus*), phosphomevalonate kinase (*Streptococcus. pneumoniae*), and diphosphomevalonate decarboxylase (*Streptococcus. pneumoniae*) were expressed and purified as described previously [46,49].

Enzyme expression and purification

37°C LB/ampicillin media was inoculated with *E. coli* BL21(DE3) freshly transformed with the expression plasmid of interest. The cells were cultured to an OD_{595} of 0.8, protein expression was induced by the addition of isopropyl-1-thio-β-D-galactopyranoside (IPTG, 0.75 mM), and the incubation was continued for 4 h at 37°C. The culture temperature was then shifted to 18°C and incubation was continued for 16 h. The cells were then harvested by centrifugation (30 min, RCF 5,000 g, 4°C). The MVK [46], PMK [46] and DPM-DC (diphosphome-

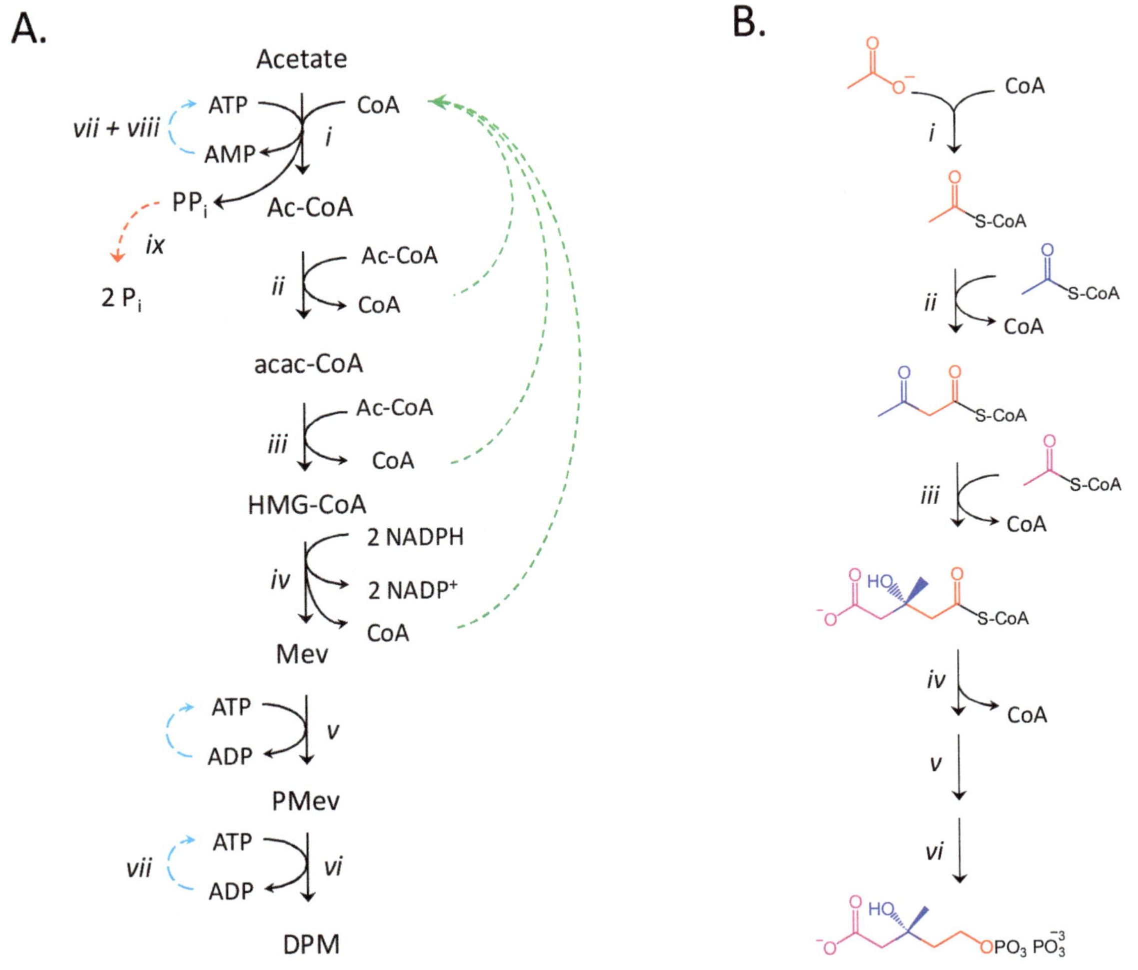

Figure 1. Schematics for the in-situ enzymatic synthesis of DPM and its isotopomers. Panel A. The enzymatic synthesis of DPM from acetate and CoA. The synthesis occurs in six steps (*i - vi*). CoA is consumed at reaction *i*, and regenerated at steps *ii-iv*. To prevent product inhibition and thermodynamically bias the system toward DPM formation, ADP (*vii*) and AMP (*vii* and *viii*) are recycled and pyrophosphate is hydrolysed (*ix*). **Panel B.** *The incorporation of acetate into DPM.* Acetate fragments are enzymatically concatenated to form the 6-carbon skeleton of DPM. Isotopic labels can be introduced at various points in the DPM synthesis to achieve a particular labeling outcome. The enyzmes used in the synthesis are as follows: *i*, acetyl-CoA synthetase; *ii*, *acetoacetyl-coA thiolase; iii*, hydroxymethylglutaryl-CoA synthase; *iv*, hydroxymethylglutaryl-CoA reductase; *v*, mevalonate kinase; *vi*, phosphomevalonate kinase; *vii*, pyruvate kinase; *viii*, adenylate kinase; *ix*, inorganic pyrophosphatase.

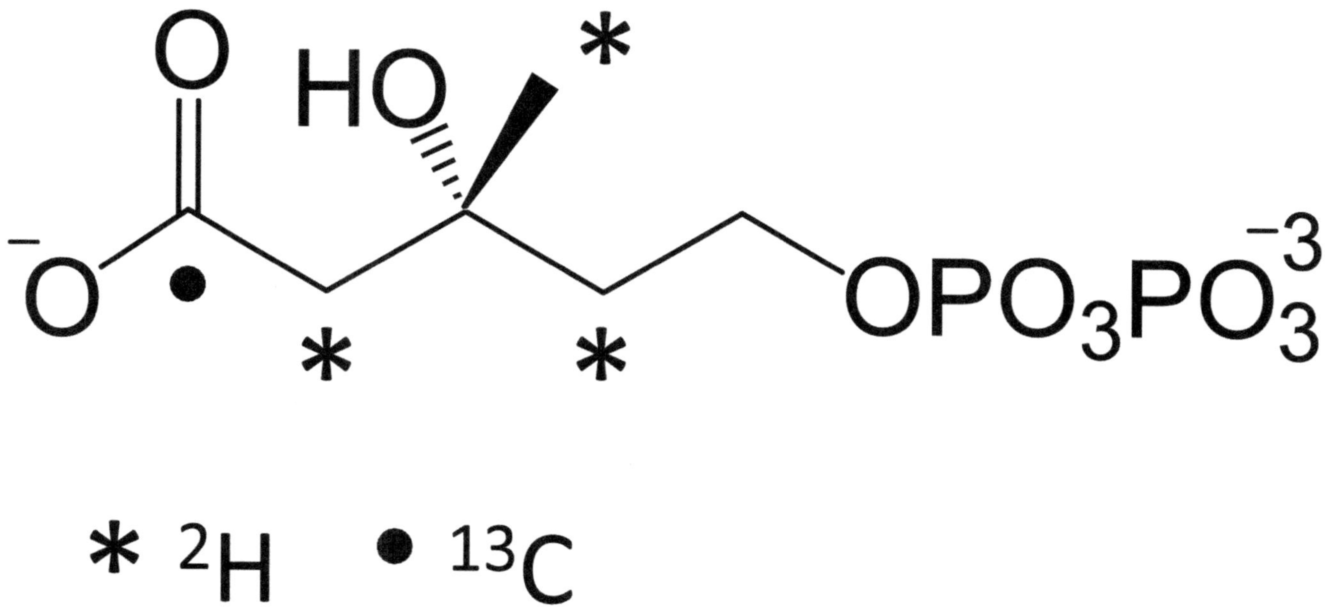

Figure 2. The isotopomers of (R)-diphosphomevalonate. Dots (●, [^{13}C]) and asterisks (*, [^{2}H]) mark the positions of heavy atoms in the synthesized compounds. Each mark represents a separate, singly-labeled compound. A triply-labeled compound, enriched at all of the [^{2}H]-positions, was also synthesized.

valonate decarboxylase) [46] expression vectors fuse a His_9-GST-tag to the N-terminus of the enzyme; whereas, the acetoacetyl-CoA thiolase [47], HMG-CoA synthase [50] and HMG-CoA reductase [47] vectors fuse a His_6-tag to the N-terminus. Dual-tag proteins were purified using a GST resin followed by a His resin. All buffers and columns were equilibrated at 4°C prior to use. Purification began by suspending cell pellets (5.0 ml/g cell paste) in **Buffer A** [H_2KPO_4 (50 mM), NaCl (140 mM), KCl (2.7 mM), pH 7.3] supplemented with lysozyme (0.10 mg/ml), PMSF (290 μM), and pepstatin A (1.5 μM). EDTA (1.0 mM) was added to Buffer A when purifying dual-tag systems. After suspension for 1 hr at 4°C, cells were disrupted by sonication and debris was removed by centrifugation (50 min, RCF 15,000 g, 4°C). Supernatants containing dual-tag proteins were loaded onto a Glutathione Sepharose 4 FF column equilibrated with Buffer A + EDTA, the column was then washed with three column volumes of the same buffer, and protein was eluted using Tris/Cl (250 mM, pH 8.0), KCl (500 mM) and reduced glutathione (10 mM). Supernatants containing singly tagged proteins, or GST-resin eluants containing dual-tag proteins, were loaded onto a Ni-NTA column equilibrated with **Buffer B** [H_2KPO_4 (50 mM), NaCl (300 mM), imidazole (10 mM), pH 8.0]. The column was washed with **Buffer C** [H_2KPO_4 (50 mM), NaCl (300 mM), imidazole (20 mM), pH 8.0], and fusion protein was eluded with **Buffer D** [H_2KPO_4 (50 mM), NaCl (300 mM), imidazole (300 mM), β-mercaptoethanol (β-ME) (10 mM), pH 8.0]. Glycerol was then added to the singly-tagged eluants (5% v/v) and stored (see below). Tags were removed from the dual-tag proteins by incubation with PreScission protease [51] during overnight dialysis at 4°C against Hepes/K^+ (50 mM, pH 8.0) containing DTT (10 mM, dithiothreitol) and KCl (100 mM). Following proteolysis, the dialysate was passed over a GSTrap column to remove the GST-tagged protease. The purity of the single- and double-tags proteins was estimated, using SDS-PAGE, at >85 and >95%, respectively. Eluants containing purified proteins were frozen rapidly and stored at −80°C.

Enzymatic assays

To establish conditions for the synthesis of DPM, the activity of each enzyme was assessed under the synthesis conditions. Apparent kinetic constants were extracted from reaction progress curves [11] and were in good agreement with published values (Table 1). *Acetyl-CoA synthetase activity* was monitored by coupling the production of AMP to the oxidation of NADH [52]. The assay conditions were as follows: inorganic pyrophosphatase (4.0 U/ml), myokinase (4.0 U/ml), PK (4.0 U/ml), LDH (lactate dehydrogenase, 8.0 U/ml), NADH (3.0 mM, $\varepsilon_{398} = 0.136\ mM^{-1}\ cm^{-1}$), acetate (2.0 mM), CoA (2.0 mM), ATP (4.0 mM), PEP (6.0 mM), $MgCl_2$ (1.0 mM + [nucleotide]), KCl (50 mM), β-ME (10 mM). *Acac-Coa thiolase activity* was monitored by following the appearance of acac-CoA (acetoacetyl-CoA) at 302 nm [53]. The conditions were: Ac-CoA (acetyl-CoA) (6.0 mM), $MgCl_2$ (2.0 mM). *HMG-CoA synthase activity* was monitored at 386 nm ($\varepsilon_{386}^{NADPH} = 0.61\ mM^{-1}\ cm^{-1}$) by coupling the production of 3-hydroxy-3-methyl glutaryl-CoA to the oxidation of NADPH using HMG-CoA reductase. The conditions were: HMG-CoA reductase (1.0 μM), acac-CoA (1.0 mM), Ac-CoA (1.0 mM), NADPH (1.5 mM), KCl (50 mM), β-ME (10 mM). *HMG-CoA reductase activity* was monitored by following oxidation of NADPH. The conditions were: 3-hydroxy-3-methyl glutaryl-CoA (0.50 mM), NADPH (0.20 mM), KCl (50 mM), β-ME (10 mM). *Mevalonate kinase activity* was monitored by coupling the production of ADP to the oxidation of NADH [46,49]. The conditions were: PK (4.0 U/ml), LDH (8.0 U/ml), NADH (200 μM, $\varepsilon_{339} = 6.22\ mM^{-1}\ cm^{-1}$), PEP (7.0 mM), mevalonate (135 μM), ATP (5.0 mM), $MgCl_2$ (1.0 mM + [nucleotide]), KCl (50 mM), β-ME (10 mM). *Phosphomevalonate kinase activity* was monitored by coupling the production of ADP to the oxidation of NADH [46,49]. The conditions were identical to those used for mevalonate kinase except phosphomevalonate (50 μM) replaced mevalonate. *DPM Decarboxylase activity* was monitored by coupling the production of ADP to the oxidation of NADH [46,49]. The conditions were

Table 1. Enzymes used in the synthesis of DPM.

[a]Enzyme	EC #	Gene	Source	Substrate	[d]K_m (mM)	[d]k_{cat} (sec^{-1})
[b]ACS	6.2.1.1	*Acs1*	*S. cerevisiae*	Acetate CoA	0.28 0.24	10
[c]ACT	2.3.1.9	*mvaE*	*E. faecalis*	Ac-CoA	0.60	2.3
[c]HMGS2	2.3.3.10	*mvaS*	*E. faecalis*	acac-CoA Ac-CoA	0.015 0.35	1.0
[c]ThRed	1.1.1.34	*mvaE*	*E. faecalis*	HMG-CoA	0.023	0.55
[c]HMGR	1.1.1.34	*mvaE*	*E. faecalis*	HMG-CoA	0.020	0.67
[c]MVK	2.7.1.36	*mvaK1*	*S. aureus*	Mev	0.027	19
[c]PMK	2.7.4.2	*mvaK2*	*S. pneumoniae*	P-mev	0.0042	5.0
[b]PK	2.7.1.40	*PKM2*	*O. cuniculus*	PEP	0.040	160
[b]MK	2.7.4.3	*AK1*	*O. cuniculus*	AMP	[e]0.50	[e]410
[b]PP_iase	3.6.1.1	*Ppa1*	*S. cerevisiae*	PP_i	[e]0.0050	[e]260

[a]Abbreviations: ACS, acetyl-CoA synthetase; ACT, acetoacetyl-CoA thiolase; HMGS2, HMG-CoA synthase; ThRed, acetoacetyl-CoA thiolase/HMG-CoA reductase (dual-function enzyme); HMGR, HMG-CoA reductase; MVK, mevalonate kinase; PMK, phosphomevalonate kinase; PK, pyruvate kinase; MK, myokinase; PP_iase, inorganic pyrophosphatase.
[b]Obtained from commercial sources.
[c]Expressed in *E. coli* and purified.
[d]Standard errors are <5% in all cases (see *Materials and Methods*).
[e]Parameters taken from literature (MK [77,78], PP_iase [79,80]).

identical to those used for mevalonate kinase except diphosphomevalonate (50 μM) replaced mevalonate. In all cases, reactions were buffered with Hepes/K^+ (50 mM), pH 8.0, and T = 25±2°C.

The synthesis of (*R*)-diphosphomevalonate

DPM was synthesized in a one-pot reaction using the following conditions: Ac-CoA synthetase (2.0 μM), acac-CoA thiolase (2.0 μM), HMG-CoA synthase (4.0 μM), HMG-CoA reductase (2.0 μM), mevalonate kinase (2.0 μM), phosphomevalonate kinase (3.0 μM), pyruvate kinase (5.0 U/mL), myokinase (2.0 U/ml), inorganic pyrophosphatase (2.0 U/ml), ATP (5.0 mM), PEP (10 mM), acetate (12 mM), CoA (5.0 mM), NADPH (10 mM), KCl (50 mM), $MgCl_2$ (1.0 mM + [ATP]), β-ME (10 mM), Hepes/K^+ (50 mM), pH 8.0, T = 25±2°C. Reactions progress was monitored by following the oxidation of NADPH associated with the HMG-CoA reductase reaction. DPM formation was assayed by adding an aliquot the DPM-synthesis reaction into a DPM decarboxylase assay mixture (DPM-DC (0.10 μM), PK (4.0 U/ml), LDH (8.0 U/ml), NADH (200 μM), PEP (4.0 mM) ATP (2.0 mM), $MgCl_2$ (1.0 mM + [nucleotide]), KCl (50 mM), β-ME (10 mM), Hepes/K^+ (50 mM), pH 8.0, T = 25±2°C) and monitoring NADH oxidation at 340 nm. The assay-reaction dilution was sufficient (330-fold dilution) to prevent the HMG-CoA reductase reaction from contributing significantly to the measurement. The reactions yielded essentially quantitative conversion of acetate to the endproduct, DPM.

The synthesis of labeled acetyl-CoA precursors

The synthesis of regiospecifically labeled DPM requires appropriately labeled Ac-CoA. Labeled Ac-CoA precursors were synthesized using the following conditions: acetyl-CoA synthetase (2.0 μM), pyrophosphatase (2.0 U/ml), labeled acetate (4.0 mM), CoA (4.0 mM), ATP (4.0 mM), $MgCl_2$ (5.0 mM), Hepes/K^+ (50 mM), pH 8.0. The reactants were mixed gently for 10 hr at T = 25±2°C. Reaction progress was monitored by assaying aliquots of the reaction for AMP synthesis using the Ac-CoA synthetase assay described above. The conversion of CoA to labeled Ac-CoA was >95%.

Synthesis of acac-CoA

The synthesis of acac-CoA was achieved using the conditions identical to those described for the synthesis of Ac-CoA with the exception that acac-CoA thiolyase (2.0 μM) and DTNB (10 mM, 5, 5'-Dithio-bis(2-nitrobenzoic acid) were present. DNTB reacts with CoA and was used to draw the acac-thiolase reaction forward. The DTNB reaction was monitored at 412 nM [54]. Acac-CoA formation was monitored at 302 nm (*see, Enzymatic assays, Materials and Methods*). The reaction reached completion after approximately 17 hr, after which >98% acetyl-CoA had converted to acac-CoA. The reaction was filtered (10 kDa membrane) to remove enzymes prior to using the acac-CoA in subsequent syntheses.

The synthesis of [1-^{13}C]DPM or [2-$^2H_2$]DPM

Labeled Ac-CoA (^{13}C or ^{2}H) was prepared from CoA and labeled acetate as described above (see, *Synthesis of labeled acetyl-CoA precursors*). Labeled DPM was synthesized by adding the following reagents to the labeled Ac-CoA reaction mixture: PK (10 U/mL) (U, μmoles product formed min^{-1} at a saturating substrate), HMG-CoA synthase (4.0 μM), HMG-CoA reductase (2.0 μM), MVK (2.0 μM), PMK (1.0 μM), PEP (10 mM), NADPH (5.0 mM), unlabelled acac-CoA (2.0 mM), ATP (5.0 mM), KCl (50 mM), and β-ME (10 mM). The unlabeled acac-CoA was prepared as describe above (see, *Synthesis of acac-CoA*). The reaction was stirred gently overnight (~16 h, 25±2°C), at which point >97% of the labeled Ac-CoA had incorporated into DPM. The quantitation of DPM is described above (see, *The synthesis of (R)-diphosphomevalonate*).

The synthesis of DPM from acetate at high concentration

DPM synthesis was accomplished in a one-pot reaction using the following conditions: Ac-CoA synthetase (5.0 μM), acac-CoA thiolase (7.0 μM), HMG-CoA synthase (10 μM), HMG-CoA reductase (7.0 μM), mevalonate kinase (5.0 μM), phosphomevalonate kinase (3.0 μM), pyruvate kinase (10 U/mL), myokinase (5.0 U/ml), inorganic pyrophosphatase (5.0 U/ml), ATP (100 mM), PEP (800 mM), acetate (340 mM), CoA (5.0 mM), NADPH (300 mM), $MgCl_2$ (110 mM), β-ME (10 mM), Hepes/

K^+ (50 mM), pH 8.0, T = 25±2°C. Reaction progress was monitored as described above (see, Enzymatic Assays). Under the high ionic strength conditions of this reaction, the conversion of acetate to DPM decreased to sixty-three percent.

The synthesis of IPP from acetate at high concentration

IPP synthesis was accomplished in a one-pot reaction using the following conditions: Ac-CoA synthetase (7.0 μM), acac-CoA thiolase (10 μM), HMG-CoA synthase (12 μM), HMG-CoA reductase (10 μM), mevalonate kinase (7.0 μM), phosphomevalonate kinase (5.0 μM), diphosphomevalonate decarboxylase (3.5 μM), pyruvate kinase (20 U/mL), myokinase (7.0 /ml), inorganic pyrophosphatase (7.0 U/ml), ATP (200 mM), PEP (800 mM), acetate (340 mM), CoA (5.0 mM), NADPH (300 mM), $MgCl_2$ (220 mM), β-ME (10 mM), Hepes/K^+ (50 mM), pH 8.0, T = 25±2°C. The conversion of acetate to IPP yields nine IPP-equivalents of pyruvate (two equivalents for each of the three Ac-CoAs required to synthesize HMG-CoA (hydroxymethylglutaryl-CoA), two for conversion of Mev to DPM and one for the decarboxylation of DPM to IPP). Pyruvate was quantitated by adding an aliquot of the IPP-synthesis reaction to a lactate dehydrogenase assay mixture: (LDH (8.0 U/ml), NADH (200 μM), KCl (50 mM), β-ME (10 mM), Hepes/K^+ (50 mM), pH 8.0, T = 25±2°C). Dilution of the synthesis reaction was sufficient (>500-fold) to prevent enzymes from the reaction from contributing significantly to the pyruvate measurements. Sixty-nine percent of the acetate was converted to DPM.

The synthesis of DPM from (R/S)-mevalonate at high concentration

DPM synthesis was accomplished in a reaction using the following conditions: mevalonate kinase (5.0 μM), phosphomevalonate kinase (3.0 μM), pyruvate kinase (10 U/mL), (*R/S*)-mevalonate (370 mM), ATP (50 mM), PEP (350 mM), $MgCl_2$ (60 mM), β-ME (7.0 mM), Hepes/K^+ (50 mM), pH 8.0, T = 25±2°C. Reaction progress was monitored as described above (see, Enzymatic Assays). It should be noted that mevalonate kinase converts only the *R*-isomer of mevaloante to phosphomevalonate [55], and the enantiomeric composition of commercial (*R/S*)-mevalonate is 1:1 [11]; hence, a maximum of 50% of the commercial product can be converted to DPM. Seventy-one percent of the (*R*)-mevalonate in the (*R/S*)-mixture was converted to DPM.

The synthesis of IPP from (R/S)-mevalonate at high concentration

IPP synthesis was accomplished in a reaction using the following conditions: mevalonate kinase (5.0 μM), phosphomevalonate kinase (3.0 μM), diphosphomevalonate decarboxylase (1.6 μM), pyruvate kinase (10 U/mL), (*R/S*)-mevalonate (375 mM), ATP (50 mM), PEP (450 mM), $MgCl_2$ (60 mM), β-ME (7.0 mM), Hepes/K^+ (50 mM), pH 8.0, T = 25±2°C. The conversion of (*R/S*)-mevalonate to IPP was monitored by following the formation of pyruvate using lactate dehydrogenase (see, *The synthesis of IPP from acetate at high concentration*). Seventy-seven percent of the (*R*)-mevalonate in the (*R/S*)-mixture was converted to IPP.

The purification (*R*)-diphosphomevalonate

To maximize the purity and recovery of DPM, PEP (which chromatographs near DPM) was converted to pyruvate by adding one PEP-equivalent of ADP to the synthesis reaction mixture. Small and large molecules were separated by ultrafilitration (10-kDa cutoff). The small-molecule filtrate was passed through a 35 mL bed of anion exchange resin (AG MP-1) equilibrated with Hepes/K^+ (10 mM, pH 7.5), and the column was "washed" with five volumes of equilibration buffer. The compounds were eluted using a 750 ml, linear salt gradient (0–1.0 M KCl) at 2.0 mL/min. DPM eluted at 0.32 mM KCl and contained <1% nucleotide. To remove excess KCl and concentrate the DPM, the purified compound was loaded onto a 5.0 ml bed of AG MP-1 equilibrated with NH_4HCO_3 (10 mM, pH 7.5). The column was then "washed" with five volumes of NH_4HCO_3 (10 mM, pH 7.5) before eluting the DPM with 1.8 volumes of NH_4HCO_3 (350 mM, pH 7.5). Excess NH_4HCO_3 was removed by rotary evaporation at 45°C. The desalted compounds were dissolved in ultra pure water (2.0 mL) and the solution was adjusted to pH 7.5 with KOH. NH_4HCO_3 in the desalted, purified DPM was measure using an enzymatic assay that couples the reduction of $NADP^+$ to the synthesis of glutamate from NH_4^+ and α-ketoglutarate [56]. The assay conditions were as follows: α-ketoglutarate (5.0 mM), $NADP^+$ (0.20 mM), glutamate dehydrogenase (14 U/mL), Hepes/K^+ (45 mM) pH 8.0 at T = 25±2°C. The NH_4^+/DPM stoichiometry was ~4:1. The DPM concentration and purity, presence of mevalonate and phosphomevalonate, were determined spectrophotometrically using the assay described above (see, *Enzymatic assay*), and the purified compounds were stored in Hepes/K^+ (10 mM, pH 8.0) at −80°C.

NMR protocols

One dimensional NMR was used to confirm the structure and isotopic labeling of the DPM isotopomers. A Bruker DRX 300 MHz spectrometer equipped with a 5 mm broadband probe was used to acquire data. Sample temperature was 25±2°C. Proton spectra were the average of 32 scans (64K points each) acquired over 20 ppm using a 1.0 s recycle delay. The residual water signal was suppressed by presaturation of the HOD resonance. Spectra were processed with 1.0 Hz line broadening, and proton chemical shifts were referenced to 3-(trimethylsilyl) propionate [57]. Proton-decoupled carbon spectra were the average of 100 scans (61K points each) acquired over 315 ppm using a 3.0 s recycle delay. Spectra were processed with a 1.5 Hz line broadening, and chemical shifts were referenced indirectly [57]. Proton-decoupled phosphorus spectra were the average of 256 scans (64K points each) acquired over 50 ppm using a 6.0 s recycle delay. Spectra were processed with a 3.0 Hz line broadening, and chemical shifts were referenced to phosphocreatine [58].

Results and Discussion

The enzymatic synthesis of DPM

Diphosphomevalonate is synthesized from acetate, ATP and NADPH in six consecutive enzymatic steps (*i* - *vi*, Fig 1A). The first four reactions produce mevalonate from 3 acetate, 3 ATP, and 2 NADPH [55,59,60]. CoA, which acts as an acetyl-carrier, is consumed in reaction *i*, and regenerated in reactions *ii*, *iii* and *iv* (see, dashed green arrows, Fig 1A). Reactions *v* and *vi* are catalyzed by kinases that phosphorylate mevalonate to produce the pyrophosphoryl-group of DPM. To bias the reactions toward the endproduct and avoid product inhibition, ADP and AMP were recycled to ATP using pyruvate kinase and myokinase, and PP_i was hydrolyzed to P_i using inorganic pyrophosphatase. In total, nine enzymes were used in the synthesis [43–48].

Enzymes *ii* - *vi* were cloned, expressed in *E. coli* and purified (see, *Materials and Methods*); *i* and *vii* - *ix* were obtained from commercial sources. The purified enzymes were 80–95% pure, as judged by Comassie staining [61] of SDS PAGE [62] gels, and

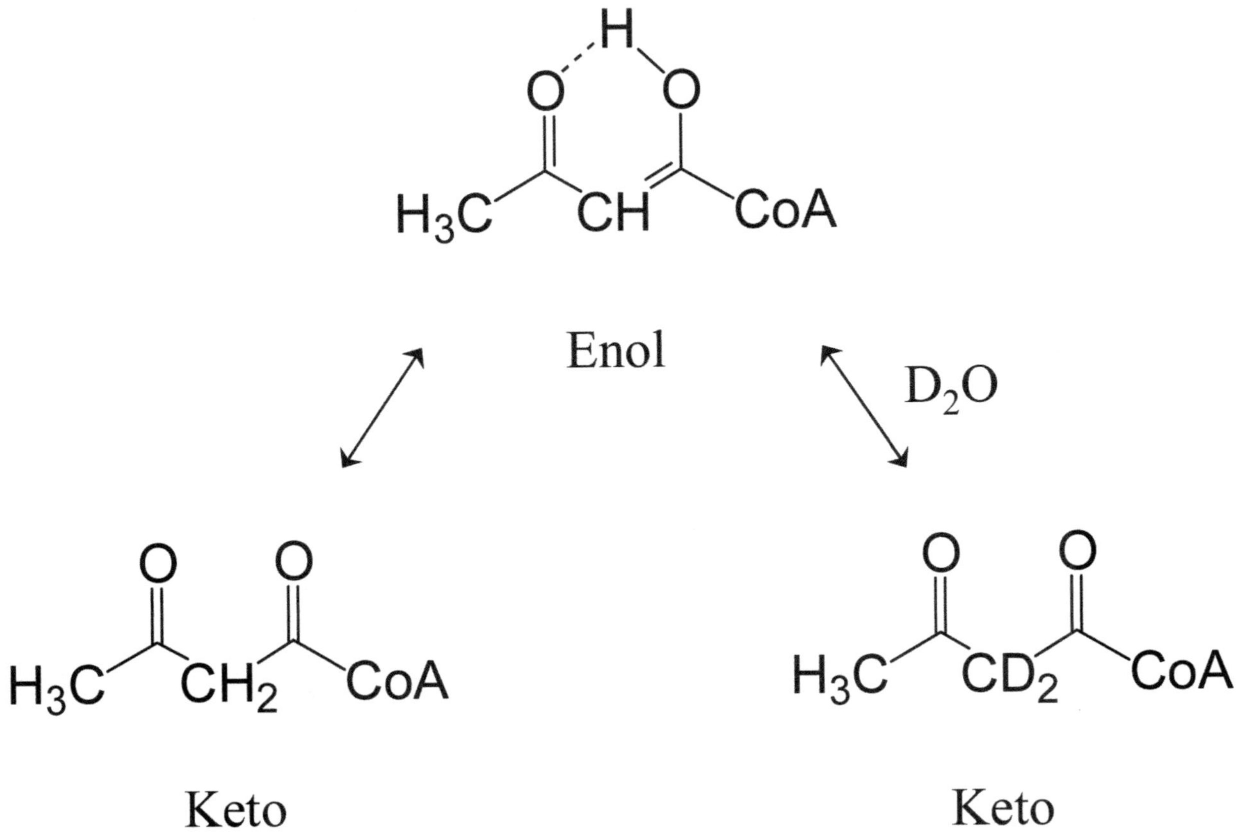

Figure 3. Tautomerization of acetoacetyl-CoA. Tautomerization allows H/^{2}H-exchange at the C_2-position of the 3-oxobutyryl-moiety of acac-CoA, and thus provides the precursor for the synthesis of [4-$^2H_2$]DPM.

were obtained in yields of 30–40 mg pure protein/liter of *E. coli*. The kinetic constants of the purified enzymes were determined under the conditions used for the synthesis, and were in good agreement with literature values (Table 1). The assays are described in *Enzymatic Assays* (see, *Materials and Methods*). The enzymes showed no significant loss of activity over an 8 month period when frozen rapidly and stored at −80°C in Hepes (50 mM, pH 8.0), 150 mM KCl, 5% glycerol (v/v).

The relative enzyme concentrations used in the DPM syntheses were determined empirically by adjusting concentrations such that flux through the pathway was not rate-limited by any single step. This was accomplished by setting PMK (*vi*) at a fixed concentration and titrating each preceding enzyme successively until the DPM-synthesis rate was 80–90% of the maximum rate achievable at each step. For example, MVK (*v*) was titrated at a fixed concentration of PMK until the rate of DPM synthesis became independent of MVK concentration – the maximum rate. The MVK concentration was then adjusted to allow 80–90% of the maximum rate, and an analogous procedure was performed with HMG-CoA reductase (*iv*). The procedure was performed in succession for each enzyme in the pathway to determine the relative enzyme concentrations to be used in the synthesis. Once relative concentrations were established, the absolute concentrations were set to achieve the desired reaction times, which ranged from 8–72 hr. Mevalonate kinase from *S. aureus* was selected because, unlike the *S. pneumoniae* enzyme, it is not allosterically inhibited by DPM [11].

Substrates were set at saturating, sub-inhibiting concentrations. ATP, a substrate for five of the enzymes (*i*, *v*, *vi*, *vii* and *viii*), was set at 5.0 mM, which ranges from 5.8–68×K_m. Typical substrate concentrations of the other reactants were as follows: acetate (12 mM, 42×K_m); CoA (2.0 mM , 8.0×K_m); NADPH (10 mM, 320×K_m); and PEP (10 mM, 250×K_m). Under these conditions, and using the enzyme concentrations detailed in *Synthesis of (R)-diphosphomevalonate* (*Materials and Methods*), ~98% of the acetate was incorporated into DPM in this single-pot reaction.

The incorporation of isotopes into DPM

The regiospecific incorporation of isotopes has proven extremely valuable in the elucidation of metabolism [63–65] and determining enzyme mechanism [66]. Indeed, this was the basis for the discovery of the non-mevalonate pathway of isoprenoid biosynthesis [40]. The enzymatic scheme shown in Fig. 1 offers a flexible and efficient means of synthesizing numerous radiolabeled and stable isotopomers of mevalonate, many of which are not commercially available. The six-carbon backbone of DPM is constructed in the first three enzymatic steps of the scheme (*i* – *iii*). Each step adds a single acetate to the CoA thioester R-group. The pattern of acetate incorporation into the R-group, and ultimately DPM, is shown in Figure 1B. Acetate is first esterfied onto the CoA thiol, and subsequent two-carbon units are added by forming carbon-carbon bonds with the existing R-group. Isotopes can be incorporated into specific positions in DPM (Fig 2) using labeled acetate or acetyl-CoA, or *via* solvent exchange with exchange-sensitive intermediates. Achieving certain labeling patterns required removal of enzymes by ultrafiltration at intermediate stages of the synthesis, and/or that reactions were run in D_2O (see below).

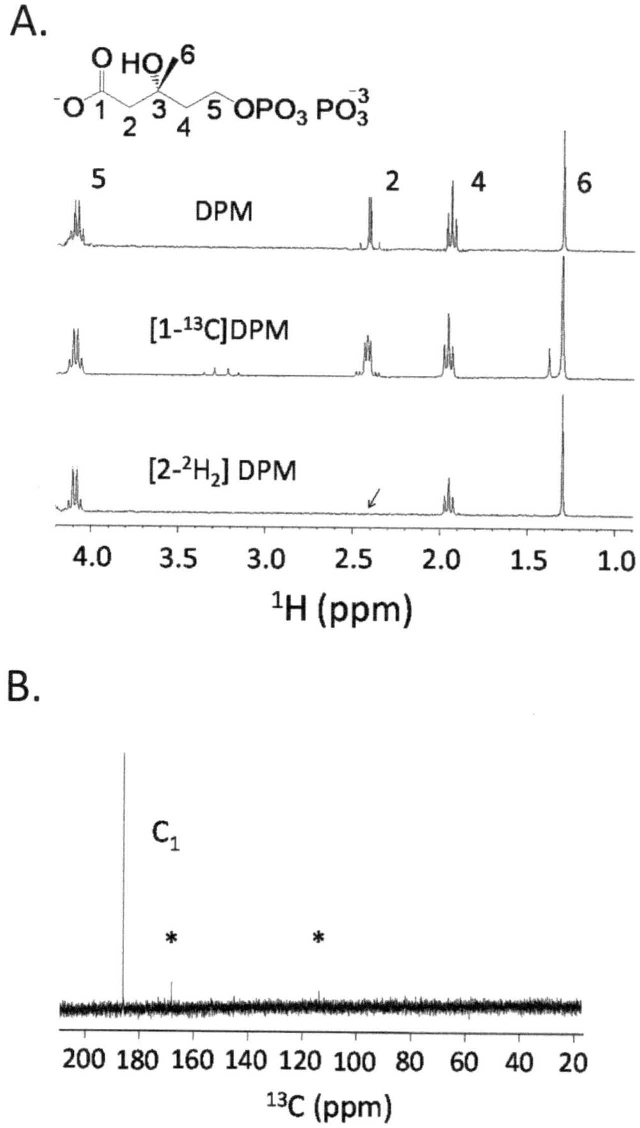

Figure 4. ^{1}H and ^{13}C NMR spectra of (R)-diphosphomevalonate isotopomers. Spectra indicate specificity and efficiency of labeling. **Panel A.** ^{1}H NMR spectra of labeled and unlabeled DPM. The efficiency and specificity of [$^2H_2$]-labeling were estimated at >98% and >95%, respectively. **Panel B.** ^{13}C NMR spectrum of [1-^{13}C]DPM. The resonance at 181 ppm corresponds to C_1. The efficiency of labeling at C_1 is estimated at >92% (see *Results and Discussion*). Based on the absence of non-C_1 signals and the S/N, the labeling specificity is calculated at >98%. Asterisks indicate instrumental artifacts.

The synthesis of [2, 4, 6,-$^2H_7$]- and [6-$^2H_3$]DPM

The compounds were synthesized in approximately 50 mg quantites in one-pot reactions using commercial [2-$^2H_3$]acetate or (*R, S*)-[6-$^2H_3$]mevalonolactone as starting material (see, *Supplementary Material*). Reactions were complete after 22 hrs, and virtually quantitative conversion of starting material to DPM was achieved in all cases. Approximately ~86% of the maximum theoretical maximum yield of DPM was obtained after purification (see, *Material and Methods*). The labeling of DPM was confirmed using ^{1}H NMR (Fig. S1).

Synthesis of [4-$^2H_2$]DPM

The synthesis of [4-$^2H_2$]DPM was carried out in several steps. First, unlabeled Ac-CoA was synthesized using acetyl-CoA synthetase (*i*) and inorganic pyrophosphatase (*ix*) (see, *Synthesis of acetyl-CoA*, *Materials and Method*). Acetyl-CoA thiolyase (*ii*) and DTNB (in excess over Ac-CoA) were then added to form acac-CoA. DTNB reacts quantitatively with CoA [54] and was used to draw the unfavorable acac-CoA-forming reaction to completion [47]. Acac-CoA tautomerizes [67], and its enol-form exchanges protons with solvent (Fig. 3). To streamline the synthesis, both enzymatic reactions were run in D_2O. ^{1}H NMR confirmed that exchange was complete and occurred exclusively at the C_4-position of DPM (Fig. S1). It is notable that this exchange suggests the possibility of using equilibrium isotope exchange to produce Ac-CoA in which the methyl-protons have been exchanged with solvent. To attach the third acetate without forming unlabelled acac-CoA, which would dilute isotopic enrichment, the Ac-CoA thiolyase was removed by ultrafiltration before adding the reactants that complete the synthesis of DPM (see, *Synthesis of [4-$^2H_2$]-DPM, Supplementary Material*). The reactions were essentially quantitative and the production of DPM was ~96% of the theoretical maximum.

The synthesis of [1-^{13}C]- and [2-$^2H_2$]DPM

The strategy used to synthesize these compounds was similar to that used in the synthesis of [4-$^2H_2$]DPM. Unlabeled acac-CoA was synthesized, the enzymes used in the synthesis were "strained" from the reaction by ultrafiltration, labeled Ac-CoA was then added along with the enzymes, and reagents needed to complete the synthesis of DPM (see, *Materials and Methods*). Care was taken to ensure that the Ac-CoA remaining in the acac-CoA synthesis reaction was ~0.2% of the labeled Ac-CoA used in the subsequent conversion to DPM. The yield was quantitative; DPM levels reached ~95% of the theoretical maximum.

Confirming the structure and labeling patterns of the compounds

The specificity and efficiency of isotopic labeling were assessed using ^{1}H and ^{13}C NMR. Deuterium incorporation at a given position was assessed by quantitating the loss of proton signal at that position. The ^{1}H NMR spectra of the synthesized compounds are compiled in Figures 4A and S1. In all cases, proton signal at the targeted position(s) was below detection (i.e., >97% incorporation efficiency) and the integrated intensities of the remaining proton peaks were identical within error (±3%); thus, deuterium did not incorporate significantly into positions other than the target sites. Comparison of the ^{1}H spectra of [1-^{13}C]- and natural abundance C_1-DPM reveals that the AB quartet associated with the C_2 of DPM (2.41 ppm) is split to into an ABX pattern by the incorporation of ^{13}C (Fig 4A). This splitting is consistent only with ^{13}C incorporation at C_1. If the synthesis had resulted in a significant fraction of natural abundance C_1-DPM, the AB and ABX resonances are expected to overlap. Close inspection of the upfield doublet of the ABX pattern gives no indication of the AB species (Fig 4 inset) indicating that the incorporation efficiency is quite high (>95%). The labeling specificity of [1-^{13}C]DPM is given by the ^{13}C spectrum (Fig 4B), which shows the expected C_1-resonance [68] and no detectible signal at the positions associated with the other carbon atoms in the molecule (dotted arrows). The integrity of the pyrophosphoryl moiety was confirmed using ^{31}P NMR (Figure S2).

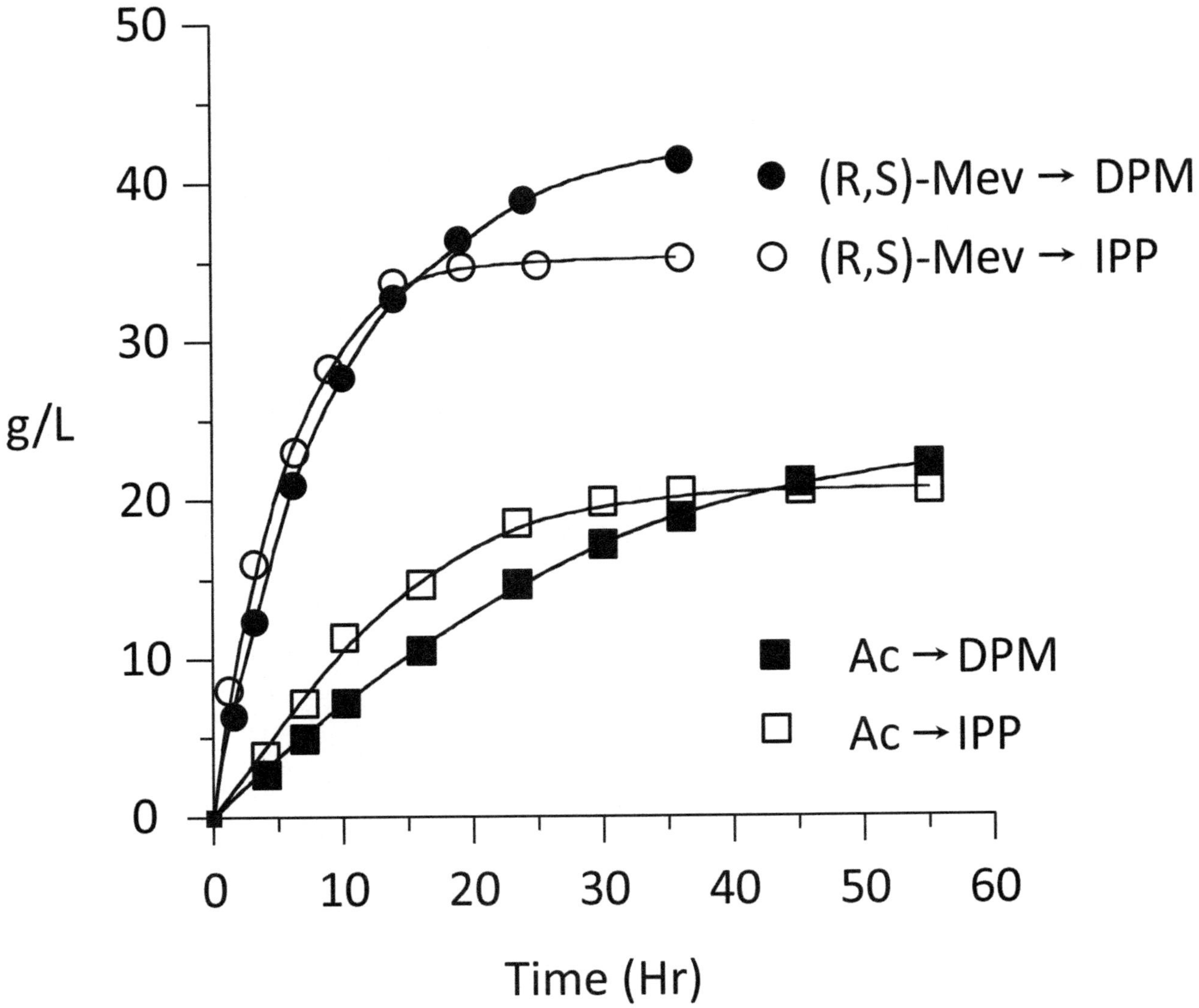

Figure 5. DPM and IPP synthesis at high reactant concentration. DPM and IPP were synthesized in separate, single-pot reactions. Reactions were initiated from acetate, or (R,S)-mevalonate. The conditions are described in *Materials and Methods*. Reactions initiated with acetate yielded 69% conversion of acetate to IPP (○), or 63% conversion of acetate to DPM (●). Reactions starting with (R,S)-mevalonate yielded 77% conversion of (*R*)-mevalonate to IPP (□), or 73% conversion of (R)-mevalonate to DPM (■). The data points represent the average of results from three independent experiments.

The synthesis of highly concentrated DPM and isopentenyldiphosphate

Given the considerable societal value of isoprenoids, the difficulties obtaining them, and the current efforts to bio-synthesize these compounds at commercial scale, it was of interest to assess the potential of the *in-situ* enzymatic synthesis to produce large quantities of product. Toward this end, the velocity of the acetate-to-DPM conversion was studied as a function of initial-reactant concentration with the goal of determining the highest, useful concentrations. The system proved remarkably robust. Only slight inhibition (~30%) was observed at 0.50 M acetate. PEP and NADPH could be increased to near saturation (~500 and 200 mM, respectively) without significant decrease in velocity, and ATP could be added to 0.15 M without inhibition or noticeable precipitation. The concentration-optimized system contained acetate, ATP, PEP and NADPH at 0.35, 0.10, 0.40, 0.30 M, respectively, and yielded DPM and IPP at 22 and 18 g/liter, respectively −63% and 69% conversions of acetate to product (Fig 5). Product formation was limited by the solubility of nucleotide and high ionic strength of these reactions.. To assess whether the enzymatic system was capable of producing even higher product concentrations, DPM and IPP synthesis was initiated from mevalonate. Reactions contained (*R/S*) mevalonate, ATP, and PEP at 0.370, 0.05, 0.35 M, respectively, and yielded DPM and IPP at 42 and 35 g/liter, respectively −73% and 76% product yields (Fig 5). The reactions conditions are further described in *Materials and Methods* (see, *Reactions that yield highly concentrated product*).

The syntheses outlined in the preceding paragraph are highly scalable. Reaction yields were independent of volume from 0.10 ml–1 liter and are expected to be similar at larger volumes. Under the high ionic strength conditions of these assays, the enzymes proved to be quite stable. The majority lost ≤20% of their activity over 2 days at room temperature. The exceptions were acetoacetyl-CoA thiolyase and inorganic pyrophosphatase, which lost 63% and 72% of their activity, respectively, over this time period.

Attempts to genetically engineer plants and bacteria to produce commercial quantities of isoprenoids have met with variable success. Artemisinin, a potent antimalarial, is currently isolated from the *quinghao* plant (*Artemisia annua*) at ~3 mg/g dry weight. In contrast, genetically engineered tobacco produces artemisinin at ~0.8 mg/g dry weight [69], and transgenic yeast secrete artemisinic acid (an artemisinin precursor [70] at 100 mg/liter [71]. Using *E. coli* as the host, pathway optimization has yielded ~0.3 g/liter of artemisinic acid in shaking flasks [34,72], high-density batch fermentation of engineered *E. coli* has produced yields as high as 23g/liter [73] and recent breakthroughs in understanding of the pathway have produced artemisinic acid at ~25 g/liter in moderately high-density *E. coli* cultures [74]. Similar efforts in *E. coli* have produce taxadiene (a precursor of Taxol, an anticancer therapeutic) at ~1g/liter in shaking flasks [75]. Production of farnesol, a relatively simple isoprenoid and potential biofuel [29], has reached 130 mg/liter in engineered *E. coli* grown in shaker flasks [76].

While these efforts have helped define the complexities associated with expressing and controlling the isoprenoid biosynthetic pathway in living organisms, only fermentation in conjunction with genetic engineering is yielding product quantities required for successful commercial application. The cell-free approach described here yields product quantities that are comparable to, or exceed those achieved in high-density fermentation and have the advantage that product is formed in a simple aqueous system from which it can be recovered readily.

Conclusions

The enzymes that comprise the HMG CoA reductase and mevalonate pathways have been used along with enzymatic substrate-recycling and product-removal systems to efficiently synthesize intermediates and end products of these pathways in high yield. Strategies for using these enzymes to regio-specifically position isotopes in the isoprenoid backbone are described and used to synthesize and purify isotopomers of DPM, the immediate end product of the mevalonate pathway. The enzymatic platform is robust and produces isoprenoid precursors, DPM and IPP, in quantites ranging from 20–40 g/liter. These values meet or exceed all published values for isoprenoid production using genetically engineered organisms. The platform produces product in simple aqueous solutions, which, in most cases, will make isoprenoid isolation far simpler than extraction from high-density fermentums. These favorable attributes recommend the enzymatic platform as a valuable alternative to cell-culturing methods as a clean, sustainable, non-biocompetitive method for the production of isoprenoids.

Supporting Information

Figure S1 ^{1}H NMR spectra of DPM isotopomers. The specificity and efficiency of labeling of the isotopomers were estimated based on the integration of the ^{1}H signals. The results were as follows: [4-$^{2}H_2$]DPM (96%, 95%), [6-$^{2}H_3$]DPM (97%, 95%) and [2, 4, 6-$^{2}H_7$]DPM (96%, 97%).

Figure S2 ^{31}P NMR spectrum of DPM. The resonance positions, splitting pattern and nearly identical integrated intensities of the <- and -resonances indicate an intact pyrophosphoryl-moiety. The asterisk identifies the resonance of phosphocreatine, which was added as an internal standard. The absence of a peak at ~5 ppm indicates that phophomevalonate is undetectable.

Acknowledgments

We thank Dr. Victor W. Rowell for providing the plasmids pET28-efTR, pET28efS2 A100G, and pET28-efR. We thank Dr. Sean Cahill for helpful discussions on the NMR data. This work was supported, in whole or in part, by National Institutes of Health Grants GM 54469, GM 106158 AI 068989.

Author Contributions

Conceived and designed the experiments: SBR TSL. Performed the experiments: SBR. Analyzed the data: SBR TSL. Contributed reagents/materials/analysis tools: SBR TSL. Contributed to the writing of the manuscript: SBR TSL.

References

1. Albery WJ, Knowles JR (1976) Evolution of enzyme function and the development of catalytic efficiency. Biochemistry 15: 5631–5640.
2. Burbaum JJ, Raines RT, Albery WJ, Knowles JR (1989) Evolutionary optimization of the catalytic effectiveness of an enzyme. Biochemistry 28: 9293–9305.
3. Eijsink VG, Gaseidnes S, Borchert TV, van den Burg B (2005) Directed evolution of enzyme stability. Biomol Eng 22: 21–30.
4. Gerlt JA, Babbitt PC (2009) Enzyme (re)design: lessons from natural evolution and computation. Curr Opin Chem Biol 13: 10–18.
5. Sen S, Venkata Dasu V, Mandal B (2007) Developments in directed evolution for improving enzyme functions. Appl Biochem Biotechnol 143: 212–223.
6. Cheng Q, Xiang L, Izumikawa M, Meluzzi D, Moore BS (2007) Enzymatic total synthesis of enterocin polyketides. Nat Chem Biol 3: 557–558.
7. Huang KT, Wu BC, Lin CC, Luo SC, Chen C, et al. (2006) Multi-enzyme one-pot strategy for the synthesis of sialyl Lewis X-containing PSGL-1 glycopeptide. Carbohydr Res 341: 2151–2155.
8. Melnick JS, Sprinz KI, Reddick JJ, Kinsland C, Begley TP (2003) An efficient enzymatic synthesis of thiamin pyrophosphate. Bioorg Med Chem Lett 13: 4139–4141.
9. Schultheisz HL, Szymczyna BR, Scott LG, Williamson JR (2008) Pathway engineered enzymatic de novo purine nucleotide synthesis. ACS Chem Biol 3: 499–511.
10. Sugiyama M, Hong Z, Liang PH, Dean SM, Whalen LJ, et al. (2007) D-Fructose-6-phosphate aldolase-catalyzed one-pot synthesis of iminocyclitols. J Am Chem Soc 129: 14811–14817.
11. Lefurgy ST, Rodriguez SB, Park CS, Cahill S, Silverman RB, et al. (2010) Probing ligand-binding pockets of the mevalonate pathway enzymes from Streptococcus pneumoniae. J Biol Chem 285: 20654–20663.
12. Kuberan B, Lech MZ, Beeler DL, Wu ZL, Rosenberg RD (2003) Enzymatic synthesis of antithrombin III-binding heparan sulfate pentasaccharide. Nat Biotechnol 21: 1343–1346.
13. Kajiwara Y, Santander PJ, Roessner CA, Perez LM, Scott AI (2006) Genetically engineered synthesis and structural characterization of cobalt-precorrin 5A and -5B, two new intermediates on the anaerobic pathway to vitamin B12: definition of the roles of the CbiF and CbiG enzymes. J Am Chem Soc 128: 9971–9978.
14. Ohdan K, Fujii K, Yanase M, Takaha T, Kuriki T (2007) Phosphorylase coupling as a tool to convert cellobiose into amylose. J Biotechnol 127: 496–502.
15. Van Den Heuvel RH, Fraaije MW, Laane C, van Berkel WJ (2001) Enzymatic synthesis of vanillin. J Agric Food Chem 49: 2954–2958.
16. Yamaguchi S, Komeda H, Asano Y (2007) New enzymatic method of chiral amino acid synthesis by dynamic kinetic resolution of amino acid amides: use of stereoselective amino acid amidases in the presence of alpha-amino-epsilon-caprolactam racemase. Appl Environ Microbiol 73: 5370–5373.
17. Koeller KM, Wong CH (2001) Enzymes for chemical synthesis. Nature 409: 232–240.
18. Schmid A, Dordick JS, Hauer B, Kiener A, Wubbolts M, et al. (2001) Industrial biocatalysis today and tomorrow. Nature 409: 258–268.
19. Martin del Campo JS, Rollin J, Myung S, Chun Y, Chandrayan S, et al. (2013) High-Yield Production of Dihydrogen from Xylose by Using a Synthetic Enzyme Cascade in a Cell-Free System. Biocatalysis 52: 4587–4590.

20. Krutsakorn B, Honda K, Ye X, Imagawa T, Bei X, et al. (2013) In vitro production of n-butanol from glucose. Metab Eng 20: 84–91.
21. You C, Myung CH, Sathitsuksanoh N, Ma H, Zhang XZ, et al. (2013) Enzymatic trnsformation of nonfood biomass to starch. PNAS 110: 7182–7187.
22. Zhu Z, Kin Tam T, Sun F, You C, Percival Zhang YH (2014) A high-energy-density sugar biobattery based on a synthetic enzymatic pathway. Nat Commun 5: 3026.
23. Arsenault PR, Wobbe KK, Weathers PJ (2008) Recent advances in artemisinin production through heterologous expression. Curr Med Chem 15: 2886–2896.
24. Hsu E (2006) The history of qing hao in the Chinese materia medica. Trans R Soc Trop Med Hyg 100: 505–508.
25. Oldfield E, Lin FY (2012) Terpene biosynthesis: modularity rules. Angew Chem Int Ed Engl 51: 1124–1137.
26. Osbourn A, Goss RJ, Field RA (2011) The saponins: polar isoprenoids with important and diverse biological activities. Nat Prod Rep 28: 1261–1268.
27. Puri M, Sharma D, Tiwari AK (2011) Downstream processing of stevioside and its potential applications. Biotechnol Adv 29: 781–791.
28. Caputi L, Aprea E (2011) Use of terpenoids as natural flavouring compounds in food industry. Recent Pat Food Nutr Agric 3: 9–16.
29. Rude MA, Schirmer A (2009) New microbial fuels: a biotech perspective. Curr Opin Microbiol 12: 274–281.
30. Peralta-Yahya PP, Ouellet M, Chan R, Mukhopadhyay A, Keasling JD, et al. (2011) Identification and microbial production of a terpene-based advanced biofuel. Nat Commun 2: 483.
31. Zhang F, Rodriguez S, Keasling JD (2011) Metabolic engineering of microbial pathways for advanced biofuels production. Curr Opin Biotechnol 22: 775–783.
32. Dugar D, Stephanopoulos G (2011) Relative potential of biosynthetic pathways for biofuels and bio-based products. Nat Biotechnol 29: 1074–1078.
33. Farhi M, Marhevka E, Ben-Ari J, Algamas-Dimantov A, Liang Z, et al. (2011) Generation of the potent anti-malarial drug artemisinin in tobacco. Nat Biotechnol 29: 1072–1074.
34. Wu T, Wu S, Yin Q, Dai H, Li S, et al. (2011) [Biosynthesis of amorpha-4,11-diene, a precursor of the antimalarial agent artemisinin, in Escherichia coli through introducing mevalonate pathway]. Sheng Wu Gong Cheng Xue Bao 27: 1040–1048.
35. Hale V, Keasling JD, Renninger N, Diagana TT (2007) Microbially derived artemisinin: a biotechnology solution to the global problem of access to affordable antimalarial drugs. Am J Trop Med Hyg 77: 198–202.
36. Lacaze C, Kauss T, Kiechel JR, Caminiti A, Fawaz F, et al. (2011) The initial pharmaceutical development of an artesunate/amodiaquine oral formulation for the treatment of malaria: a public-private partnership. Malar J 10: 142.
37. Hemmerlin A, Rivera SB, Erickson HK, Poulter CD (2003) Enzymes encoded by the farnesyl diphosphate synthase gene family in the Big Sagebrush Artemisia tridentata ssp. spiciformis. J Biol Chem 278: 32132–32140.
38. Thulasiram HV, Erickson HK, Poulter CD (2008) A common mechanism for branching, cyclopropanation, and cyclobutanation reactions in the isoprenoid biosynthetic pathway. J Am Chem Soc 130: 1966–1971.
39. Wouters J, Oudjama Y, Barkley SJ, Tricot C, Stalon V, et al. (2003) Catalytic mechanism of Escherichia coli isopentenyl diphosphate isomerase involves Cys-67, Glu-116, and Tyr-104 as suggested by crystal structures of complexes with transition state analogues and irreversible inhibitors. J Biol Chem 278: 11903–11908.
40. Rohmer M (1999) The discovery of a mevalonate-independent pathway for isoprenoid biosynthesis in bacteria, algae and higher plants. Nat Prod Rep 16: 565–574.
41. Rohmer M, Knani M, Simonin P, Sutter B, Sahm H (1993) Isoprenoid biosynthesis in bacteria: a novel pathway for the early steps leading to isopentenyl diphosphate. Biochem J 295 (Pt 2): 517–524.
42. Zhao L, Chang WC, Xiao Y, Liu HW, Liu P (2013) Methylerythritol phosphate pathway of isoprenoid biosynthesis. Annu Rev Biochem 82: 497–530.
43. Sutherlin A, Rodwell VW (2004) Multienzyme mevalonate pathway bioreactor. Biotechnol Bioeng 87: 546–551.
44. Miziorko HM (2011) Enzymes of the mevalonate pathway of isoprenoid biosynthesis. Arch Biochem Biophys.
45. Middleton B (1974) The kinetic mechanism and properties of the cytoplasmic acetoacetyl-coenzyme A thiolase from rat liver. Biochem J 139: 109–121.
46. Andreassi JL, 2nd, Dabovic K, Leyh TS (2004) Streptococcus pneumoniae isoprenoid biosynthesis is downregulated by diphosphomevalonate: an antimicrobial target. Biochemistry 43: 16461–16466.
47. Hedl M, Sutherlin A, Wilding EI, Mazzulla M, McDevitt D, et al. (2002) Enterococcus faecalis acetoacetyl-coenzyme A thiolase/3-hydroxy-3-methylglutaryl-coenzyme A reductase, a dual-function protein of isopentenyl diphosphate biosynthesis. J Bacteriol 184: 2116–2122.
48. Miziorko HM, Lane MD (1977) 3-Hydroxy-3-methylgutaryl-CoA synthase. Participation of acetyl-S-enzyme and enzyme-S-hydroxymethylgutaryl-SCoA intermediates in the reaction. J Biol Chem 252: 1414–1420.
49. Pilloff D, Dabovic K, Romanowski MJ, Bonanno JB, Doherty M, et al. (2003) The kinetic mechanism of phosphomevalonate kinase. J Biol Chem 278: 4510–4515.
50. Sutherlin A, Hedl M, Sanchez-Neri B, Burgner JW, 2nd, Stauffacher CV, et al. (2002) Enterococcus faecalis 3-hydroxy-3-methylglutaryl coenzyme A synthase, an enzyme of isopentenyl diphosphate biosynthesis. J Bacteriol 184: 4065–4070.
51. Romanowski MJ, Bonanno JB, Burley SK (2002) Crystal structure of the Streptococcus pneumoniae phosphomevalonate kinase, a member of the GHMP kinase superfamily. Proteins 47: 568–571.
52. Pilloff DE, Leyh TS (2003) Allosteric and catalytic functions of the PPi-binding motif in the ATP sulfurylase-GTPase system. J Biol Chem 278: 50435–50441.
53. Clinkenbeard KD, Sugiyama T, Lane MD (1975) Cytosolic acetoacetyl-CoA thiolase from chicken liver. Methods Enzymol 35: 167–173.
54. Kredich NM, Tomkins GM (1966) The enzymic synthesis of L-cysteine in Escherichia coli and Salmonella typhimurium. J Biol Chem 241: 4955–4965.
55. Cornforth RH, Fletcher K, Hellig H, Popjak G (1960) Stereospecificity of enzymic reactions involving mevalonic acid. Nature 185: 923–924.
56. Ozer N (1985) A new enzyme-coupled spectrophotometric method for the determination of arginase activity. Biochem Med 33: 367–371.
57. Wishart DS, Bigam CG, Yao J, Abildgaard F, Dyson HJ, et al. (1995) 1H, 13C and 15N chemical shift referencing in biomolecular NMR. J Biomol NMR 6: 135–140.
58. Gorenstein DG (1984) Phosphorus-31 NMR : principles and applications. Orlando, Fla.: Academic Press. xiv, 604 p. p.
59. Cornforth JW, Phillips GT, Messner B, Eggerer H (1974) Substrate stereochemistry of 3-hydroxy-3-methylglutaryl-coenzyme A synthase. Eur - J Biochem 42: 591–604.
60. Popjak G (1970) Stereospecificity of Enzymic Reactions. In: P. D . Boyer,editor editors. The Enzymes. New York: Academic Press, Inc. pp. 115–215.
61. Neuhoff V. SR, Eibl H. (1885) Clear background and highly sensitive protein staining with Coomassie Blue dyes in polyacrylamide gels: A systematic analysis. Electrophoresis 6: 427–448.
62. Shapiro AL, Vinuela E, Maizel JV, Jr. (1967) Molecular weight estimation of polypeptide chains by electrophoresis in SDS-polyacrylamide gels. Biochem Biophys Res Commun 28: 815–820.
63. Fokt I, Skora S, Conrad C, Madden T, Emmett M, et al. (2012) D-Glucose and D-mannose-based metabolic probes. Part 3: Synthesis of specifically deuterated D-glucose, D-mannose, and 2-deoxy-D-glucose. Carbohydr Res 368: 111–119.
64. Huang YQ, Liu JQ, Gong H, Yang J, Li Y, et al. (2011) Use of isotope mass probes for metabolic analysis of the jasmonate biosynthetic pathway. Analyst 136: 1515–1522.
65. Xiao Y, Guo L, Wang Y (2013) Isotope-coded ATP probe for quantitative affinity profiling of ATP-binding proteins. Anal Chem 85: 7478–7486.
66. Klinman JP (2014) The power of integrating kinetic isotope effects into the formalism of the Michaelis-Menten equation. FEBS J 281: 489–497.
67. Cederstav AK (1994) Investigations into the Chemistry of Thermodynamically Unstable Species. The Direct Polymerization of Vinyl Alcohol, the Enolic Tautomer of Acetaldehyde. Journal of the American Chemical Society 116: 4073–4076.
68. Levy GC, Nelson GL (1972) Carbon-13 nuclear magnetic resonance for organic chemists. New York,: Wiley-Interscience. xiii, 222 p. p.
69. Fox JL (2011) Interest groups jostle to influence PDUFA V. Nat Biotechnol 29: 1062.
70. Brown GD (2010) The biosynthesis of artemisinin (Qinghaosu) and the phytochemistry of Artemisia annua L. (Qinghao). Molecules 15: 7603–7698.
71. Ro DK, Paradise EM, Ouellet M, Fisher KJ, Newman KL, et al. (2006) Production of the antimalarial drug precursor artemisinic acid in engineered yeast. Nature 440: 940–943.
72. Anthony JR, Anthony LC, Nowroozi F, Kwon G, Newman JD, et al. (2009) Optimization of the mevalonate-based isoprenoid biosynthetic pathway in Escherichia coli for production of the anti-malarial drug precursor amorpha-4,11-diene. Metab Eng 11: 13–19.
73. Tsuruta H, Paddon CJ, Eng D, Lenihan JR, Horning T, et al. (2009) High-level production of amorpha-4,11-diene, a precursor of the antimalarial agent artemisinin, in Escherichia coli. PLoS One 4: e4489.
74. Paddon CJ, Westfall PJ, Pitera DJ, Benjamin K, Fisher K, et al. (2013) High-level semi-synthetic production of the potent antimalarial artemisinin. Nature 496: 528–532.
75. Ajikumar PK, Xiao WH, Tyo KE, Wang Y, Simeon F, et al. (2010) Isoprenoid pathway optimization for Taxol precursor overproduction in Escherichia coli. Science 330: 70–74.
76. Wang C, Yoon SH, Shah AA, Chung YR, Kim JY, et al. (2010) Farnesol production from Escherichia coli by harnessing the exogenous mevalonate pathway. Biotechnol Bioeng 107: 421–429.
77. Noda L (1958) Adenosine triphosphate-adenosine monophosphate transphosphorylase. III. Kinetic studies. J Biol Chem 232: 237–250.
78. Noda L (1962) Nucleoside triphosphate-nucleoside monophosphokinases. In: P. D Boyer, Lardy, H and Myrbäck, K, editor editors. The Enzymes New York: Academic Press. pp. 139–149.
79. Zyryanov AB, Shestakov AS, Lahti R, Baykov AA (2002) Mechanism by which metal cofactors control substrate specificity in pyrophosphatase. Biochem J 367: 901–906.
80. Pohjanjoki P, Lahti R, Goldman A, Cooperman BS (1998) Evolutionary conservation of enzymatic catalysis: quantitative comparison of the effects of mutation of aligned residues in Saccharomyces cerevisiae and Escherichia coli inorganic pyrophosphatases on enzymatic activity. Biochemistry 37: 1754–1761.

4

Comparative Proteomics of Milk Fat Globule Membrane Proteins from Transgenic Cloned Cattle

Shunchao Sui[1], Jie Zhao[1], Jianwu Wang[2,3], Ran Zhang[1], Chengdong Guo[1], Tian Yu[3], Ning Li[1]*

1 State Key Laboratory of Agrobiotechnology, China Agricultural University, Beijing, PR China, **2** Beijing GenProtein Biotech Company Ltd., Beijing, PR China, **3** Wuxi Kingenew Biotechnology Co., Ltd., Wuxi, Jiangsu Province, China

Abstract

The use of transgenic livestock is providing new methods for obtaining pharmaceutically useful proteins. However, the protein expression profiles of the transgenic animals, including expression of milk fat globule membrane (MFGM) proteins, have not been well characterized. In this study, we compared the MFGM protein expression profile of the colostrum and mature milk from three lines of transgenic cloned (TC) cattle, i.e., expressing recombinant human α-lactalbumin (TC-LA), lactoferrin (TC-LF) or lysozyme (TC-LZ) in the mammary gland, with those from cloned non-transgenic (C) and conventionally bred normal animals (N). We identified 1, 225 proteins in milk MFGM, 166 of which were specifically expressed only in the TC-LA group, 265 only in the TC-LF group, and 184 only in the TC-LZ group. There were 43 proteins expressed only in the transgenic cloned animals, but the concentrations of these proteins were below the detection limit of silver staining. Functional analysis also showed that the 43 proteins had no obvious influence on the bovine mammary gland. Quantitative comparison revealed that MFGM proteins were up- or down-regulated more than twofold in the TC and C groups compared to N group: 126 in colostrum and 77 in mature milk of the TC-LA group; 157 in colostrum and 222 in mature milk of the TC-LF group; 49 in colostrum and 98 in mature milk of the TC-LZ group; 98 in colostrum and 132 in mature milk in the C group. These up- and down-regulated proteins in the transgenic animals were not associated with a particular biological function or pathway, which appears that expression of certain exogenous proteins has no general deleterious effects on the cattle mammary gland.

Editor: Silvia Mazzuca, Università della Calabria, Italy

Funding: This project was supported by the "863" High-Tech Research Development Project (project grant numbers 2010AA10A105 and 2011AA100601). The funders had no role in study design, data collection and analysis, decision to publish, or preparation of the manuscript.

Competing Interests: Jianwu Wang of the authors is employed by the commercial company Beijing GenProtein Biotech Co., Ltd., who has a role in performing the experiment, analyzing the data and preparing the manuscript. Tian Yu of the authors is employed by the commercial company Wuxi Kingenew Biotechnology Co., Ltd., who has a role in performing the experiment, analyzing the data. There are no further patents, products in development or marketed products to declare

* Email: ninglicaulab@gmail.com

These authors contributed equally to this work.

Introduction

The technology of using genetically modified bovine lines to produce recombinant proteins in milk has flourished since the 1990s, based on the historic breakthrough of somatic cell cloning technology [1,2]. Milk has been proven to be an excellent vehicle for producing and delivering recombinant human proteins [3]. The cow mammary gland can be modified to synthesize foreign proteins in transgenic cloned bovine milk, which may be useful for human consumption and assimilation because of its similarities of composition to human milk [4]. The advantages of milk-based system also include the easiness of animal housing and maintenance and lower cost of harvesting proteins than the cell bioreactor [5]. However, we have little information about how transgenic modification and cloning process influence bovine endogenous gene expression, milk characteristics and health. A further consideration is how these changes of products from these animals might affect the health of consumers. To address these issues, we examined both the protein expression profiles and composition of milk from transgenic cloned cows compared with conventionally bred animals. Many works have been done to analyze the compositions of meat and milk from transgenic cloned cattle, but the majority of these studies have focused on a few major constituents rather than a comprehensive analysis of proteins expression, especially of low-abundance proteins [6–8].

The concerns about consumption of food from transgenic cloned animals were the hot spot in recent research. The FDA has raised the instructions for the food from cloned animal and the studies have showed that the milk and meats from cloned cattle were as safe as the conventionally bred cattle. However, there were no instructions for food products from transgenic cloned animals, we have to use the instructions of cloned animals to examine the food products from transgenic cloned cattle. Milk proteins can be divided into three classes, namely casein, whey proteins, and MFGM proteins. Compared to other classes, MFGM proteins are the least abundant, making up 1–4% of total milk proteins, but they include thousands of different proteins [9], which were important composition of the milk proteins.

The MFGM is a protein-rich lipid bilayer that surrounds lipids in milk. Although the MFGM has been studied for 50–60 years, previous studies have focused primarily on membrane globule formation, intracellular transition, and secretion [10]. It is believed that the MFGM originates from the apical plasma membrane of mammary epithelial cells [11,12]. Precursor microlipid droplets from the endoplasmic reticulum fuse to each other and travel to the apical cytoplasm where they are surrounded by the apical plasma membrane and then are secreted into the alveolar lumen [10,13–15]. Therefore, we can obtain information about mammary epithelial cell health by analyzing the protein composition of the MFGM.

In addition, MFGM proteins have many beneficial bioactivities, e.g., as antibiotics and anticancer agents, MUC1 is one of the most abundant proteins in the MFGM and is thought to protect exposed surfaces from physical damage and invasive pathogenic microorganisms [16,17]. For example, bovine lactophorin C-terminal fragment and PAS6/7, both MFGM constituents, inhibit replication of human rotavirus and prevent gastroenteritis [18]. MFGM proteins also have commercial value in cheese and butter production.

Technological advances in proteomics, especially quantitative proteomics, have enabled an increased understanding and characterization of milk proteins [19]. So far, proteomics has been successfully applied to studies of human milk whey proteins, bovine milk whey proteins, human MFGM proteins, bovine MFGM proteins, and goat MFGM proteins [20–23].

We genetically engineered three transgenic bovine lines that specifically express recombinant human α-lactalbumin (TC-LA group), lactoferrin (TC-LF group), and lysozyme (TC-LZ group) [24–26], from which the compositions of the milk include the human proteins with higher expression level compared to that from conventionally bred animals. Prior research for milk whey proteins and the nutritional components from transgenic cloned cattle have indicated that the expression of exogenous proteins did not significantly change the milk whey protein profile, and the mean values for the majority of the measured parameters were within the normal range [27].

In this work, we compared the MFGM proteins of colostrum and mature milk from these three healthy transgenic cloned lines to those from cloned and conventionally bred animals. The MFGM proteins of the five groups were characterized using 2D-PAGE and 2D LC-MS/MS coupled with the iTRAQ proteomic strategy. Our objectives were to: (1) characterize the proteins of the MFGM in transgenic cloned cattle; (2) compare the relative expression of MFGM proteins among transgenic cloned cattle milk; and (3) investigate the health of mammary epithelial cells in the transgenic cloned cattle. The study will supplement the former study by completing the evaluation for the milk products from transgenic cloned cattle and revealing the situation of the mammary epithelial cells from transgenic cloned cattle. This study provides the complement evidence of the milk composition analysis for the transgenic cloned bovine, and the results may indirectly reflect the health of the mammary epithelial cells of the transgenic cloned bovine.

Materials and Methods

Milk Sample Collection

The protocol was approved by the Institutional Animal Care and Use Committee of China Agricultural University (ID: SKLAB-2010-05-01). The three transgenic cattle lines TC-LZ (n = 10), TC-LA (n = 4), and TC-LF (n = 3) have been described [24–26] and were compared to cloned cattle (C, n = 3) which were also cloned by somatic cell nuclear transfer (SCNT), and conventionally bred cattle (N, n = 9) of a similar genetic background. The cattle were similar in age and lactation period and were housed under the same conditions. The milk collection were carried out as described previously [27]. The colostrum was obtained during the initial three days of lactation and mature milk was obtained on the 30th, 60th and 90th day after lactation and the milk was collected from the cows twice daily and was pooled to form one daily sample. Milk samples were collected by milking machine, and samples were immediately stored at −20°C until further analysis.

Protein Extraction

Milk sample (50 ml) from each cow was centrifuged at 4000 × g to separate milk fat globule cream from whole milk. The separated cream was rinsed in 25 ml phosphate buffered saline twice and 25 ml deionized water once. Then the milk fat globule proteins were obtained by methanol/trichloromethane precipitation as follows. The cream (2 ml) was then stored at −80°C, and upon thawing 10 ml methanol and 10 ml trichloromethane were added. The samples were centrifuged at 50,000 × g for 30 min at 4°C (Avanti J-26XP, Beckman Coulter, Indianapolis, USA), and the supernatant was discarded. The precipitated proteins were solubilized in a solution containing 7 M urea, 2 M thiourea, 0.025 M DTT, 2% (w/v), and 1% CHAPS [20]. The relative concentration of protein in each extract was determined by measuring the amount of whey protein with the 2D-Quant quantitate kit (GE Healthcare). MFGM protein samples from each cow were pooled together to generate one individual sample and then individual samples in the same groups (TC-LZ, TC-LA, TC-LF, C and N) were pooled again to generate one group sample according to the equal protein mass.

In-solution digestion

Soluble proteins (200 μg) were digested in-solution using the filter aided sample preparation (FASP) method [28]. Each protein extract (30 μl) was mixed with 200 μl of buffer containing 8 M urea in 0.1 M Tris-HCl (pH 8.5) in the filter unit and centrifuged at 14,000 × g for 40 min and repeated one time. The flow-through was discarded, and iodoacetamide solution was added to the filter and mixed at 600 rpm for 1 min then incubated without mixing for 5 min. The filtered units were centrifuged at 14,000 × g for 30 min, then 100 μl of buffer containing 8 M urea in 0.1 M Tris-HCl (pH 8.0) was added to the filtered unit, then the solution was centrifuged again at 14,000 × g for 40 min and repeated twice. We then added 40 μl of the above-mentioned buffer containing endoproteinase Lys-C (enzyme to protein ratio 1:50, w/w) and mixed at 600 rpm in thermo-mixer for 1 min. The units were incubated overnight then transferred to new collection tubes, and then 120 μl of 0.05 M NH_4HCO_3 in water with trypsin (enzyme to protein ratio 1:100 w/w) was added and mixed at 600 rpm for 1 min. The units were incubated at room temperature for 4 h and then centrifuged at 14,000 × g for 40 min. Finally, 50 μl of 0.5 M NaCl was added and the units were centrifuged at 14,000 × g for 20 min.

The extracted peptides were desalted using a 1.3-ml C18 solid-phase extraction column (Sep-Pak® cartridge; Waters Corporation, Milford, USA). The peptides were dried using a vacuum centrifuge and then resuspended in 5 mM ammonium formate containing 5% acetonitrile, pH 3.0.

LC-MS/MS Analysis

LC-MS/MS was performed on a nano Acquity UPLC system (Waters Corporation) connected to an LTQ Orbitrap XL mass spectrometer (Thermo Electron Corp., Bremen, Germany) equipped with an online nano-electrospray ion source (Michrom Bioresources, Auburn, USA). The MS was operated in data-dependent mode to switch automatically between MS and MS/MS acquisition. Survey full-scan MS spectra (m/z 300–1800) were acquired in the Obitrap with one microscan and with a mass resolution of 60,000 at m/z 400, followed by MS/MS of the eight most-intense peptide ions in the LTQ analyzer. LC-MS/MS spectra were acquired using SEQUEST (v.28 revision 12, Thermo Electron Corp) against the International Protein Index bovine database version 3.54. A decoy database containing the reverse sequences was appended to the database. The search parameters were set as follows: partial trypsin cleavage with two missed cleavage sites was considered; the peptide mass tolerance was 1.4 Da; and the fragment ion tolerance was 1 Da. Trans Proteomic Pipeline software (revision 4.20; Institute of Systems Biology, Seattle, WA) was then used to identify proteins based on corresponding peptide sequences with ≥95% confidence. The peptide results were filtered by PeptideProphet with a p value> 0.90 and a ProteinProphet probability = 0.95 [29,30].

Quantitative Analysis of MFGM Proteins by iTRAQ LC-MS/MS

The digested peptides of the five groups were transferred to vials containing individual iTRAQ reagents (Applied Biosystems) following the iTRAQ standard protocol for the 8-plex kit. The N group was labeled with iTRAQ113, the TC-LF with iTRAQ114, TC-LZ with iTRAQ115, TC-LA with iTRAQ116, and the C group with iTRAQ117.

The iTRAQ-labeled samples were pooled and the SCX HPLC experiment was performed on a Shimadzu 20AD 5-μm SCX polysulfoethyl column (2.1 mm×100 mm, The Nest Group, Inc., MA) as the first dimension. Each collected components of the processed SCX fractions ran via RP-LC ESI-MS/MS on an Applied Biosystems Q-Star Elite XL mass spectrometer in which the RPLC column was a ZORBAX 300SB-C18 (5 μm, 0.1 mm×150 mm, Microm, USA). The Q-Star Elite XL mass spectrometer was operated in the smart information-dependent acquisition activated mode with automatic collision energy and automatic MS/MS accumulation. Survey full-scan MS spectra (m/z 400–1800) were acquired with one microscan and a mass resolution of 60,000 at m/z 400, followed by MS/MS of the four most-intense peptide ions in the analyzer. The relative abundance of the MFGM proteins in the different samples was derived from the ionic peak areas of the iTRAQ reporter.

iTRAQ identification and quantification analysis of the MFGM proteins were obtained using Protein Pilot 3.0 (Applied Biosystems, USA) with the following user-defined parameters: Sample Type, iTRAQ 8-plex (Peptide Labeled); Cysteine alkylation, MMTS; Digestion, Trypsin; Specify Processing, Quantitate; Database, IPI v3.62 bovine; Search Effort, thorough ID; Results Quality, Detected Protein Threshold [Unused ProtScore (Conf)] >1.3 (95%); Run False Discovery Rate Analysis.

Results

Overview of MFGM Proteome in the TC Groups

MFGM proteins were characterized in three TC bovine lines and in C group and N group using 2D nano-LC-MS/MS and a quantitative proteomics method (iTRAQ). LC-MS/MS identified 1225 proteins among the five groups of MFGM samples, 939 proteins in colostrum, and 910 proteins in mature milk (Table 1). There were 637 proteins that were present in TC-LA group, 721 in the TC-LF group, 720 in the TC-LZ group, 527 in the C group, and 668 in the N group. iTRAQ identified 851 MFGM proteins in colostrum and 775 MFGM proteins in mature milk of all the bovine lines. The identified proteins from the three TC groups were compared to those from the C and N groups. The TC-LA group contained 166 proteins that were not found in the control lines, the TC-LF group contained 265 proteins, and the TC-LZ group contained 184 proteins (Figure 1). We found that 43 of these proteins were shared by all the three TC groups (Table 2). The molecular mass of this subset was between 10 kDa and 90 kDa, and the pI values were between 4 and 10. DAVID Bioinformatics Resources was used to analyze the function of the 43 common proteins in GO terms [31]. We found that these proteins were enriched in the biological processes of protein transport, cellular component of the Golgi apparatus, and the molecular function of GTP binding. However, the functions of these specific expressed proteins did not cluster in any particular pathway (Figure 2).

Quantitative Comparison of MFGM Proteins

We obtained relative MFGM protein quantity information from the five groups using iTRAQ. Only proteins with consistent quantitative results in two independent iTRAQ runs were used for further analysis of expression and function. A total of 459 proteins from colostrum MFGM and 426 proteins from mature MFGM were identified. Statistical analysis of expression levels in the three TC groups and C group compared with the N group revealed more than twofold differences ($p \leq 0.05$) in the expression of many of these proteins (Figure 3). In this regard, 126 proteins in the colostrum and 77 proteins in the mature milk of the TC-LA group had a greater than twofold difference in expression, likewise, in the TC-LF group there were 157 proteins in colostrum and 222 proteins in mature milk, in the TC-LZ group there were 49 proteins in colostrum and 98 proteins in mature milk, and in the cloned control group there were 98 proteins in colostrum and 132 proteins in mature milk.

Comparison between the MFGM proteins among the five groups also revealed many differences. The unsupervised hierarchical cluster analysis (HCA) and the principal component analysis (PCA) were used to assess the variability among the 5 groups. We found that the TC-LA, TC-LZ, and C groups had similar protein compositions in colostrum and mature milk. The TC-LF group was distinguishable from the other two TC groups (Figure 4). The colostrum of the TC-LF and TC-LA groups clustered together as that of the TC-LZ and C groups did. The results indicated that in the colostrum the TC-LZ was closer to the C group and in the mature milk the TC-LZ and TC-LA were closer to the C group. Both in colostrum and mature milk, the TC-LF groups were showed a greater difference from the other TC groups. In colostrum, 11 proteins were differentially expressed in the TC-LA group compared to the control groups, 25 in the TC-LF group, and 6 in the TC-LZ group. In mature milk, there were 8 proteins that were differentially expressed compared to the controls in the TC-LA group, 25 in the TC-LF group, and 5 in the TC-LZ group. In TC groups, the proteins with expression changed simultaneously were not appeared. Several of these differentially expressed proteins, such as fibrinogen α chain, α-S2-casein, κ-casein, β-lactoglobulin, and lactoferrin have been explored in other studies of whey and MFGM [20,32–34].

Our identified MFGM proteins were subjected to cluster analysis to evaluate the relationship between TC animals and conventionally bred controls. In the TC-LA group, six proteins were notably down-regulated, including α-2-antiplasmin (−4.9 fold), β-lactoglobulin (−4.8 fold), serum albumin (−4.9 fold), platelet glycoprotein 4 (−4.1 fold), lactoferrin (−4.2 fold), serine hydroxymethyltransferase, mitochondrial (−4.1 fold), and a

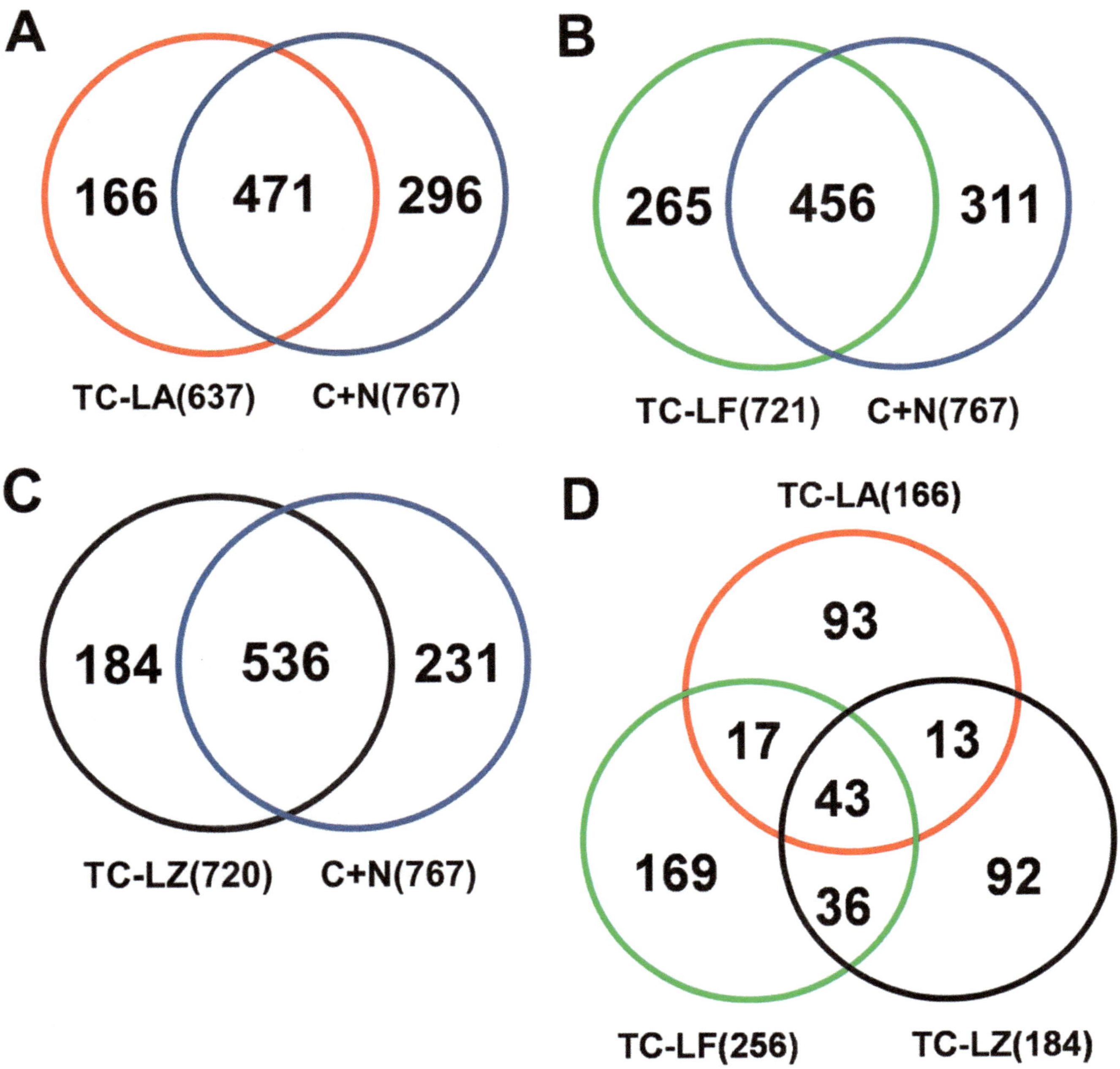

Figure 1. Comparison of MFGM proteins from the different TC and control lines. Cloned and conventionally bred control animals **[(C+N)]** were compared with (A) MFGM proteins from the TC-LA line, (B) MFGM proteins from the TC-LF line, and (C) MFGM proteins from the TC-LZ line. (D) The intersection of differentially expressed MFGM proteins from the three TC groups yielded 43 proteins in common.

Table 1. Proteins identified in the MFGM of colostrum and mature milk of the TC-LZ, TC-LA, and TC-LF transgenic cloned lines and of the cloned and conventionally bred normal controls.

	TC-LZ	TC-LA	TC-LF	Cloned	Normal	Total
Colostrum	620	516	411	220	562	939
Mature milk	444	393	567	464	353	910
In total	720	637	721	527	668	1225

Table 2. Proteins specifically expressed in transgenic cloned cattle MFGM.

No	Accession	Description	MW	PeptideLength	Isoelectric Point
1	IPI00686092.1	Peroxiredoxin-1	22194.98	199	8.6084
2	IPI00687416.2	Rras2 Protein	23384.41	204	5.8076
3	IPI00687550.1	3′(2′),5′-Bisphosphate Nucleotidase 1	33306.63	308	5.1936
4	IPI00689149.2	Oxysterol-Binding Protein	84236.69	746	6.2374
5	IPI00692733.2	Transmembrane Protein 30A	40656.91	361	8.6073
6	IPI00693697.3	Similar To Adiponutrin	50423.45	455	7.5812
7	IPI00695308.5	Mucin-16 (Fragment)	31335.98	282	9.0396
8	IPI00695670.1	Upf0585 Protein C16Orf13 Homolog	22699.17	204	7.9092
9	IPI00695741.2	62 Kda Protein	62274.23	569	5.3891
10	IPI00696647.1	Similar To Pincher Isoform 1	60985.31	540	7.6711
11	IPI00697565.3	Ras-Related Protein Rab-21	24130.66	222	8.0481
12	IPI00698430.1	Eukaryotic Translation Initiation Factor 5 Isoform 7	48939.49	429	5.2609
13	IPI00700391.1	Similar To V-Crk Sarcoma Virus Ct10 Oncogene Homolog Isoform 1	33809.31	304	5.2451
14	IPI00703351.2	Similar To Aldehyde Dehydrogenase 3B1 Isoform 1	53661.38	486	8.4692
15	IPI00703423.1	B-Cell Receptor-Associated Protein 31	27883.64	245	9.765
16	IPI00704666.1	Trna Methyltransferase 112 Homolog	14309.9	125	5.8553
17	IPI00704752.1	Ras-Related Protein Rab-7A	23528.72	207	6.5881
18	IPI00706451.3	Adp-Ribosylation Factor 4	20514.36	180	5.9819
19	IPI00706632.1	Elongation Factor 1-Gamma	50345.45	440	6.5236
20	IPI00706968.3	Nuclear Factor Of Kappa Light Polypeptide Gene Enhancer In B-Cells Inhibitor, Beta	37543.32	357	4.2979
21	IPI00707103.4	Synaptosomal-Associated Protein 29	28471.85	258	5.1001
22	IPI00707616.3	Cidea Protein	24524.03	219	9.7273
23	IPI00708591.1	Znrf2 Protein	23870.75	238	7.1119
24	IPI00710727.1	Transitional Endoplasmic Reticulum Atpase	89272.24	806	4.8908
25	IPI00711304.4	Similar To Inositol Polyphosphate-5-Phosphatase A	70475.79	637	8.6043
26	IPI00713743.1	Inosine-5′-Monophosphate Dehydrogenase 2	55726.54	514	7.339
27	IPI00714515.1	Fgr Protein	59292.95	527	5.3694
28	IPI00714621.3	Uncharacterized Protein C1Orf93 Homolog	21482.83	201	6.2609
29	IPI00714818.1	Eukaryotic Translation Initiation Factor 3 Subunit H	39879.94	352	6.653
30	IPI00714992.4	Cgmp-Dependant Type Ii Protein Kinase	87033.26	762	8.6701
31	IPI00715153.2	Fas-Associated Factor 2	52629.57	445	5.4387
32	IPI00716638.7	Wd Repeat-Containing Protein 1	66214.06	606	6.7147
33	IPI00717376.3	Similar To Cytoskeleton-Associated Protein 4	64489.56	589	5.9276
34	IPI00718291.4	Rab14 Protein	23881.49	215	6.1266
35	IPI00718671.5	Synaptosomal-Associated Protein	23212.11	211	4.4574
36	IPI00729755.2	Acyl-Coa Synthetase Long-Chain Family Member 1	78228.82	699	7.2276
37	IPI00730045.3	Similar To Rififylin Isoform 4	39655.22	356	6.1033
38	IPI00733615.1	Dnaj Homolog Subfamily A Member 2	45731.44	412	6.4421
39	IPI00867128.1	Phosphoserine Aminotransferase	40502.07	370	7.7533

Table 2. Cont.

No	Accession	Description	MW	PeptideLength	Isoelectric Point
40	IPI00867311.1	Scamp2 Protein	36678.97	328	6.1107
41	IPI00906554.1	Guanine Nucleotide-Binding Protein Subunit Alpha-11	42042.75	359	6.029
42	IPI00924133.1	Similar To Ciliary Neurotrophic Factor Receptor Alpha Precursor (Cntfr Alpha) Isoform 3	40661.3	372	6.9436
43	IPI00944416.1	Cyb5B Protein	16267.5	146	4.8233

putative uncharacterized protein (−4.4 fold). In the TC-LF group, 10 proteins were notably different from the control group, including apolipoprotein A-II (−4.4 fold), ATP-binding cassette sub-family G (WHITE) member 2 (+4.3 fold), α-1-acid glycoprotein (−4.9 fold) similar to solute carrier family 39 (zinc transporter) member 8 (+4.1 fold), α-S2-casein (+4.2 fold), MUC 1 (+4.9 fold), acyl-CoA synthetase long-chain family member 1 (+4.8 fold), and vitamin D–binding protein (−5.2 fold). In the TC-LZ, four proteins were notably different from the control group, namely fibrinogen α chain (+2.8 fold), β-lactoglobulin (−2.7 fold), cidea protein (−3.1 fold), and similar to ATPase type 13A4 (−2.9 fold). In the C group, six proteins were notably different from the conventionally bred group, namely fibrinogen α chain (+3.1 fold), α-S2-casein (+5.3 fold), α-S1-casein (+4 fold), Rit1 protein (−3.6 fold), α-lactalbumin (−4.8 fold), and nucleobindin-1 (+3.3 fold).

Function and Pathway Analysis of Differentially Expressed Proteins

The Ingenuity Pathway Analysis (IPA, Ingenuity Systems, Inc. Redwood City) software was used to analyze the functions and networks of the differentially expressed proteins. To further examine the specifically expressed proteins, they were categorized according to their biological process, cellular component or molecular function as annotated in the Gene Ontology (GO) database. Differentially expressed MFGM proteins from TC-LA colostrum were mainly enriched in the biological processes of response to wounding and the acute inflammatory response, in the cellular components of the extracellular region, and in the molecular function of endopeptidase inhibitor activity. Differentially expressed MFGM proteins from TC-LA mature milk were mainly enriched in cytoskeleton organization, membrane-bounded vesicles, and unfolded protein binding. Differentially expressed MFGM proteins from TC-LF colostrum were mainly enriched in response to organic substance, the extracellular region, and endopeptidase inhibitor activity. Differentially expressed MFGM proteins from TC-LF mature milk were mainly enriched in small GTPase–mediated signal transduction, pigment granules, and GTPase activity. Differentially expressed MFGM proteins from TC-LZ colostrum were mainly enriched in response to wounding, the extracellular region, and cell-surface binding. Finally, differentially expressed MFGM proteins from TC-LZ mature milk were mainly enriched in cell motility, the actin cytoskeleton, and actin binding (Figure 5).

In the TC-LA group, there were 18 proteins that were changed above 3 folds in colostrum and mature milk. The pathway analysis showed that the changed MFGM proteins in the TC-LA colostrum most involved in the network which have the functions of cellular movement, immune cell trafficking, cell-to-cell signaling and interaction. And in the mature milk are cell death, cell-to-cell signaling and interaction, hematological system development and function (Figure 6A, B).

In TC-LF group, there were 46 proteins that were changed in colostrum and mature milk significantly. The IPA software showed that the changed MFGM proteins in mature milk involved in the network of which top functions are cellular assembly and organization, cellular function and maintenance, dermatological diseases and conditions. And the MFGM proteins from colostrum involved in the network of which top functions are lipid metabolism, molecular transport, and small molecule biochemistry (Figure 6C, D).

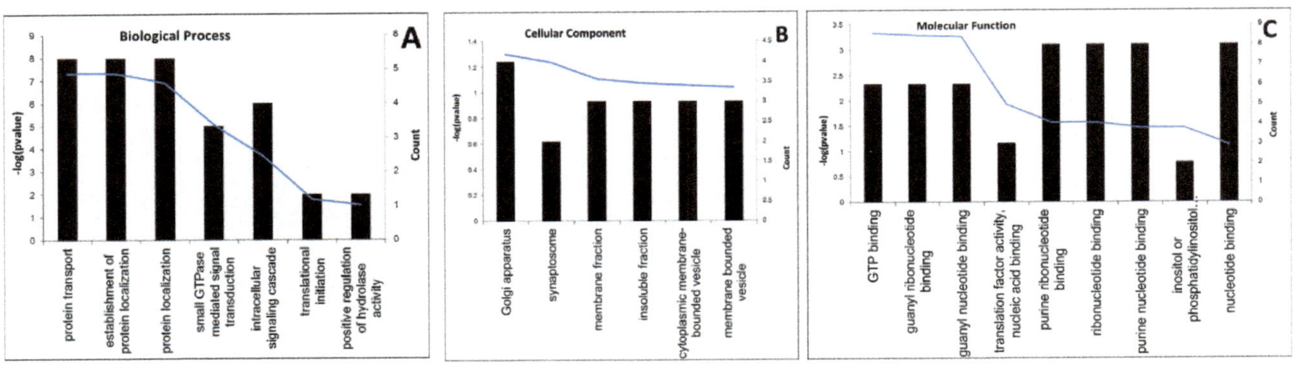

Figure 2. Functional analysis of MFGM proteins specifically expressed in the transgenic cloned animals. Analysis of proteins enriched in the transgenic cloned groups according to: (A) biological process, (B) cellular component, and (C) molecular function. Blue curves indicate the degree of enrichment of gene function.

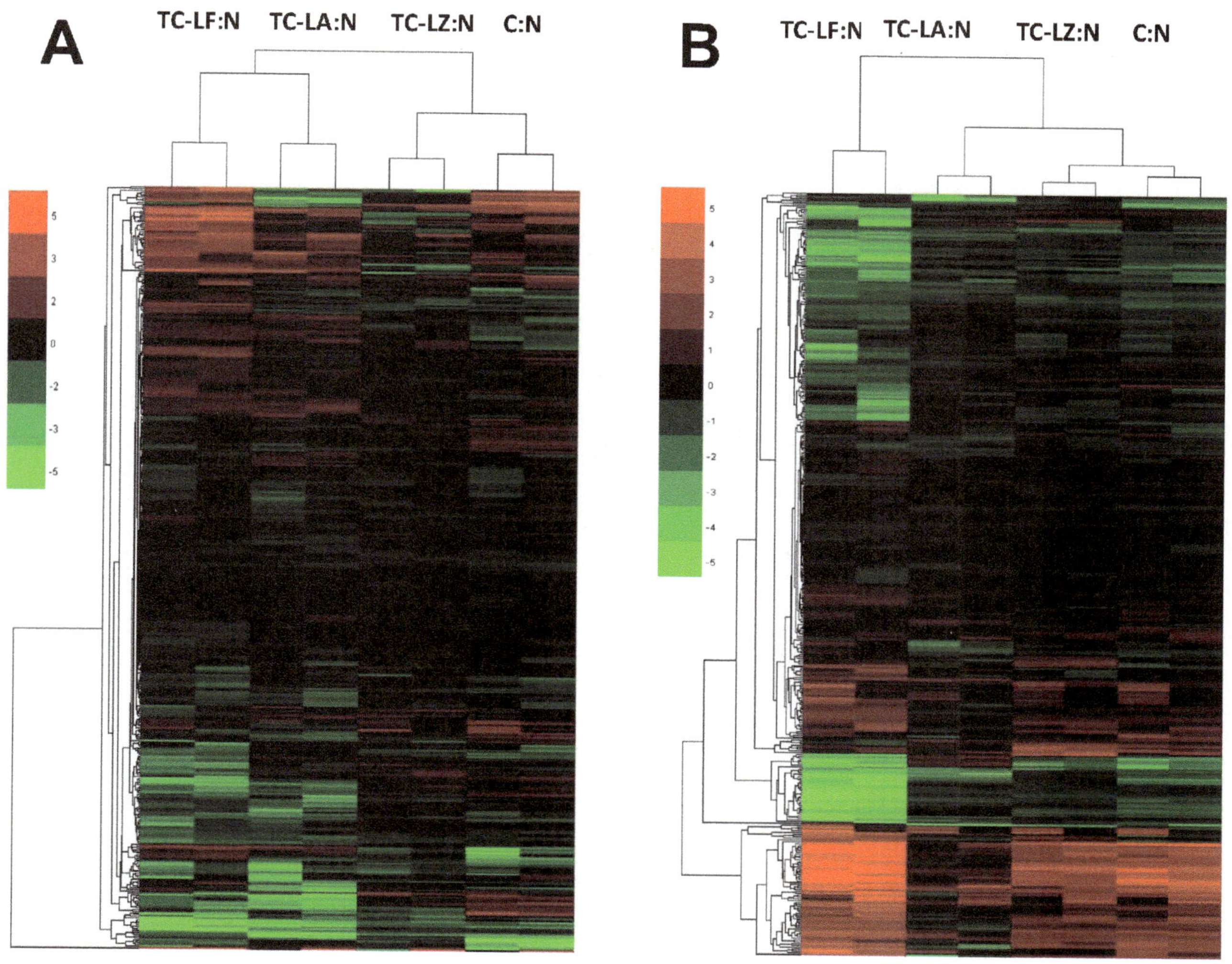

Figure 3. Cluster analysis of MFGM proteins. (A) Analysis of MFGM proteins in colostrum of the three TC groups compared to cloned **[(C)]** and conventionally bred **[(N)]** controls. (B) Analysis of MFGM proteins in mature milk of the three transgenic groups compared to controls.

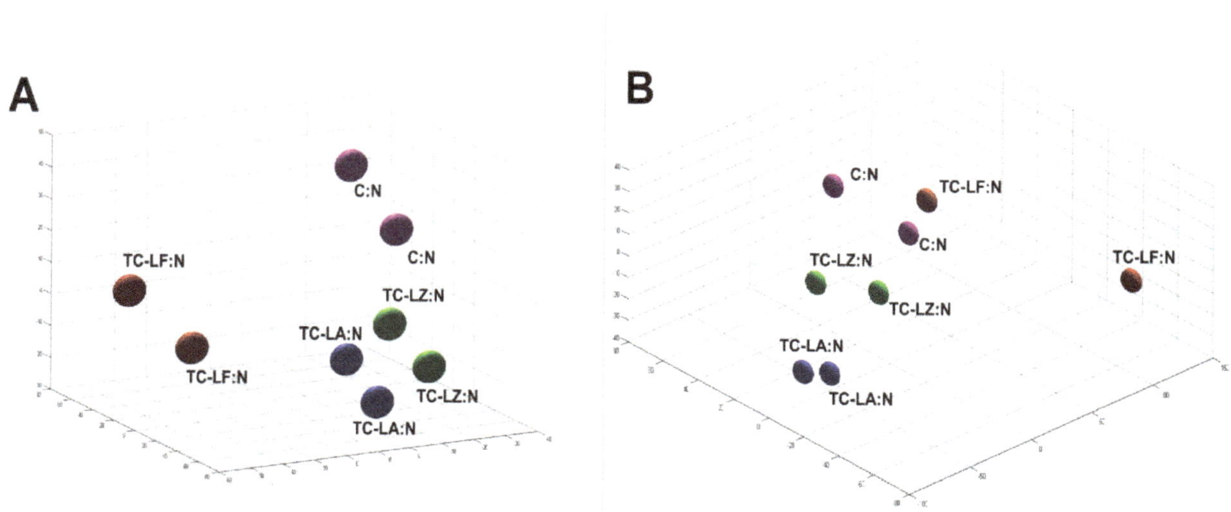

Figure 4. Principal component analysis of MFGM proteins from the three transgenic groups, the cloned group, and the conventionally bred control group. (A) Analysis of MFGM proteins in colostrum. (B) Analysis of MFGM proteins in mature milk.

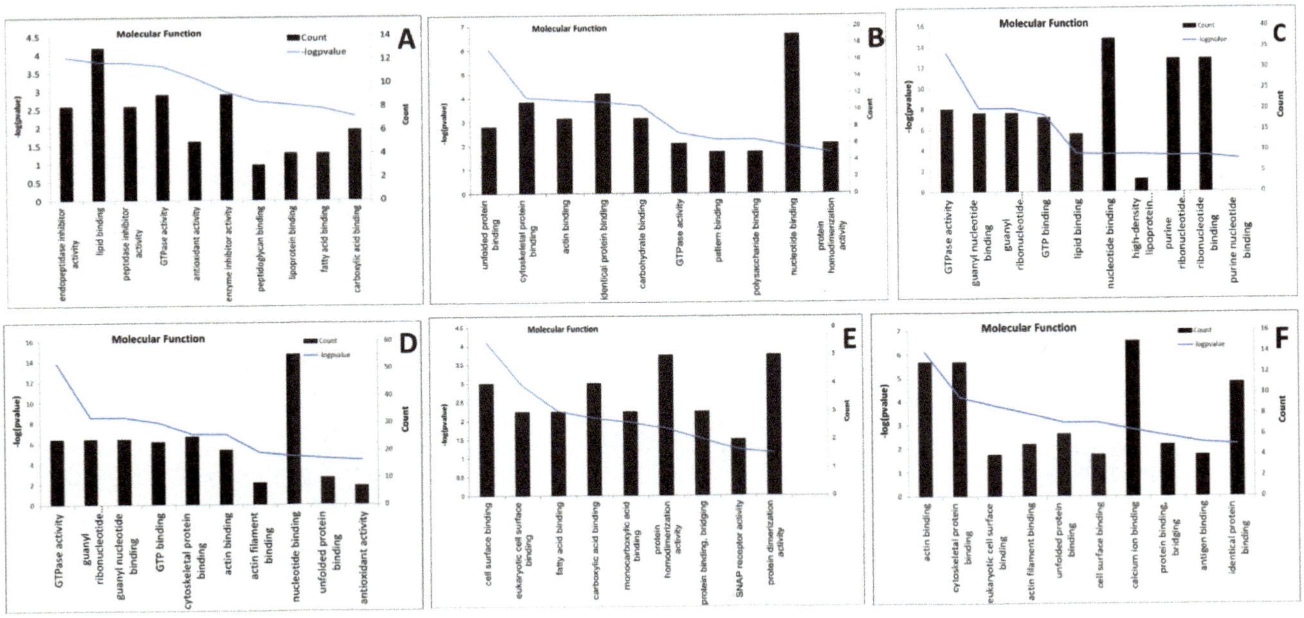

Figure 5. Molecular function analysis of differentially expressed MFGM proteins in the transgenic groups. Enrichment analysis of differentially expressed MFGM proteins in (A) TC-LA colostrum, (B) TC-LA mature milk, (C) TC-LF colostrum, (D) TC-LF mature milk, (E) TC-LZ colostrum, and (F) TC-LZ mature milk. Blue curves indicate the degree of enrichment of gene function.

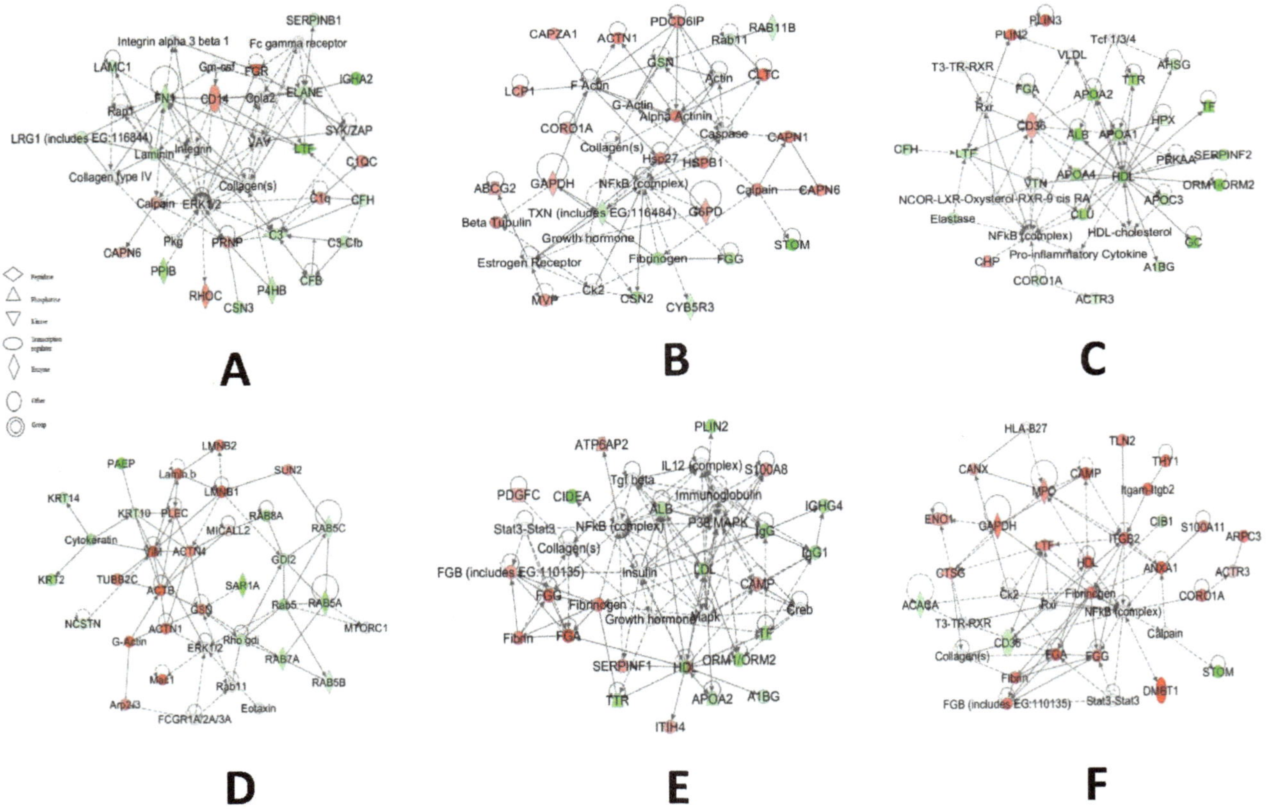

Figure 6. Pathway analysis of differentially expressed proteins from the transgenic groups. IPA defined pathways of the differentially expressed MFGM proteins from: (A) TC-LA colostrum, (B) TC-LA mature milk, (C) TC-LF colostrum, (D) TC-LF mature milk, (E) TC-LZ colostrum, and (F) TC-LZ mature milk.

In group of TC-LZ, there were 10 proteins that were altered obviously in colostrum and mature milk. The IPA software showed that the changed MFGM proteins in TC-LZ mature milk involved in the network of which the top functions are cell-to-cell signaling and interaction, tissue development, cell death. And in colostrum MFGM proteins, the top functions are inflammatory response, cell-to-cell signaling and interaction, cell signaling (Figure 6E, F).

The MFGM proteins of C group were also compared with the N group, the results showed that the changed proteins in the colostrum involved in the functions of cell-to-cell signaling and interaction, tissue development, hematological system development and function. In mature MFGM, the changed proteins were in the network of functions: cellular movement, cellular assembly and organization, cellular function and maintenance. Differentially expressed MFGM proteins in colostrum from the C group were enriched in the biological processes of response to wounding and the acute inflammatory response, in the cellular components of the extracellular region, and in the molecular function of GTPase activity. Differentially expressed MFGM proteins in mature milk of the cloned control animals were enriched in the biological process of glucose catabolism, in the cellular component of the actin cytoskeleton, and in the molecular function of cytoskeletal protein binding (Figure 7).

Discussion

Proteomics is one of the most popular methods to analyze complex samples. Recent advances in proteomics make the identification of low-abundance proteins and analysis of complex proteins such as high-molecular-weight transmembrane proteins possible, such analyses are not possible with standard 2D-PAGE methods alone. The combination of LC-MS/MS and iTRAQ or methods with the iTRAQ technologies are suitable for analysis of hydrophobic macro-molecules like plasma membrane proteins and are highly sensitive, fast, and reliable. Comparing with traditional 2D-PAGE method, the methods based on mass spectrometry are more suitable for MFGM proteins analysis. Data from several previous studies of MFGM proteins in humans [23], cattle [35] and goats [36] demonstrated that the amount of the identified proteins by this method was larger than other methods and the iTRAQ method was suitable to analyze the MFGM proteins with the same sensitivity compared to LC-MS/MS.

Applying these methods, we examined the effects of exogenous gene expression and cloning techniques on bovine MFGM proteins, an neglected area of research. Comparing three TC bovine lines expressing different exogenous proteins with C and N group, we identified 43 proteins that were specifically expressed in the transgenic animals below the detection limit for 2D-PAGE and silver staining. Considering both the concentration and functional

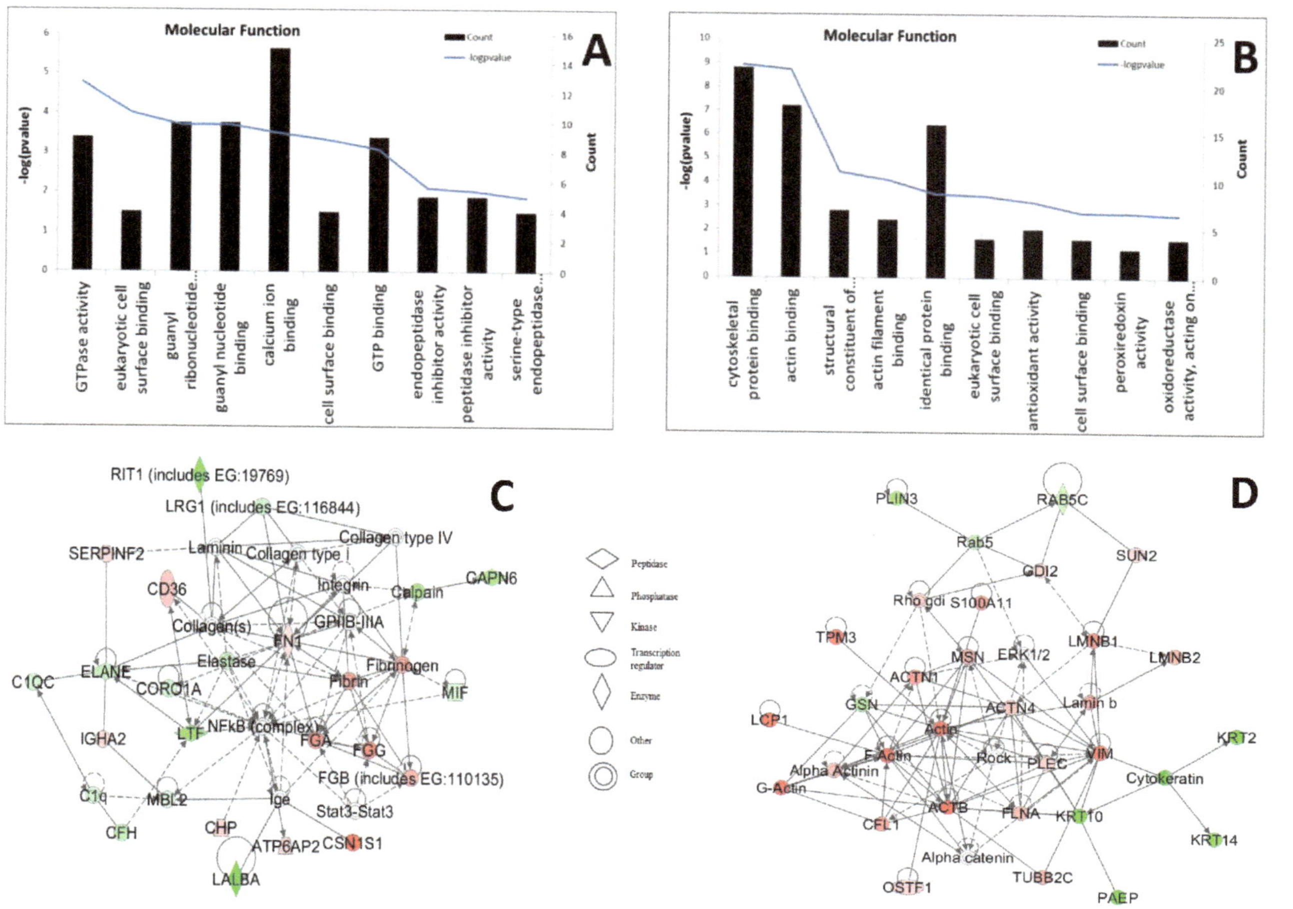

Figure 7. Molecular function and pathway analysis of differentially expressed MFGM proteins from the cloned control group. (A, B): Molecular function of MFGM proteins from colostrum (A) and mature milk (B). (C, D): IPA defined pathways of MFGM proteins from colostrum (C) and mature milk (D).

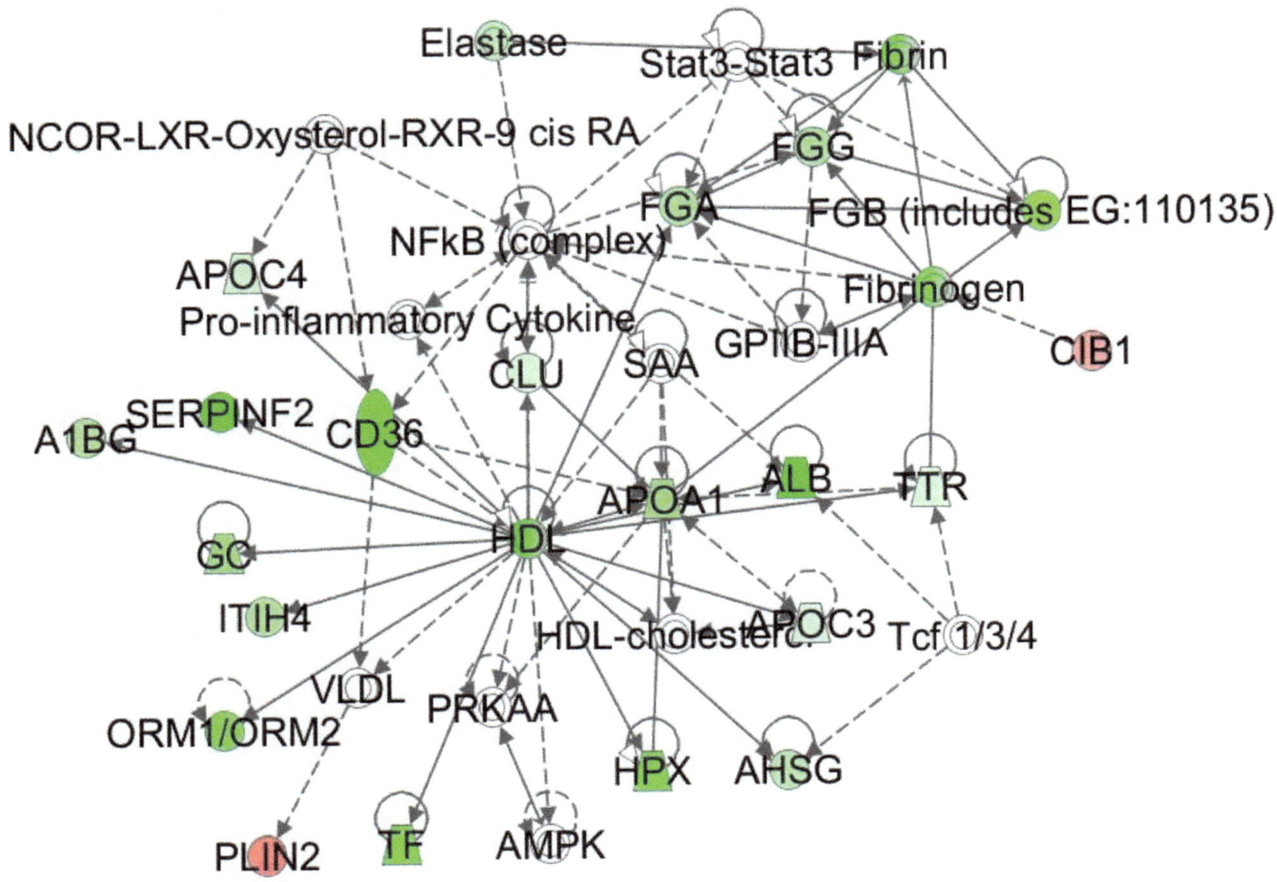

Figure 8. Pathway analysis involved in the Lipid Metabolisms of the TC-LA MFGM proteins.

analysis, we believe that these specific expressed proteins did not adversely influence the composition of milk or the health of these transgenic animals. Peroxiredoxin-1 was hypothesized to responsible for the anti-oxidative effect of laminar share stress [37]. 3′ (2′), 5′-bisphosphate nucleotidase 1, inosine-5′-monophosphate dehydrogenase 2, elongation factor 1-gamma, sdp-ribosylation factor 4 and phosphoserine aminotransferase involved in the metabolism of the cell, the protein ras-related protein rab-21 and guanine nucleotide-binding protein subunit alpha-11 involved in the mediated signal transduction. The function of b-cell receptor-associated protein 31 involves in apoptotic cleavage of cellular proteins, the function of ras-related protein rab-7A involves in protein transport, the function of acyl-coa synthetase long-chain family member 1 involves in the fatty acid degradation and metabolism, the function of Eukaryotic Translation Initiation Factor 3 Subunit H was in regulation of translational initiation.

There were distinguishable differences between the three transgenic lines in terms of exogenous protein expression. The TC-LF group expressed recombinant human lactoferrin at high levels, i.e., 2.5–3.8 g/l [25], the TC-LA group, expressed recombinant human α-lactalbumin at intermediate levels, i.e., 1.5 g/l [26], and the TC-LZ group expressed recombinant human lysozyme at low levels, i.e., 13–28 mg/l [24]. The expression profile of TC-LF group displayed a marked difference with that of other TC groups but the range of variation was comparable to the C group. Comprehensive analysis of the relative quantities of MFGM proteins in the transgenic and control groups led to the hypothesis that MFGM protein expression was influenced by the quantity of the exogenous proteins. In the TC-LF group, lactoferrin expression was higher than that of the exogenous proteins in the other two transgenic groups, and the number of the affected MFGM proteins was also higher. One possible mechanism for this could entail the S100 proteins, which are involved in inflammatory and antimicrobial responses. We found low abundance of the S100A8 and S100A9 proteins in colostrum of the transgenic animals relative to the N group. This may be a consequence of the fact that LF, LA, and LZ (i.e., which were expressed exogenously in our study animals) are all involved in the inflammatory response.

The MFGM proteins are important in cattle cultivation and quality of livestock products, especially in cheese industry. The MFGM proteins were the major composition in MFGM, they play a core role in mammary epithelial cell against bacteria, so these proteins and immune globulin proteins are highly expressed in colostrum MFGM, and the colostrum was beneficial to the acquired immunity in new born.

In TC-LA group, we analyzed the functions of the proteins and found out that plenty of them were not defined, such as similar to lipocalin, α-2-antiplasmin, Cd5L protein, putative uncharacterized protein, canx protein. Hemopexinand serotransferrin coming from blood majorly influenced by the cattle individually. Actin-related protein 2/3 complex subunit 3, rab35 protein, serine hydroxymethyltransferase, mitochondrial, l-lactate dehydrogenase B Chain are involved in ubiquitin ligase [38]. Apolipoprotein E plays a key role in regulating plasma levels of lipoproteins and its level was not associated with high-density lipoprotein fractions [39]. The former studies of the composition of milk from transgenic cloned cattle indicated that the fatty acids of TC-LA colostrum were less than that of the N group especially in monounsaturated fatty acids [27]. The pathway analysis found the down-regulated lipid metabolism (Figue 8), which presumably was involved in monounsaturated fatty acids synthesis.

In TC-LF group, some proteins have high abundance in milk such as κ-Casein, α-S2-Casein, serotransferrin, α-lactalbumin,-serum albumin and lactoperoxidase, so a small amount of residual proteins will cause detectable change of concentrations. Plenty proteins with the function of lipid metabolism were differentially expressed. Bovine MUC1 prevents binding of bacteria to human intestinal cells and has a role in preventing the binding of common enteropathogenic bacteria to human intestinal epithelial surfaces [40]. The expression of fatty acids of mature TC-LF milk was less than that of other four groups, especially in saturated fatty acids [27]. Plenty proteins which involve in lipid metabolism were expressed lowly in TC-LF mature MFGM, such as apolipoprotein A-II, which is one of the major proteins in HDL and its main function is to modulate the lipid binding and lecithin-cholesterol acyltransferase activities of HDL by promoting the dissociation of apo A-I from HDL [41]. ATP-binding cassette, sub-family G (White), member 2 have the function of maintaining lipid homeostasis in the mammary gland [42]. Fatty acid-binding protein (FABP) regulates the channeling of fatty acids toward copious milk fat synthesis in bovine mammary gland [43]. The expression level of CIDE-A protein was regulated by insulin and/or fatty acids in mammary epithelial cells, and thereby played an important role in lipid and energy metabolism [44]. Acyl-coa synthetase long-chain family member 1 (ACSL1) which involves in fatty acid analysis [43] was differentially expressed in lactation,. ACSL1, FABP and CD 36 were suggested to have the function of channel long chain fatty acids toward esterification into milk triacylglycerol [43,45].

The studies showed that both the bovine and human α-lactalbumin had the functions of feeds back on the mammary gland to regulate involution [46]. In our study, the bovine α-Lactalbumin and the exogenous human α-Lactalbumin were also expressed in mammary gland and the concentration was higher than the normal bovine. But we did not notice that the α-lactalbumin could induce the apoptotic activity, and our former study have confirmed the biological activity of recombination human α-lactalbumin [26]. In this study, over expression of the α-lactalbumin did not influence the expression of MFGM proteins, which indicated that the mammary gland epithelial cell would not be affected from the apoptotic activity.

In summary, the study uses proteomics methods to analyze transgenic cloned cattle MFGM proteins for the first time. The expression of exogenous proteins did not significantly change the MFGM protein profiles, and the relative quantity expression of MFGM proteins were all within the normal ranges. The differences among the TC groups were not greater than those between the N group and C group. The data from this study improves the understanding of the bovine milk proteome and provides data for the assessment of the food safety of transgenic cloned animals.

Acknowledgments

We thank Jin He for the valuable comments and suggestions during the paper preparation.

Author Contributions

Conceived and designed the experiments: NL. Performed the experiments: SS JZ JW TY CG. Analyzed the data: SS JZ JW TY. Contributed reagents/materials/analysis tools: NL RZ. Contributed to the writing of the manuscript: SS JZ.

References

1. Simons JP, McClenaghan M, Clark AJ (1987) Alteration of the quality of milk by expression of sheep beta-lactoglobulin in transgenic mice. Nature 328: 530–532.
2. Gordon K, Lee E, Vitale JA, Smith AE, Westphal H, et al. (1992) Production of human tissue plasminogen activator in transgenic mouse milk. 1987. Biotechnology 24: 425–428.
3. Rudolph NS (1999) Biopharmaceutical production in transgenic livestock. Trends Biotechnol 17: 367–374.
4. Houdebine LM (2000) Transgenic animal bioreactors. Transgenic Res 9: 305–320.
5. Wheeler MB, Walters EM, Clark SG (2003) Transgenic animals in biomedicine and agriculture: outlook for the future. Anim Reprod Sci 79: 265–289.
6. Laible G, Brophy B, Knighton D, Wells DN (2007) Compositional analysis of dairy products derived from clones and cloned transgenic cattle. Theriogenology 67: 166–177.
7. Baldassarre H, Hockley DK, Olaniyan B, Brochu E, Zhao X, et al. (2008) Milk composition studies in transgenic goats expressing recombinant human butyrylcholinesterase in the mammary gland. Transgenic Res 17: 863–872.
8. Baldassarre H, Schirm M, Deslauriers J, Turcotte C, Bordignon V (2009) Protein profile and alpha-lactalbumin concentration in the milk of standard and transgenic goats expressing recombinant human butyrylcholinesterase. Transgenic Res 18: 621–632.
9. Quaranta S, Giuffrida MG, Cavaletto M, Giunta C, Godovac-Zimmermann J, et al. (2001) Human proteome enhancement: high-recovery method and improved two-dimensional map of colostral fat globule membrane proteins. Electrophoresis 22: 1810–1818.
10. Keenan TW (2001) Milk lipid globules and their surrounding membrane: a brief history and perspectives for future research. J Mammary Gland Biol Neoplasia 6: 365–371.
11. Keenan TW (2001) Assembly and secretion of the lipid globules of milk. Adv Exp Med Biol 501: 125–136.
12. Auty MA, Twomey M, Guinee TP, Mulvihill DM (2001) Development and application of confocal scanning laser microscopy methods for studying the distribution of fat and protein in selected dairy products. J Dairy Res 68: 417–427.
13. Mather IH, Keenan TW (1998) Origin and secretion of milk lipids. J Mammary Gland Biol Neoplasia 3: 259–273.
14. Heid HW, Keenan TW (2005) Intracellular origin and secretion of milk fat globules. Eur J Cell Biol 84: 245–258.
15. Wu CC, Howell KE, Neville MC, Yates JR 3rd, McManaman JL (2000) Proteomics reveal a link between the endoplasmic reticulum and lipid secretory mechanisms in mammary epithelial cells. Electrophoresis 21: 3470–3482.
16. Patton S, Gendler SJ, Spicer AP (1995) The epithelial mucin, MUC1, of milk, mammary gland and other tissues. Biochim Biophys Acta 1241: 407–423.
17. Guri A, Griffiths M, Khursigara CM, Corredig M (2012) The effect of milk fat globules on adherence and internalization of Salmonella Enteritidis to HT-29 cells. J Dairy Sci 95: 6937–6945.
18. Inagaki M, Nagai S, Yabe T, Nagaoka S, Minamoto N, et al. (2010) The bovine lactophorin C-terminal fragment and PAS6/7 were both potent in the inhibition of human rotavirus replication in cultured epithelial cells and the prevention of experimental gastroenteritis. Biosci Biotechnol Biochem 74: 1386–1390.
19. Moore JB, Weeks ME (2011) Proteomics and systems biology: current and future applications in the nutritional sciences. Adv Nutr 2: 355–364.
20. Bianchi L, Puglia M, Landi C, Matteoni S, Perini D, et al. (2009) Solubilization methods and reference 2-DE map of cow milk fat globules. J Proteomics 72: 853–864.
21. Picariello G, Ferranti P, Mamone G, Klouckova I, Mechref Y, et al. (2012) Gel-free shotgun proteomic analysis of human milk. J Chromatogr A 1227: 219–233.
22. Spertino S, Cipriani V, De Angelis C, Giuffrida MG, Marsano F, et al. (2012) Proteome profile and biological activity of caprine, bovine and human milk fat globules. Mol Biosyst 8: 967–974.
23. Liao Y, Alvarado R, Phinney B, Lonnerdal B (2011) Proteomic characterization of human milk fat globule membrane proteins during a 12 month lactation period. J Proteome Res 10: 3530–3541.
24. Yang B, Wang J, Tang B, Liu Y, Guo C, et al. (2011) Characterization of bioactive recombinant human lysozyme expressed in milk of cloned transgenic cattle. PLoS One 6: e17593.
25. Yang P, Wang J, Gong G, Sun X, Zhang R, et al. (2008) Cattle mammary bioreactor generated by a novel procedure of transgenic cloning for large-scale production of functional human lactoferrin. PLoS One 3: e3453.
26. Wang J, Yang P, Tang B, Sun X, Zhang R, et al. (2008) Expression and characterization of bioactive recombinant human alpha-lactalbumin in the milk of transgenic cloned cows. J Dairy Sci 91: 4466–4476.
27. Zhang R, Guo C, Sui S, Yu T, Wang J, et al. (2012) Comprehensive assessment of milk composition in transgenic cloned cattle. PLoS One 7: e49697.

28. Wisniewski JR, Zougman A, Nagaraj N, Mann M (2009) Universal sample preparation method for proteome analysis. Nat Methods 6: 359–362.
29. Nesvizhskii AI, Keller A, Kolker E, Aebersold R (2003) A statistical model for identifying proteins by tandem mass spectrometry. Anal Chem 75: 4646–4658.
30. Keller A, Nesvizhskii AI, Kolker E, Aebersold R (2002) Empirical statistical model to estimate the accuracy of peptide identifications made by MS/MS and database search. Anal Chem 74: 5383–5392.
31. Huang da W, Sherman BT, Lempicki RA (2009) Bioinformatics enrichment tools: paths toward the comprehensive functional analysis of large gene lists. Nucleic Acids Res 37: 1–13.
32. D'Amato A, Bachi A, Fasoli E, Boschetti E, Peltre G, et al. (2009) In-depth exploration of cow's whey proteome via combinatorial peptide ligand libraries. J Proteome Res 8: 3925–3936.
33. Aziz A, Zhang W, Li J, Loukas A, McManus DP, et al. (2011) Proteomic characterisation of Echinococcus granulosus hydatid cyst fluid from sheep, cattle and humans. J Proteomics 74: 1560–1572.
34. Affolter M, Grass L, Vanrobaeys F, Casado B, Kussmann M (2010) Qualitative and quantitative profiling of the bovine milk fat globule membrane proteome. J Proteomics 73: 1079–1088.
35. Lu J, Boeren S, de Vries SC, van Valenberg HJ, Vervoort J, et al. (2011) Filter-aided sample preparation with dimethyl labeling to identify and quantify milk fat globule membrane proteins. J Proteomics 75: 34–43.
36. Addis MF, Pisanu S, Ghisaura S, Pagnozzi D, Marogna G, et al. (2011) Proteomics and pathway analyses of the milk fat globule in sheep naturally infected by Mycoplasma agalactiae provide indications of the in vivo response of the mammary epithelium to bacterial infection. Infect Immun 79: 3833–3845.
37. Mowbray AL, Kang DH, Rhee SG, Kang SW, Jo H (2008) Laminar shear stress up-regulates peroxiredoxins (PRX) in endothelial cells: PRX 1 as a mechanosensitive antioxidant. J Biol Chem 283: 1622–1627.
38. Lee KA, Hammerle LP, Andrews PS, Stokes MP, Mustelin T, et al. (2011) Ubiquitin ligase substrate identification through quantitative proteomics at both the protein and peptide levels. J Biol Chem 286: 41530–41538.
39. Takahashi Y, Itoh F, Oohashi T, Miyamoto T (2003) Distribution of apolipoprotein E among lipoprotein fractions in the lactating cow. Comp Biochem Physiol B Biochem Mol Biol 136: 905–912.
40. Parker P, Sando L, Pearson R, Kongsuwan K, Tellam RL, et al. (2010) Bovine Muc1 inhibits binding of enteric bacteria to Caco-2 cells. Glycoconj J 27: 89–97.
41. Mahley RW, Innerarity TL, Rall SC Jr, Weisgraber KH (1984) Plasma lipoproteins: apolipoprotein structure and function. J Lipid Res 25: 1277–1294.
42. Viturro E, Farke C, Meyer HH, Albrecht C (2006) Identification, sequence analysis and mRNA tissue distribution of the bovine sterol transporters ABCG5 and ABCG8. J Dairy Sci 89: 553–561.
43. Bionaz M, Loor JJ (2008) ACSL1, AGPAT6, FABP3, LPIN1, and SLC27A6 are the most abundant isoforms in bovine mammary tissue and their expression is affected by stage of lactation. J Nutr 138: 1019–1024.
44. Yonezawa T, Haga S, Kobayashi Y, Katoh K, Obara Y (2009) Saturated fatty acids stimulate and insulin suppresses CIDE-A expression in bovine mammary epithelial cells. Biochem Biophys Res Commun 384: 535–539.
45. Spitsberg VL, Matitashvili E, Gorewit RC (1995) Association and coexpression of fatty-acid-binding protein and glycoprotein CD36 in the bovine mammary gland. Eur J Biochem 230: 872–878.
46. Sharp JA, Lefevre C, Nicholas KR (2008) Lack of functional alpha-lactalbumin prevents involution in Cape fur seals and identifies the protein as an apoptotic milk factor in mammary gland involution. BMC Biol 6: 48.

5

Differential Bees Flux Balance Analysis with OptKnock for *In Silico* Microbial Strains Optimization

Yee Wen Choon[1], Mohd Saberi Mohamad[1]*, Safaai Deris[1], Rosli Md. Illias[2], Chuii Khim Chong[1], Lian En Chai[1], Sigeru Omatu[3], Juan Manuel Corchado[4]

1 Artificial Intelligence and Bioinformatics Group, Faculty of Computing, Universiti Teknologi Malaysia, Johor, Malaysia, **2** Department of Bioprocess Engineering, Faculty of Chemical Engineering, Universiti Teknologi Malaysia, Skudai, Johor, Malaysia, **3** Department of Electronics, Information and Communication Engineering, Osaka Institute of Technology, Osaka, Japan, **4** Biomedical Research Institute of Salamanca/BISITE Research Group, University of Salamanca, Salamanca, Spain

Abstract

Microbial strains optimization for the overproduction of desired phenotype has been a popular topic in recent years. The strains can be optimized through several techniques in the field of genetic engineering. Gene knockout is a genetic engineering technique that can engineer the metabolism of microbial cells with the objective to obtain desirable phenotypes. However, the complexities of the metabolic networks have made the process to identify the effects of genetic modification on the desirable phenotypes challenging. Furthermore, a vast number of reactions in cellular metabolism often lead to the combinatorial problem in obtaining optimal gene deletion strategy. Basically, the size of a genome-scale metabolic model is usually large. As the size of the problem increases, the computation time increases exponentially. In this paper, we propose Differential Bees Flux Balance Analysis (DBFBA) with OptKnock to identify optimal gene knockout strategies for maximizing the production yield of desired phenotypes while sustaining the growth rate. This proposed method functions by improving the performance of a hybrid of Bees Algorithm and Flux Balance Analysis (BAFBA) by hybridizing Differential Evolution (DE) algorithm into neighborhood searching strategy of BAFBA. In addition, DBFBA is integrated with OptKnock to validate the results for improving the reliability the work. Through several experiments conducted on *Escherichia coli*, *Bacillus subtilis*, and *Clostridium thermocellum* as the model organisms, DBFBA has shown a better performance in terms of computational time, stability, growth rate, and production yield of desired phenotypes compared to the methods used in previous works.

Editor: Ying Xu, University of Georgia, United States of America

Funding: The authors would like to thank Malaysian Ministry of Science, Technology and Innovation for supporting this research by two e-science research grants (Grant numbers: 06-01-06-SF1029 and 01-01-06-SF1234). This research is also funded by an Exploratory Research Grant Scheme (Grant number: RJ130000.7807.4L096) and a Fundamental Research Grant Scheme (Grant number: RJ130000.7807.4F190) from Malaysian Ministry of Higher Education. The funders had no role in study design, data collection and analysis, decision to publish, or preparation of the manuscript.

Competing Interests: The authors have declared that no competing interests exist.

* Email: saberi@utm.my

Introduction

In the recent years, more accurate genome annotation and more complete pathway information, promising achievements have been made on *in silico* metabolic network reconstruction [1]. In addition, one of the most comprehensive stoichiometric models built upon the genome of *E. coli* had been available [2]. Currently, the new emerging study is aiming to computationally optimize the microbial strains for the overproduction of particular chemical and biochemical compounds, diversely ranging from industrial interests to environmental usages [3]. Computational models are of the central importance for the investigation of general biological functions and applications in the area of biomedicine and biotechnology, this accelerates the process of developing computational models to simulate the actual processes inside the cells [4]. Retrofitting of the cellular metabolism is essential, as naturally, due to the cellular metabolism responses in the history of selective pressures, microorganisms evolve by optimizing their growth rather than having overproduction of specific chemical compounds. Classical strain improvement such as random mutagenesis and screening are among traditional methods to retrofit the microbial metabolism. However, the effects of genetic modification on the desirable phenotypes are difficult to predict due to the complexities of the metabolic networks that lead to data ambiguity. Furthermore, the combinatorial problem that is caused by a large number of reactions in cellular metabolism makes the process to obtain optimal gene deletion strategies challenging. Basically, a genome-scale metabolic model is usually large in size. Another problem in the current metabolic engineering field is as the size of the problem increases, the computation time increases exponentially. In later years, metabolic engineering is introduced to retrofit microbial metabolism. In metabolic engineering, the main objective is to increase the production of particular chemical and biochemical compounds through genetic engineering. Gene knockout is one of the most common genetic engineering techniques for the overproduction of particular chemical and biochemical compounds, in which one of the organism's genes is made inoperative. Currently, this technology has been successfully applied in many organisms, from unicellular eukaryotes to mammals, including human cells.

In the recent years, the study on computational algorithms to identify optimal gene knockout strategies for obtaining improved phenotypes is growing rapidly. The first rational modeling framework (known as OptKnock) is introduced by Burgard *et al.*, for predicting gene knockout strategies aiming to the overproduction of a desired metabolite. Without affecting the operation of the internal flux distribution, OptKnock identifies a set of gene (reaction) deletions to maximize the flux of a desired metabolite so that the growth or another objective function is optimized [5].

OptKnock uses mixed integer linear programming (MILP) to formulate a bi-level linear optimization that is promising to find the global optimal solution. OptGene is extended from OptKnock which formulates the *in silico* design problem by using Genetic Algorithm (GA) [6]. Meta-heuristic methods are usually known to be able to produce near-optimal solutions with reasonable computation time, which is favourable, and the objective function that can be optimized is flexible. OptGene is available in two representation schemes which are binary and integer. The binary representation is more complex and leads to a solution with a large number of knockouts even though it is closer to the natural evolution of microbial genomes. The integer representation gives a more compact genome but the problem is that a priori number of gene knockouts need to be defined [7]. To overcome the problems, Rocha et al. [7] proposed two optimization algorithms: Simulated Annealing (SA) and Set-based Evolutionary Algorithms (SEAs) to allow the automatic finding of the best number of gene deletions to achieve a given productivity goal. The performance to identify optimal gene knockout strategies still needs improvement.

A hybrid of BA and FBA (BAFBA) was proposed by Choon *et al.* [8], which showed a better performance in predicting optimal gene knockout strategies in terms of growth rate and production yield. Bees Algorithm (BA) was first introduced by Pham *et al.* [9], which is a typical meta-heuristic optimization approach that has been applied to several problems, such as controller formation [10], image analysis [11], and job multi-objective optimization [12]. BA mimicks the intelligent behaviors of honeybees. It locates the most promising solutions, and selectively explores its neighborhoods for the global maximum of the objective function. According to a series of recent publications, BA is efficient in solving optimization problems [9–12]. Nevertheless, BA is relatively weak in local search activities due to its dependency on random search [13]. In this paper, DBFBA, a hybrid of DE algorithm into the neighborhood searching strategy of BAFBA has been proposed to improve the performance of BAFBA. DE algorithm is a heuristic approach mainly having three advantages; finding the true global minimum regardless of the initial parameter values, fast convergence, and using few control parameters. In addition, to improve the reliability of the work, the OptKnock was integrated into DBFBA to validate the results rather than validating the results only through literature. OptKnock is a well-recognized modelling framework, which is widely used in assisting biologists for gene knockout experiments in laboratory. It is proven that predicted strains from OptKnock can lead to successful production strains and that adaptive evolution of the engineered strains can lead to improved production capabilities [14]. Besides, the previous work, OptGene, integrated with OptKnock to validate the results too, hence OptKnock is used to validate the results in this paper [7]. This paper shows that DBFBA is capable of solving large size problems in short computational time as well as improves the performance in predicting optimal gene knockout strategies. The results obtained

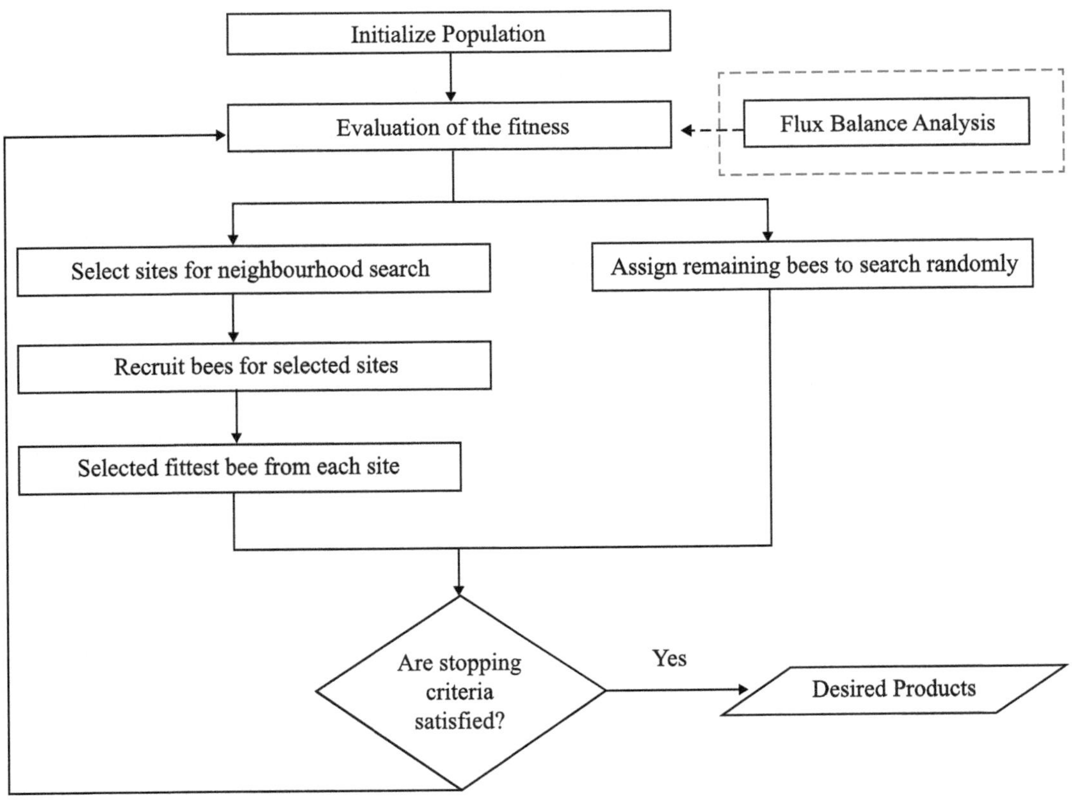

Figure 1. BAFBA Flowchart. Note: Red-dotted box is Flux Balance Analysis which is hybridized into standard BA as an objective function in order to predict the effect of gene knockout.

by DBFBA in four case studies are also presented where *E.coli* (*Escherichia Coli*) iJR904 model, *B.subtilis* (*Bacillus subtilis*) model, and *C. thermocellum* (*Clostridium thermocellum*) model are the target microorganisms [2,15–16]. In addition, a computational benchmarking analysis was also conducted to test the performance of the hybrid of Bee algorithm and DE algorithm.

This paper is organized as follows: Firstly, problem formulation is introduced, and the details of the BAFBA and the proposed DBFBA are described. The validation analyses by using OptKnock are also explained. Then, the datasets and experimental setup are described. Next, experimental results are presented. Then, the discussion on the obtained results is addressed, which deliberates the contributions of this work. Lastly, this paper is summarized by providing main conclusion and future developments.

Materials and Methods

Problem Formulation

The problem to identify optimal gene knockout strategies from the biological models can be formulated as follows: Suppose a model which contains the stoichiometric matrix **S** provides the linear relationship of the model between the flux rates of the reactions (**v**) and the derivatives of the reactant concentrations (**x**). The matrix is a constant, while the flux vector is a variable. Assume that there are m reactants and n reactions between them.

Flux vector:

$$\mathbf{v}=(v_1, v_2, \cdots, v_n)^T \tag{1}$$

Concentration vector:

$$\mathbf{x}=(x_1, x_2, \cdots, x_m)^T \tag{2}$$

Dynamic mass balance equation:

$$\frac{d\mathbf{X}}{dt}=\mathbf{Sv} \tag{3}$$

where $\boldsymbol{T}$ represents the time.

The chemical elements, ionic charge, and biochemical moieties must be balanced in the stoichiometric matrix. The objective is to find the optimal gene knockout strategy which can improve the product yields of industrially important chemicals, while sustaining the growth rate of the microorganism. This is commonly

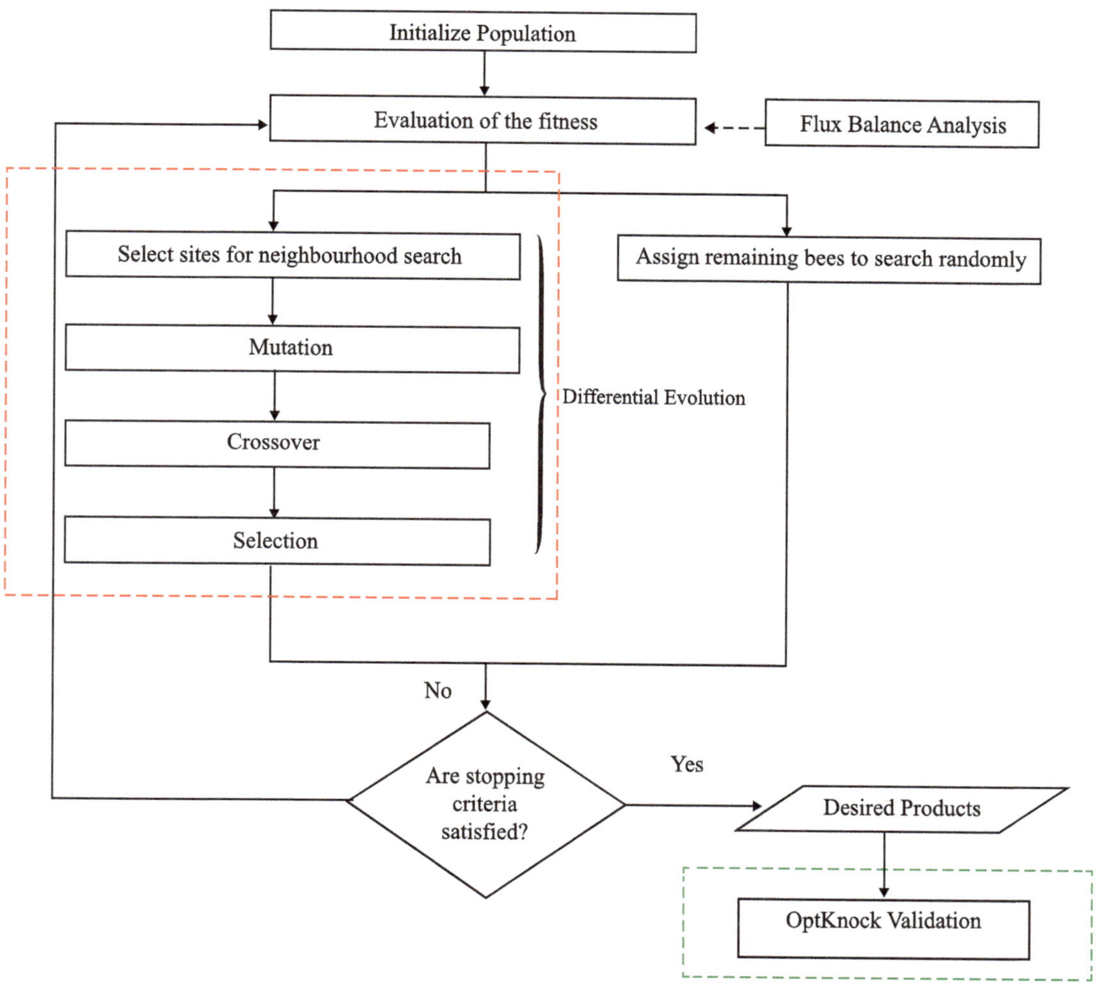

Figure 2. Overall framework of DBFBA with OptKnock. Note: Red-dotted box is DE algorithm that is hybridized into BAFBA in order to improve the local search performance of BAFBA. Green-dotted box is OptKnock validation that has been integrated into DBFBA in this paper.

```
Step 1. Initialise the population with n scout bees
Step 2. Evaluate the fitness of the population (Flux Balance Analysis)
maximize          cTx
subject to        Sv = 0
and               lowerbound ≤ x ≤ upperbound
v represents the vector of fluxes
S is the stoichiometric matrix
Expression (cTx) is the objective function
c is a vector of weights
fitness = BPCY
Step 3. While (stopping criterion not met)
Step 4. Select sites (m) for neighbourhood search (Differential Evolution)
Population ← m
Evaluate Population(m)
Sbest ← GetBestSolution(m)
While (¬StopCondition())
NewPopulation ←Ø
For (Pi ∈ Population)
Si ←NewSample(Pi, Population, Problemsize, Weighting factor, Crossover rate)
If (Cost(Si) ≤ Cost(Pi))
        NewPopulation ←Si
      Else
        NewPopulation ← Pi
      End
   End
   Population ←NewPopulation
   EvaluatePopulation(Population)
     Sbest ←GetBestSolution(Population)
End
Return (Sbest)
Step 5. Assign remaining bees to search randomly and evaluate their fitnesses. // (n-m) bees
assigned randomly to search the whole solution space
Step 6. End while
```

Figure 3. Pseudo code of DBFBA.

performed by using the linear programming, defined as follows:

$$\begin{aligned} &\textit{maximize} \quad \mathbf{c}^{\mathrm{T}}\mathbf{x} \\ &\textit{subject to} \quad \mathbf{Sv} = 0 \\ &\textit{and} \quad \mathbf{lowerbound} \leq \mathbf{x} \leq \mathbf{upperbound} \end{aligned} \tag{4}$$

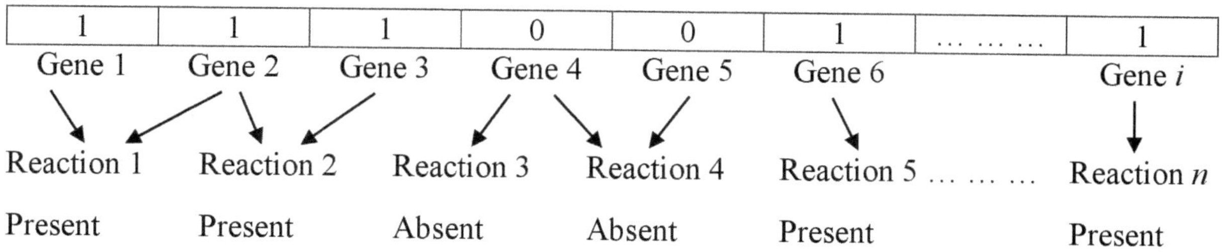

Figure 4. Bee representation of metabolic genotype [6].

where **v** represents the vector of fluxes, **S** is the stoichiometric matrix. The expression ($\mathbf{c}^{\mathbf{T}}\mathbf{x}$) to be maximized or minimized is known as the objective function, where **c** is a vector of weights, indicating contributions of each reaction to the objective function. The inequalities of lower bound and upper bound define the maximal rates of flux for every reaction corresponding to the columns of the stoichiometric matrix.

A hybrid of BA and FBA (BAFBA)

Figure 1 shows the flow of BAFBA. BAFBA is initialized by mimicking a population of bees. In identifying gene knockout strategies, a bee is represented by a binary variable to indicate the absence or the presence of genes in the reaction. In this study, the BAFBA is started with the bees being placed randomly in the search space. The fitness of the sites visited by the bees is evaluated using FBA. Bees with the highest fitness would be denoted as 'selected bees' and the sites visited by them would be chosen for neighbourhood search. Small amount of 'selected bees' was expected to encourage local exploitation. After performing many tests, it was found that appropriate maximum 'selected bees' was $(1/4)\times n$. The amount of selected bees was limited within the range $[1,(1/4)\times n]$ to prevent unnecessary selection of sites for neighborhood search. Each bee was required to go through this repetitive local search neighborhood procedure until the best possible answer obtained. Meanwhile, the remaining bees were assigned randomly to search for new potential solutions.

Before attempting to propose DBFBA, it is crucial to find the limitations of BAFBA [9]. BA dependence on random search makes it relatively weak in local search activities and it suffers of slow convergence due to the repetitive iteration of the algorithm.

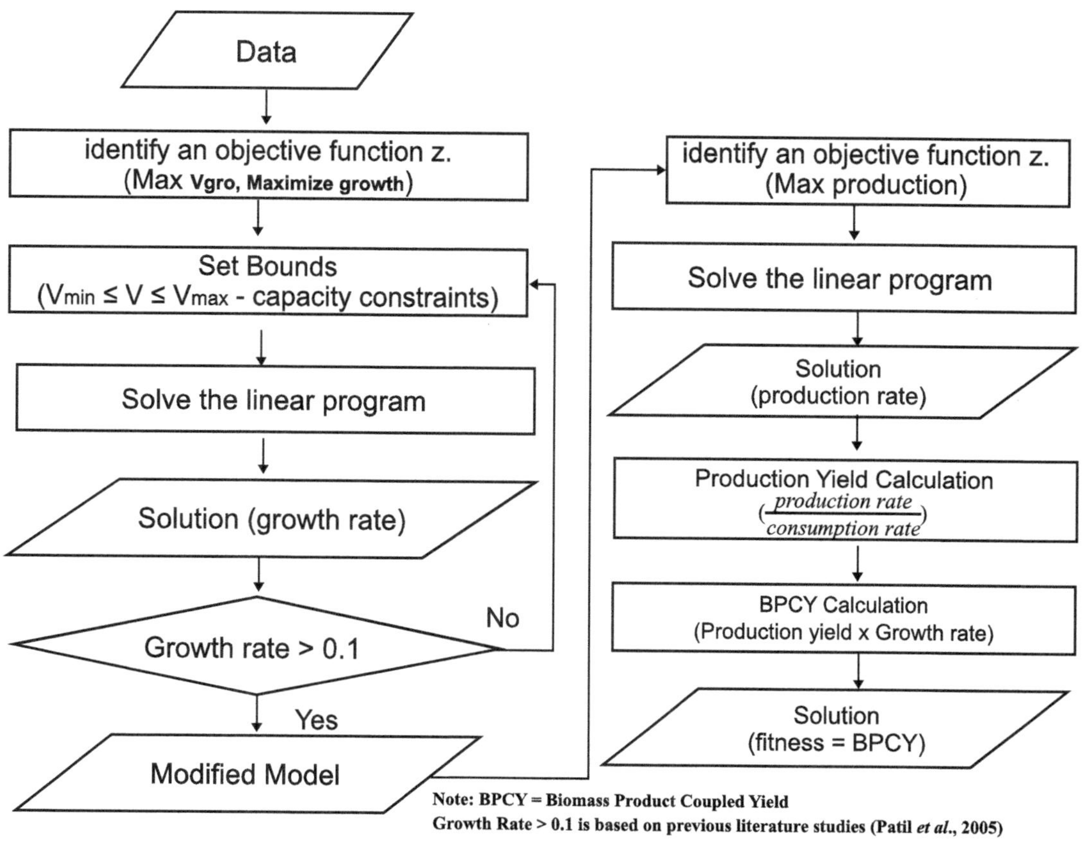

Figure 5. The flow of fitness calculation.

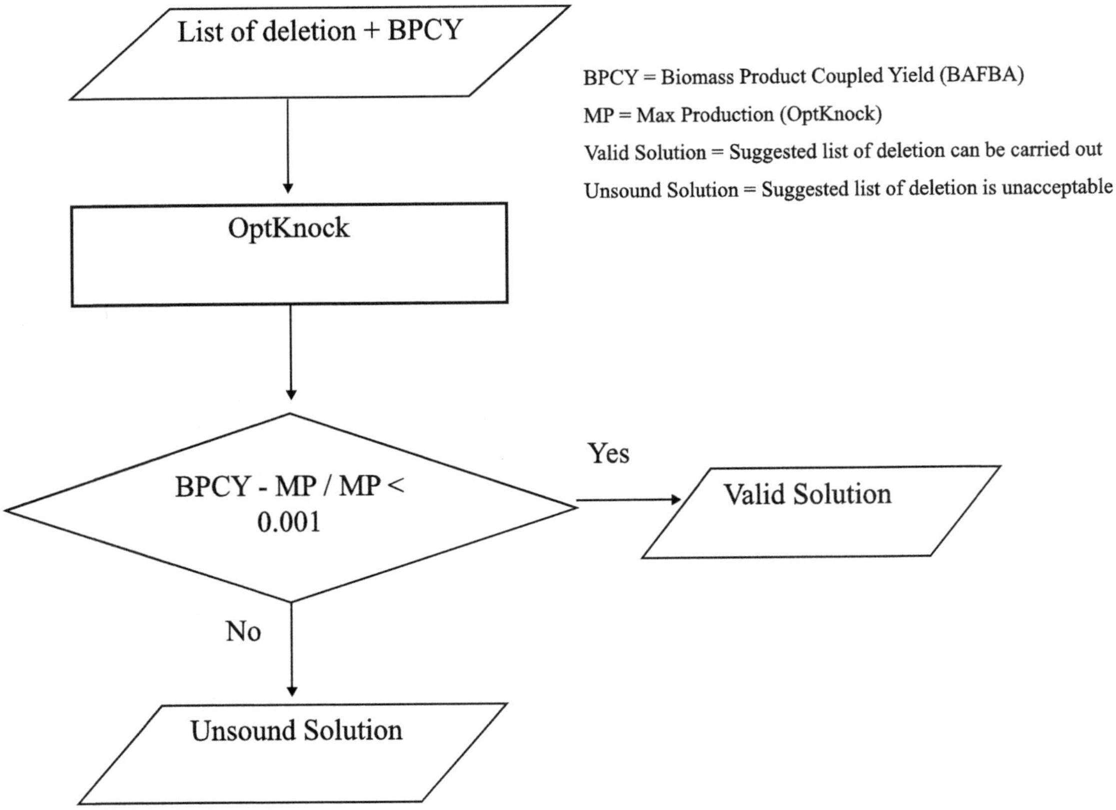

Figure 6. The flow of OptKnock validation.

Repetition of unnecessary similar process in neighborhood search causes an additional computational time in generating solution.

Differential Bees Flux Balance Analysis (DBFBA) with OptKnock

In this paper, DBFBA with OptKnock for identifying optimal gene knockout strategy was proposed to overcome the limitations of BAFBA and previous works [5–7,9]. DBFBA in this work differs from the BAFBA in the neighborhood search activities. The proposed DBFBA improved the operation by hybridizing DE algorithm into BAFBA. In addition, OptKnock was integrated into DBFBA to improve the reliability of the results. Figure 2 shows the overall framework of DBFBA and Figure 3 is the pseudo code of DBFBA. The important steps are explained in the following subsections.

Bee Representation of Metabolic Genotype. One or more genes can be discovered in each reaction in a metabolic model. In this paper, each of those genes is represented by a binary variable, where 0 represents the absence of the gene and 1 represents the presence of the gene in the reaction. These variables form a 'bee' representing a specific mutant that lacks some metabolic reactions when compared with the wild type (Figure 4).

Table 1. Mathematical representation of De Jong Martin & Gaddy, Schwefel, and Griewangk functions.

Name	Mathematical representation
De Jong	$\max F = (3905.93) - 100({x_1}^2 - x_2)^2 - (1 - x_1)^2$
Martin & Gaddy	$\min F = (x1 - x2)^2 + \left(\frac{x1 + x2 - 10}{3}\right)^2$
Branin	$\min F = a(x_2 - bx_1^2 + cx_1 - d)^2 + e(1 - f)\cos(x1) + e$ $a = 1, b = \frac{5.1}{4}\left(\frac{7}{22}\right)^2, c = \frac{5}{22} \times 7, d = 6, e = 10, f = \frac{1}{8} \times \frac{7}{22}$
Griewangk	$\min F = \frac{1}{\left(0.1 + \sum_{i=1}^{n}\left(\frac{x_i^2}{4000}\right) - \prod_{i=1}^{n}\cos\left(\frac{x_i}{\sqrt{i}} + 1\right)\right)}$

Table 2. Obtained fitness value of all benchmark functions.

Function	Mean		STD	
	BA	DB	BA	DB
De Jong	3.91e+03	3.91E+03	0.000504	**2.13793E-05**
Martin & Gaddy	11.1083	11.1111	0.002797	**0**
Branin	26.5619	26.5537	0.02917	**0.02771**
Griewangk	−0.5263	−0.5263	5.76765E-09	**0**

Note: The bold numbers represent the best result.

Initialization of the Population. The algorithm starts with an initial population of n scout bees. Each bee is initialized as by assuming a reaction with n genes. Bees in the population are initialized by setting present or absent status to each gene randomly. Initialization of the population is done randomly so that all bees in the population have an equal chance to be selected. Specific rules like co-regulated genes tending to be clustered did not affect the result of selecting the genes. The result might not truly reflect the population if it is done with bias setting.

Flux Balance Analysis. Each site is given a fitness score that determines whether to recruit more bees or should be abandoned. In this work, FBA has been used to calculate the fitness score for each site (refer to Equation (4)). In the study made by García Sánchez *et al.* [17], the best predictions were obtained using "maximization of growth", and with some combinations that included this objective. Hence, in this paper, maximization of growth is applied. After maximizing the cellular growth, mutant with a growth rate of more than 0.1 continues the process by maximizing the desired product flux at fixed optimal cellular growth value. After conducting a small number of trial, the optimal cellular growth value was fixed at 90% from the value obtained from FBA, since the production yield of the desired metabolite is always 0 when the growth is at maximum. Production yield is the maximum amount of product that can be generated per unit of substrate. The following shows the calculation for production yield:

$$\text{Production yield} = \frac{\text{production rate}_{\text{production}}}{\text{consumption rate}_{\text{substrate}}}\left(\frac{mmol}{mmol}\right)\left(\frac{gm}{gm}\right) \quad (5)$$

where mmol = millimole and gm is gram.

We used Biomass-product coupled yield (BPCY) as the fitness score in this work. According to Soons *et al.* (2013), metabolic networks can function in living cells under various biological objectives, depending on the relevant organism and its genetic and environmental context. However, biological objectives have been clarified as only applicable for analyzing a number of organisms in terms of the microbial metabolic engineering. It is desirable to couple the formation of the desired product to growth [18]. The calculation for BPCY is as follows:

$$BPCY = production\ yield \times growth\ rate\left(\frac{mmol}{mmol \times hr}\right)\left(\frac{gm}{gm \times hr}\right) \quad (6)$$

Where mmol is millimole, hr is hour and gm is gram.

The flow of the fitness calculation is showed in Figure 5.

Neighborhood Search (Differential Evolution Algorithm). This algorithm carries out neighbourhood searches in the favored sites (m) by using DE algorithm. DE algorithm operates by maintaining a population of candidate solutions and creating new candidate solutions through the mutation and crossover operation of DE, and keeps the fitness candidate solution. In this paper, the candidate solutions are the m favored sites from the population initialized by using BA. The algorithm starts with the solution, then goes through the mutation and

Table 3. Comparison between different methods for growth rate and BPCY Succinic acid by *E.coli.*

Method	Growth Rate (1/hr)	BPCY	List of knockout genes
DBFBA	**0.74529**	**13.58**	ENO, PFL, PYK**
BAFBA [9]	0.58512	3.84893	FUM**, PTAr**, RPE
SA + FBA [7]	N/A	0.35785	MALS, ORNDC, FUM**, GLYCL, GHMT2, ADPT, DCYTD, DUTPDP, URIDK2r, NTD8, PUNPI, THD2, GND, PFL, SUCFUMt
OptKnock [5]	0.31	N/A	PYK, ACKr, PTAr**, Phosphotransferase system

Note: The bold numbers represent the best result. N/A – Not Applicable.
*Common genes for all methods.
**Common genes in either 2 methods. BPCY is in gram (gram-glucose.hour)$^{-1}$.

Table 4. Comparison between different methods for growth rate and BPCY of Lactic acid by *E.coli.*

Method	Growth Rate (1/hr)	BPCY	List of knockout genes
DBFBA	**0.86719**	**16.1905**	ACALD*, MDH, TALA
BAFBA [9]	0.58586	3.5656	GAPD, L_LACD2, PTAr**
SA + FBA [7]	N/A	0.39850	ACLD19**, DRPA, GLYCDx, F6PA, TPI, LDH_D2, EDA, TKT2, LDH_D-
OptKnock [5]	0.28	N/A	ACKr, PTAr**, ACALD**

Note: The bold numbers represent the best result. N/A – Not Applicable.
*Common genes for all methods.
**Common genes in either 2 methods. BPCY is in gram (gram-glucose.hour)$^{-1}$.

crossover operation to create new candidate solutions. Since the candidate solutions are initialized by BA randomly, hence the specific rules on the distribution of gene on chromosome did not affect the result for neighbourhood gene selection too. The mutation factor F is a constant from [0,2] and elements of the candidate solutions enter the new candidate solutions with probability *CR*. In this paper, *m* is equal to 15, F is equal to 1, and *CR* is equal to 0.5. The values are obtained by conducting a small number of trials with the range of 10 to 25, 0 to 2, and 0.1–0.9, respectively. This step is important as there might be better solutions than the original solution in the neighborhood area.

Random Assignment and Termination. The remaining bees in the population are sent randomly around the search space to scout for new feasible solutions. This step is done randomly to avoid overlooking the potential results that are not in the range. These steps are repeated until either the maximum loop value is met or the fitness function has converged. In the end, the colony generates two parts to its new population – representatives from each selected patch and other scout bees assigned to perform random searches. There is no necessities to shrink the final result from DBFBA, there will be no additional genes to the solutions, because the fitness calculation is based on the bees that consist of a set of genes and it is not calculate the effect of each gene, which may contribute to the production of desired phenotypes.

OptKnock Validation. In this paper, OptKnock [5] has been used to evaluate the result obtained through the list of gene knockout from DBFBA. If the difference between the BPCY obtained from DBFBA and the maximum production rate obtained by OptKnock are less than 0.001, the list would be considered as a valid solution. Figure 6 shows the flow of the validation.

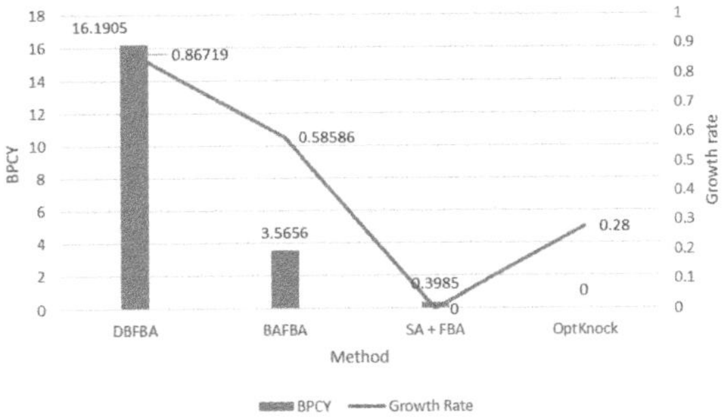

Figure 7. Comparison between different methods for growth rate and BPCY of Succinic acid by *E.coli.* Note: BPCY is in gram (gram-glucose.hour)$^{-1}$. Unit for growth rate is hr^{-1}.

Datasets and Experimental Setup

There are three models used to test the operation of DBFBA in this work, which are *E. coli*, *B. subtilis*, and *C. thermocellum* models, respectively. The three models are well-established models, the models are developed and used to study the bacterium's metabolism and phenotypic behaviour. Therefore the distribution of the genes (clustered/location) has no impact in our works since our works are not dealing with the expression patterns of the genes. All the models are pre-processed through several steps based on biology assumptions and computational approaches before it was applied to DBFBA. Lethal reactions such as genes that are found to be lethal *in vivo*, but not *in silico*, should be removed to improve the quality of the results. The results are invalid if a lethal reaction is deleted. In addition, dead-end reactions that cannot be activated within a feasible flux distribution when considering the entire universal reactions set are removed. The first model is *E. coli* model that contains 904 genes, 931 unique biochemical reactions, and 761 metabolites [2] (http://bigg.ucsd.edu/). *E. coli i*JR904 has been used in this work to test the reliability of DBFBA as this model was also used in previous works [5,7]. After the pre-processing step, the size of the model is reduced to 667 reactions. The second model is *B. subtilis i*Bsu1103 model [15] (http://genomebiology.com/content/supplementary/gb-2009-10-6-r69-s4.xml) that includes 1437 reactions associated with 1103 genes. This model has been preprocessed and the size is reduced to 763 reactions. The last model is *C.thermocellum* (ATCC 27405) *i*SR432 model [16] (http://www.biomedcentral.com/content/supplementary/1752-0509-4-31-s3.xml) that contains 577 reactions representing the function of 432 genes. This model underwent the pre-processing step and the size has reduced to 351 reactions. This work has generated three results, which are the list of knockout genes, growth rate, and BPCY (Biomass Product Coupled Yield). Unit for growth rate is hour^{-1} while unit for BPCY is in milli-gram (gram-glucose.hour)$^{-1}$. The experiments were conducted using a 2.3 GHz Intel Core i7 processor with 8 GB RAM workstation.

The results were compared with the previous works as reported in the literature studies [5,7,9]. The experiment carries out 100 individual runs to test on DBFBA with OptKnock, in which the result shown is the best result among the runs.

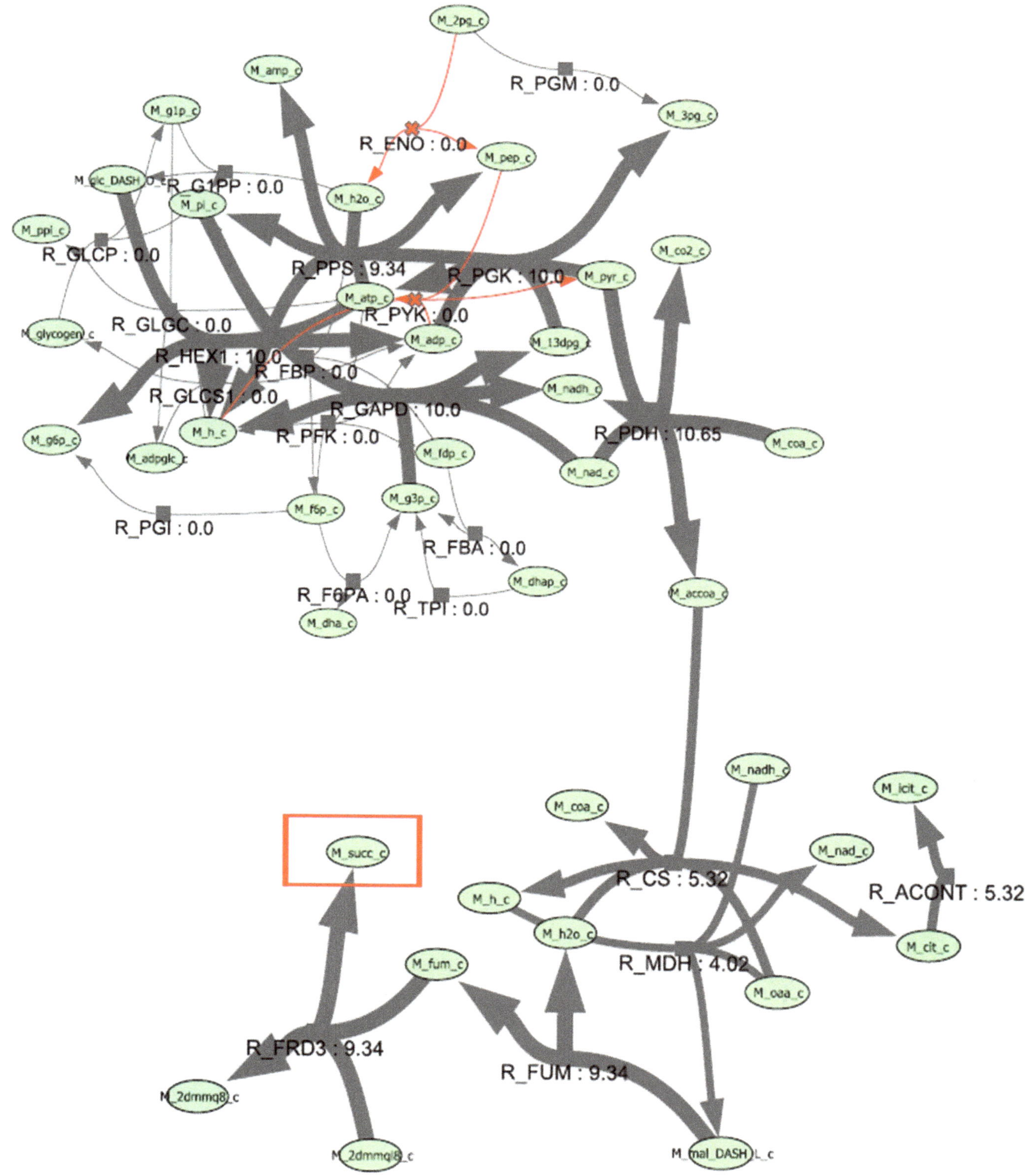

Figure 8. Flux map of the production of succinic acid.

Results and Discussion

Benchmark Functions

In this paper, an improved method, DBFBA, was proposed to test the performance of DBFBA. For evaluation, a benchmarking analysis was conducted. However, benchmark functions could only be tested on DB and BA, as FBA is an objective function. Hence, the benchmark functions on DB and BA in this study were tested. As BA is used to attain the maximum, the functions are inverted before the algorithm is applied. The De Jong, Martin & Gaddy, Branin, and Griewangk functions have been used in this

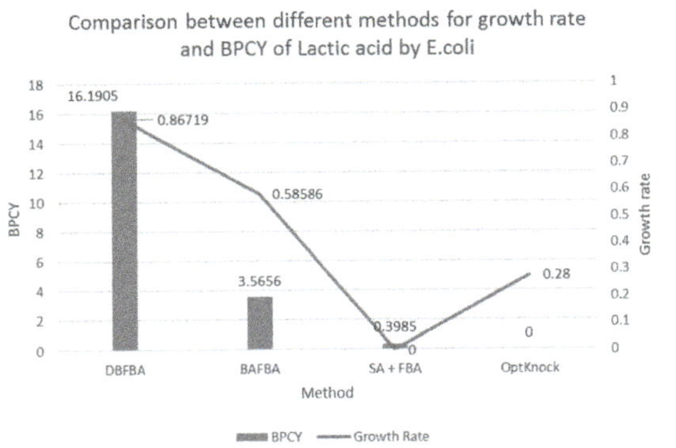

Figure 9. Comparison between different methods for growth rate and BPCY of Lactic acid by *E.coli.* Note: BPCY is in gram (gram-glucose.hour)$^{-1}$. Unit for growth rate is hr^{-1}.

study. A total of 100 individual runs have been carried out to test on DB and BA.

Table 1 shows the mathematical representation of the functions. Table 2 shows the mean and standard deviation (STD) of all the functions, De Jong, Martin & Gaddy, Branin, and Griewangk, tested on both original BA and DB. The results showed that DB performs better than BA. All functions obtained low STD, indicating that the result from each run is very close to the mean. As a conclusion, the stability of the proposed method is high as the difference between the result for each individual run is small. In addition, the means for both algorithms are similar, indicating that DB is indeed reliable, as the results obtained from DB are consistent with the results from previous works.

Production of Succinic Acid and Lactic Acid by *E.coli*

Tables 3 and 4 summarize the results obtained from DBFBA for succinic acid and lactic acid production in *E.coli*. As shown from the results, this method produces better results than the previous works in terms of growth rate and BPCY, and can also identify potential genes that can be removed [5,7,9].

Table 3 shows DBFBA for the removal of three reactions from the network results in succinic acid growth rate reaching 0.85512 and BPCY reaching 14.2907, which is better than previous works. The removal of enolase (ENO) affects the phosphotransferase system, which causes the network to rely exclusively on glucokinase for the uptake of glucose. In addition, the removal of pyruvate kinase (PYK), makes PEP carboxykinase as the only central metabolic reaction capable of draining a significant amount of PEP supplied by glycolysis [5]. Figure 7 presents the comparison between different methods for growth rate and BPCY of succinic acid. Figure 8 shows the flux map for the production of succinic acid. The flux map shows that there is an increase in the flux to succinic acid.

DBFBA is then applied to identify the knockout strategy for producing lactic acid. Table 4 shows the growth rate of this method, which is 0.86719 and 16.1905 for BPCY. Figure 9 shows the comparison between different methods for growth rate and BPCY of lactic acid. The deletion of transaldolase (TALA) decreased the efficiency of gluconeogenesis which resulted in the increased concentration of phosphoenolpyruvate. Phosphoenol-pyruvate is then converted into pyruvate and continues to convert into lactic acid. Knocking out acetaldehyde dehydrogenase (ACALD) which catalyzes the conversion of acetaldehyde into

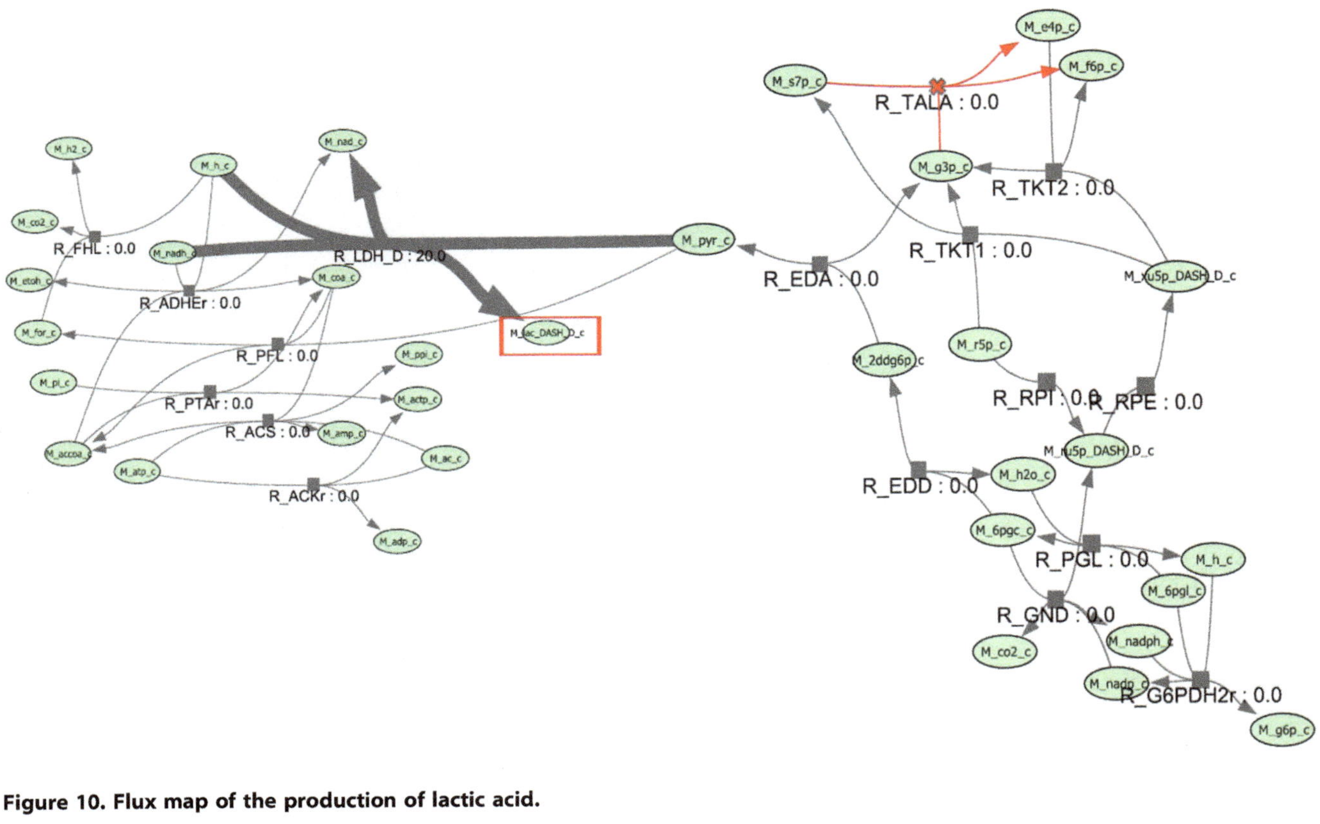

Figure 10. Flux map of the production of lactic acid.

Table 5. Comparison between different methods for growth rate and BPCY of ethanol by *B.subtilis.*

Method	Growth Rate (1/hr)	BPCY	List of knockout genes
DBFBA	**122.9657**	**1.1573e+05**	ALAD_L*, GK, sn-Glycero-3-phosphoethanolamine glycerophosphohydrolase
BAFBA[9]	122.8861	1.1154e+05	ALAD_L*, LDH_L, XYLI1, inosose 2,3-dehydratase

Note: The bold numbers represent the best result.
*Common genes for all methods. BPCY is in gram (gram-glucose.hour)$^{-1}$.

acetic acid eliminated the competing product, acetic acid. In addition, the removal of malate dehydrogenase (MDH) affects the oxaloacetate concentration which resulted in the increased concentration of pyruvate. As a consequence, the yield of lactic acid is improved. Figure 10 shows the flux map for the production of lactic acid. The flux map shows the flux increases to lactic acid.

Production of Ethanol by *B. subtilis*

Table 5 shows the results of DBFBA and previous works to enhance the production of ethanol in *B. subtilis*. Ethanol is a volatile, flammable, colorless liquid, and it is a promising biofuel. Ethanol is currently used as an alternative fuel for gasoline worldwide. Therefore, ethanol was found to be the most suitable case study for this research.

DBFBA obtained a better growth rate and BPCY which are 122.9657 and 1.1573e+05 respectively. Glycerol kinase (GK) and sn-Glycero-3-phosphoethanolamine glycerophosphohydrolase participate in glycerophospholipid metabolism, which involves the formation of glycerol. As stated by Pagliardini *et al.* (2013), glycerol is one of the main by-products in ethanol fermentation hence a substantial reduction of glycerol yield may lead to a significant increase in ethanol yield. In Pagliardini *et al.* (2013) experiment, a drastic reduction of the glycerol yield improved the yield of ethanol [19]. In addition, Figure 11 shows the comparison between different methods for growth rate and BPCY of ethanol, DBFBA obtains better results in both growth rate and BPCY compared to previous methods.

Production of Ethanol by *C.thermocellum*

Table 6 shows the results of DBFBA and previous works to enhance the production of ethanol in *C.thermocellum*. DBFBA

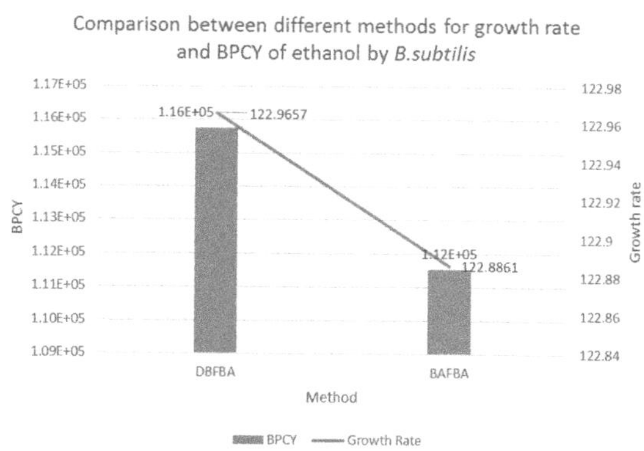

Figure 11. Comparison between different methods for growth rate and BPCY of Ethanol by *B.subtilis.* Note: BPCY is in gram (gram-glucose.hour)$^{-1}$. Unit for growth rate is hr^{-1}.

shows a better result for *C.thermocellum* model with the growth rate of 11.4826 and the BPCY 1.07e+004. Figure 12 shows the comparison between different methods for the growth rate and BPCY of ethanol. DBFBA shows that it could reach a higher value in both growth rate and BPCY than previous methods in the figure. The list of knockout genes shows the deletion of glycerate kinase (GLYCK), L-lactatedehydrogenase (LDH_L), and 2-Oxobutanoateformatelyase (OBTFL). Glycerate kinase participates in glycerolipid metabolism, which involves the formation of glycerol. In the experiment of Pagliardini *et al.* (2013), a drastic reduction of the glycerol yield improved the yield of ethanol due to which glycerol becomes one of the main by-products in ethanol fermentation [19]. As stated in Kim *et al.* (2012), lactate dehydrogenase (LDH_L) plays a key role in the fermentative metabolism in the metabolic engineering for ethanol production. The deletion of LDH_L inhibited the conversion from pyruvate to lactate, therefore more pyruvate was decarboxylated to acetaldehyde and further converted to ethanol [20]. 2-Oxobutanoateformatelyase is a key reaction for propionate metabolism. However, there is no specific report on the effect of 2-Oxobutanoateformatelyase for ethanol production currently. Figure 13 and Figure 14 show the convergence speed of DBFBA on ethanol production for *B.subtilis* and *C.thermocellum*.

Discussion

In addition, Table 7 shows the computational time comparison between DBFBA and BAFBA for 1000 iterations. The average computational time of DBFBA improved by 80%, 81%, and 87% compared to BAFBA result for 1000 iterations, respectively.

The results showed that both DBFBA and DB performed better than other algorithms. It can be concluded that the capability of DE algorithm in finding local optimum improved the performance of the original BA. The original BA had the problem of repetitive iterations of the algorithm in local search, where each bee keeps searching until the best possible answer was obtained. The proposed DBFBA solved the problem by implementing DE algorithm into the local search part. DE algorithm operates by

Table 6. Comparison between different methods for growth rate and BPCY of ethanol by *C.thermocellum.*

Method	Growth Rate (1/hr)	BPCY	List of knockout genes
DBFBA	**11.4826**	**1.07e+04**	GLYCK, LDH_L, OBTFL
BAFBA [9]	9.9313	8.329e+003	MDH, G3PD1, PTAr

Note: The bold numbers represent the best result.

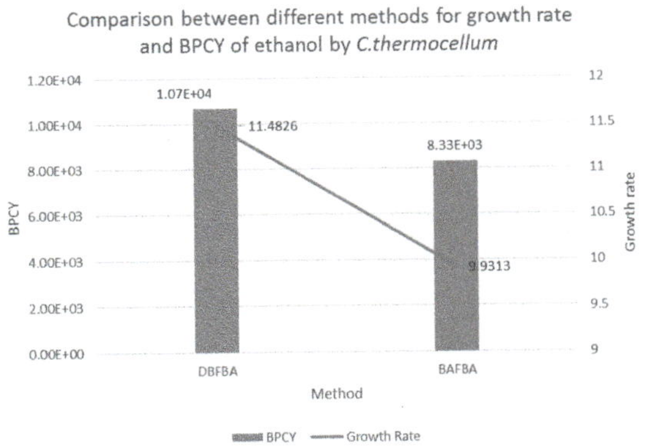

Figure 12. Comparison between different methods for production of Ethanol by *C. thermocellum*. Note: BPCY is in gram (gram-glucose.hour)$^{-1}$. Unit for growth rate is hr^{-1}.

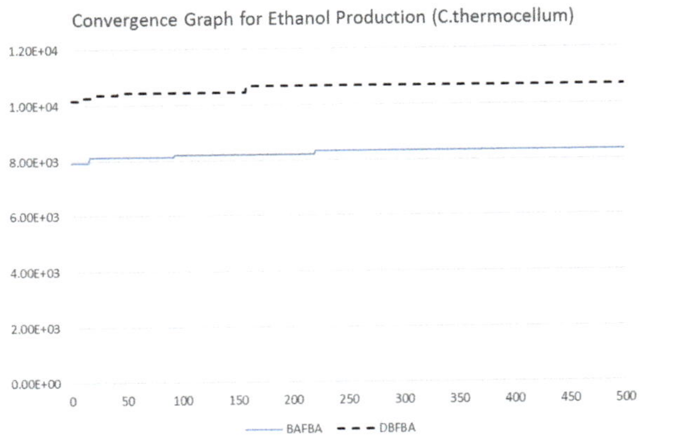

Figure 14. Convergence graph of different methods for ethanol production in *C.thermocellum*.

maintaining a population of candidate solutions and creating new candidate solutions through the mutation and crossover operation of DE, and keeping the fitness candidate solution. Furthermore, DE algorithm is a heuristic approach mainly having three advantages; finding the true global minimum regardless of the initial parameter values, fast convergence, and using few control parameters. In addition, the OptKnock validation, which is integrated into BAFBA in this work, improves the reliability of BAFBA as the identified lists of potential genes have proven crucial in improving the desired phenotypes through literature [5,19–20].

Conclusion and Future Works

The development of accurate and efficient modeling methods as well as optimization methods in the field of metabolic engineering is crucial, as they will contribute significantly to the field of biotechnology. Consequently, it will lead to substantial economic gains in the production of pharmaceuticals, fuels and food ingredients. In this work, DBFBA has been proposed to be able to predict optimal sets of gene knockout strategies to maximize the production of desired phenotype and in the meantime, sustains the cellular growth. DBFBA improves the performance of BAFBA by implementing DE algorithm, which is indeed a promising algorithm in finding local optimum. Experimental results on *E. coli* model, *B. subtilis* model, and *C.thermocellum* model showed a hybrid approach, where DBFBA works as an effective tool in generating optimal solutions for the gene knockout identification, and therefore could be a useful tool in the field of metabolic engineering. The ability of DBFBA in predicting a set of genes that might affect the production of desired phenotypes is helpful to the biologists. Instead of testing single gene knockout effect on the production, the biologists can test on the effect of multiple genes knockout. Lastly, since user-friendly and publicly accessible web-servers represent the future direction for developing practically more useful models, simulated methods, or predictors [21], we shall make efforts in our future work to provide a web-server for the method presented in this paper.

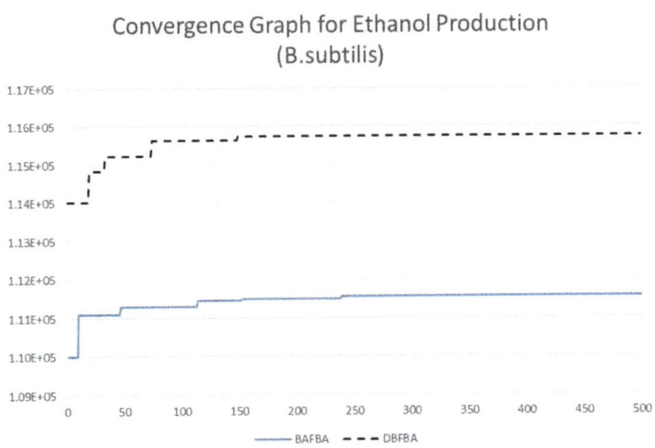

Figure 13. Convergence graph of different methods for ethanol production in *B.subtilis*.

Table 7. Comparison between average computational time of DBFBA and BAFBA for 1000 iterations.

Model	Method	Computation Time (seconds)
E.coli	DBFBA	**2076**
	BAFBA [11]	10253
	OptKnock [5]	N/A
	SA + FBA [7]	N/A
B.subtilis	DBFBA	**4173**
	BAFBA [11]	22515
	OptKnock [5]	N/A
	SA + FBA [7]	N/A
C. thermocellum	DBFBA	**1350**
	BAFBA [11]	10282
	OptKnock [5]	N/A
	SA + FBA [7]	N/A

Note: The bold numbers represent the best result. N/A represents that the results are not reported in literature.

Acknowledgments

We would like to express gratitude to the editor and reviewers for helpful suggestions and Universiti Teknologi Malaysia for sponsoring this research.

Author Contributions

Conceived and designed the experiments: YWC MSM SD RMI CKC LEC SO JMC. Performed the experiments: YWC MSM. Analyzed the data: YWC. Contributed reagents/materials/analysis tools: YWC MSM. Wrote the paper: YWC MSM SD RMI CKC LEC SO JMC.

References

1. Van Dien SJ, Lidstrom ME (2002) Stoichiometric model for evaluating the metabolic capabilities of the facultative methylotrophMethylobacteriumextorquens AM1, with application to reconstruction of C(3) and C(4) metabolism. Biotechnol Bioeng 78: 296–312.
2. Reed JL, Vo TD, Schilling CH, Palsson BO (2003) An expanded genome-scale model of Escherichia coli K-12 (iJR904 GSM/GPR). Genome Biology 4, 54.
3. Feist AM, Herrgård MJ, Thiele I, Reed JL, Palsson BØ (2009) Reconstruction of biochemical networks in microorganisms. Nat Rev Microbiol 72, 129–143.
4. Alt W, Deutsch A, Preziosi L (2009) Computational Cell Biology: Second Theme Issue on "Computational Biology". J Math Biol 58, 1–5.
5. Burgard AP, Pharkya P, Maranas CD (2003) OptKnock: A bilevel programming framework for identifying gene knockout strategies for microbial strains optimization. BiotechnolBioeng 84, 647–657.
6. Patil KR, Rocha I, FÖrster J, Nielsen J (2005) Evolutionary programming as a platform for in silico metabolic engineering. BMC Bioinformatics 6, 308.
7. Rocha M, Maia P, Mendes R, Pinto JP, Ferreira EC, et al. (2008) Natural computation meta-heuristics for the in silico optimization of microbial strains. BMC Bioinformatics 9, 499.
8. Choon YW, Mohamad MS, Deris S, Chong CK, Chai LE, et al. (2012) Identifying Gene Knockout Strategies Using a Hybrid of Bees Algorithm and Flux Balance Analysis For in silico Optimization of Microbial Strains. Advances in Intelligent and Soft-Computing. 151: 371–378.
9. Pham DT, Ghanbarzadeh A, Koç E, Otri S, Zaidi M (2006) The bees algorithm – a novel tool for complex optimization problems. In: Proceedings of the Second International Virtual Conference on Intelligent Production Machines and Systems. July 3–14, 2006.
10. Pham DT, Darwish AH, Eldukhri EE (2006) Optimisation of a fuzzy logic controller using the bees algorithm. International Journal of Computer Aided Engineering and Technology 1(2): 250–264.
11. Olague G, Puente C (2006) The honeybee search algorithm for three-dimensional reconstruction. In: Proceeding EuroGP'06 Proceedings of the 2006 international conference on Applications of Evolutionary Computing, pp. 427–437. Springer-Verlag Berlin, Heidelberg (2006).
12. Pham DT, Ghanbarzadeh A (2007) Multi-objective optimisation using the bees algorithm. Proceedings of the Third International Virtual Conference on Intelligent Production Machines and Systems, July 2–13.
13. Cheng MY, Lien LC (2011) A Hybrid Swarm Intelligence Based Particle Bee Algorithm For Benchmark Functions And Construction Site Layout Optimization. In: Proceedings of the 28th ISARC, pp. 898–904. Seoul.
14. Fong SS, Burgard AP, Herring CD, Knight EM, Blattner FR, et al. (2005) In silico design and adaptive evolution of Escherichia coli for production of lactic acid. Biotechnol Bioeng, 91(5): 643–8.
15. Henry CS, Zinner JF, Cohoon MP, Stevens RL (2009) iBsu1103: a new genome-scale metabolic model of Bacillus subtilis based on SEED annotations. Genome Biology, 10: 69.
16. Roberts SB, Gowen CM, Brooks JP, Fong SS (2010) Genome-scale metabolic analysis of Clostridium thermocellum for bioethanol production. BMC System Biology, 4: 31.
17. García Sánchez CE, Vargas García CA, Torres Sáez RG (2012) Predictive Potential of Flux Balance Analysis of Saccharomyces cerevisiae Using as Optimization Function Combinations of Cell Compartmental Objectives. PLoS ONE 7(8): e43006. doi:10.1371/journal.pone.0043006.
18. Soons ZITA, Ferreira EC, Patil KR, Rocha I (2013) Identification of Metabolic Engineering Targets through Analysis of Optimal and Sub-Optimal Routes. PLoS ONE 8(4): e61648. doi:10.1371/journal.pone.0061648.
19. Pagliardini J, Hubmann G, Alfenore S, Nevoigt E, Bideaux C, et al. (2013) The metabolic costs of improving ethanol yield by reducing glycerol formation capacity under anaerobic conditions in Saccharomyces cerevisiae. Microbial Cell Factories 12(29). doi:10.1186/1475-2859-12-29.
20. Kim JW, Chin YW, Park YC, Seo JH (2012) Effects of deletion of glycerol-3-phosphate dehydrogenase and glutamate dehydrogenase genes on glycerol and ethanol metabolism in recombinant Saccharomyces cerevisiae. Bioprocess BiosystEng 35, 49–54.
21. Chou KC, Shen HB (2009) Review: recent advances in developing web-servers for predicting protein attributes. Natural Science, 2: 63–92.

6

Dlf1, a WRKY Transcription Factor, Is Involved in the Control of Flowering Time and Plant Height in Rice

Yuhui Cai[1], Xujun Chen[1], Ke Xie[3], Qikai Xing[1], Yawen Wu[2], Jing Li[3], Caihong Du[3], Zongxiu Sun[2]*, Zejian Guo[1,3]*

1 Key Laboratory of Plant Pathology, China Agricultural University, Beijing, China, **2** China National Rice Research Institute, Hangzhou, China, **3** Zhejiang University, Hangzhou, China

Abstract

Flowering time and plant height are important agronomic traits for crop production. In this study, we characterized a semi-dwarf and late flowering (*dlf1*) mutation of rice that has pleiotropic effects on these traits. The *dlf1* mutation was caused by a T-DNA insertion and the cloned *Dlf1* gene was found to encode a WRKY transcription factor (OsWRKY11). The *dlf1* mutant contains a T-DNA insertion at the promoter region, leading to enhanced accumulation of *Dlf1* transcripts, resulting in a semidominant mutation. The *dlf1* mutation suppressed the transcription of *Ehd2/RID1/OsId1* and its downstream flowering-time genes including *Hd1*, *Ehd1* and *Hd3a* under both long-day (LD) and short-day (SD) conditions. Knock-down of *Dlf1* expression exhibited early flowering at LD condition related to the wild-type plants. Accumulation of *Dlf1* mRNA was observed in most tissues, and two splicing forms of *Dlf1* cDNAs were obtained (*OsWRKY11.1* and *OsWRKY11.2*). These two proteins showed transactivation activity in yeast cells. Dlf1 protein was found to be localized in the nucleus. Enhanced expression of *OsWRKY11.2* or its 5′ truncated gene showed similar phenotypes to the *dlf1* mutant, suggesting that it might function as a negative regulator. We conclude that Dlf1 acts as a transactivator to downregulate *Ehd2/RID1/OsId1* in the signal transduction pathway of flowering and plays an important role in the regulation of plant height in rice.

Editor: Jin-Song Zhang, Institute of Genetics and Developmental Biology, Chinese Academy of Sciences, China

Funding: This research was financially supported by National Nature Science of Foundation (31171833 and 31272025), the Program for Changjiang Scholars and Innovative Research Team in University (IRT042), and the 111 project (B13006). The funders had no role in study design, data collection and analysis, decision to publish, or preparation of the manuscript.

Competing Interests: The authors have declared that no competing interests exist.

* Email: sunzx405@163.com (ZS); guozj@cau.edu.cn (ZG)

These authors contributed equally to this work.

Introduction

Increasing cereal output has been a fundamental goal to meet the soaring demand for food. Plant height, potential yield and flowering time are important traits for cereal production. Plant height, one of the main selection trait in rice (*Oryza sativa*) breeding, is controlled mostly by genes related to the synthesis and regulation of phytohormones, such as gibberellin (GA) and brassinolide [1–3]. Potential yield per rice plant is determined by grain weight, and numbers of grains per panicle and tillers per plant [4–6]. Flowering time of plants is controlled by both environmental and developmental factors, with photoperiod as an important environmental signal. Molecular genetic analysis in *Arabidopsis thaliana*, a long-day (LD) plant, has identified a set of key regulators functioning in the photoperiod-mediated flowering pathway. For example, the nuclear protein CONSTANS (CO) positively regulates the flowering activator *FLOWERING LOCUS T* (*FT*), which further interacts with the bZIP transcription factor *FLOWERING LOCUS D* (*FD*) to control flowering time in *Arabidopsis* [7–9]. *SUPPRESSOR OF OVEREXPRESSION OF CONSTANS1* (*SOC1*), encoding a MADS box transcription factor, is activated by CO through FT and repressed by FLC (FLOWERING LOCUS C) via direct binding to the promoter [10].

Analysis of natural variants and mutants in rice, a short-day (SD) plant, has revealed the existence of a genetic pathway similar to that in *Arabidopsis* in photoperiodic flowering. *Heading date 1* (*Hd1*), *Heading date 3a* (*Hd3a*), *Heading date 6* (*Hd6*) and *OsGI* in rice are orthologs of *Arabidopsis CO*, *FT*, the α-subunit of kinase CK2, and *GIGANTEA* (*GI*), respectively. *OsGI*, a gene involved in the circadian clock control, regulates *Hd1* and *Hd3a* in photoperiodic flowering, which promotes flowering under SD conditions and suppresses it under LD conditions [11–14]. In addition, *early heading date 1* (*Ehd1*), a B-type response regulator that is specific to floral induction in rice, regulates the expression of *Hd3a*, *FTL1* and *OsMADS14* [15]. *Ehd1* functions upstream of *Hd3a* and *RFT1* through the *Hd1*-independent pathway. *Ehd2/RID1/OsId1* were isolated separately by three groups and the locus was found to encode a Cys2/His2-type zinc finger transcription factor orthologous to the *INDETERMINATE 1* (*ID1*) gene in maize [16–18]. Loss-of-function *ehd2/rid1/Osid1* mutants were never- or extremely late-flowering in comparison with wild-type plants under both SD and LD conditions. Functional analysis revealed that *Ehd2/RID1/OsId1* acts as a master switch and promoter of phase transition mainly by regulating *Ehd1* and the downstream genes. Further, a CCT (CO, CO-LIKE and TIMING OF CAB1)-domain protein encoding gene *Ghd7*, which was uncovered from a natural variant rice, suppresses the expression of *Ehd1* and *Hd3a* genes under LD conditions but does not affect *Hd3a* expression under SD conditions [19]. Recently, *Hd17/Ef7*, an ortholog of *Arabidopsis EARLY FLOWERING 3* (*ELF3*), has been characterized to promote rice flowering by repressing *Ghd7* expression under both LD and SD conditions [20,21]. In addition, *Ehd3*, which encodes a protein containing two plant homeodomain (PHD) finger motifs, functions also as a LD suppressor of *Ghd7* [22]. The mutation in *DTH8/Ghd8/Hd5* shows pleiotropic effects on grain number, plant height and heading date, and causes delayed flowering by down-regulation of *Ehd1* under LD conditions [23–25]. On the other hand, *OsMADS51* is a SD promoter functioning downstream of *OsGI1* and upstream of *Ehd1* [9,26], whereas *Ehd4*, encoding a CCCH-zinc finger regulator, promotes rice flowering by stimulating the expression of *Ehd1* and its downstream genes [27].

Much progress has been achieved in the genetic dissection of photoperiodic flowering of rice, but the molecular regulation is still largely unknown. In this study, we characterized a semi-dwarf and late flowering (*dlf1*) mutant and identified *Dlf1* gene that encodes a WRKY transcription factor. Our results showed that overexpression of *Dlf1* suppressed flowering by inhibiting the expression of *Ehd2/RID1/OsId1* under both LD and SD conditions and influenced plant height in rice.

Results

Isolation and morphological characterization of the *dlf1* mutant

A semi-dwarf and late flowering (*dlf1*) mutant was identified from a T_1 population of T-DNA insertion lines of cultivar *Oryza sativa* L. *japonica* Zhonghua 11 (ZH11). The flowering time of *dlf1* plants was delayed for about two weeks compared with wild-type plants in the experimental field in late May, 2003 in Hangzhou, China (Fig. 1A). DNA blot analysis showed that there was only one copy of T-DNA insertion in the mutant line (data not shown).

To determine whether the heading time of the *dlf1* mutant differed under different photoperiod conditions, the mutant and wild-type plants were grown under SD conditions (10/14 h light/dark) and LD conditions (14/10 h light/dark). Heading time of the *dlf1* mutant plants was 87.2±1.0 d, which was delayed ca. 16 d in comparison with wild-type (71.5±0.8 d) under SD conditions (Fig. 1E). Under LD conditions, heading time of the *dlf1* mutant plants was 140.9±2.2 d and increased by ca. 23 d compared with ZH11 (117.8±2.2 d). Other phenotypes differed significantly under natural LD conditions, including homozygous *dlf1* mutant lowered plant height (Figs. 1A and F) caused by reducing cell size (Fig. S1A), and decreased number of spikelets per panicle (Fig. S1B). The *dlf1* mutant plants also had less 1000-grain weight (Fig. S1C) and showed leaf rolling phenotypes (Figs. S1D and E) under natural LD conditions.

Dlf1 encodes a WRKY transcription factor

Using the thermal asymmetric interlaced PCR (Tail-PCR) method, we isolated the genomic DNA flanking of the T-DNA insertion site from the *dlf1* mutant. A BLAST search of the flanking sequence revealed that the T-DNA was inserted at 67 bp upstream of the initial ATG (the 'A' was defined as +1) of the predicted coding sequence of the gene *OsWRKY11* (LOC_Os01g43650, AK108745; and named as *Dlf1* in the present study) (Fig. 2A). The flanking sequence of the other T-DNA border was amplified and revealed a 42 bp deletion at the insertion site.

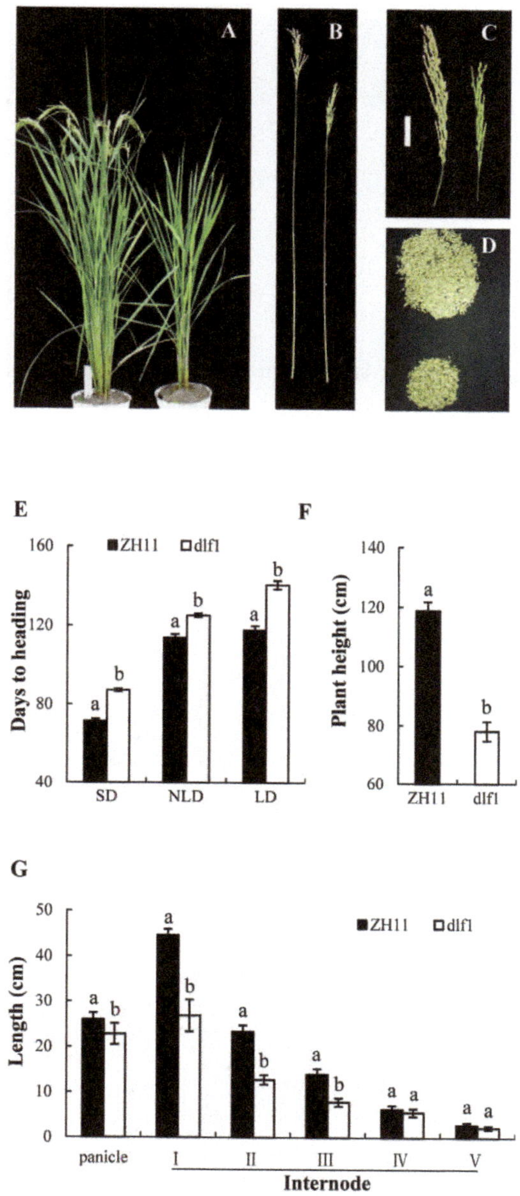

Figure 1. Phenotypes of Zhonghua 11 (ZH11) and *dlf1* mutant. (**A**) Photos of the wild-type ZH11 (left) and mutant *dlf1* (right), taken when ZH11 reached maturity. (**B**) Photos of main culms of ZH11 (left) and *dlf1* (right). (**C**) Main panicles of ZH11 (left) and *dlf1* (right). (**D**) Grains from main panicles of ZH11 (top) and *dlf1* (bottom). (**E**) Days to heading of ZH11 and *dlf1* under SD, LD and NLD (natural LD) conditions. Data are represented as mean values ± standard derivation (SD) of 20 plants. (F) Plant heights of ZH11 and *dlf1* under natural LD conditions. Values are means ± SD, n = 20. Experiment were performed three times, showing similar results. (**G**) Difference of internode lengths between matured ZH11 and *dlf1* plants. The plants were grown in the experimental field under natural LD conditions. Values are means ± SD, n = 20. The same experiments were repeated three times, and the similar results were obtained. For SD and LD treatments, the plants were grown in greenhouse under natural light conditions and shaded at the time designated. a and b in figure indicate ranking by Duncan test at $P<0.05$, starting from a. b is significantly different from a.

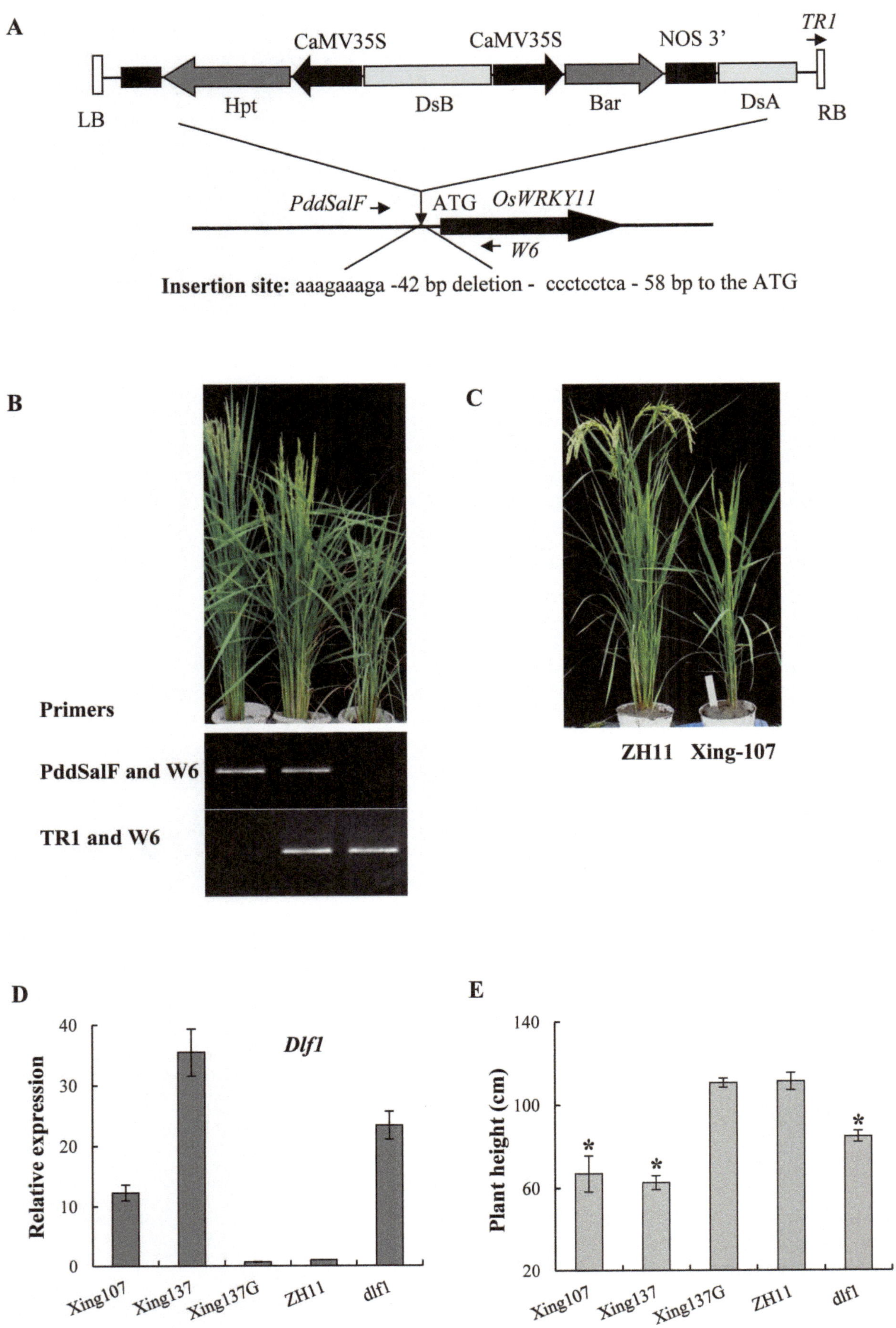
A
CaMV35S
CaMV35S
NOS 3'
TR1
LB
Hpt
DsB
Bar
DsA
RB
PddSalF
ATG
OsWRKY11
W6
Insertion site: aaagaaaga -42 bp deletion - ccctcctca - 58 bp to the ATG
B
Primers
PddSalF and W6
TR1 and W6
C
ZH11 Xing-107
D
Dlf1
Relative expression
0
10
20
30
40
Xing107
Xing137
Xing137G
ZH11
dlf1
E
Plant height (cm)
20
60
100
140
Xing107
Xing137
Xing137G
ZH11
dlf1

Figure 2. Molecular features of *Dlf1*. (**A**) Structure of T-DNA insertion site. The perpendicular arrow head indicates the insertion site. The T-DNA was inserted into the upstream of *OsWRKY11*. Arrows indicate the primers used for analyzing the insertion site. LB and RB represent the left and right borders of T-DNA. (**B**) PCR genotyping *Dlf1* segregants in F_2 with primers as indicated in **A** (showed in *italics*). Primers PddSalF and W6 amplify a product from wild-type. PCR-positive plants with primers of TR1 and W6 indicate T-DNA insertion in the examined site and a homozygote for the insertion if without an amplification product from PddSalF and W6. (**C**) Photos of ZH11 (left) and *Cp-Ins-Dlf1* transgenic T_1 plant (named as Xing-, right), taken when ZH11 reached maturity. (**D**) Expression of *Dlf1* in transgenic plants of T_1 progenies under natural LD conditions. The first and second youngest leaves were sampled from 90-d-old plants for RNA isolation. Transcription levels were quantified by quantitative reverse-transcription PCR (qPCR) and the gene expression was normalized to rice *ubiquitin* gene (*Ubq*) for each sample. Transcription levels are expressed as ratio to the level of transcript in ZH11. (**E**) Plant heights. Xing-, the transgenic plants, the suffix G for segregated non-transgenic plants; ZH11, the control and *dlf1*, the mutant. The plants were grown in the experimental field under natural LD conditions. Values are shown as means ± SD of two independent experiments. Asterisks indicate significant difference between ZH11 and other lines ($P<0.05$, Duncan test).

To ascertain whether the *dlf1* phenotypes are associated with the T-DNA insertion, genetic analysis was performed using two filial populations of reciprocal crosses between the *dlf1* mutant and wild-type plants. The flowering time and the plant heights co-segregated in a manner of fit the 1:2:1 population (data not shown). Correspondingly, the genotypes of the T-DNA insertion in the same populations were determined by PCR and showed a tight linkage with the phenotypes (Fig. 2B). These results indicated that the *Dlf1* is semi-dominant and the T-DNA insertion event upstream of the *dlf1* gene is responsible for the mutant phenotype.

To examine the effect of the T-DNA insertion on the gene expression, the *Dlf1* total expression level was assayed by quantitative real-time PCR (qPCR). The transcript levels of *Dlf1* in rice leaves were enhanced significantly in the *dlf1* mutant compared with the wild-type ZH11 (Figs. 2D). To verify that the mutant phenotypes were caused by the high expression of *Dlf1* that was related to the T-DNA insertion, a fragment from the T-DNA insertion site to the end of the *Dlf1* coding region (*Cp-Ins-Dlf*) were transformed into the wild-type background. Among 14 plants regenerated (named as Xing-), 10 showed variation in dwarfism, and late flowering phenotypes in the T_0 progeny (data not shown). Two lines were used for further analysis in the T_1 generation. The levels of *Dlf1* expression were higher in dwarf plants compared with the wild-type and segregated non-transgenic plants (Figs. 2C, D and E). The dwarf plants also showed delayed flowering for about two weeks late under natural LD conditions, indicating that the increase of *Dlf1* expression caused the *dlf1* phenotypes.

Dlf1 is ubiquitously expressed and alternatively spliced in rice

To examine the expression pattern of *Dlf1*, qPCR analysis was performed with total RNA prepared from leaves, sheaths, young panicles and roots of ZH11 wild-type plants grown under natural LD conditions. *Dlf1* was expressed in all tissues examined (Fig. 3A). Further, we stained the transgenic plants harboring the *Dlf1* promoter- driven *Gus* fusion gene (−2010 – +134 bp; *Cp-WP:Gus*, Fig. 3B). Gus staining was observed in leaves, roots and panicles (Fig. 3C), which confirmed the qPCR results. Since the plants harboring *Cp-Ins-Dlf* construct showed the *dlf1* phenotypes, the fragment from the insertion site (−67 – + 134 bp; *Cp-InR3:Gus*) of *Dlf1* was also fused with *Gus* gene. Surprisingly, *Cp-InR3:Gus* plants exhibited the similar level of expression as the *Cp-WP:Gus* plants in the young seedlings (Figs. 3D and E) though the *InR3* fragment was a region transcribed. However, the expression of *Gus* gene and Gus activity were low in the *Cp-In2R:Gus* (−67 – +5 bp) transgenic plants, which revealed that the deduced translation region in *InR3* was required for the promoter activity in comparison with the *In2R* fragment.

To obtain the *Dlf1* cDNA, we amplified the open reading frame region from total RNA isolated from ZH11 using RT-PCR with the primers W1 and W5. Two amplified products were obtained (Fig. 3F), the longer one (assigned as *OsWRKY11.1*) encoded the deduced 379-aa Dlf1 protein and the shorter sequence (assigned as *OsWRKY11.2*) contained a 161 bp deletion in the third exon causing a premature stop of translation (Fig. S2). The deduced amino acid sequence of OsWRKY11.2 encodes a protein of 270 residues, which still contains the WRKY domain (aa 205–262). Northern blot analysis revealed two hybridization bands using the *Dlf1* coding region, (Fig. 3G), confirming the existence of alternative splicing of the *Dlf1* mRNAs.

Dlf1 has transcriptional activation activity in yeast

Dlf1 contains an acidic N-terminus that may function as a transcriptional activation domain. To investigate this possibility, the coding region of the *Dlf1* full length cDNA and its truncated derivatives were fused in frame to the GAL4 DNA-binding domain in the *pGBKT7* vector. The transactivation activity assay in yeast showed that the region of 91–120 aa was required for its transactivation (Fig. 4). We also determined the activation activity of OsWRKY11.2 (*pBD-WRKY11.2*), showing slight increase in activity compared with full length *Dlf1* transcript (*pBD-WRKY11.1*). This result was confirmed by deletion of the 3′ terminal of full length transcript (*pBD-dC2* construct).

Nuclear localization signal (NLS) of Dlf1 was predicted using cNLS Mapper (http://nls-mapper.iab.keio.ac.jp/). A NLS (amino acid 151–181) was identified with a high score of 7.5. The OsWRKY11.2 protein also contains the sequence of nuclear localization signal. To confirm the subcellular localization of Dlf1, we fused *OsWRKY11.1* with the enhanced green fluorescent protein (*GFP*) gene. The chimeric gene was put under the control of maize ubiquitin (*Ubi*) promoter. The resulting plasmid (*Ubi:Dlf1-GFP*) was then bombarded into onion epidermal cells. Localization of the WRKY11.1-GFP fusion protein was visualized exclusively in the nucleus (Fig. 5), whereas the control GFP (*Ubi:GFP*) was distributed both in the cytoplasm and the nucleus, indicating Dlf1 is a nuclear protein.

The expression of *Ehd2*, *Ehd1*, *Hd1* and *Hd3a* was repressed in *dlf1* mutant

To determine whether the late flowering phenotype of the *dlf1* mutant was due to the changes in flowering-related gene expression, qPCR analysis was performed in the wild-type and *dlf1* plants. Leaf samples were collected from 40 or 90-d-old plants grown under SD or LD conditions. The developmental stage of the plants was about one month before flowering in ZH11. The total expression levels of *Dlf1* in the mutant were higher than in the wild-type plants under both SD and LD conditions (Fig. 6). Moreover, *Dlf1* mRNA accumulation showed diurnal changes in the wild-type and *dlf1* mutant plants. *Ehd2/RID1/OsId1* is a key regulator in the genetic network that controls photoperiodic flowering in rice, promoting floral transition by upregulating *Ehd1* and then the downstream *Hd3a* [15–17]. *Ehd2/RID1/OsId1*

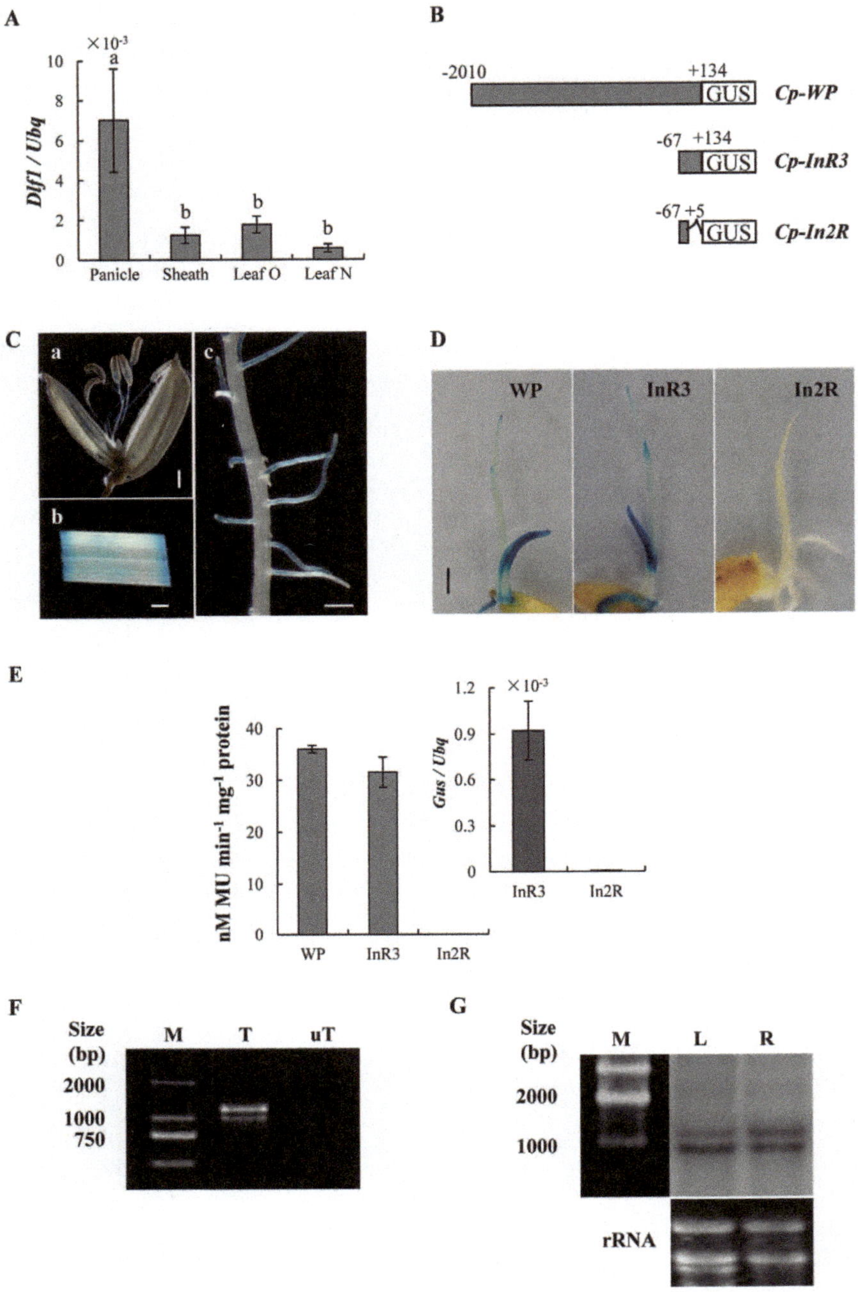

Figure 3. Expression pattern of *Dlf1*. (**A**) Expression of *Dlf1* in different tissues was analyzed by qPCR. RNA was isolated from young panicles, sheathes, older leaves (leaf O) and younger leaves (leaf N). Tissue samples were collected at 4 h after dawn. Values are shown as means ± SD of two biological experiments. Values marked with different letters are significantly different ($P<0.05$, Duncan test). (**B**) Gus-fused constructs with different lengths of *Dlf1* promoter. (**C**) *Dlf1* promoter-driven *Gus* expression (*Cp-WP*) in (a) young panicles (b) leaves and (c) roots of three-week-old plants. (**D**) Gus histochemical staining of 6-day-old transgenic lines. (**E**) Gus enzyme activity was measured in six-day-old seedlings harboring different constructs. Transcription analysis of *Gus* gene in the transgenic plants. The gene expression was normalized to rice *ubiquitin* gene (*Ubq*) for each sample. Means and their standard deviations are shown from three independent experiments. (**F**) Different transcripts of *Dlf1*. Total RNAs were isolated from leaves of three-week-old ZH11 plants. PCR products were obtained by amplification using templates of reverse-transcribed RNA (T) and RNA (uT) and separated by electrophoresis. (**G**) Northern blot analysis of *Dlf1* expression in leaves (L) and roots (R), using the total RNAs isolated from three-week-old plants. The probes used were the *Dlf1* coding region. rRNA of ethidium bromide staining was used as the loading control. Bar = 1 mm (**C**) and 2 mm (**D**).

mRNA accumulation was reduced in the *dlf1* mutant compared with wild-type plants under SD and LD conditions (Fig. 6). Subsequently, the levels of *Hd3a* and *Ehd1* expression were decreased in the mutant plants. The expression of *Hd1* was also suppressed in the *dlf1* mutant under both SD and LD conditions (Fig. 6). However, the expression of photoperiod-related genes *OsGI*, *Se5* and *Ghd7* were not significantly affected in the *dlf1* mutant (Fig. S3), indicating that Dlf1 specifically suppressed the expression of *Ehd2* and downstream genes.

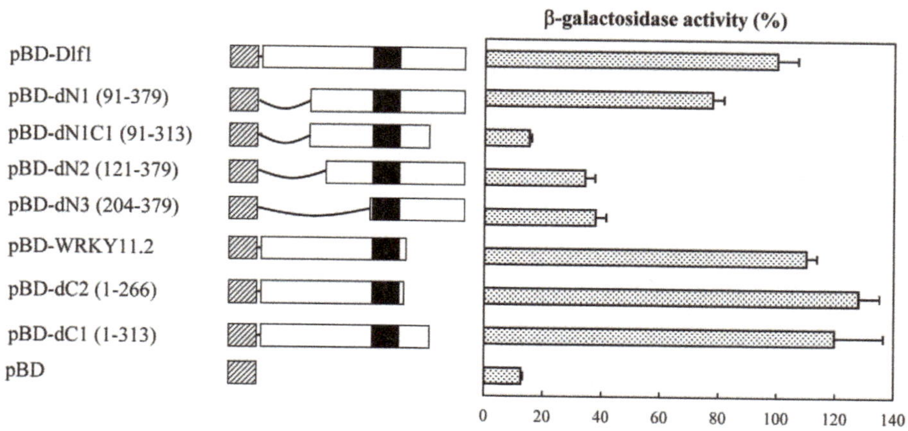

Figure 4. Dlf1 is transactivator in yeast. The full encoding sequence and deletion mutants of *Dlf1* were fused in frame to the *GAL4*-binding domain (BD) in *pBDKT7* to generate various vectors for yeast transformation. The constructed vectors were transformed into yeast AH109 strain, and grew on the selective medium at 30°C for 3 d. The β-galactosidase activity of transformants was determined using *o*-nitrophenyl β-D-galatopyranoside as a substrate. An empty vector *pGBKT7* (*pBD*) was used as the negative control. The values from three independent experiments were shown as means ± SD. Slash boxes represent BD in *pGBKT7* and the black boxes for the WRKY domain of Dlf1, whereas the white boxes represent the rest part of Dlf1, and the line indicates the deleted region. The numbers in the brackets are the start and end positions of each translation product of Dlf1 in the construct.

Transcription levels of *Dlf1* and the flowering-related genes were also examined at different developmental stages under LD conditions every 20 days. The accumulation of *Dlf1* mRNA slightly increased and reached a peak at 70 d after germination. Subsequently, the transcript level gradually decreased and remained at low levels even after flowering in the wild-type plants (Fig. 7). In the *dlf1* mutant, *Dlf1* transcript accumulated in a pattern quite similar to that of the wild type ZH11 plants, but the expression levels of *Dlf1* were at least 10-fold higher throughout the experiment periods.

Decrease of *Dlf1* expression showed early flowering under LD conditions

To further dissect the function of *Dlf1*, several constructs were generated to examine possible roles of the different transcripts by means of overexpression and RNAi. The full-length *OsWRKY11.1* and the alternatively spliced *OsWRKY11.2* transcript were put under control of the *Ubi* promoter, respectively, generating the *Ubi:W11.1* and *Ubi:W11.2* constructs for rice transformation. To knockdown *Dlf1*, a 276-bp fragment was used for the RNAi construct under the control of the CaMV35S promoter. More than 15 independent overexpressing transgenic lines were obtained for *Ubi:W11.1* in the genetic background of ZH11 and ZH17, respectively. However, only two *Ubi:W11.1* lines of ZH17 genetic background (named as C-) were found to increase *OsWRKY11* expression level and to delay flowering under LD condition comparing with ZH17 plants (Figs. 8A and B). Most of the transgenic lines in ZH11 background (named as OE-) showed slight variations in plant height (Fig. S4B) and insignificant changes in total transcripts of *OsWRKY11* (data not shown). Further, RNAi lines of dch53 and dch57 displayed suppressed transcription of *OsWRKY11* and flowered earlier than the ZH11 controls under LD condition (Figs. 8C and D). These results collectively suggested that *Dlf1* negatively regulate flowering in rice.

The C-terminus of *Dlf1* plays a role in the regulation of its expression level

In contrast to *Ubi:W11.1* construct, most of the *Ubi:W11.2* transgenic lines (named as Ka-) exhibited severe dwarfism (Fig. 9B)

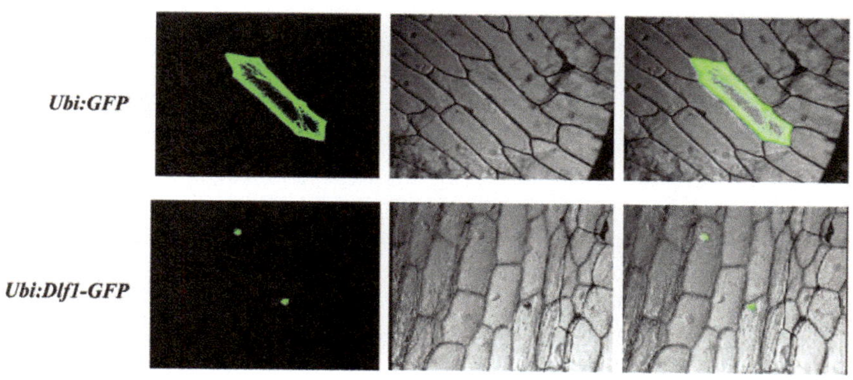

Figure 5. Nuclear localization of Dlf1-GFP fusion protein in onion epidermal cells. Onion epidermal cells were transformed with plasmids expressing GFP (top panel), or the Dlf1-GFP fusion protein (bottom panel) by bombardment and examined after 24 h. GFP fluorescence (left panel) and differential interference contrast image (middle panel) were compared to show the subcellular localization of GFP (cytoplasmic and nuclear) and Dlf1-GFP (nuclear). The images of the right panel were merged for each.

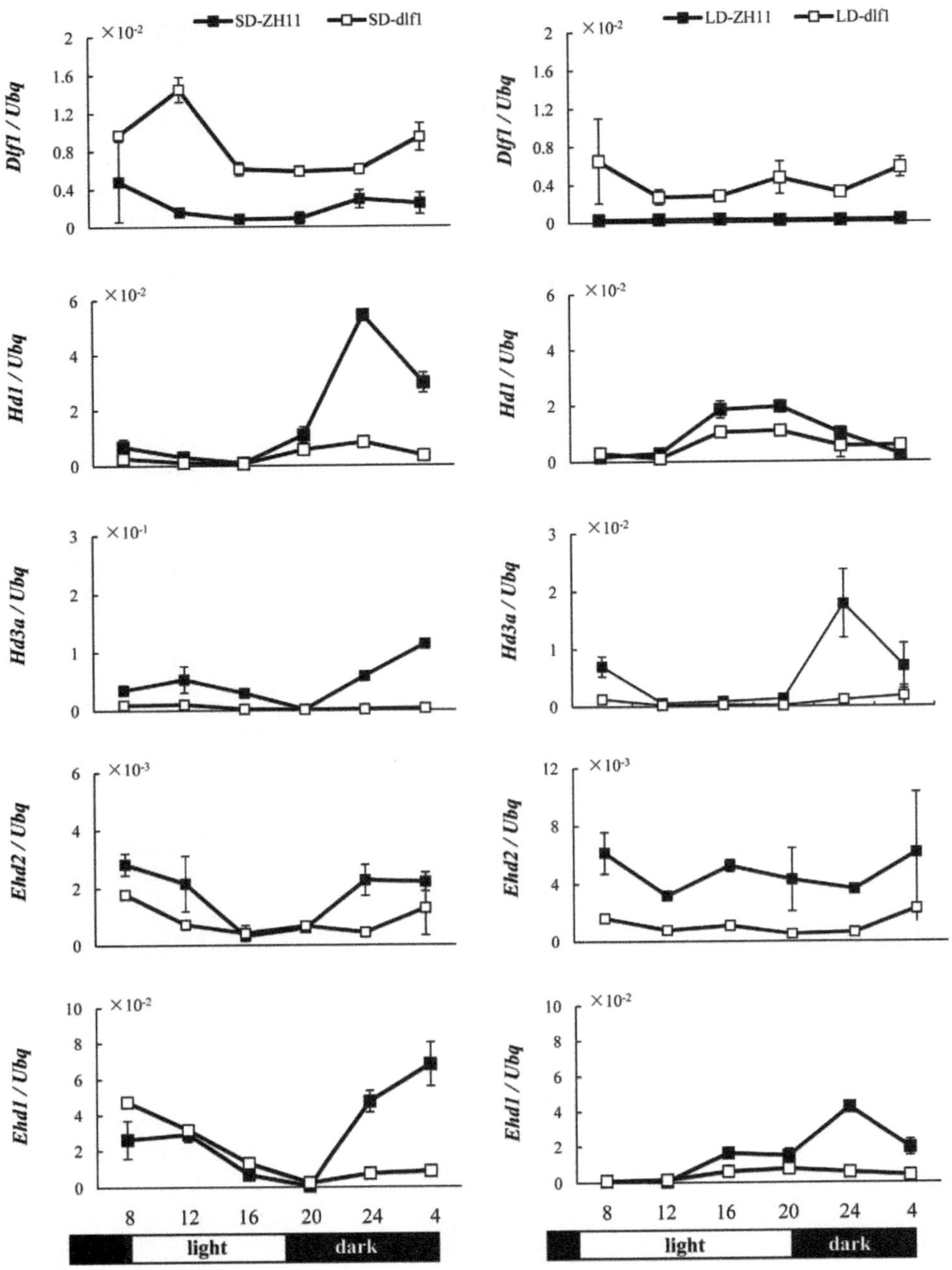

Figure 6. Dlf1 suppresses the expression of *Hd3a, Hd1, Ehd1* and *Ehd2*. Diurnal expression patterns of *Dlf1, Hd1, Hd3a, Ehd2* and *Ehd1* in wild-type ZH11 (filled squares) and *dlf1* (open squares) plants under SD (10/14 h light/dark) and LD (14/10 h light/dark) conditions by qPCR analysis. The expression levels are relative to the *ubiquitin* (*Ubq*) mRNA. The plants were grown at conditions as described in Fig. 1. Values are shown as means ± SD of two independent experiments. The open and filled bars at the bottom represent the light and dark periods, respectively.

and delayed flowering of about 2–3 weeks compared with ZH11 under natural LD conditions. The expression of total *OsWRKY11* in transgenic plants was much higher than in ZH11 control and the segregated non-transgenic plants (Fig. 9C), suggesting that the C-terminus of Dlfl plays a role in controlling its expression level.

Due to the promoter activity of *InR3* fragment, transgenic plants were also obtained with the constructs excluding the N-terminal 37 aa of Dlfl (assigned as *Ubi:d4W11.1* and *Ubi:d4W11.2*). The *Ubi:d4W11.2* transgenic lines (named as 4S-) presented dwarf and late flowering phenotypes under both LD and SD conditions (Figs. 9D and E), whereas the *Ubi:d4W11.1* plants had no such phenotype (data not shown). Analysis of the expression level of the transgenic plants revealed that the accumulation of *OsWRKY11* total mRNA was increased over 400-fold in the *Ubi:d4W11.2* progenies (Fig. 9F). These results suggest that the C-terminal region of Dlfl influenced its expression and high level of expression of truncated *OsWRKY11.2* might function in a similar manner as *OsWRKY11.2*.

To get a clue of the effect on *Dlf1* expression, we examined the possible degraded mRNA of *Dlf1* using the RNA ligase-mediated amplification of 5′ cDNA ends (RLM-RACE) [28]. A stronger PCR band was obtained in the *dlf1* mutant than the ZH11 (Fig. 9G). Sequencing verified the degraded positions at 1093 and 1099 bp of *Dlf1*. These observations support the idea that RNA

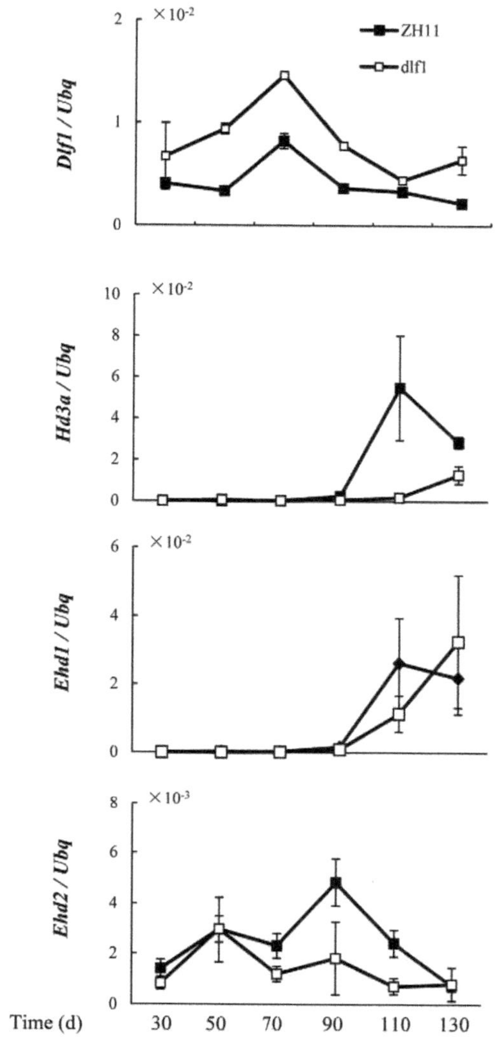

Figure 7. Expression of *Dlf1* and other flowering-time genes during development. Changes of *Dlf1*, *Hd3a*, *Ehd1* and *Ehd2* transcription levels in wild-type (filled squares) and *dlf1* (open squares) plants during development under LD conditions. The expression levels are relative to the *ubiquitin* (*Ubq*) mRNA. The values of *Dlf1* expression in ZH11 were scaled up 10 times. The plants were grown at conditions as described in Fig. 1. Developing leaves were harvested 4 h after dawn. Values are shown as means ± SD of two independent experiments.

processing proteins or microRNAs may regulate the expression of *Dlf1*, if the level of *Dlf1* mRNA containing the 3′ end reached a limit.

Discussion

In this study, we demonstrated that the *Dlf1* gene had pleiotropic effects on a variety of traits, including flowering time, plant height, grain number and leaf rolling. Yield, plant height and heading date are the most important agronomic traits in rice, and a number of genes have been isolated that control each of these traits. For example, *Gn1a*, a gene for cytokinin oxidase/dehydrogenase, regulates the number of grains per panicle [5], and a RING-type E3 ligase (*GW2*) controls grain width and weight [6]. The height of a rice plant is regulated by the gibberellin-insensitive gene *Dwarf 1*, encoding the alpha-subunit of GTP-binding protein, and the brassinosteroid biosynthesis gene *Dwarf 11* [1,3]. *Ghd7*, encoding a CCT-domain protein, was shown to have multiple effects on grain number, heading date and plant height [19]. *DTH8/Ghd8/Hd5*, encoding a HAP3 subunit of the CCAAT-box binding protein, is also reported to suppress rice flowering under LD conditions and regulate plant height and yield potential [23,24]. Our data indicate that the rice WRKY transcription factor Dlf1 also widely affects rice development. *Dlf1* regulates plant height by altering cell size in the internodes, similar to the effect of *DTH8* but different from that of *Ghd7* (Fig. S1) [19,23]. The phenotypes of short internode length and leaf rolling in the *dlf1* mutant are supported by a recent report of *OsWRKY11* transgenics, which is controlled by the promoter of heat shock-inducible *HSP101* gene [29]. Among the four transgenic rice plants reported, three had bent leaves or dwarf phenotype, and two had significantly enhanced heat and drought tolerance under heat induction conditions.

The T-DNA insertion site in the *dlf1* mutant was 67 bp upstream of the predicted translational starting site of *OsWRKY11* (Fig. 2A). However, the *Dlf1* expression was significantly increased in comparison with the wild-type plants. Transgenic plants harboring the genomic DNA of *Dlf1*, starting from the T-DNA insertion site to the end of *Dlf1* coding region, recapitulated the *dlf1* phenotypes (Fig. 2C), suggesting that the region has promoter function. This is confirmed by fusion with the *Gus* reporter gene (Fig. 3B, D and E). However, the Gus activities of *Cp-WP:Gus* and *Cp-InR3:Gus* were at a similar level, inconsistent with a higher level of expression *Dlf1* in the *dlf1* mutant than that of the wild-type, implying a suppressor element existed outside of the *WP* fragment used in the experiment. We also generated *Dlf1* overexpressing and RNAi transgenic lines. Unexpectedly, most of the *OsWRKY11.1* transgenic plants did not show morphological differences to controls (Fig. S4). Nevertheless, two RNAi lines with decrease in *OsWRKY11* expression showed early flowering under LD conditions (Fig. 8C and D). On the other hand, the accumulation of total *OsWRKY11* mRNA (including the endogenous and transferred gene) was extremely high in the lines harboring the *Ubi:W11.2* construct, which is 109 aa shorter than *Ubi:W11.1* in the C-terminus (Fig. S2). These results suggested that the C-terminus of *Dlf1* was involved in controlling accumulation level of its mRNA. This notion is further supported by comparison of the transgenic plants containing *Ubi:d4W11.1* or *Ubi:d4W11.2* constructs. Interestingly, most of the *Ubi:W11.2* and *Ubi:d4W11.2* transgenic plants exhibited dwarf and late flowering phenotypes, similar to the *dlf1* mutant (Fig. 9). Likely, OsWRKY11.2 retained the transactivation activity (Fig. 4) and the sequence of nuclear localization signal (position 179–187 aa). The results suggested that the high level of OsWRKY11.2, or its N-terminus-truncated protein might function as a negative regulator. This information also implies that different splicesomes of *Dlf1* might work together to regulate downstream gene expression, although further study is required to test the existence of alternative splicing *in planta*.

RNA processing plays an important role in control of plant flowering time. FLOWERING LOCUS C (FLC), a repressor of the transition to flowering in *Arabidopsis*, functions to delay flowering by down-regulating expression of genes promoting the floral state. Processing of *FLC* mRNA is regulated by autonomous pathway components of FCA and FY, which encode a pre-mRNA processing protein and a homolog of the yeast RNA 3′-processing factor Pfs2p, respectively [30,31]. *FCA* expression is also regulated through alternative processing of its pre-mRNA into four different transcripts, in which only the fully spliced *FCA* transcripts can promote flowering [30]. Furthermore, *FCA* negatively regulates its own expression by increasing cleavage and polyadenylation within

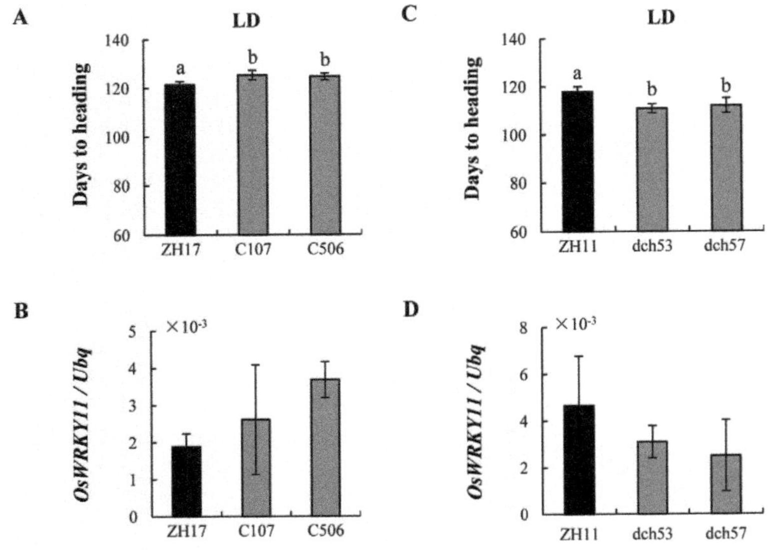

Figure 8. Alternation of *Dlf1* expression changes the heading day of rice. (**A**) The heading time of *Ubi:W11.1* transgenic lines in T_2 progenies (named as C-) and the ZH17 control under LD conditions. (**B**) Expression of total *OsWRKY11* (including the transferred and endogenous genes) in the *Ubi:W11.1* transgenic and ZH17 plants. (**C**) The heading day of ZH11 and the RNAi lines in T_2 progenies (named as dch-) under LD condition. (**D**) Expression of total *OsWRKY11* in the RNAi transgenic and ZH11 control plants. Leaves of the first and second youngest were sampled from 90-d old plants. Values are shown as means ± SD of two independent experiments. a and b in figure indicate ranking by Duncan test at $P<0.05$, starting from a. b is significantly different from a.

intron 3, thus limiting the production of active FCA protein to keep the balance of pathways controlling flowering time. They also found that active *FCA* can be overexpressed only when the *cis*-element within intron 3 required for the negative feedback is removed [32,33]. As mentioned above, the accumulation of *OsWRKY11.1* mRNA was not much increased in the overexpressing plants. An explanation is that the C-terminal part of OsWRKY11.1 interacts with a protein involved in regulation of the *OsWRKY11.1* mRNA level. When the OsWRKY11.1 protein reaches a threshold level it will activate the protein–protein interaction and decrease accumulation of the *OsWRKY11.1* transcript. This or another interaction might also possibly involve in the alternative splicing of *Dlf1*.

Rice is a facultative SD plant which flowers under LD conditions. As a counterpart of the GI–CO–FT signaling pathway in *Arabidopsis*, the rice orthologous proteins consist of OsGI–Hd1–Hd3a. The clock-associated protein OsGI upregulates *Hd1* expression and in turn Hd1 induces the expression of *Hd3a* during SD conditions in rice [9,12,13]. The expression of *Hd3a* is also induced by the Ehd1 activator, which is an evolutionarily unique gene in rice with no counterpart in *Arabidopsis* [15]. Ehd2/RID1/OsId1 was found to promote flowering under both SD and LD conditions by upregulating *Ehd1* expression [16–18]. Since *Ehd2/RID1/OsId1* expression was suppressed under both SD and LD conditions in the *dlf1* mutant (Fig. 6), the late flowering phenotype of the mutant is easily explained by loss of the promoting action of *Ehd2/RID1/OsId1* (Fig. 9).

Dlf1 was expressed in leaves, roots and panicles (Fig. 3A). Expression of *Dlf1* in leaves is consistent with the role of genes in flowering-time regulation, such as *Ehd2* and *Ghd7* [16,19]. *Ehd2/RID1/OsId1* is considered the master switch for the transition from vegetative to reproductive phase, a crucial process in higher plants. We found that increased *Dlf1* expression delayed the phase transition and initiation of floral induction, leading to late flowering in the *dlf1* mutant. This was further supported by studies of gene expression in the whole developmental process, which showed that *Ehd2/RID1/OsId1* is suppressed in the *dlf1* mutant with a higher level of *Dlf1* mRNA accumulation under LD conditions (Fig. 7). Diurnal expression of *Dlf1* was observed under both LD and SD conditions; however, the expression of photoperiod-related genes *OsGI* and *Se5* was not significantly changed between the mutant and wild-type plants (Fig. 6; Fig. S3), suggesting that *Dlf1* is unlikely to be upstream of these genes in the pathways of photoperiodic flowering in rice. WRKY proteins are a super family of plant transcription factors, which are characterized by binding specifically with W-box (a core sequence of TGAC). Genetic and molecular analyses have revealed that *WRKY* genes play roles in diverse biotic and abiotic stresses, as well as in development [34–36]. It is commonly observed that a single transcription factor may regulate multiple plant processes and that some may work in a redundant manner. AtWRKY6 is reported to be associated with senescence- and defense-related processes, and was shown to respond to low Pi stress in *Arabidopsis* [37,38]. Recently, Roboni et al. [39] have shown that GI plays a key role in photoperiodic cues and drought escape response via ABA-dependent activation of florigens and SOC1. The *gi* mutants exhibit hyper tolerance to oxidative stress, enhanced salt tolerance, and elevated starch content, highlighting the importance of carbohydrate metabolism in the regulation of flowering [40–43]. Meanwhile, SOC1 and FLC are mediated in crosstalk between cold response and flowering time regulation [44].

Our data combined with others indicate that Dlf1 play important roles on plant height, heading date, yield potential and responses to abiotic stress [29]. The CCT-domain protein Ghd7 and DTH8 protein have been demonstrated to have pleiotropic effects on heading time, height and yield potential [19,23]. Further investigation of their relationship should help illuminate the complexities of these important agronomic traits, as well as aid in manipulation of the traits for yield improvement.

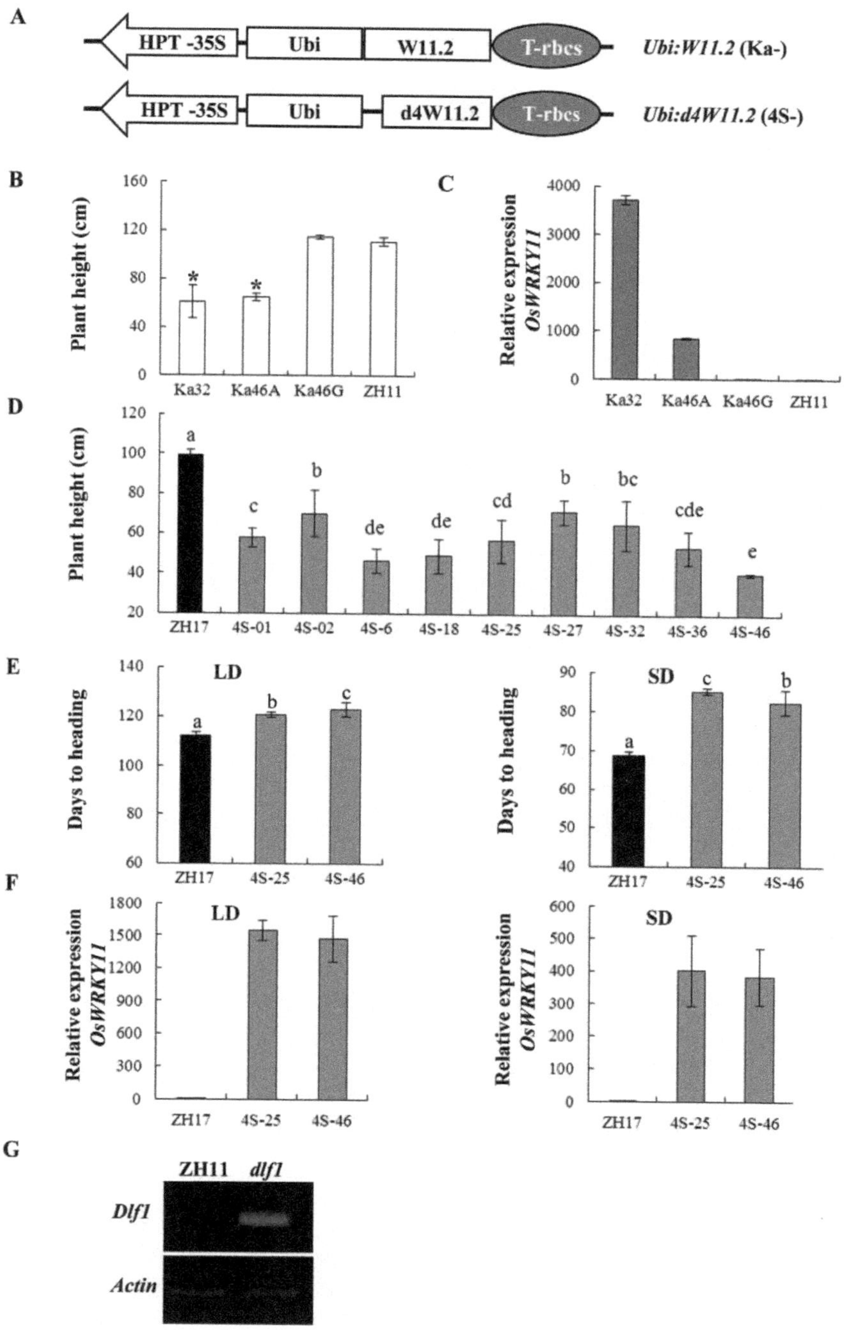

Figure 9. High level expression of the *OsWRKY11.2* leads to dwarfism and late flowering. (**A**) Schematic diagram of *Ubi:W11.2* (Ka-) and *Ubi:d4W11.2* (4S-, with 37-aa deletion at the N-terminus of W11.2) constructs. (**B**) and (**D**) Plant heights of those transformed with *Ubi:W11.2* or *Ubi:d4W11.2* in ZH11 or ZH17 genetic background, respectively. (**C**) Expression of total *OsWRKY11* (including the transferred and endogenous genes) in ZH11 and the *Ubi:W11.2* transgenic lines of T1 progenies under natural LD conditions. The first and second youngest leaves were sampled from 90-d-old plants for RNA isolation. Transcription levels were quantified by qPCR and the gene expression was normalized to rice *ubiquitin* gene (*Ubq*) for each sample. Transcription levels are expressed as ratio to the level of transcript in ZH11. The suffix A for dwarf and G for segregated non-transgenic plants. (**E**) Days to heading of the *Ubi:d4W11.2* plants of T_2 progenies under SD and LD conditions (the same treatments as in Fig. 6). (**F**) Expression of total *OsWRKY11* in ZH17 and the *Ubi:d4W11.2* transgenic plants of T_2 progenies under both LD and SD conditions (the same treatments as in Fig. 6). Transcription levels are expressed as ratio to the level of transcript in ZH17. (**G**) Analysis of the possible degraded mRNA of *Dlf1* using the RNA ligase-mediated amplification of 5′ cDNA ends (RLM-RACE). Total RNAs were isolated from the *dlf1* mutant (*dlf1*) and the wild-type (ZH11) of three-week-old plants. Two rounds of PCR were performed: 1) by using an outer primer from the adaptor and a *Dlf1*-specific reverse primer W5; and 2) by using the inner primer from the adaptor and a *Dlf1* gene primer W4 for 30 cycles. *Actin* gene was used as an internal standard. Asterisks indicate significant difference between ZH11 and the overexpression lines ($P<0.05$, Duncan test). a, b, c, d, and e indicate ranking by Duncan test at $P<0.05$, starting from a. Different letters indicate significantly difference from each other.

Materials and Methods

Plant materials and growth conditions

Rice seeds of the wild-type (*Oryza sativa* L. *japonica*, Zhonghua 11 or Zhonghua 17, ZH11 or ZH17), mutant and transgenic progenies were sterilized and germinated on half-strength Murashige and Skoog medium for 7 d and then transferred to soil in a greenhouse. For flowering time measurements, plants were grown either in 10/14 h light/dark for SD or 14/10 h light/dark for LD. Rice flowering time was measured in days from germination until emergence of the first panicle. For diurnal expression analyses, young leaves were harvested from wild-type ZH11 and the *dlf1* mutant of 40-d-old (SD) or 90-d-old (LD) plants at 4-h intervals for a total of 24 h. To analyze gene expression during development, the first and second youngest leaves from three plants were harvested from 30, 50, 70, 90, 110 and 130-d-old plants at 4 h after dawn under LD condition. The rice plants examined under natural field conditions were sown at late April and transplanted at early June in the experimental field of China Agricultural University, Beijing (39°54′N, 116°23′E), China.

Genotyping of mutant plants and Tail-PCR

For genotyping analysis, the *dlf1* segregating population was assayed by PCR using the primers of PddSalF, TR1 and W6 (Table S1). The primer pair of TR1 and W6 was used for amplification of the T-DNA insertion. Thermal asymmetric interlaced PCR (Tail-PCR) method was used to isolate genomic fragment flanking of the T-DNA insertion site from the *dlf1* mutant plant. The primers TR1, TR2, TR3, AD1, AD2 and AD3 are shown in Table S1.

Vector construction and transformation

The full-length coding region of *Dlf1* was obtained by PCR amplification using the primers W1 and W5, along with a shorter product (assigned as *OsWRKY11.2*). The *OsWRKY11.1* and *OsWRKY11.2* cDNAs were put under the control of a maize ubiquitin promoter in a modified *pCambia 1301* vector to generate *Ubi:W11.1* and *Ubi:W11.2* for overexpression [34]. Similarly, the PCR products, amplified with the primer pairs of d40BIF/W10H3r and d40BIF/W10SH3r were used to construct *Ubi:d4W11.1* and *Ubi:d4W11.2* vectors, with deletions of the 5′-ends in comparison with *OsWRKY11.1* and *OsWRKY11.2*, respectively. Each overexpressing construct contained a Flag tag in the 5′-end of the gene. The *Dlf1* fragment of 276 bp (from −39 to +237 bp) was used to generate the *Dlf1*-RNAi plasmid using procedures similar to the previous description [34]. The hairpin structure was put under the control of the CaMV35S promoter (*Cam35S:dsW11*). For promoter constructs, the PCR products were fused to *β-glucuronidase* (*Gus*) reporter gene as following: the *Dlf1* promoter from −2010 to +134 bp as *Cp-WP:Gus*, −67 – + 134 bp as *Cp-InR3:Gus*, and −67 – +5 bp as *Cp-In2R:Gus*. For complementation, the genomic DNA fragment from −67 to the end of *Dlf1* was put into a modified *pCambia 1301* vector, designed as *Cp-Ins-Dlf*. All vectors were verified by sequencing and transformed into rice through *Agrobacterium*-mediated transformation [35]. The transgenic lines obtained were determined by PCR amplifications.

Transactivation activity assay

The coding region of *Dlf1* was amplified with the primers W10EI and W10BgSal (Table S3). The PCR product was fused to the GAL4 DNA binding-domain vector *pGBKT7* (*pBD*, Clotech) to generate the plasmid *pBD–Dlf1*. Similarly, a fragment of *Dlf1* encoding amino acids 91–379 (*pBD–dN1*), 121–379 (*pBD–dN2*), 204–379 (*pBD–dN3*), 91–313 (*pBD–dN1C1*), the first 313 amino acids (*pBD–dC1*), the first 266 amino acids (*pBD–dC2*), or OsWRKY11.2 (*pBD-WRKY11.2*) were fused to the GAL4 DNA binding-domain. These constructs or empty vector *pBD* were individually transformed into yeast cells of AH109 strain and grown on SD–Trp selective medium at 30°C for 3 d. An assay of β-galactosidase activity was performed with transformed cell lines grown in liquid SD-Trp medium using *o* -nitrophenyl β-D-galactopyranoside as a substrate, according to the manufacturer's protocol.

Subcellular localization of Dlf1

The coding sequence of *Dlf1* was amplified and fused in frame to the upstream of green fluorescent protein (*GFP*) gene to generate the *CamUbi:Dlf1-GFP* construct. The resultant and the control *CamUbi:GFP* vectors were transformed into onion (*Allium cepa*) inner epidermal cells by bombardment using the PDS-1000/He system (Bio-Rad) with DNA-coated gold particles. The transformed cells were cultured on 1/2 MS medium at 28°C for 2 d and observed under a confocal microscope (Bio-Rad MRC 1024).

Gus assay and histochemical staining

Gus activity assay and histochemical staining were performed as described [45] and photographed using a Nikon SMZ 1000 stereoscope with a Nikon digital sight DS-SM camera or Nikon camera.

RNA gel-blot

Trizol reagent was used to extract the RNA from rice tissues. Total RNA was fractionated in 1.5% agarose containing formaldehyde, blotted onto Hybond-N$^+$ nylon membrane and probed with the coding sequence of *Dlf1*. Hybridization was performed as previously described [34] and the membrane was autoradiographed by using a phosphoimaging system (Amersham Pharmacia Biotech, UK).

Real-time quantitative RT-PCR

Total RNA was isolated from different rice tissues using Trizol reagent following the manufacturer's procedures. DNaseI-treated RNAs were reverse transcribed with oligo (dT) and random primers. Real-time quantitative RT-PCR (qPCR) was performed in a final volume of 20 μL, including 10 μL SYBR Premix EX Taq (Takara), 2 μL of the diluted first-strand cDNA as templates and 0.2 μM of each primer. The reactions were carried out with an ABI PRISM 7000 or ABI StepOne in the following program: 95°C for 2 min, 40 cycles of 95°C for 5 s, 60°C for 20 s, and 72°C for 31 s. Every experiment was repeated more than twice. The primers of *Ehd2, Hd1, Ehd1, Hd3a, Ubq, Ghd7, OsGI, Se5 FTL6* and *Dlf1* are listed in Table S2. The level of *ubiquitin* (*Ubq*) expression was used to normalize the expression ratio of each gene.

RLM-RACE PCR

The RNA ligase-mediated amplification of 5′ cDNA ends (RLM-RACE) was performed according to the manufacturer's manual with modifications. Two μg total RNA was directly ligated to a RNA adaptor using T4 RNA ligase and transcribed by random primers. PCR was performed by using an outer primer from the adaptor and a *Dlf1*-specific reverse primer W5 at conditions of 95°C for 3 min (1 cycle), 95°C for 30 s, 55°C for 30 s, 72°C for 1 min (20 cycles), then 72°C for 10 min (1 cycle). A second round of nested PCR was amplified by using the inner primer and a *Dlf1* gene primer W4 for 30 cycles. *Actin* gene was used as an internal standard. The RACE PCR products were sequenced to identify the mRNA degraded ends.

Supporting Information

Figure S1 Phenotypes of Zhonghua 11 (ZH11) and *dlf1* mutant. (**A**) Longitudinal section of the stems approximately 2 cm above the upper-most nodes from the tiller culms of plants. (**B**) Number of spikelets per panicle. Values are means ± SD, n = 20. (**C**) 1000-grain weight. Values are means ± SD, n = 10. (**D**) Leaf rolling. (**E**) Transverse sections of the middle part of the first leaf from tillering plants. The plants were grown in the experimental field under natural LD conditions.

Figure S2 Multiple sequence alignment of *OsWRKY11* cDNAs and proteins. (**A**) The full-length cDNA of *Dlf1* (*OsWRKY11.1*) and the alternatively spliced transcript *OsWRKY11.2* were aligned using the CLUSTAL W program. (B) The alignment of Dlf1 and OsWRKY11.2 proteins using CLUSTAL W program.

Figure S3 Expression of *OsGI, Ghd7, Se5*, and *FTL6*. Diurnal expression patterns of *OsGI, Ghd7, Se5*, and *FTL6* in the ZH11 control (filled circle) and the mutant *dlf1* (open circle) plants under SD (10 h light/14 h dark) and LD (14 h light/10 h dark) conditions by qPCR analysis. The expression levels are relative to the *ubiquitin* (*Ubq*) mRNA. Values are shown as means ± SD of two independent experiments. The open and filled bars at the bottom represent the light and dark periods, respectively.

Figure S4 Heights of *Ubi:W11.1* transgenic plants. (**A**) Schematic diagram of *Ubi:W11.1* construct (OE-). (**B**) Plant heights of some *Ubi:W11.1* lines in T_2 progenies and the ZH11 control. Values are means ± SD.

Acknowledgments

The authors thank Professors Yiji Xia (Hong Kong Baptist University) and Jun Fan (China Agricultural University) for critical reading of the manuscript.

Author Contributions

Conceived and designed the experiments: YC XC ZG. Performed the experiments: YC XC KX QX YW JL CD. Analyzed the data: YC XC ZG. Contributed reagents/materials/analysis tools: ZS. Contributed to the writing of the manuscript: YC XC ZG.

References

1. Ashikari M, Wu J, Yano M, Sasaki T, Yoshimura A (1999) Rice gibberellin-insensitive dwarf mutant gene *Dwarf 1* encodes the α-subunit of GTP-binding protein. Proc Natl Acad Sci USA 96: 10284–10289.
2. Itoh H, Tatsumi T, Sakamoto T, Otomo K, Toyomasu T, et al. (2004) A rice semi-dwarf gene, Tan-Ginbozu (D35), encodes the gibberellin biosynthesis enzyme, *ent*-kaurene oxidase. Plant Mol Biol 54: 533–547.
3. Tanabe S, Ashikari M, Fujioka S, Takatsuto S, Yoshida S, et al. (2005) A novel cytochrome P450 is implicated in brassinosteroid biosynthesis via the characterization of a rice dwarf mutant, *dwarf11*, with reduced seed length. Plant Cell 17: 776–790.
4. Li X, Qian Q, Fu Z, Wang Y, Xiong G, et al. (2003) Control of tillering in rice. Nature 422: 618–621.
5. Ashikari M, Sakakibara H, Lin S, Yamamoto T, Takashi T, et al. (2005) Cytokinin oxidase regulates rice grain production. Science 309: 741–745.
6. Song XJ, Huang W, Shi M, Zhu MZ, Lin HX (2007) A QTL for rice grain width and weight encodes a previously unknown RING-type E3 ubiquitin ligase. Nat Genet 39: 623–630.
7. Corbesier L, Vincent C, Jang S, Fornara F, Fan Q, et al. (2007) FT protein movement contributes to long-distance signaling in floral induction of *Arabidopsis*. Sci Signaling 316: 1030–1036.
8. Kobayashi Y, Weigel D (2007) Move on up, it's time for change-mobile signals controlling photoperiod-dependent flowering. Genes Dev 21: 2371–2384.
9. Tsuji H, Taoka KI, Shimamoto K (2011) Regulation of flowering in rice: two florigen genes, a complex gene network, and natural variation. Curr Opin Plant Biol 14: 45–52.
10. Lee J, Lee I (2010) Regulation and function of SOC1, a flowering pathway integrator. J Exp Bot 61: 2247–2254.
11. Yano M, Katayose Y, Ashikari M, Yamanouchi U, Monna L, et al. (2000) Hd1, a major photoperiod sensitivity quantitative trait locus in rice, is closely related to the *Arabidopsis* flowering time gene CONSTANS. Plant Cell 12: 2473–2483.
12. Izawa T, Oikawa T, Sugiyama N, Tanisaka T, Yano M, et al. (2002) Phytochrome mediates the external light signal to repress FT orthologs in photoperiodic flowering of rice. Genes Dev. 16: 2006–2020.
13. Hayama R, Yokoi S, Tamaki S, Yano M, Shimamoto K (2003) Adaptation of photoperiodic control pathways produces short-day flowering in rice. Nature 422: 719–722.
14. Hayama R, Coupland G (2004) The molecular basis of diversity in the photoperiodic flowering responses of *Arabidopsis* and rice. Plant Physiol 135: 677–684.
15. Doi K, Izawa T, Fuse T, Yamanouchi U, Kubo T, et al. (2004) Ehd1, a B-type response regulator in rice, confers short-day promotion of flowering and controls FT-like gene expression independently of *Hd1*. Genes Dev 18: 926–936.
16. Matsubara K, Yamanouchi U, Wang ZX, Minobe Y, Izawa T, et al. (2008) *Ehd2*, a rice ortholog of the maize *INDETERMINATE1* gene, promotes flowering by up-regulating *Ehd1*. Plant Physiol 148: 1425–1435.
17. Park SJ, Kim SL, Lee S, Je BI, Piao HL, et al. (2008) *Rice Indeterminate 1* (*OsId1*) is necessary for the expression of *Ehd1* (*Early heading date 1*) regardless of photoperiod. Plant J 56: 1018–1029.
18. Wu C, You C, Li C, Long T, Chen G, et al. (2008) *RID1*, encoding a Cys2/His2-type zinc finger transcription factor, acts as a master switch from vegetative to floral development in rice. Proc Natl Acad Sci USA 105: 12915–12920.
19. Xue W, Xing Y, Weng X, Zhao Y, Tang W, et al. (2008) Natural variation in *Ghd7* is an important regulator of heading date and yield potential in rice. Nat Genet 40: 761–767.
20. Matsubara K, Ogiso-Tanaka E, Hori K, Ebana K, Ando T, et al. (2012) Natural variation in Hd17, a homolog of *Arabidopsis* ELF3 that is involved in rice photoperiodic flowering. Plant Cell Physiol 53: 709–716.
21. Saito H, Ogiso-Tanaka E, Okumoto Y, Yoshitake Y, Izumi H, et al. (2012) Ef7 encodes an ELF3-like protein and promotes rice flowering by negatively regulating the floral repressor gene Ghd7 under both short- and long-day conditions. Plant Cell Physiol 53: 717–728.
22. Matsubara K, Yamanouchi U, Nonoue Y, Sugimoto K, Wang ZX, et al. (2011) *Ehd3*, encoding a plant homeodomain finger-containing protein, is a critical promoter of rice flowering. Plant J 66: 603–612.
23. Wei X, Xu J, Guo H, Jiang L, Chen S, et al. (2010) DTH8 suppresses flowering in rice, influencing plant height and yield potential simultaneously. Plant Physiol 153: 1747–1758.
24. Yan WH, Wang P, Chen HX, Zhou HJ, Li QP, et al. (2011) A major QTL, Ghd8, plays pleiotropic roles in regulating grain productivity, plant height, and heading date in rice. Mol Plant 4: 319–330.
25. Fujino K, Utako Yamanouchi U, Yano M (2013) Roles of the *Hd5* gene controlling heading date for adaptation to the northern limits of rice cultivation. Theor Appl Gen 126: 611–618.
26. Kim SL, Lee S, Kim HJ, Nam HG, An G (2007) OsMADS51 is a short-day flowering promoter that functions upstream of Ehd1, OsMADS14, and Hd3a. Plant Physiol 145: 1484–1494.
27. Gao H, Zheng XM, Fei G, Chen J, Jin M, et al. (2013) *Ehd4* encodes a novel and *Oryza*-genus-specific regulator of photoperiodic flowering in rice. PLoS Genet 9: e1003281.
28. Franco-Zorrilla JM, Del Toro FJ, Godoy M, Pérez-Pérez J, López-Vidriero I, et al. (2009) Genome-wide identification of small RNA targets based on target enrichment and microarray hybridizations. Plant J 59: 840–850.

29. Wu X, Shiroto Y, Kishitani S, Ito Y, Toriyama K (2009) Enhanced heat and drought tolerance in transgenic rice seedlings overexpressing *OsWRKY11* under the control of *HSP101* promoter. Plant Cell Rep 28: 21–30.
30. Quesada V, Macknight R, Dean C, Simpson GG (2003) Autoregulation of FCA pre-mRNA processing controls *Arabidopsis* flowering time. EMBO J 22: 3142–3152.
31. Simpson GG, Dijkwel PP, Quesada V, Henderson I, Dean C (2003) FY is an RNA 3′ end-processing factor that interacts with FCA to control the *Arabidopsis* floral transition. Cell 113: 777–787.
32. Macknight R, Bancroft I, Page T, Lister C, Schmidt R, et al. (1997) FCA, a gene controlling flowering time in *Arabidopsis*, encodes a protein containing RNA-binding domains. Cell 89: 737–745.
33. Macknight R, Duroux M, Laurie R, Dijkwel P, Simpson G, et al. (2002) Functional significance of the alternative transcript processing of the *Arabidopsis* floral promoter FCA. Plant Cell 14: 877–888.
34. Wang H, Hao J, Chen X, Hao Z, Wang X, et al. (2007) Overexpression of rice *WRKY89* enhances ultraviolet B tolerance and disease resistance in rice plants. Plant Mol Biol 65: 799–815.
35. Zhang J, Peng Y, Guo Z (2008) Constitutive expression of pathogen-inducible *OsWRKY31* enhances disease resistance and affects root growth and auxin response in transgenic rice plants. Cell Res 18: 508–521.
36. Pandey SP, Somssich IE (2009) The role of WRKY transcription factors in plant immunity. Plant Physiol 150: 1648–1655.
37. Robatzek S, Somssich IE (2001) A new member of the *Arabidopsis* WRKY transcription factor family, AtWRKY6, is associated with both senescence- and defence- related processes. Plant J 28: 123–133.
38. Chen YF, Li LQ, Xu Q, Kong YH, Wang H, et al. (2009) The WRKY6 transcription factor modulates PHOSPHATE1 expression in response to low Pi stress in *Arabidopsis*. Plant Cell 21: 3554–3566.
39. Riboni M, Galbiati M, Tonelli C, Conti L (2013) GIGANTEA enables drought escape response via abscisic acid-dependent activation of the florigens and suppressor of overexpression of CONSTANS1. Plant Physiol 162: 1706–1719.
40. Eimert K, Wang SM, Lue W, Chen J (1995) Monogenic recessive mutations causing both late floral initiation and excess starch accumulation in *Arabidopsis*. Plant Cell 7: 1703–1712.
41. Kurepa J, Smalle J, Van Montagu M, Inzé D (1998) Effects of sucrose supply on growth and paraquat tolerance of the late-flowering *gi-3* mutant. Plant Growth Regul 26: 91–96.
42. Kim WY, Ali Z, Park HJ, Park SJ, Cha JY, et al. (2013) Release of SOS2 kinase from sequestration with GIGANTEA determines salt tolerance in *Arabidopsis*. Nature Comm 4: 1352–1364.
43. Wahl V, Ponnu J, Schlereth A, Arrivault S, Langenecker T, et al. (2013) Regulation of flowering by trehalose-6- phosphate signaling in *Arabidopsis thaliana*. Science 339: 704–707.
44. Seo E, Lee H, Jeon J, Park H, Kim J, et al. (2009) Crosstalk between cold response and flowering in *Arabidopsis* is mediated through the flowering-time gene *SOC1* and its upstream negative regulator *FLC*. Plant Cell 21: 3185–3197.
45. Jefferson RA, Kavanagh TA, Bevan MW (1987) GUS fusions: beta-glucuronidaseas a sensitive and versatile gene fusion marker in higher plants. EMBO J 6: 3901–3907.

Low Levels of the Reverse Transactivator Fail to Induce Target Transgene Expression in Vascular Smooth Muscle Cells

Nikenza Viceconte, Tomás McKenna, Maria Eriksson*

Department of Biosciences and Nutrition, Center for Biosciences, Karolinska Institutet, Novum, Huddinge, Sweden

Abstract

Hutchinson-Gilford progeria syndrome (HGPS) is a genetic disease with multiple features that are suggestive of premature aging. Most patients with HGPS carry a mutation on one of their copies of the *LMNA* gene. The *LMNA* gene encodes the lamin A and lamin C proteins, which are the major proteins of the nuclear lamina. The organs of the cardiovascular system are amongst those that are most severely affected in HGPS, undergoing a progressive depletion of vascular smooth muscle cells, and most children with HGPS die in their early teens from cardio-vascular disease and other complications from atherosclerosis. In this study, we developed a transgenic mouse model based on the tet-ON system to increase the understanding of the molecular mechanisms leading to the most lethal aspect of HGPS. To induce the expression of the most common HGPS mutation, *LMNA* c.1824C>T; p.G608G, in the vascular smooth muscle cells of the aortic arch and thoracic aorta, we used the previously described reverse tetracycline-controlled transactivator, sm22α-rtTA. However, the expression of the reverse sm22α-transactivator was barely detectable in the arteries, and this low level of expression was not sufficient to induce the expression of the target human lamin A minigene. The results from this study are important because they suggest caution during the use of previously functional transgenic animal models and emphasize the importance of assessing transgene expression over time.

Editor: Carlos E. Ambrósio, Faculty of Animal Sciences and Food Engineering, University of São Paulo, Pirassununga, SP, Brazil, Brazil

Funding: This study was in part performed at the Live Cell Imaging unit, Department of Biosciences and Nutrition, Karolinska Institutet, Huddinge, Sweden, supported by grants from the Knut and Alice Wallenberg Foundation, the Swedish Research Council and the Center for Biosciences. This research was supported by grants to ME from the Swedish Research Council, the OE & Edla Johansson foundation, the in memory of Lars Hierta foundation, the Svenska Läkare Sällskapet, and the Åke Wiberg foundation. The funders had no role in study design, data collection and analysis, decision to publish, or preparation of the manuscript.

Competing Interests: The authors have declared that no competing interests exist.

* Email: Maria.Eriksson.2@ki.se

Introduction

Hutchinson-Gilford progeria syndrome (HGPS) is a very rare genetic disorder characterized by multiple features and pathologies typical of advanced age. Initial symptoms start to appear approximately one year after birth and include postnatal growth retardation, decreased subcutaneous fat, alopecia and failure to thrive [1,2]. The disorder is usually caused by a *de novo* point mutation c.1824C>T, p.G608G in exon 11 of the *LMNA* gene. The *LMNA* gene encodes the lamin A and lamin C proteins, which are the major proteins of the nuclear lamina located beneath the inner nuclear membrane [3].

The *LMNA* c.1824C>T, p.G608G mutation and other less-prevalent mutations, including the *LMNA* c.1822G>A, p.G608S mutation [3,4], result in the partial activation of a cryptic splice site and in the removal of the 150 carboxy-terminal nucleotides of exon 11 [3,4]. This internal deletion leads to the expression of a truncated lamin A protein with an internal deletion of 50 amino acids, called progerin [3,5,6]. Since none of the mutations create a sequence with a perfect match to a splice site the full-length lamin A protein will also be expressed from the mutant allele. The accumulation of progerin is thought to affect nuclear structure and function, which in turn leads to the process of premature aging [5,6]. Previous studies have shown that the cardiovascular phenotype in HGPS involves the progressive medial loss of vascular smooth muscle cells and replacement with proteoglycans and collagens [7,8]. Further studies should be performed to explore the molecular dysfunction and activation of signaling pathways caused by progerin accumulation, particularly in the cardiovascular system, because such accumulation leads to death from myocardial infarction and/or stroke, usually during the second decade of life [1,9].

Several mouse models have been developed to explore the functions of the lamin gene in health and disease [10]. One of the most commonly used inducible protein expression systems is regulated by tetracycline and its derivatives [11]. This system has regulatory and responsive element components. The regulatory component consists of one of the transactivators, tTA or rtTA, that is associated with a tissue specific promoter. The responsive element contains the target gene downstream of the *tetO* sequence (tetop). Various Tet-regulated expression systems have been developed to enable the induction or repression of the transcription of a target gene in response to tetracycline, Tet, or a derivative of tetracycline, usually doxycycline. These tet-ON or tet-OFF expression systems allow for the control of target gene expression in a specific tissue at different time points and are reversible [11,12]. Several mouse models have been developed to study the underlying causes of HGPS [10], and two animal models are

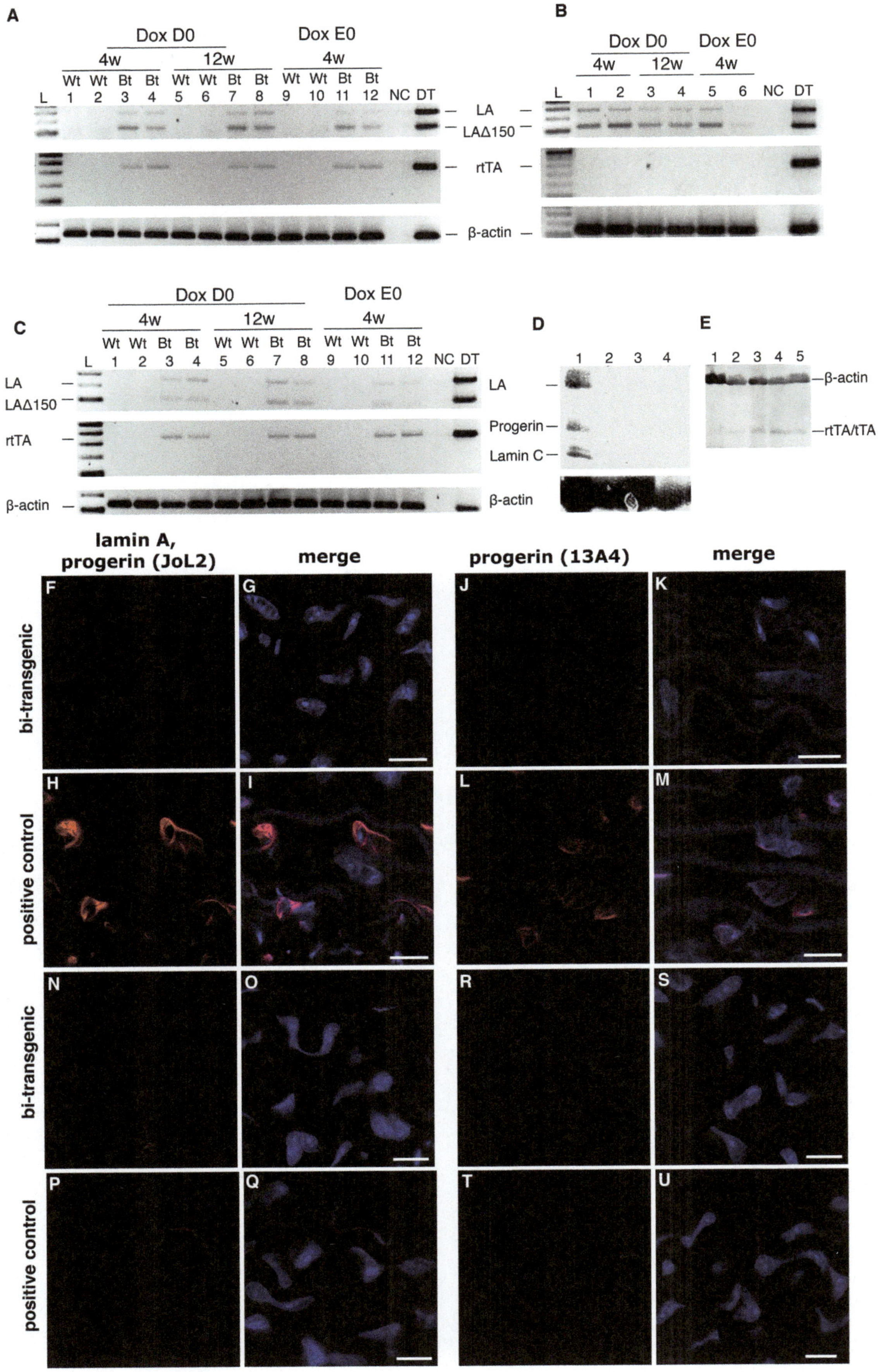
A
Dox D0
Dox E0
4w
12w
4w
Wt Wt Bt Bt Wt Wt Bt Bt Wt Wt Bt Bt
L 1 2 3 4 5 6 7 8 9 10 11 12 NC DT
LA
LAΔ150
rtTA
β-actin
B
Dox D0
Dox E0
4w
12w
4w
L 1 2 3 4 5 6 NC DT
C
Dox D0
Dox E0
4w
12w
4w
Wt Wt Bt Bt Wt Wt Bt Bt Wt Wt Bt Bt
L 1 2 3 4 5 6 7 8 9 10 11 12 NC DT
LA
LAΔ150
rtTA
β-actin
D
1 2 3 4
LA
Progerin
Lamin C
β-actin
E
1 2 3 4 5
β-actin
rtTA/tTA
lamin A, progerin (JoL2)
merge
progerin (13A4)
merge
bi-transgenic
positive control
bi-transgenic
positive control
F
G
J
K
H
I
L
M
N
O
R
S
P
Q
T
U

Figure 1. Low levels of transgene expression in the aortic arch. (A–C) RT-PCR using mRNA from the aortic arch showed very weak amplification products for human lamin A, lamin Adel150 and the reverse transactivator after 35 cycles of PCR. (A, C) RT-PCR on samples from bi-transgenic mice encoding both the reverse transactivator (sm22α-rtTA^{+}) and the lamin A minigene (tetop-LA^{G608G+}) that were supplied with doxycycline from the date of birth until postnatal week 4 (Dox D0, lanes 3–4) or 12 (Dox D0, lanes 7–8), or supplied with dox during embryogenesis and postnatally for 4 weeks (Dox E0, lanes 11–12). (A) C57BL/6J genetic background. (C) C57BL/6J; FVB/NCrl mixed genetic background. (B) RT-PCR for human lamin A and lamin Adel150 in samples from transactivator negative control animals (tetop-LA^{G608G+}; sm22α-rtTA^{-}, lanes 1–6) supplied with doxycycline from the date of birth until postnatal week 4 (Dox D0, lanes 1 and 2, C57BL/6J and C57BL/6J; FVB/NCrl genetic background, respectively) or week 12 (Dox D0, lanes 3 and 4, C57BL/6J and C57BL/6J; FVB/NCrl genetic background, respectively), or that were supplied with doxycycline during embryogenesis and postnatally for 4 weeks (Dox E0, lanes 5 and 6, C57BL/6J and C57BL/6J; FVB/NCrl genetic background, respectively). Bt, bi-transgenic. Wt, wild-type. NC, control with no template. DT, sample from a different transactivator strain was used as a control for the PCR assay and showed amplification for human lamin A and lamin Adel150 with cDNA from the bone of bi-transgenic tetop-LA^{G608G+}; Sp7-tTA^{+} mice [18]. Genomic DNA from a tetop-LA^{G608G+}; sm22α-rtTA^{+} bi-transgenic animal was used as a positive control for the amplification of the reverse transactivator (350 base pair product) [15]. The RT-PCR results for β-actin served as a control. (D) Western blot analysis on protein extracts from pooled aortic regions from bi-transgenic tetop-LA^{G608G+}; sm22α-rtTA^{+} (lane 2) and tetop-LA^{G608G+}; NSE-tTA^{+} (lane 3) animals did not show transgenic expression of human lamin A and progerin. A single transgenic tetop-LA^{G608G-}; sm22α-rtTA^{+} animal carrying only the transactivator was used as a negative control (lane 4). Protein extract from HGPS patient cell line AG11513A was used as positive control (lane 1). (E) Western blot analysis on protein extracts from pooled aortic regions from bi-transgenic tetop-LA^{G608G+}; sm22α-rtTA^{+} (lanes 2 and 5) and single transgenic animals, tetop-LA^{G608G-}; NSE-tTA^{+} (lane 3) and tetop-LA^{G608G-}; sm22α-rtTA^{+} (lane 4). Protein extracts from wild-type tissue, tetop-LA^{G608G-}; NSE-tTA^{-}, was used as a negative control (lane 1). (F–G and J–K) Very few transgene positive cells, <1%, were detected in the aortic arches of bi-transgenic animals at postnatal week 12 using antibodies specific for human lamin A/C and progerin (JoL2) (F–G), and human progerin (13A4) (J–K). (H–I and L–M) Positive staining was obtained using the same antibodies, on sections of the aortic arch from tetop-LA^{G608G+}; NSE-tTA+ bi-transgenic mice [17]. (N–O and R–S) Very few transgene protein positive cells, <1%, were detected in the aortic arches of adult bi-transgenic animals, not supplied with doxycycline for the last 4 weeks prior to sacrifice, using antibodies for human lamin A/C and progerin (JoL2) (N–O), and human progerin (13A4) (R–S). (P–Q and T–U) Almost no positive staining was obtained using the same antibodies, on sections of the aortic arch from tetop-LA^{G608G+}; NSE-tTA+ bi-transgenic mice supplied with doxycycline for 3 weeks (indicating a significant down-regulation of the transgenic expression with the doxycycline supplement). G, I, O, Q: merge of the transgenic lamin A and progerin with DAPI fluorescence signals. K, M, S, U: merge of the progerin and DAPI fluorescence signals. Scale bars: 10 μm.

currently available to study the cardiovascular pathology of HGPS [8,13]; however, none of these models is inducible or tissue-specific. Our goal was to induce the pathology of HGPS in arteries by developing an inducible, tissue-specific expression system based on the tet-ON system [11], under the control of the vascular smooth muscle cell specific promoter (sm22α) that is expressed during mouse embryogenesis (from embryonic day 9.5) and continuously active in postnatal vascular smooth muscle cells [14–15]. Here, we report one of the problems with the tet-ON system and emphasize the need to assess the expression of transgenes over time.

Materials and Methods

Ethics Statement

This study was performed in accordance with the guidelines for working with experimental animals set by the Karolinska Institute and all efforts were made to minimize animal suffering. All animal studies were approved by the Stockholm South Ethical review board (Dnr. S101–12 to Maria Eriksson).

Transgenic animals

Transgenic mice were housed in a pathogen-free animal facility within the Karolinska University Hospital, Huddinge, Sweden, and maintained in a 12-hour light/dark cycle, at 20–22°C temperature and 50–65% air humidity. Heterozygous animals carrying the tissue-specific promoter-regulated reverse transactivator sm22α-rtTA (on the C57BL/6J genetic background) [15] were crossed with heterozygous tetop-LA^{G608G} animals (on the C57BL/6J or FVB/NCrl genetic background) carrying the human lamin A minigene containing the most common HGPS mutation, *LMNA* c.1824c>T; p.G608G (tetop-LA^{G608G}; F1-line VF1-07) [16]. PCR genotyping was performed according to previously described procedures [15,16]. Offspring that were positive for both transgenes, tetop-LA^{G608G+}; sm22α-rtTA^{+}, were referred to as bi-transgenic animals. The single transgenic offspring, tetop-LA^{G608G+}; sm22α-rtTA^{-}, were referred to as control mice. The non-transgenic offspring, tetop-LA^{G608G-}; sm22α-rtTA^{-}, were referred to as wild-type mice. The animals were given drinking water supplemented with 5% sucrose and doxycycline. A total of twenty-one litters were treated with doxycycline during embryogenesis, out of these one litter was given 1.5 mg/ml doxycycline and the remainder were given 2 mg/ml doxycycline. Sixteen litters were given 2 mg/ml doxycycline postnatally. Tissues from tetop-LA^{G608G+}; NSE-tTA^{+}, tetop-LA^{G608G-}; NSE-tTA^{-}, tetop-LA^{G608G-}; NSE-tTA^{+} mice, and tetop-LA^{G608G+}; Sp7-tTA^{+} [17,18] were used as a controls.

RNA extraction and RT-PCR analysis

Animals were sacrificed by administering an overdose of isoflurane (Baxter, Deerfield, IL, USA) at postnatal week 4 or 12, and the aortic arch, thoracic aorta and abdominal aorta were collected for analysis. RNA was extracted using the TriZol Reagent (Invitrogen, Carlsbad, CA, USA). Following RNA extraction, the samples were treated with DNase (Promega, Madison, WI, USA) and column purified (Qiagen, Valencia, CA, USA). The cDNA synthesis was performed on 0.5 μg total RNA with random hexamers using the SuperScript II Reverse Transcriptase kit (Invitrogen, Carlsbad, CA, USA), according to the manufacturer's instructions. The expressions of transgenic human lamin A and lamin Adel150 (progerin) were analyzed by PCR using the primers 5′-AGTTCTGGGGGCTCTGGGT-3′, 5′-ACTGCAGCAGCTCGGGG-3′, and 5′-TCTGGGGGCTCTGGGC-3′ [16]. The fragment sizes for human lamin A and human lamin Adel150 were 276 and 123 base pairs, respectively [16]. PCR amplification of the cDNA using primers for β-actin (5′- CCTAGGCACCAGGGTGTGAT-3′ and 5′-CCATGTCGTCCCAGTTGGTAA-3′) was performed on all samples as a control. The expression of the transactivator rtTA was analyzed using the following primers: 5′-GACGCGC-TAGACGATTTCGA-3′ and 5′-GGAAAGACCCCTAG-GAATGC-3′[15].

Protein extraction and Western blot

Protein was extracted from arteries of bi-transgenic and single transgenic animals as previously described [16]. To be able to

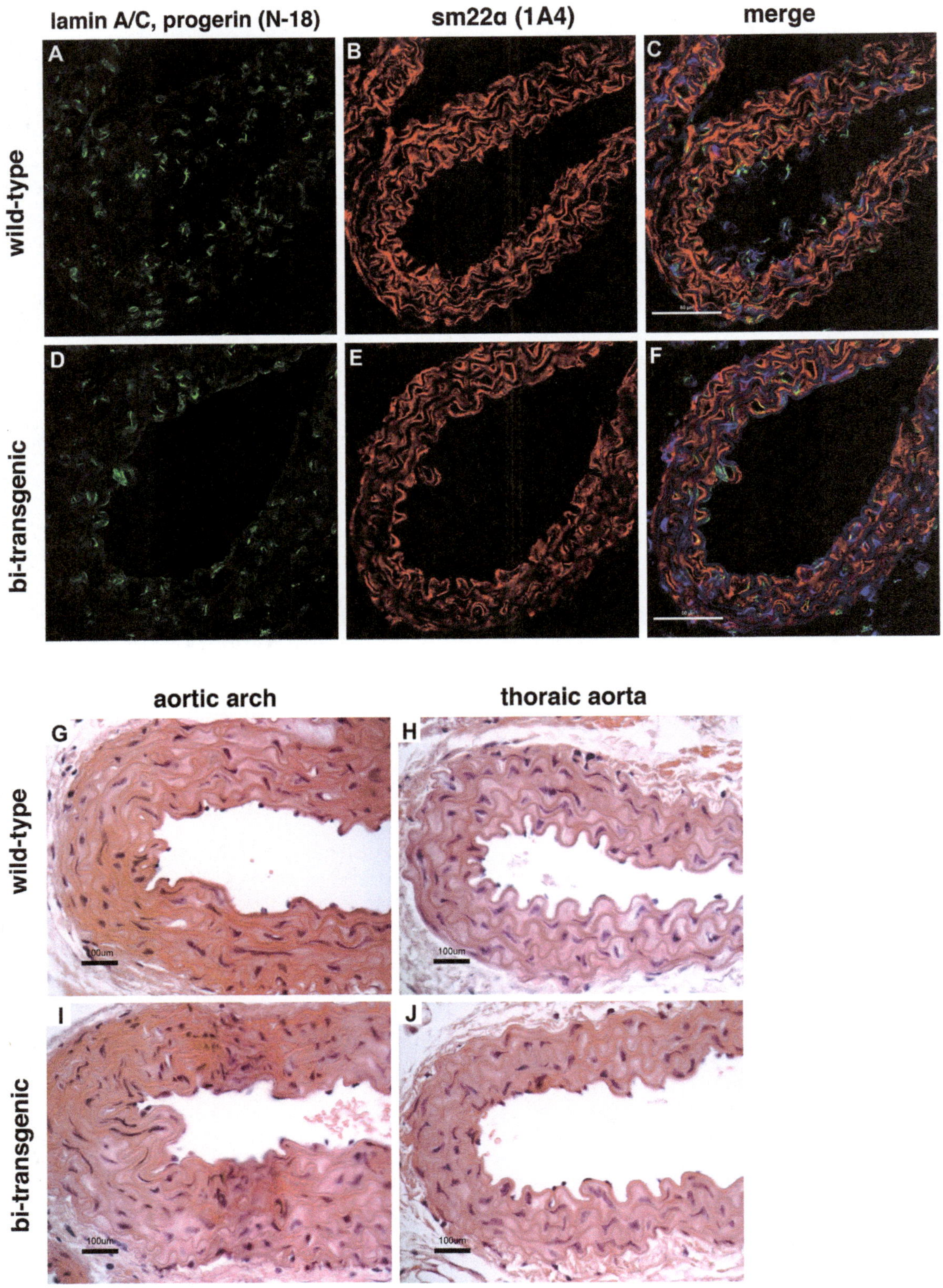

Figure 2. Expression of mouse lamin A/C and sm22α-actin is unaffected in the aortic arch. Immunofluorescence staining with an anti-human-lamin A/C antibody (N-18), which also binds to progerin of human origin and lamin A/C of mouse origin, and an antibody for vascular smooth muscle cells (1A4) in wild-type (A–C) and bi-transgenic tetop-LA^{G608G+}; sm22α-rtTA^{+} animals (D–F). Representative images from the sections of aortic arches of mice with the C57BL/6J; FVB/NCrl genetic background supplied with doxycycline from the date of birth until postnatal week 12. Scale bars: 50 μm. C, F: merge of the lamin A/C, sm22α-actin, and DAPI fluorescence signals. (G–J) Histological examination of aortic sections with haematoxylin eosin staining shows normal structure of the aorta. G, I: aortic arch. H, J: thoracic aorta. Scale bars: 100 μm.

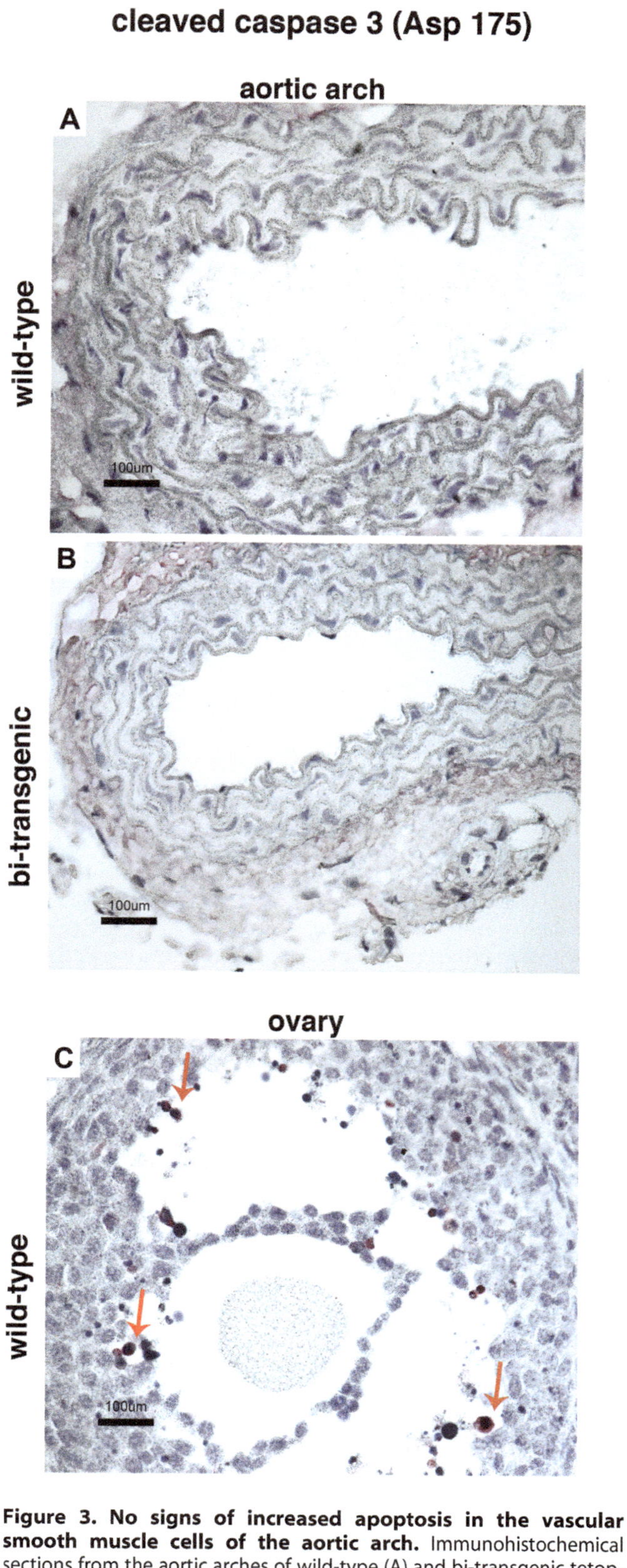

Figure 3. No signs of increased apoptosis in the vascular smooth muscle cells of the aortic arch. Immunohistochemical sections from the aortic arches of wild-type (A) and bi-transgenic tetop-LA^{G608G+}; sm22α-rtTA^{+} (B) mice supplied with doxycycline from the date of birth to postnatal week 12, stained with an anti-Cleaved Caspase 3 antibody (Asp 175) to highlight apoptotic cells. (C) A section of the ovary from a wild-type animal used as a positive control tissue for apoptotic cells. Arrows indicate apoptotic cells. Scale bars: 100 μm.

obtain enough protein for the Western blot, tissue samples from aortic arch, thoracic aorta and abdominal aorta from each animal were pooled for protein extraction. Protein extracts from HGPS patient cell line AG11513A (Coriell Cell Respositories), tetop-LA^{G608G-}; NSE-tTA^{-}, tetop-LA^{G608G+}; NSE-tTA^{+} and tetop-LA^{G608G-}; NSE-tTA^{+} mice [17] were used as a controls. Enhanced protein separation Western blot analysis was performed using the Odyssey system (LI-COR Biosciences) and as previously described [16]. Primary antibodies used for Western blot were: anti-TetR (9G9, Clontech Laboratories Inc.), anti-TetR (631108, Clontech Laboratories Inc.), anti-human lamin A/C (JoL2, Chemicon), and anti-*B*-actin (A5441, Sigma).

Immunofluorescence, immunohistochemistry and imaging

Animals were sacrificed by administering an overdose of isoflurane and the aortic arch, thoracic aorta and abdominal aorta were collected from animals at postnatal weeks 4 or 12 from 5 wild-type and 5 bi-transgenic mice. The samples were fixed in 4% paraformaldehyde (pH 7.4) overnight and then dehydrated in 70% ethanol. After fixation, the tissues samples were embedded in paraffin, cut into 4-μm sections and dried at 60°C for 30 minutes. The sections were then stained with haematoxylin and eosin (H&E) according to standard procedures. For immunostaining analysis, the sections were re-hydrated, followed by antigen retrieval using a pressure cooker. The primary antibodies used were anti-human-lamin A/C (1:30, JoL2, mouse, Chemicon) (cross reacts with lamin A/C and progerin of human origin [16]), anti-human-progerin (1:100, 13A4, mouse, Enzo Life Science) (cross reacts with progerin of human origin [19], anti-human-lamin A/C (1:75, N-18, goat, Santa Cruz) (cross reacts with lamin A/C and progerin of human origin and lamin A/C of mouse origin [16]), and anti-sm α actin conjugated to Cy3 (1:200, 1A4, mouse, Sigma). The samples were incubated with the primary antibodies overnight at 4°C. The secondary antibodies used were Alexa 555-conjugated goat anti-mouse (1:100, A-2122, Life Technologies) and Alexa 488-conjugated donkey anti-goat (1:100, A-11055, Life Technologies). Blocking was performed with normalized goat or donkey serum, BSA or mouse-to-mouse blocking reagent (Scytek, Logan, UT, USA). The sections were mounted in Vectashield mounting media containing DAPI (Vector laboratories, Burlingame, CA, USA). Sections from aortic arch from tetop-LA^{G608G+}; NSE-tTA^{+} [17] were used as controls. Imaging was performed using a Nikon A1+ imaging system, (Nikon Corporation, Japan), and images were analyzed using NIS elements (Nikon Corporation, Japan). Immunohistochemistry using the anti-Cleaved Caspase 3 (1:200, Asp 175, Cell Signaling) primary antibody was performed on deparaffinized sections. The endogenous peroxidase activity was blocked using a 2.5% hydrogen peroxide solution. The tissue sections were subjected to heat-induced epitope retrieval by incubation in sodium citrate buffer (10 mM, pH 6.0) for 30 min in a 95 centigrade water bath, followed by blocking with 3% goat serum for 30 min. The primary antibody was then applied overnight, and the secondary antibody, biotin-goat anti-rabbit IgG (Zymed 65–6140, Invitrogen, CA, USA) was applied for 30 min, followed by the label antibody (ABC Elite, Vector Laboratories, Burlingame, CA, USA) for 30 min. The DAB chromagen (Dako Cytomation, Carpinteria, CA, USA) was applied for 3 min, followed by 2 rinses in distilled water. Mayers Hematoxylin (Histolab) was used as a counterstain.

Statistical analysis

Statistical analyses were performed using Chi-squared test.

Table 1. Genotype frequencies.

Doxycycline treatment	Background strain	Obtained genotype frequencies (%)			
		sm22α-rtTA⁺; tetop-LA^{G608G-}	sm22α-rtTA⁻; tetop-LA^{G608G+}	sm22α-rtTA⁺; tetop-LA^{G608G+}	sm22α-rtTA⁻; tetop-LA^{G608G-}
D0	C57BL/6J	30.2	26.4	34.0	9.4
	C57BL/6J; FVB/NCrl	14.6	29.2	33.3	22.9
E0	C57BL/6J	36.7	24.1	17.7	21.5
	C57BL/6J; FVB/NCrl	22.9	40.0	17.1	20.0

D0, mice supplied with doxycycline from the date of birth; E0, mice supplied with doxycycline during embryogenesis and postnatally; +, presence of transgene; –, absence of transgene. The expected frequency for each individual genotype was 25%.

Results and Discussion

In this study, we used the previously functional transgenic animal models, sm22α-rtTA [15] and tetop-LA^{G608G} [16], to develop a tissue-specific and inducible transgenic animal model for HGPS with expression in the vascular smooth muscle cells of the aorta to further the study the most lethal aspect of this disease.

Previously published results using sm22α-rtTA have shown that it could induce target transgene expression in the presence of doxycycline in the aortic arch, abdominal aorta and thoracic aorta, with the strongest induction observed in the aortic arch [15]. The aortic arch, thoracic aorta and abdominal aorta were isolated from bi-transgenic, control and wild-type animals, and the tissues were analyzed by RT-PCR, immunofluorescence and immunohistochemistry. Several conditions and time points were analyzed. To analyze the transgene expression at the RNA level, we used RT-PCR with primers that were specific for human lamin A, lamin Adel150 (progerin), and the reverse transactivator (Figures 1A–1C). Since the tetop-LA^{G608G} transgene contained a minigene with the human *LMNA* c.1824c>T; p.G608G mutation, which has previously been shown to generate only a partial activation of a cryptic splice site, we expected both wild-type lamin A and lamin Adel150 to be expressed from the transgene. However, the RT-PCR amplification of human lamin A and lamin Adel150 yielded very weak signals, even after 35 cycles of amplification (Figures 1A–1C). The weak signal was also observed in samples that did not carry the reverse transactivator transgene (Figure 1B), which suggested that the weak amplification product occurred due to the leakiness of the system, which has been reported in previous studies that used the tet-ON/OFF model systems [20–22]. A positive control reaction was performed using cDNA synthesized from RNA extracted from the bone of a previously reported animal model containing the tet-OFF system with the same lamin A target gene, available in the lab [18]. The positive control reaction yielded strong amplification of both the human lamin A and lamin Adel150 transcripts, confirming that the experimental procedure was functional (Figures 1A–1C). To analyze the expression of the reverse transactivator in the arteries, PCR with primers specific for the rtTA was performed on cDNA derived from samples from the bi-transgenic tetop-LA^{G608G+}; sm22α-rtTA⁺ animals (Figures 1A, 1C). The RT-PCR confirmed the presence of the transactivator in the arteries but only resulted in a weak amplification signal. Western blot analysis on protein extracts from pooled aortic regions from bi-transgenic tetop-LA^{G608G+}; sm22α-rtTA⁺ animals did not show transgenic expression of human lamin A and progerin (Figure 1D). Western blot experiments using an antibody to the transactivator (rtTA) showed a weak signal in protein extracts from the arteries of bi-transgenic animals and single transgenic animals for the sm22α-rtTA or the NSE-tTA (Figure 1E). One reason for the low expression of the transactivator could be the difference in genetic backgrounds since the original report of the sm22α-rtTA was based on offspring from sm22α-rtTA hybrids on C57BL/6 X CBA that were intercrossed with a target gene on C57BL/6 background [15]. However, this was unlikely because similarly weak amplification products were obtained using the aortic tissue samples extracted from animals with both pure C57BL/6J (Figure 1A) and mixed C57BL/6J; FVB/NCrl genetic backgrounds (Figure 1C). Similar weak amplification products were also observed using samples from mice that were administered doxycycline for different time periods, during embryogenesis to postnatal week 4, and from the date of birth to postnatal week 4 or 12 (see lanes 11–12, 3–4 and 7–8, respectively, in Figures 1A and 1C). Previous results have indicated that the tet-ON system is preferable to the tet-OFF system because induction by doxycycline is more rapid [12]. Thus, doxycycline addition, particularly during early/prenatal development, was expected to result in high transgene expression, which was not evident at the RNA level.

To monitor the expression of transgenic human lamin A and progerin at the protein level in the different aortic sections, we performed immunofluorescence staining using the anti-human-lamin A/C antibody (JoL2) which cross reacts with lamin A/C and progerin of human origin, but does not cross-react with mouse lamins (Figures 1F–1I). Unfortunately, only a few weakly positive cells, <1%, were found in samples from bi-transgenic animals (in both genetic backgrounds C57BL/6J and C57BL/6J; FVB/NCrl) that were supplied with doxycycline for 12 weeks after birth (data not shown). Aortic sections from bi-transgenic animals incubated with only the secondary antibody, used for transgene detection, did not show any fluorescence signal (data not shown). Aortic sections from wild-type littermate controls, stained with antibodies used for transgene detection, also showed no staining indicating that the weakly positive cells were exhibiting specific staining (data not shown). There were no positive cells in 4-week-old animals that were administered doxycycline treatment during embryogenesis or postnatally (data not shown). It is well known that with the progression of disease in HGPS, the smooth muscle cell population in arteries progressively decreases and the thickness of the vascular medial layer from the aortic arch to the abdominal aorta decreases [7,9]. In a successful model system, we expected to have a high positive fluorescence signal for progerin and structural defects due to progerin toxicity, specifically in the aortic arch and the first part of thoracic aorta, compared to the abdominal aorta. However, in agreement with the results from the RT-PCR analysis and the immunofluorescence analysis using the transgene-specific lamin A/C antibody (JoL2), there were no progerin-positive immunofluorescent cells detected using an antibody against progerin (13A4) in any of the three aortic regions in bi-transgenic animals from all of the studied experimental groups (Figures 1J–1K, and data not shown). Aortic tissue from bi-transgenic animals with the same lamin A target gene driven by a different transactivator, NSE-tTA, was included as a positive control [17] (Figures 1D–E, 1H–I, 1L–M). The failure to detect human lamin A and progerin expression from aortic tissues with the NSE-tTA transactivator on Western blot (Figure 1D) (even though positive cells were seen on the immunofluorescence analysis using the transgene-specific lamin A/C and progerin antibodies (Figure 1H–I, 1L–M)) could be caused by heterogenous tissue and the fact that the protein extraction included cells from the aortic arch and all the way down to the abdominal aorta. Our previous analysis, on immunofluorescence from aortic regions of these control animals, have indicated a higher fraction of cells that express the transgene in the aortic arch compared to the thoracic aorta (data not shown). A control group of bi-transgenic animals that were not supplied with doxycycline were analyzed by immunofluorescent which showed a few positive cells, in the aortic arch sections, for the transgene-specific lamin A/C (Figure 1N–O) and progerin antibodies (Figure 1R–S), confirming the leakiness of the system. Treatment with doxycycline in the NSE-tTA+; tetop-LA^{G608G+} animal model the addition of doxycycline for three weeks was sufficient to downregulate the transgenic expression (Figure 1P–Q, 1T–U).

To analyze whether the expression of transgenic human lamin A and progerin, despite being nearly undetectable at the RNA level at the time points analyzed, resulted in pathological changes such as the severe loss of vascular smooth muscle cells, we analyzed the expression of the lamin A and C proteins (with the N-18 antibody) in the vascular smooth muscle cells of the aortic arch, thoracic aorta and abdominal aorta (Figures 2A–2F and data not shown). Immunofluorescence analysis revealed a normal expression pattern of the A-type lamins (lamins A and C) in the vascular smooth muscle cells of bi-transgenic mice (Figures 2D–2F) similar to wild-type animals (Figures 2A–2C). In agreement with the immunofluorescence results, there were no signs of pathology in haematoxylin and eosin stained sections and no indication of loss of vascular smooth muscle cells in the different aortic regions (Figures 2G–2J, and data not shown). The expression of progerin in HGPS has previously been associated with increased apoptosis [23]. We therefore decided to investigate whether there was an increased population of apoptotic cells in the aortic arch, thoracic aorta, and abdominal aorta of bi-transgenic animals. However, immunohistochemistry using an antibody for cleaved caspase 3 (Asp 175) did not show increased apoptosis in the aortic sections from bi-transgenic animals when compared to wild-type animals (Figure 3).

Despite the absence of reports of toxicity associated with the administration of 2 mg/ml of doxycycline during embryogenesis, we carefully analyzed the numbers of individual genotypes for signs of embryonic lethality. Statistical analyses were performed to evaluate the deviation from Mendels law (see Table 1 for Genotype frequencies). Mendelian inheritance showed no significant deviation from expected values, (for pure C57BL6/J mice with doxycycline supplied during embyrogenesis ($N = 79$), $P = 0.093$ and for C57BL6/J and FVB/NCrl hybrids with doxycycline supplied during embryogenesis ($N = 35$), $P = 0.219$), for Hardy-Weinberg equilibrium based on genotype, when assessed with the Chi-squared test. Similarly no deviation from expected values from Mendels law was obtained for mice that were supplied with doxycycline postnatally (for pure C57BL6/J mice ($N = 53$), $P = 0.059$ and for C57BL6/J and FVB/NCrl hybrids ($N = 48$), $P = 0.280$).

Several investigators have reported problems using the tet-ON/OFF system and low expression of the transactivator [20–22,24]. Unfortunately, we were unable to induce transgene expression of lamin A and progerin in the different regions of aorta using the sm22α-rtTA transactivator. The sm22α-rtTA has previously been used successfully by Mizrachi et al., wherein mice carrying the reverse transactivator-regulated promoter sm22α-rtTA expressed the target gene Ucp1 in the aortic arch, thoracic aorta and the abdominal aorta, with expression clearly detectable after 10 days of doxycycline treatment [15]. In our study, despite analyzing mice with different genetic backgrounds and allowing different time periods for doxycycline induction, we were unable to obtain expression of the HGPS allele above background levels. Currently, we can only speculate on the reasons for the lack of transgene expression. One possibility, is that epigenetic modifications of the promotor sequence of the sm22α-rtTA transgene may have arisen over multiple mouse generations, resulting in suppression of the rtTA expression [22,25–26].

Acknowledgments

We are thankful to Dr. Paolo Parini for teaching us the technique of artery extraction and Neshmil Rada for technical assistance. We would like to acknowledge Raoul Kuiper for pathological consultation and Sylvie Le Guyader for consultation on microscopy. We thank Ylva Rosengardten for her contribution in the maintenance of transgenic animals.

Author Contributions

Conceived and designed the experiments: NV TK ME. Performed the experiments: NV. Analyzed the data: NV TK ME. Wrote the paper: NV TK ME. Read and approved the final manuscript: NV TM ME.

References

1. DeBusk FL (1972) The Hutchinson-Gilford progeria syndrome. Report of 4 cases and review of the literature. J Pediatr 80: 697–724.
2. Merideth MA, Gordon LB, Clauss S, Sachdev V, Smith AC, et al. (2008) Phenotype and course of Hutchinson-Gilford progeria syndrome. N Engl J Med 358: 592–604.
3. Eriksson M, Brown WT, Gordon LB, Glynn MW, Singer J, et al. (2003) Recurrent de novo point mutations in lamin A cause Hutchinson-Gilford progeria syndrome. Nature 423: 293–298.
4. De Sandre-Giovannoli A, Bernard R, Cau P, Navarro C, Amiel J, et al. (2003) Lamin a truncation in Hutchinson-Gilford progeria. Science 300: 2055.
5. Baek J-H MT, Eriksson M (2012) Hutchinson-Gilford Progeria Syndrome. In: Puiu M, editor. Genetic Disorders: Intech. 65–87.
6. Goldman RD, Shumaker DK, Erdos MR, Eriksson M, Goldman AE, et al. (2004) Accumulation of mutant lamin A causes progressive changes in nuclear architecture in Hutchinson-Gilford.
7. Stehbens WE, Wakefield SJ, Gilbert-Barness E, Olson RE, Ackerman J (1999) Histological and ultrastructural features of atherosclerosis in progeria. Cardiovasc Pathol. 8: 29–39.
8. Varga R, Eriksson M, Erdos MR, Olive M, Harten I, et al. (2006) Progressive vascular smooth muscle cell defects in a mouse model of Hutchinson-Gilford progeria syndrome. Proc Natl Acad Sci U S A 103: 3250–3255.
9. Capell BC, Collins FS, Nabel EG (2007) Mechanisms of cardiovascular disease in accelerated aging syndromes. Circ Res 101: 13–26.
10. Zhang H, Kieckhaefer JE, Cao K (2013) Mouse models of laminopathies. Aging Cell 12: 2–10.
11. Gossen M, Bujard H (1992) Tight control of gene expression in mammalian cells by tetracycline-responsive promoters. Proc Natl Acad Sci U S A 89: 5547–5551.
12. Schonig K, Bujard H, Gossen M (2010) The power of reversibility regulating gene activities via tetracycline-controlled transcription. Methods Enzymol 477: 429–453.
13. Osorio FG, Navarro CL, Cadinanos J, Lopez-Mejia IC, Quiros PM, et al. (2011) Splicing-directed therapy in a new mouse model of human accelerated aging. Sci Transl Med 3: 106ra107.
14. Li L, Miano JM, Mercer B, Olson EN (1996) Expression of the SM22alpha promoter in transgenic mice provides evidence for distinct transcriptional

regulatory programs in vascular and visceral smooth muscle cells. J Cell Biol 132: 849–859.
15. Bernal-Mizrachi C, Gates AC, Weng S, Imamura T, Knutsen RH, et al. (2005) Vascular respiratory uncoupling increases blood pressure and atherosclerosis. Nature 435: 502–506.
16. Sagelius H, Rosengardten Y, Hanif M, Erdos MR, Rozell B, et al. (2008) Targeted transgenic expression of the mutation causing Hutchinson-Gilford progeria syndrome leads to proliferative and degenerative epidermal disease. J Cell Sci 121: 969–978.
17. Chen J, Kelz MB, Zeng G, Sakai N, Steffen C, et al. (1998) Transgenic animals with inducible, targeted gene expression in brain. Mol Pharmacol 54: 495–503.
18. Schmidt E, Nilsson O, Koskela A, Tuukkanen J, Ohlsson C, et al. (2012) Expression of the Hutchinson-Gilford progeria mutation during osteoblast development results in loss of osteocytes, irregular mineralization, and poor biomechanical properties. J Biol Chem 287: 33512–33522.
19. McKenna T, Rosengardten Y, Viceconte N, Baek J-H, Grochova D, et al. (2013) Embryonic expression of the common progeroid lamin A splice mutation arrests postnatal skin development. Aging Cell Dec 4.
20. Bao-Cutrona M, Moral P (2009) Unexpected expression pattern of tetracycline-regulated transgenes in mice. Genetics 181: 1687–1691.
21. Lee S, Agah R, Xiao M, Frutkin AD, Kremen M, et al. (2006) In vivo expression of a conditional TGF-beta1 transgene: no evidence for TGF-beta1 transgene expression in SM22alpha-tTA transgenic mice. J Mol Cell Cardiol 40: 148–156.
22. Schmidt E, Eriksson M (2011) A previously functional tetracycline-regulated transactivator fails to target gene expression to the bone. BMC Res Notes 4: 282.
23. Bridger JM, Kill IR (2004) Aging of Hutchinson-Gilford progeria syndrome fibroblasts is characterised by hyperproliferation and increased apoptosis. Exp Gerontol 39: 717–724.
24. Lee S, Agah R, Xiao M, Frutkin AD, Kremen M, et al. (2005) Regulated transgene expression in vascular smooth muscle. Circ Res 97: e85; author reply e85–87.
25. Boger H, Gruss P (1999) Functional determinants for the tetracycline-dependent transactivator tTA in transgenic mouse embryos. Mech Dev 83: 141–153.
26. Fedorov LM, Tyrsin OY, Sakk O, Ganscher A, Rapp UR (2001) Generation dependent reduction of tTA expression in double transgenic NZL-2/tTA(CMV) mice. Genesis 31: 78–84.

8

The EAR Motif Controls the Early Flowering and Senescence Phenotype Mediated by Over-Expression of SlERF36 and Is Partly Responsible for Changes in Stomatal Density and Photosynthesis

Rakesh Kumar Upadhyay[1,3,4], Asmita Gupta[1], Sanjay Ranjan[2], Ruchi Singh[2], Uday V. Pathre[2], Pravendra Nath[1], Aniruddha P. Sane[1]*

1 Plant Gene Expression Lab, CSIR-National Botanical Research Institute (Council of Scientific and Industrial Research), Lucknow, India, **2** Department of Plant Physiology, CSIR-National Botanical Research Institute (Council of Scientific and Industrial Research), Lucknow, India, **3** Sustainable Agricultural Systems Laboratory, USDA-ARS, Beltsville Agricultural Research Center, Beltsville, Maryland, United States of America, **4** Department of Biology, Pennsylvania State University, Harrisburg, Pennsylvania, United States of America

Abstract

The EAR motif is a small seven amino acid motif associated with active repression of several target genes. We had previously identified *SlERF36* as an EAR motif containing gene from tomato and shown that its over-expression results in early flowering and senescence and a 25–35% reduction of stomatal density, photosynthesis and stomatal conductance in transgenic tobacco. In order to understand the role of the EAR motif in governing the phenotypes, we have expressed the full-length *SlERF36* and a truncated form, lacking the EAR motif under the CaMV35S promoter, in transgenic Arabidopsis. Plants over-expressing the full-length *SlERF36* show prominent early flowering under long day as well as short day conditions. The early flowering leads to an earlier onset of senescence in these transgenic plants which in turn reduces vegetative growth, affecting rosette, flower and silique sizes. Stomatal number is reduced by 38–39% while photosynthesis and stomatal conductance decrease by about 30–40%. Transgenic plants over-expressing the truncated version of SlERF36 (lacking the C-terminal EAR motif), show phenotypes largely matching the control with normal flowering and senescence indicating that the early flowering and senescence is governed by the EAR motif. On the other hand, photosynthetic rates and stomatal number were also reduced in plants expressing SlERF36ΔEAR although to a lesser degree compared to the full- length version indicating that these are partly controlled by the EAR motif. These studies show that the major phenotypic changes in plant growth caused by over-expression of SlERF36 are actually mediated by the EAR motif.

Editor: Peter Meyer, University of Leeds, United Kingdom

Funding: The work was supported by a grant (GAP244525) from the Department of Biotechnology, Government of India, to A.P.S. Senior Research Fellowship was provided by the Council of Scientific and Industrial Research, Government of India, to R.K.U., A.S., R.S. The funders had no role in study design, data collection and analysis, decision to publish, or preparation of the manuscript

Competing Interests: The authors have declared that no competing interests exist.

* Email: saneanil@rediffmail.com

Introduction

The AP2-ERF (APETALA2-ETHYLENE RESPONSE FACTOR) domain family of transcription factors is one of the largest families of transcription factors comprising about 140–280 members in various plants [1]. The family governs plant responses to various biotic and abiotic stresses largely by controlling responses to different plant hormones such as ABA, ethylene and jasmonic acid (JA). A small sub set of this family is characterized by the presence of a seven amino acid repression motif designated as the ERF-associated amphiphilic repression (EAR) motif that is present at the C-terminal end of the protein. The EAR motif with an L/FDLNL/FxP sequence functions in concert with the AP2 domain but is not restricted to the ERF family. It has been found to be associated with about 21 different types of transcriptional regulator families that also include the AUX/IAA family (with the similar LxLxL motif), C2H2 Zn finger family and the JAZ family [2,3]. The domain can confer the ability to repress transcription even when it is hooked on to proteins that otherwise function as transcriptional activators [4,5]. The EAR motif and the related LxLxL motif actively repress target genes by recruitment of co-repressors such as AtSAP18/SIN3 [6,7] or those of the TOPLESS family such as TPL/TPR [8]. These co-repressors in turn interact with histone deacetylases and remodel chromatin to repress transcription [9,3]. The presence of the EAR motif is essential for the repression activity with mutations in the D and L residues of the motif affecting the repressor function [4,10,11].

At least eight EAR motif containing ERF genes are known in Arabidopsis and rice [1] and at least seven are present in tomato [12,13]. They show diverse roles such as in herbivory and wounding [14,15], cold and drought stresses [16], salt stress signalling [17], ABA responses [6] ethylene response [13,18,19], JA responses [20] and senescence [21].While the EAR motif has been shown to function as a repressor motif *in vitro*, its role and the extent of its contribution in governing the processes mediated by its expression are still not clear. We had previously identified SlERF36/SlERF.F.1 (Accession No. SGN-U564952) as an EAR motif containing AP2 domain gene from tomato, the expression of which accelerated flowering and senescence, and reduced stomatal density by 25–35% in transgenic tobacco plants with direct or indirect effects on photosynthesis, stomatal conductance and transpiration [22]. In this paper, we show that the phenotypic effects of early flowering and senescence imparted by *SlERF36* over-expression are largely governed by the EAR motif while the reduction in stomatal density appears to be partly dependent on the EAR motif.

Materials and Methods

Development of constructs and transgenic *Arabidopsis* plants

The SlERF36 gene was identified in a previous study as an EAR motif containing AP2 domain gene encoding a protein of 221 amino acids [22]. Constructs containing the SlERF36 gene under the CaMV35S promoter in a pBI121 background (used for transgenic tobacco plants [22]) were also used to transform the Col-0 ecotype of *Arabidopsis*. In order to study the effects of lack of EAR repressor motif, a reverse primer SlERF36ΔEAR-R GTCAAGGCAGTGGATTTCTGAGAGATGA was designed just upstream of the EAR motif that incorporated a termination codon. This primer was used in combination with the forward primer ATGGATCCTATGAGAAGAGGCAGAGC (containing the initiation codon) to amplify a fragment of 598 nucleotides that contained an ORF of 585 nt. This ORF encoded a protein of 195 amino acids and lacked the last 26 amino acids that also included the EAR motif. The fragment was cloned in pBI121 under the CaMV35S promoter. *Arabidopsis* plants were transformed with these constructs using the floral dip method [23].

Of the various transformants generated, two each (lines 1-8 and 2-1 for SlERF36 and lines 6-1 and 9-1 for SlERF36ΔEAR, lacking the EAR motif) were taken to the third generation to obtain homozygous lines. Progeny of these homozygous lines was used for all phenotypic and physiological studies as well as for gene expression. The plants were grown in culture racks in cups in soilrite mix at a light intensity of 150 μmoles and a photoperiod of 16 h light/8 h dark at 22°C unless otherwise mentioned. For short day treatments plants were grown in a 10 h light/14 h dark cycle. The plants were monitored and compared for various growth parameters that included leaf number, rosette diameter, time of flowering, time of senescence, chlorophyll content, plant height, flower and silique sizes as well as photosynthetic parameters and stomatal density.

Gene expression

Gene expression studies were carried out using RNA isolated from seedlings of control (Col-0) and transgenic plants (with and without the EAR motif). RNA was isolated using the Spectrum Plant total RNA isolation kit (Sigma) and cDNA prepared using the REVERTAID MMLV kit (Fermentas). Expression of *SlERF36* in the transgenic lines was carried out using the primers SlERF36-FRT: TTCCGTCGTTGACCACGGCG and SlERF36-RRT: TGCAGTGAAGA TCGTCGGCGC and normalized against actin amplified using the primers Act-F: ATGACATGGAGAAGATCTGGCATCA and Act-R: AGCCTGGATGGCAACATACAT AGC. Analysis of FT transcript levels was carried out using the primers FT-F: TGTTGGAGACGTTCTTGATCC and FT-R: AGCCACTCTCCCTCTGACAA [24] while analysis of SEN4 was carried out using the primers SEN4-F: TCTTCTTCACGACTCTTCTC and SEN4-R: TTGCCCAATCGTCTGCGTTC [25].

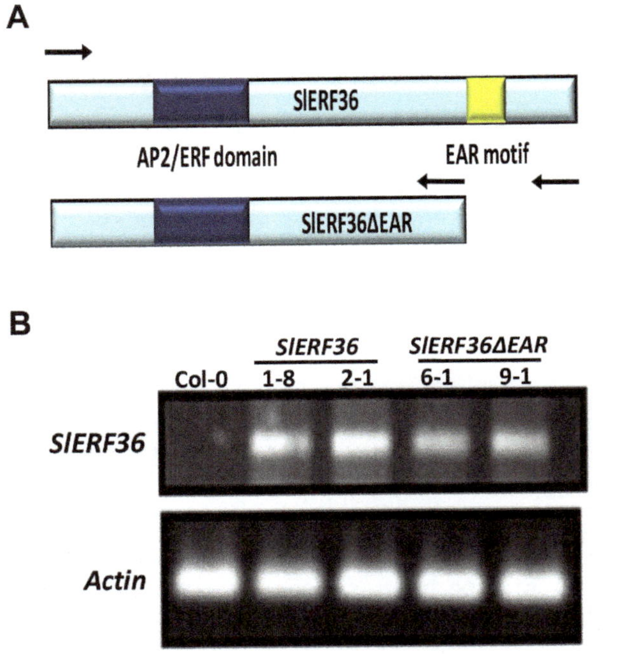

Figure 1. The full length and truncated versions of SlERF36 used for study. A. Schematic representation of SlERF36 and SlERF36ΔEAR that lacks the C-terminal region. The EAR motif is shown in yellow while the AP2 domain has been shown in dark blue. The arrows represent the primers used to amplify the two genes. **B.** Expression of the transgenes *SlERF36* and *SlERF36ΔEAR* in the transgenic lines 1-8 and 2-1 (*SlERF36*) and the lines 6-1 and 9-1 (*SlERF36ΔEAR*). Actin was used as a control.

Physiological studies

Homozygous *Arabidopsis* plants expressing *SlERF36* and *SlERF36 ΔEAR* under the CaMV35S promoter were grown in pots in the culture room under white light. Progeny of homozygous lines was used for each experiment with 5–6 plants per line. Gas exchange parameters were determined on plants enclosed in an *Arabidopsis* chamber (3010-A) under controlled conditions using the GFS-3000 (Heinz Walz Gmbh, Effeltrich, Germany) instrument attached with a fluorescence module (LED-array/PAM fluorometer 3055-FL, Walz, Germany). Measurements were made on 20-day-old plants. Measurements of steady-state photosynthesis rate (A), stomatal conductance (g_s) and transpiration rate (E) were carried out at 70% humidity, 25°C leaf temperature, CO_2 concentration of 400 μmol mol^{-1}and PPFD adjusted to 400 μmol photons m^{-2}s^{-1} as described before [22].

Chlorophyll content was measured by isolating chlorophyll from leaves of 12-day-old plants (4th leaf from bottom) and calculated according to Arnon [26].

Estimation of stomatal density

Leaf epidermis (about 1 cm^2) from the abaxial surface of fully expanded leaves (7th leaf from bottom of 30-day-old plants) was peeled off with a pair of forceps and placed immediately in water and later mounted in 10% glycerol and observed under a light microscope (Nikon Eclipse TE300 Inverted microscope). Stomata

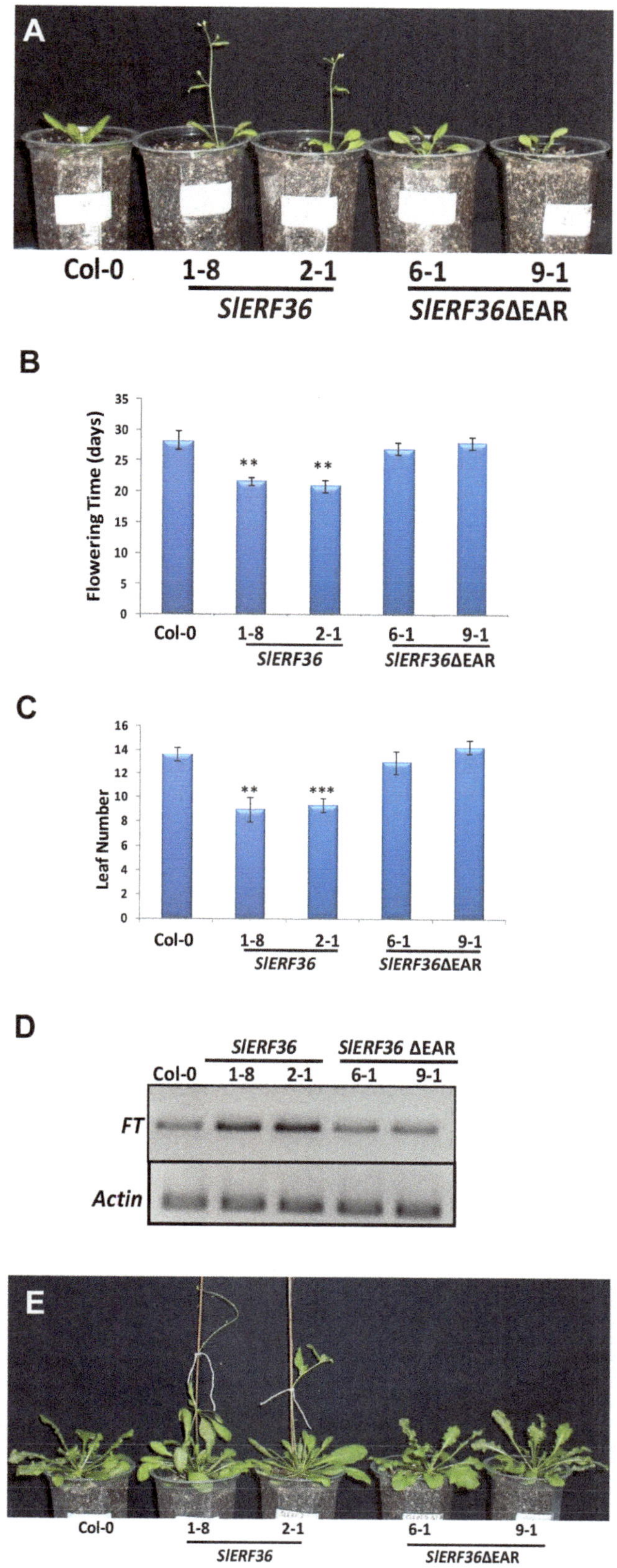

Figure 2. Early flowering in transgenic *SlERF36* over-expressing plants. A. Early flowering in transgenic *SlERF36* over-expressing plants grown under long day (16 h light/8 h dark) conditions. Col-0 wild type; Lines 1-8 and 2-1 – *SlERF36* over-expressing lines; Lines 6-1 and 9-1 – *SlERF36ΔEAR* over-expressing lines. **B.** Graphical representation of days to flowering in transgenic *SlERF36* and *SlERF36ΔEAR* over-expressing lines. Values represent the average ± SD of 3–5 homozygous plants of each independent transformant.** P<0.01; ***P<0.001. **C.** Graphical representation of number of leaves at the time of flowering in transgenic *SlERF36* and *SlERF36ΔEAR* over-expressing lines. Values represent the average ± SD of 3–5 homozygous plants of each independent transformant.** P<0.01; ***P<0.001. **D.** Transcript accumulation of the *FT* gene in control and transgenic lines in 12-day-old plants by semi-quantitative RT-PCR. Actin was used as internal control. **E.** Early flowering in transgenic *SlERF36* over-expressing plants grown under short day (10 h light/14 h dark) conditions. Note the delay in flowering and the larger rosettes compared to plants grown in long days in Fig. 3A.

were counted in an area of 0.24 μm^2 in three different regions from three independent leaves of three plants from the same position from two independent lines (1-8 and 2-1 expressing *SlERF36* and 6-1 and 9-1 expressing *SlERF36ΔEAR*).

Statistical Analysis

Stastitical analysis for all growth parameters and stomatal numbers was carried out using Student's t-test with P values of < 0.05 considered statistically significant. For physiological parameters, the significance of correlations was tested by using linear regression, with P values of <0.05 considered statistically significant. Means were compared by using one-way analysis of variance and post hoc means comparison (Scheffé Test). Data analysis and plotting were performed with SigmaPlot version 8 0.

Results

Over-expression of SlERF36/SlERF.F.1 was previously shown to affect flowering time and senescence and was responsible for a 25–35% reduction in stomatal density that affected several photosynthetic parameters in tobacco [22]. In order to perform a more detailed study regarding the role of the EAR motif in these effects, constructs containing the full length *SlERF36* and one lacking the EAR motif (*SlERF36ΔEAR*) were generated for expression under the CaMV35S promoter (Fig. 1A). Both constructs were used to transform *Arabidopsis*. Of the various independent lines generated, two designated as SlERF36-1-8 and SlERF36-2-1 (over-expressing the full length *SlERF36*) and SlERF36ΔEAR-6-1 and SlERF36ΔEAR-9-1 (over-expressing *SlERF36ΔEAR*) were selected for detailed analysis in the third generation. Progenies of homozygous lines were first checked by semi-quantitative RT-PCR (using actin as internal control) with primers specific to a region common to *SlERF36* and *SlERF36-ΔEAR* to confirm that the genes were expressed in the respective transgenic lines (Fig. 1B). The plants were then monitored for various visible growth parameters such as time of flowering, leaf shape and size, senescence, height, and physiological characters.

SlERF36 over-expressing plants show early photoperiod independent flowering

One of the most prominent features of transgenic *SlERF36* over-expressing plants was the early flowering phenotype (Fig. 2A). In comparison to control plants which produced inflorescence bolts at 28.33±1.52 days, transgenic *SlERF36* over-expressing plants produced inflorescence bolts at about 21.66±0.57 days (*SlERF36*-1-8) and 21.0±1.0 days (*SlERF36*-2-1) indicating a decrease in flowering time by about 7 days compared to the control (Fig. 2B). In contrast, transgenic *SlERF36ΔEAR* over-expressing plants lacking the EAR motif, produced inflorescence bolts at 27.0±1.0 days (SlERF36ΔEAR-6-1) and 28.0±1.0 days (SlERF36ΔEAR-9-1).

Flowering in transgenic *SlERF36* plants was initiated with fewer leaves as compared to control and *SlERF36ΔEAR* (Fig. 2C). Control plants had an average number of 13.66±0.57 leaves at the time of flowering while transgenic *SlERF36* over-expressing plants flowered at 9.0±1.00 leaves/plant (SlERF36-1-8) to 9.33±0.57 leaves/plant (SlERF36-2-1). This early flowering phenotype of *SlERF36* over-expressing plants was no longer observed after deletion of the C-terminal region containing the repressor domain. Transgenic *SlERF36ΔEAR* over-expressing plants flowered at 13.0±1.00 leaves/plant (*SlERF36ΔEAR*-6-1) to 14.33±0.57 leaves/plant (*SlERF36ΔEAR*-9-1).

We next checked for expression levels of *FT* (*FLOWERING LOCUS T*) transcript in 20-day-old plants. As shown in Fig. 2D, a higher transcript level of *FT* was observed in *SlERF36* plants compared to Col-0 and *SlERF36ΔEAR* plants indicating that the presence of the C-terminal region containing the EAR motif was essential for the increase in *FT* levels.

We then tested whether the early flowering phenotype was dependent on photoperiod. For this, plants were grown under 24 h light as well as under short day conditions. Under all conditions, flowering was earlier in *SlERF36* over-expressing plants as compared to control and *SlERF36ΔEAR* over-expressing plants. Under a 24 h light period, flowering was advanced by about 10 days in transgenic *SlERF36* plants while under short day conditions it was advanced by one and half months compared to control and *SlERF36ΔEAR* plants (Fig. 2E) indicating that the early flowering conferred by SlERF36 was independent of photoperiod.

SlERF36 over-expressing plants show early senescence

The initiation of flowering was followed by rapid senescence and leaf death in transgenic *SlERF36* over-expressing lines as shown in Fig. 3A. The early onset of senescence manifested itself in the form of reduced chlorophyll content, early yellowing and early death in lower leaves of transgenic *SlERF36* over-expressing lines as compared to the controls (Fig. 3A and B). At the 12-day-stage, control plants (fourth leaf from bottom) had a chlorophyll content of 4.55±0.13 mg/g FW. In contrast, the leaves of transgenic plants of *SlERF36* over-expressing lines had a chlorophyll content of 3.51±0.07 mg/g FW (*SlERF36*-1-8) and 2.60±0.05 mg/g FW (*SlERF36*-2-1) indicating a decrease in chlorophyll content by 25–50% of the control (Fig. 3B). In contrast, transgenic *SlERF36*-ΔEAR over-expressing plants showed normal wild type senescence with chlorophyll content ranging from a minimum of 4.17±0.23 mg/g FW (*SlERF36ΔEAR*-9-1) to 4.42±0.02 mg/g FW (*SlERF36ΔEAR*-6-1). The early senescence in *SlERF36* expressing plants was also evident at the molecular level by higher expression of the *SEN4* gene (a marker gene for senescence)[25] (Fig. 3C).

The early flowering phenotype markedly affected plant growth, height and flower and silique length. In general, transgenic *SlERF36* over-expressing plants had a smaller rosette diameter as compared to control Col-0 and *SlERF36*ΔEAR over-expressing plants (Fig. 4A). Rosettes measured in 28-day-old plants had a diameter of 7.0±0.5 cm for control plants. In contrast the rosette diameter of transgenic plants of *SlERF36* over-expressing lines ranged from a minimum of 4.1±0.85 cm (*SlERF36* -2-1) to 5.0±1.0 cm (*SlERF36* -1-8) indicating a decrease by 2–3 cm in diameter compared to the control. Deletion of the repressor domain led to formation of normal rosettes with size ranges from 6.9±0.45 cm (*SlERF36*ΔEAR-6-1) to 7.33±0.28 cm (*SlERF36*-ΔEAR-9-1).

Transgenic *SlERF36* over-expressing plants also had smaller flowers and siliques as compared to control Col-0 and *SlERF36-*

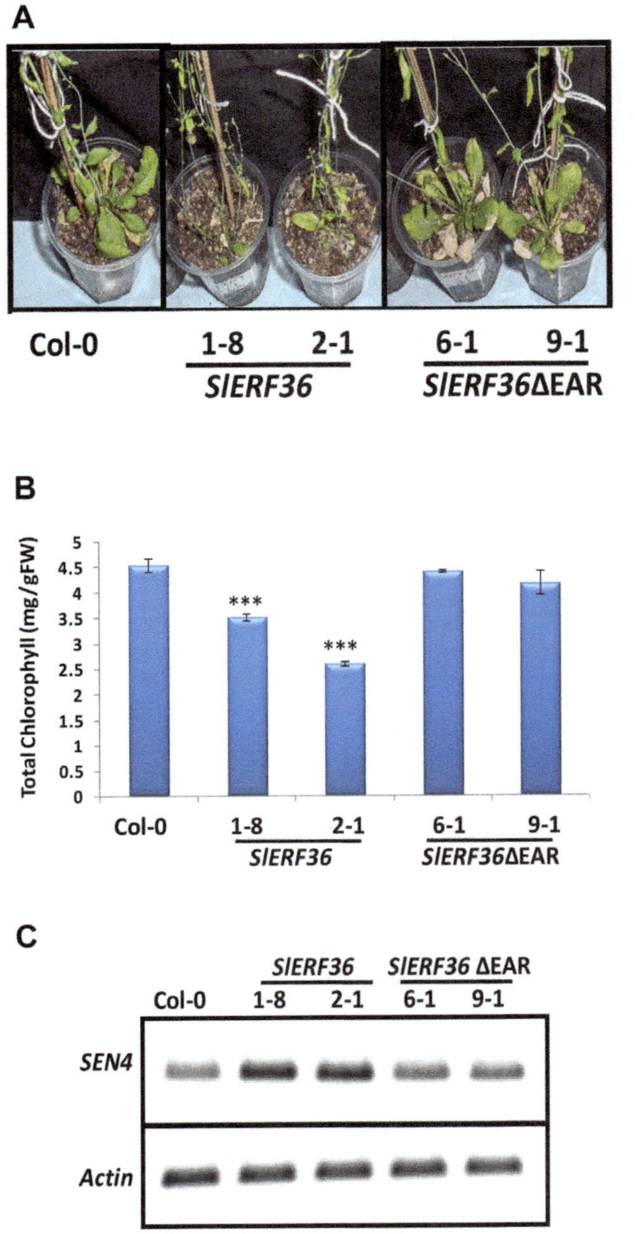

Figure 3. Early senescence in transgenic *SlERF36* over-expressing plants. One and half month old plants (grown under long day conditions from Fig. 3A) showing early leaf senescence and death. Graphical representation of differences in chlorophyll content between control, *SlERF36* over-expressing (1-8 and 2-1) and *SlERF36ΔEAR* over-expressing (6-1 and 9-1) plants. Values represent the average ± SD of 5 leaves of each independent transformant (4th leaf from bottom from 12-day-old plants). Transcript accumulation of the senescence associated *SEN4* gene in control and transgenic lines in 12-day-old plants. Actin was used for normalization.

ΔEAR over-expressing plants (Fig. 4B). As compared to a silique length of 11.6±0.89 cm in controls, transgenic *SlERF36* over-expressing plants showed siliques of sizes 7.4±0.54 cm (*SlERF36*-1-8) and 7.0±0.0 cm (*SlERF36*-2-1). In contrast, transgenic *SlERF36ΔEAR* over-expressing plants had siliques ranging in length from 9.8±0.83 cm (*SlERF36ΔEAR*-9-1) to 10.6±0.89 cm (*SlERF36ΔEAR*-6-1) (Fig. 4C). Full grown plants over-expressing *SlERF36* were shorter as compared to control Col-0 and *SlERF36ΔEAR* over-expressing plants (Fig. 4D) with a final

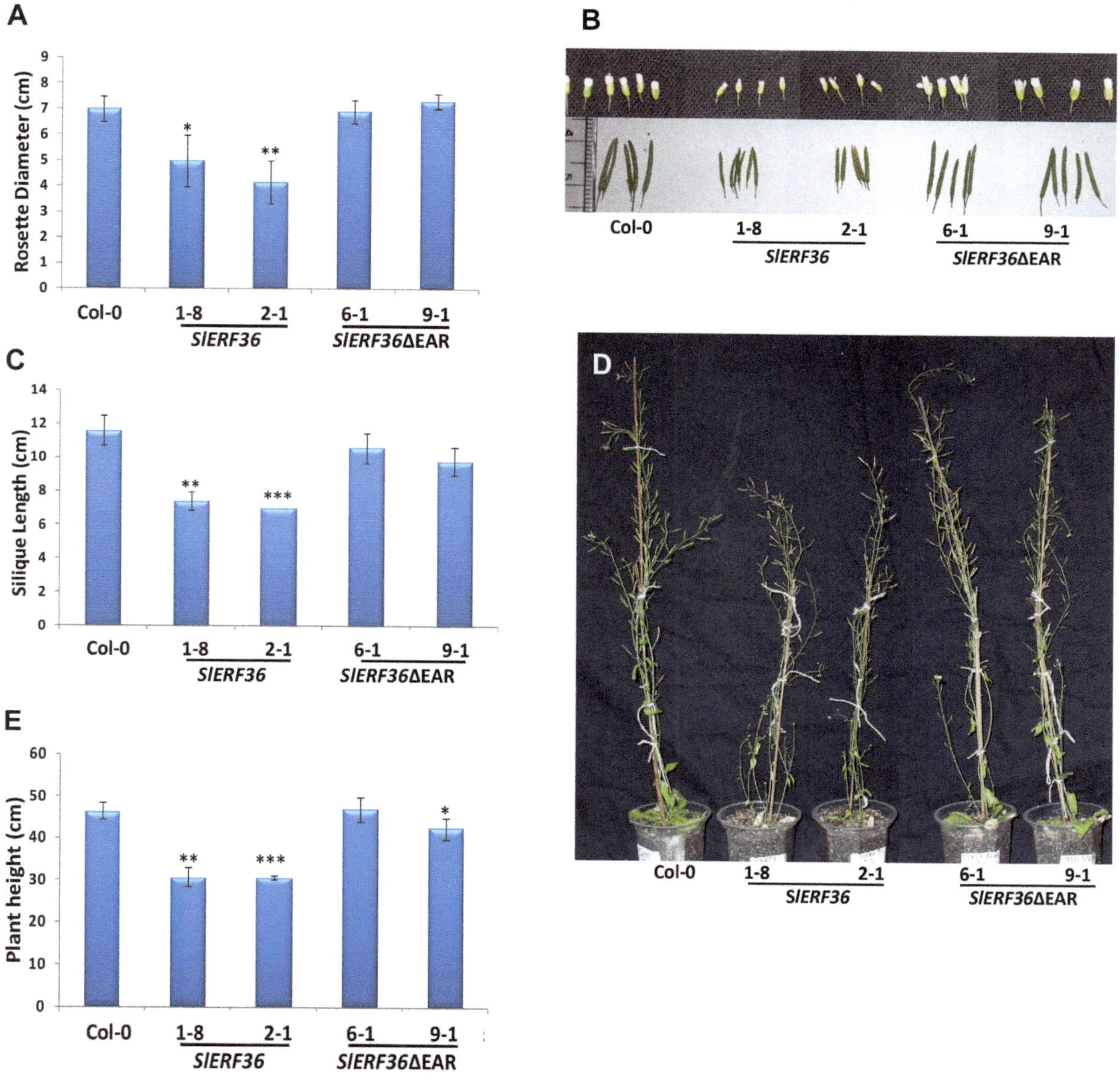

Figure 4. Reduction in organ size and plant height of transgenic *SlERF36* over-expressing plants grown under long day conditions. A. Graphical representation of rosette diameter of 28-day-old transgenic plants over-expressing *SlERF36* and *SlERF36ΔEAR*. Values represent the average ± SD of 3–5 homozygous plants of each independent transformant.** P<0.01; ***P<0.001. B. Flower (top) and silique (bottom) size variation in transgenic *SlERF36* and *SlERF36*ΔEAR *Arabidopsis* plants. C. Graphical representation of the variation in silique sizes in transgenic *SlERF36* and *SlERF36*ΔEAR *Arabidopsis* plants. Values represent the average ± SD of 10 siliques each from 3–5 homozygous plants of each independent transformant.** P<0.01; ***P<0.001. D. Comparison of transgenic *SlERF36* and *SlERF36*ΔEAR over-expressing plants showing differences in height. E. Graphical representation of plant height of transgenic *SlERF36* and *SlERF36*ΔEAR plants. Values represent the average ± SD of 3–5 homozygous plants of each independent transformant. ** P<0.01; ***P<0.001.

plant height of 30.66±2.30 cm (*SlERF36*-1-8) and 30.70±0.60 cm (*SlERF36*-2-1) as compared to about 46.33±2.08 cm in controls and 47.00±3.0 cm (*SlERF36ΔEAR*-6-1) and 44.83±5.29 cm (*SlERF36ΔEAR*-9-1). This indicated a decrease in height by 35–40% for 35S:SlERF36 plants compared to that of the control and transgenic 35S:*SlERF36*ΔEAR plants.

Transgenic 35S:SlERF36 *Arabidopsis* plants show reduced photosynthesis and stomatal conductance

To study if reduced growth and early senescence was due to an effect on photosynthesis we carried out a detailed analysis of the various photosynthetic parameters of the transgenic lines using a GFS-3000 system. As shown in Fig. 5, photosynthetic rates showed a marked reduction in transgenic lines. Compared to control plants that showed photosynthetic rates of 8.75±1.24 µmol CO_2 fixed $m^{-2}s^{-1}$ plants over-expressing *SlERF36* showed reduced photosynthetic rates ranging from 4.8±1.1 (*SlERF36*-1-8) to 5.9±0.54 µmol CO_2 fixed $m^{-2}s^{-1}$ (*SlERF36*-2-1). This indicated a reduction of 33–45% compared to the control. Interestingly, the reduction in photosynthetic rates brought about by over-expression of *SlERF36* was partially affected by the removal of the EAR motif with transgenic *SlERF36ΔEAR* over-expressing lines showing rates of 6.4±1.26 (*SlERF36ΔEAR*-6-1) to 6.1±1.24 μmol CO_2 fixed $m^{-2}s^{-1}$ (*SlERF36ΔEAR*-9-1) indicating a reduction of 27–30% compared to control. The reduction in photosynthetic rates was associated with a decrease in stomatal conductance and transpiration in transgenic lines. Stomatal conductance reduced from 385±62 mmol H_2O $m^{-2}s^{-1}$ in control plants to 240±52 (*SlERF36*-1-8) and 323±64 mmol H_2O $m^{-2}s^{-1}$ (*SlERF36*-2-1) in the transgenic lines. A decrease in stomatal conductance was also seen in transgenic *SlERF36ΔEAR* over-expressing lines with values of 332±52 and 340±81 mmol H_2O $m^{-2}s^{-1}$ in lines *SlERF36ΔEAR*-6-1 and *SlERF36ΔEAR*-9-1 respectively. Transpiration also showed a decrease of 14–40% in transgenic *SlERF36* over-expressing lines and a decrease of 14–27% in transgenic *SlERF36ΔEAR* over-expressing lines. No changes were observed in internal Ci concentration, ETR rates, yield and Fv/Fm values (data not shown).

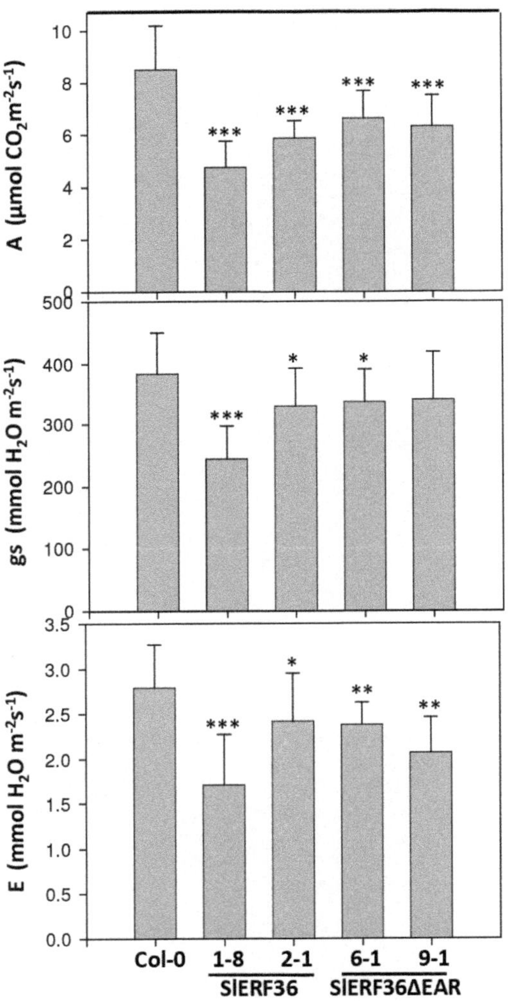

Figure 5. Net photosynthesis (A), stomatal conductance (gs) and transpiration (E) rates of transgenic *Arabidopsis* plants over-expressing *SlERF36* and *SlERF36ΔEAR*. Five plants of homozygous progeny of two independent lines for each gene (lines 1-8 and 2-1 over-expressing *SlERF36* and lines 6-1 and 9-1 over-expressing *SlERF36ΔEAR*) were grown as described in the methods section. Measurements were carried out using a GFS-3000 system under a light intensity of 400 μmole photons $m^{-2}s^{-1}$ and a CO_2 concentration of 400 μmol mol^{-1}. Values are average ± SEs of five replicates. *P<0.05, ** P<0.01; ***P<0.001, ****P<0.0001.

Transgenic *SlERF36* over-expressing lines show reduced stomatal density

Our previous results had shown that over-expression of *SlERF36* affects stomatal density and that many of the defects in photosynthetic parameters were most likely associated with adaptive responses due to reduced stomatal number [22]. To test whether the reduction in photosynthetic parameters was associated with stomatal number, we measured the stomatal density on the abaxial leaf surface (7th leaf from bottom of 30-day-old plants) of control and transgenic lines. As shown in Fig. 6 (A and B) control *Arabidopsis* plants showed a stomatal number of about 8.94±0.76/240 μm^2. In contrast transgenic *SlERF36* over-expressing lines showed a stomatal number of 5.55±0.61/240 μm^2 (*SlERF36*-1-8) and 5.61±0.61/240 μm^2 (*SlERF36*-2-1) in the same field. This indicated a decrease in stomatal number to just 61–62% of the controls in transgenic *SlERF36* over-expressing lines. The removal of the EAR motif in transgenic *SlERF36ΔEAR* over-expressing lines led to a comparative increase in stomatal numbers with values of 6.66±0.6/240 μm^2 and 6.77±0.64/240 μm^2 in the lines *SlERF36ΔEAR*-6-1 and *SlERF36ΔEAR*-9-1 respectively. Thus over-expression of the truncated gene lacking the EAR motif affected the stomatal number to a lesser degree with a decrease of 25–26% compared to the decrease of 38–39% in lines expressing the full length *SlERF36* gene. The difference between the stomatal numbers of the *SlERF36* and *SlERF36ΔEAR* over-expressing lines was significant at P<0.01. Interestingly, the decrease in stomatal number appeared to be partly due to an increase in cell size. Cells of transgenic lines appeared to be larger in size with fewer cells in the same relative area. The decrease in non-stomatal cell number was about 25% in transgenic *SlERF36* over-expressing lines and those over-expressing *SlERF36ΔEAR* (Fig. 6C).

Discussion

SlERF36 is a repressor type of ERF containing a C-terminal EAR motif that was shown to affect flowering time and senescence in transgenic tobacco plants [22]. Another prominent phenotype in tobacco was a 25–35% reduction in stomatal density that in turn reduced stomatal conductance, CO_2 uptake and utilization and photosynthesis thereby affecting development. These could either be independent effects or they could be related. In order to elucidate the role of the EAR motif in these different changes, we generated transgenic *Arabidopsis* over-expressing the complete *SlERF36* and a truncated version of the gene lacking the EAR motif. Interestingly, all the effects observed in transgenic tobacco namely early flowering, early senescence, reduced stomatal number and reduced photosynthesis were replicated in transgenic *Arabidopsis* upon over-expression of the full-length *SlERF36*. The observations have two major implications:

1. The targets of SlERF36 action for early flowering, early senescence and stomatal number are most likely conserved between tomato/tobacco and *Arabidopsis* suggesting functional conservation of these pathways.
2. The sites for SlERF36 binding, upstream within the promoter of the target genes, are also most likely conserved between *Arabidopsis* and tomato/tobacco for it to bind and produce similar effects in these plants.

The EAR motif has been associated with active repression of transcription by several repressor genes that function in different processes [27] and its presence is necessary for repression since mutations in the EAR motif reduce or abolish the repression

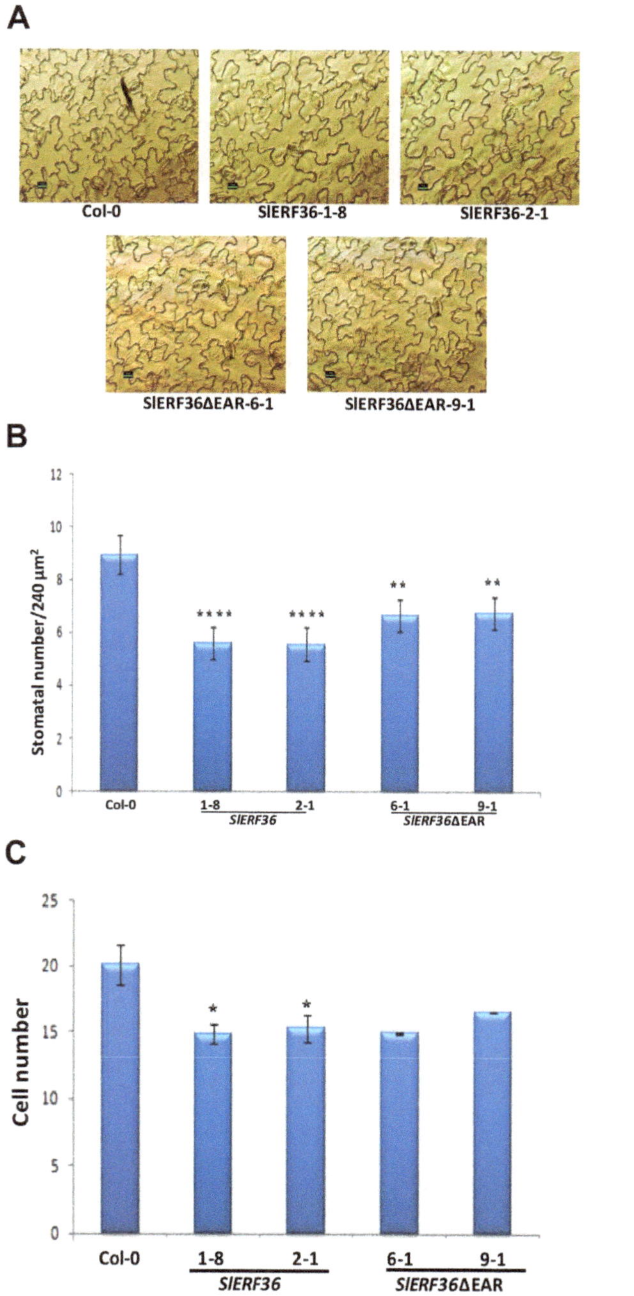

Figure 6. Reduction in stomatal density in transgenic *SlERF36* and *SlERF36ΔEAR* plants. A. Stomatal density on the leaf abaxial surface in control (C) and transgenic *Arabidopsis* plants from two independent lines (lines 1-8 and 2-1 over-expressing *SlERF36* and lines 6-1 and 9-1 over-expressing *SlERF36ΔEAR*). Stomatal density from leaf epidermal peels was estimated in the leaf sections in three different regions of three different leaves (7th leaf from bottom from 30-day-old plants) under a light microscope (Nikon Eclipse TE300 Inverted microscope). The small black bar at the base of each picture on the left hand side represents a length of 10 µm. **B.** Graphical estimation of the stomatal density of the lower leaf epidermis of control (Col-0) and transgenic *SlERF36* and *SlERF36ΔEAR* over-expressing lines from Fig. 6A. Values represent the average stomatal density ± SD in an area of 240 μm^2 of three independent leaves (from the same position). ** $P<0.01$; ***$P<0.001$, ****$P<0.0001$. Figure 6C. Graphical estimation of the non-stomatal cell number of the lower leaf epidermis of control (Col-0) and transgenic *SlERF36* and *SlERF36ΔEAR* over-expressing lines from Fig. 6A. Values represent the average cell number± SD in an area of 240 μm^2 of three independent leaves (from the same position). * $P<0.05$.

activity and gene function [4,10]. For instance, plants expressing the Arabidopsis *RAP2.1* show enhanced sensitivity to drought while mutation of the EAR motif reduces drought sensitivity [16]. Similarly, plants expressing *SlERF3* show severe growth suppression while plants expressing the gene without the EAR motif have no suppressive effect on growth [17]. Likewise, in tobacco, induction of hypersensitive cell death expression by *NtERF3* is dependent on the EAR motif [28]. In another study, transgenic rice expressing a mutated version of the EAR motif of the *OsERF3* gene, showed increased ethylene biosynthesis and greater drought tolerance compared to the non-mutated *OsERF3* expressing lines that showed suppression of ethylene biosynthesis genes [19]. In some cases however, complete removal of the EAR motif does not affect all aspects of the gene function. For example, expression of *ZAT7* reduces growth and imparts salt tolerance but deletion of the EAR motif from ZAT7 does not reduce growth suppression although plants lose their salinity tolerance [29].

Of the changes that we observed, the most prominent visible change was the early flowering phenotype. Interestingly, early flowering was seen not only under long day conditions (both 16 h light and 24 h light) but also under short day conditions (10 h light) indicating that the effects on flowering were largely photoperiod independent. Nevertheless, it should be noted that transgenic plants grown under short day conditions did show a delay in flowering compared to those grown under long day conditions indicating that photoperiod did to some extent influence the timing of flowering in transgenic lines. The early flowering was associated with higher levels of the *FT* (*FLOWERING LOCUS T*) transcript, the gene involved in initiating flowering [30]. Expression of the truncated form of *SlERF36* (lacking the EAR motif) abrogated the early flowering phenotype of full length *SlERF36* expression under both short day and long day conditions. Its expression did not affect (increase) *FT* transcript levels (Fig. 2D) in spite of the presence of the AP2 domain. This indicated that the presence of the EAR motif was essential for the higher *FT* transcript levels and the early flowering phenotype although an effect of other deleted C-terminal residues cannot be ruled out. The fact that *SlERF6* over-expression accelerates flowering regardless of photoperiod and in plants as different as *Arabidopsis* and tobacco suggests that SlERF36 (and the EAR motif) might interact in some way with the general flowering machinery and regulate a component that is common to both photoperiods. Considering that EAR motif containing proteins function as active repressors of transcription, and that *SlERF36* expression leads to increased *FT* transcript levels, one could envisage a possibility where the direct or indirect repression of a floral inhibitor by SlERF36 could activate *FT* and thereby flowering in the transgenic lines. An interesting possibility that would require further studies is whether SlERF36 affects expression of homologues of *TEMPRANILLO* (*TEM1* and *TEM2*) that are known to directly repress *FT* expression [31] or whether it in some way controls TOE1/TOE2 or miRNA172, the regulation of which affects flowering in both short and long day conditions [32]. Incidentally, both TEM and TOE members belong to the AP2/ERF/RAV domain family of transcription factors.

Of the other changes, those related to early senescence appeared to be a consequence of the early flowering phenotype and therefore developmental in nature. This is based on the observations that although senescence was early in transgenic *SlERF36* plants (as seen through reduced chlorophyll and higher expression of the *SEN4* gene), it was dependent on the photoperiodic flowering and was delayed when flowering was delayed in short day conditions. Under these conditions, rosette sizes were larger than under long day conditions and plants took a longer time to senesce (Fig. 2E). The reduced sizes of the various organs (rosettes, flowers and siliques) in transgenic *SlERF36* over-expressing plants under long day conditions was most likely an effect due to reduced vegetative growth of these plants and the fewer number of leaves that were present at the time of flowering, leading to fewer photosynthates being synthesized and translocated. In transgenic *SlERF36ΔEAR* plants, where the timing of flowering was not affected, senescence was also normal.

In contrast to the flowering and senescence phenotypes, the other major phenotype namely stomatal number was affected in both transgenic *SlERF36* plants and those lacking the EAR motif (albeit to a lesser extent). At least in plants over-expressing *SlERF36ΔEAR* the apparent reduction in stomatal number by 25% could be attributed to an increase in cell size which increased by about 25%. Stomatal density is known to be tightly controlled by a large number of negative regulators such as ERECTA, ERL1, ERL2, EPF1, EPF2, CHALLAH, YODA, TMM, and SDD1 that are responsible for determining the spacing between stomata [33]. The fact that SlERF36 reduces stomatal density by 25–35% in tobacco and 38–39% in *Arabidopsis* would indicate that a stomatal development regulator common to both tobacco and Arabidopsis might be controlled by the repressor SlERF36. Stomatal density is also controlled by environmental factors such as CO_2, humidity, light intensity and water availability with CO_2 levels being by far the most important determinants of stomatal density and conductance in angiospermic plants. An inverse relationship between CO_2 levels and stomatal density and conductance has been noted under experimental conditions as well as in fossil studies [34,35,36] with CO_2 doubling leading to an average decrease of almost 22–29% in stomatal density in *Arabidopsis* and other plants [37,38]. Changes in stomatal density can affect photosynthesis and growth [39] particularly under conditions where there is no corresponding increase in CO_2 levels. Indeed the reduction in stomatal number and density does affect photosynthesis with growth being affected in *SlERF36* over-expressing plants but not in *SlERF36ΔEAR* over-expressing plants (where the reduction in stomatal number is lower). Several studies in Arabidopsis have shown that reduced stomatal densities in the range of 20–25% such as in mutants like *gtl1, edt1, gpa1* do not affect plant growth [40,41,42] unlike in *SlERF36* expressing plants where a much a larger decrease is observed. It is likely that beyond a certain point the reduced density may limit CO_2 availability affecting plant growth. Whether the effects on stomatal density/photosynthesis and flowering are related or independent effects is not yet known. Photosynthetic rates control starch reserves and these in turn could affect C/N ratios and thereby flowering. SlERF36 has recently been shown to actively repress the ethylene responsive GCC box *in vitro* [13]. The net effect of such a function would be to reduce ethylene responses. In this context, it is interesting to note that recent microarray studies in *SlERF36* expressing plants, although not conclusive, showed reduction in transcript levels of *AtERF1, AtERF2* (involved in ethylene responses) and *AtMKK9* (involved in ethylene biosynthesis), suggesting a reduction in ethylene responses (data not shown). The net effect of a reduction in ethylene responses would be an increase in cell size since ethylene is known to repress cell elongation. Ethylene is also known to delay flowering. Both these effects of increased cell size and early flowering, possibly indicative of reduced ethylene responses, are seen in transgenic *SlERF36* over-expressing lines. However, more detailed studies especially through loss of SlERF36 function lines are required to get at a causal relationship between SlERF36, ethylene responses and the early flowering.

In conclusion, we demonstrate that the EAR motif of SlERF36 is most likely responsible for the strong early flowering phenotype and a reduction in stomatal density and photosynthesis that is common to both Arabidopsis and tobacco when SlERF36 is over-expressed. The indication that this motif may also directly or indirectly control the expression of *FT*, although not studied in detail as yet by us, adds a new dimension to the complex pathways by which flowering is controlled in plants.

Author Contributions

Conceived and designed the experiments: APS PN UVP RKU. Performed the experiments: RKU AG SR RS. Analyzed the data: RKU AG UVP PN APS. Contributed reagents/materials/analysis tools: RKU AG SR RS UVP. Wrote the paper: APS.

References

1. Nakano T, Suzuki K, Fujimura T, Shinshi H (2006) Genome-wide analysis of the ERF gene family in *Arabidopsis* and rice. Plant Physiol 140: 411–432.
2. Kagale S, Links MG, Rozwadowski K (2010) Genome-wide analysis of ethylene-responsive element binding factor-associated amphiphilic repression motif-containing transcriptional regulators in *Arabidopsis*. Plant Physiol 15: 1109–1134.
3. Kagale S, Rozwadowski K (2011) EAR motif mediated transcriptional repression in plants. Epigenetics 6:2: 141–146.
4. Ohta M, Matsui K, Hiratsu K, Shinshi H, Ohme-Takagi M (2001) Repression domains of class II ERF transcriptional repressors share an essential motif for active repression. Plant Cell 13: 1959–1968.
5. Hiratsu K, Matsui K, Koyama T, Ohme-Takagi M (2003) Dominant repression of target genes by chimeric repressors that include the EAR motif, a repression domain, in *Arabidopsis*. Plant J 34: 733–739.
6. Song CP, Agarwal M, Ohta M, Guo Y, Halfter U, et al. (2005) Role of an *Arabidopsis* AP2/EREBP-type transcriptional repressor in abscisic acid and drought stress responses. Plant Cell 17: 2384–2396.
7. Song CP, Galbraith DW (2006) AtSAP18, an orthologue of human SAP18, is involved in the regulation of salt stress and mediates transcriptional repression in *Arabidopsis*. Plant Mol Biol 60:241–257.
8. Causier B, Ashworth M, Guo W, Davies B (2012) The TOPLESS interactome: A framework for gene repression in *Arabidopsis*. Plant Physiol 158: 423–438.
9. Hill K, Wang H, Perry SE (2008) A transcriptional repression motif in the MADS factor AGL15 is involved in recruitment of histone deacetylase complex components. Plant J 53: 172–185.
10. Hiratsu K, Mitsuda N, Matsui K, Ohme-Takagi M (2004) Identification of the minimal repression domain of SUPERMAN shows that the DLELRL hexapeptide is both necessary and sufficient for repression of transcription in *Arabidopsis*. Biochem Biophys Res Commun 321: 172–178.
11. Tsukagoshi H, Saijo T, Shibata D, Morikami A, Nakamura K (2005) Analysis of a sugar response mutant of *Arabidopsis* identified a novel B3 domain protein that functions as an active transcriptional repressor. Plant Physiol 138: 675–685.
12. Sharma MK, Kumar R, Solanke AU, Sharma R, Tyagi AK, et al. (2010) Identification, phylogeny, and transcript profiling of ERF family genes during development and abiotic stress treatments in tomato. Mol Genet Genom 284: 455–475.
13. Pirrello J, Prasad BCN, Zhang W, Chen K, Mila I, et al. (2012) Functional analysis and binding affinity of tomato ethylene response factors provide insight on the molecular bases of plant differential responses to ethylene. BMC Plant Biol 12:190 (doi:10.1186/1471-2229-12-190)
14. Yang Z, Tian L, Latoszek-Green M, Brown D, Wu K (2005) *Arabidopsis* ERF4 is a transcriptional repressor capable of modulating ethylene and abscisic acid responses. Plant Mol Biol 58: 585–596.
15. Lu J, Ju H, Zhou G, Zhu C, Erb M, et al. (2011) An EAR-motif-containing ERF transcription factor affects herbivore-induced signaling, defense and resistance in rice. Plant J 68: 583–596.
16. Dong C-J, Liu J-Y (2010) The *Arabidopsis* EAR-motif-containing protein RAP2.1 functions as an active transcriptional repressor to keep stress responses under tight control BMC Plant Biol 10: 47.
17. Pan I-C, Li C-W, Su R-C, Cheng C-P, Lin C-S, et al. (2010) Ectopic expression of an EAR motif deletion mutant of SlERF3 enhances tolerance to salt stress and *Ralstonia solanacearum* in tomato. Planta 232: 1075–1086.
18. Li Z, Zhang L, Yu Y, Quan R, Zhang Z, et al. (2011) The ethylene response factor AtERF11 that is transcriptionally modulated by the bZIP transcription

factor HY5 is a crucial repressor for ethylene biosynthesis in *Arabidopsis*. Plant J 68: 88–99.

19. Zhang H, Zhang J, Quan R, Pan X, Wan L, et al. (2013) EAR motif mutation of rice OsERF3 alters the regulation of ethylene biosynthesis and drought tolerance. Planta 237: 1443–1451.
20. McGrath KC, Dombrecht B, Manners JM, Schenk PM, Edgar CI, et al. (2005) Repressor- and activator-type ethylene response factors functioning in jasmonate signaling and disease resistance identified via a genome-wide screen of *Arabidopsis* transcription factor gene expression. Plant Physiol 139: 949–959.
21. Koyama T, Nii H, Mitsuda N, Ohta M, Kitajima S, et al. (2013) A regulatory cascade involving class II ETHYLENE RESPONSE FACTOR transcriptional repressors operates in the progression of leaf senescence. Plant Physiol 162: 991–1005.
22. Upadhyay RK, Soni DK, Singh R, Dwivedi UN, Pathre UV, et al. (2013) *SlERF36,* an EAR motif containing ERF gene from tomato, alters stomatal density and modulates photosynthesis and growth. J Exp Bot 64: 3237–3247.
23. Clough SJ, Bent AF (1998) Floral dip: a simplified method for *Agrobacterium*-mediated transformation of *Arabidopsis thaliana*. Plant J 16: 735–743.
24. Adamczyk BJ, Lehti-Shiu MD, Fernandez DE (2007) The MADS domain factors AGL15 and AGL18 act redundantly as repressors of the floral transition in Arabidopsis. Plant J 50: 1007–1019
25. Wu K, Zhang L, Zhou C, Yu C-W, Chaikam V (2008) HDA6 is required for jasmonate response, senescence and flowering in Arabidopsis. J. Exp Bot 59: 225–234.
26. Arnon DI (1949) Copper enzymes in isolated chloroplasts. Polyphenoloxidase in *Beta vulgaris*. Plant Physiol 24: 1–15.
27. Kazan K (2006) Negative regulation of defence and stress genes by EAR motif-containing repressors. Trends Plant Sci. 11: 109–112.
28. Ogata T, Kida Y, Arai T, Kishi Y, Manago Y, et al. (2012) Overexpression of tobacco ethylene response factor NtERF3 gene and its homologues from tobacco and rice induces hypersensitive response-like cell death in tobacco. J Gen Plant Pathol 78:8–17.
29. Ciftci-Yilmaz S, Morsy MR, Song L, Coutu A, Krizek BA, et al. (2007) The EAR-motif of the Cys2/His2-type zinc finger protein Zat7 plays a key role in the defense response of *Arabidopsis* to salinity stress. J Biol Chem 282: 9260–9268.
30. Turck F, Fornara F, Coupland G (2008) Regulation and identity of florigen: FLOWERING LOCUS T moves center stage. Annu Rev Plant Biol 59: 573–594.
31. Castillejo C, Pelaz S (2008) The balance between CONSTANS and TEMPRANILLO activities determines FT expression to trigger flowering. Curr Biol 18: 1338–1343.
32. Jung JH, Seo YH, Seo PJ, Reyes JL, Yun J, et al. (2007) The GIGANTEA-regulated microRNA172 mediates photoperiodic flowering independent of CONSTANS in *Arabidopsis*. Plant Cell 19: 2736–2748.
33. Shimada T, Sugano SS, Hara-Nishimura I (2011) Positive and negative peptide signals control stomatal density. Cell Mol Life Sci 68: 2081–2088.
34. Royer DL (2001) Stomatal density and stomatal index as indicators of paleoatmospheric CO_2 concentration. Rev Palaeobot Palynol 114: 1–28.
35. Franks PJ, Beerling DJ (2009) Maximum leaf conductance driven by CO_2 effects on stomatal size and density over geologic time. Proc Natl Acad Sci USA 106: 10343–10347.
36. Brodribb TJ, McAdam SAM, Jordan GJ, Field TS (2009) Evolution of stomatal responsiveness to CO_2 and optimization of water-use efficiency among land plants. New Phytol 183: 839–847.
37. Woodward FI, Lake JA, Quick WP (2002) Stomatal development and CO_2: ecological consequences. New Phytol 153: 477–484.
38. Woodward FI, Kelly CK (1995) The influence of CO_2 concentration on stomatal density. New Phytol 131: 311–327.
39. Doheny-Adams T, Hunt L, Franks PJ, Beerling DJ, Gray JE (2012) Genetic manipulation of stomatal density influences stomatal size, plant growth and tolerance to restricted water supply across a growth carbon dioxide gradient. Phil Trans Royal Soc B 367: 547–555.
40. Yoo CY, Pence HE, Jin JB, Miura K, Gosney MJ, et al. (2010) The *Arabidopsis* GTL1 transcription factor regulates water use efficiency and drought tolerance by modulating stomatal density via transrepression of SDD1. Plant Cell 22: 4128–4141.
41. Yu H, Chen X, Hong YY, Wang Y, Xu P, et al. (2008). Activated expression of an *Arabidopsis* HD-START protein confers drought tolerance with improved root system and reduced stomatal density. Plant Cell 20: 1134–1151.
42. Nilson SE, Assmann SM (2010) The α-subunit of the *Arabidopsis* heterotrimeric G protein, GPA1, is a regulator of transpiration efficiency. Plant Physiol 152: 2067–2077.

9

Assessment of Cultivation Factors that Affect Biomass and Geraniol Production in Transgenic Tobacco Cell Suspension Cultures

Nikolay Vasilev[1]*, Christian Schmitz[1], Ulrike Grömping[2], Rainer Fischer[1,3], Stefan Schillberg[1,4]

1 Department Plant Biotechnology, Fraunhofer Institute for Molecular Biology and Applied Ecology IME, Aachen, Germany, **2** Department II–Mathematics, Physics and Chemistry, Beuth University of Applied Sciences, Berlin, Germany, **3** Institute for Molecular Biotechnology, RWTH Aachen University, Aachen, Germany, **4** Institute for Phytopathology and Applied Zoology, Justus-Liebig University Giessen, Giessen, Germany

Abstract

A large-scale statistical experimental design was used to determine essential cultivation parameters that affect biomass accumulation and geraniol production in transgenic tobacco (*Nicotiana tabacum* cv. Samsun NN) cell suspension cultures. The carbohydrate source played a major role in determining the geraniol yield and factors such as filling volume, inoculum size and light were less important. Sucrose, filling volume and inoculum size had a positive effect on geraniol yield by boosting growth of plant cell cultures whereas illumination of the cultures stimulated the geraniol biosynthesis. We also found that the carbohydrates sucrose and mannitol showed polarizing effects on biomass and geraniol accumulation. Factors such as shaking frequency, the presence of conditioned medium and solubilizers had minor influence on both plant cell growth and geraniol content. When cells were cultivated under the screened conditions for all the investigated factors, the cultures produced ~5.2 mg/l geraniol after 12 days of cultivation in shaking flasks which is comparable to the yield obtained in microbial expression systems. Our data suggest that industrial experimental designs based on orthogonal arrays are suitable for the selection of initial cultivation parameters prior to the essential medium optimization steps. Such designs are particularly beneficial in the early optimization steps when many factors must be screened, increasing the statistical power of the experiments without increasing the demand on time and resources.

Editor: M. Lucrecia Alvarez, Mayo Clinic Arizona, United States of America

Funding: This research was funded by the European Union Seventh Framework Programme SmartCell (Rational design of plant systems for sustainable generation of value-added industrial products, grant agreement no. 222716). The funder had no role in study design, data collection and analysis, decision to publish, or preparation of the manuscript.

Competing Interests: The authors have declared that no competing interests exist.

* Email: nikolay.vasilev@ime.fraunhofer.de

Introduction

Geraniol is an intermediate metabolite in the monoterpenoid-secoiridoid biosynthesis pathway and is used as a flavor and fragrance compound in the agricultural, food and cosmetic industries. It is an ingredient in 43% of fragrances currently on the market [1]. The annual global demand for geraniol was approximately 327 kg in 2008, with a 3–4% annual market growth rate [2]. The three major geraniol manufacturing strategies are biosynthesis (isolation from essential oils), semi-synthesis (chemical conversion of pinenes) and total synthesis from petrochemicals, and the yield varies between 60% (biosynthesis) and 98% (total synthesis) [2].

The enzyme geraniol synthase (GES) is responsible for the biosynthesis of geraniol from geranyl diphosphate [3–10]. GES expression in plants may help to reconstitute the terpenoid-indole alkaloid pathway and therefore satisfy the consumer demand for geraniol.

Plant cell and tissue cultures can be used as an alternative to whole plants for the production of valuable natural compounds [11,12]. Physical factors that have a major impact on the performance of plant-derived cultures include the inoculum state, size and age [11,13–15], the volume of medium in the culture vessel and the shaking speed, both of which influence on the oxygen supply during cultivation [16,17]. Biological factors also play an important role, including the use of conditioned medium to influence the accumulation of plant cell biomass [18] and the selection of carbohydrates as sources of energy and carbon due to the low levels of CO_2 in the closed-glass cultivation vessels during the photoperiod [19]. Light influences both growth and secondary metabolism in plants, e.g. by regulating the diurnal pattern of monoterpenoid synthase transcript levels and monoterpenoid production [20,21]. Finally, the equilibrium of production may be influenced by introducing excipients such as cyclodextrins that increase the solubility of sparingly-soluble compounds [22].

All the cultivation factors described above may have a direct influence on the productivity of plant cell suspension cultures in terms of biomass accumulation and the synthesis of secondary metabolites. Typically, a one-factor-at-a-time testing approach is used to find optimal cultivation conditions, but this does not identify interactions between factors that can improve or inhibit productivity. Statistical experimental designs are widely used in the biotechnology industry to avoid the drawbacks of testing single factors while leaving the others unchanged. Instead, statistical experimental designs determine the impact of several nutritional and physical factors simultaneously using full factorial, fractional factorial or response surface designs [18,23–25].

To our knowledge, this is the first report describing a statistical experimental design approach that includes multiple physical factors (inoculum size, shaking frequency, filling volume, light) and biological factors (conditioned medium, carbohydrate source and solubilizers) in one experiment. Our results allowed us to optimize geraniol production in transgenic tobacco suspension cultures expressing plastid-targeted GES from *Valeriana officinalis*.

Materials and Methods

Origin of transgenic material and initiation of cell suspension cultures

Cloning of *V. officinalis* GES (*VoGES*) cDNA (GenBank accession: KF951406.1) and stable transformation of tobacco plants was conducted as described previously [3]. *N. tabacum* cv. Samsun NN seeds from a homozygous T_7 line carrying a single *VoGES* expression cassette were selected for initiation of cell suspension cultures. The transgenic seeds were germinated under a 16-h light daily period in sterile conditions on hormone-free MS medium (Duchefa, Netherlands), supplemented with 100 mg/l kanamycin. Next, the intact *in vitro* plants were established on the same medium without antibiotics in plastic transparent containers. Sterile transgenic seedlings were used for the onset of callus culture in Petri dishes containing MS medium with vitamins, supplemented with 0.1 mg/l kinetin and 1 mg/l 1-naphthaleneacetic acid (NAA). Suspended cell cultures were developed subsequently by transferring callus tissue into 10 ml Gamborg's B5 medium plus vitamins, 0.1 mg/l kinetin and 1 mg/l NAA, accommodated in 50 ml TubeSpin bioreactors (Techno Plastic Products AG, Switzerland). Cell suspension cultures were maintained on a gyratory shaker (180 rpm) at 26°C under 16-h illumination (35.6 μmol/s)/8-h darkness and underwent subcultivation every two weeks prior to the start of the screening experimental design. Fresh biomass was collected at the end of the screening cultivation (day 10) and subjected to double filtration under vacuum.

Time curve of *VoGES* suspensions growth was performed in three biological replicates and the samples were collected for analysis every three days after the start of cultivation. Dissolved oxygen was measured online in 250 ml flasks by means of the BPM-60 (BioProcess Monitoring) device (Kühner AG, Switzerland). All the screening or validation experiments were conducted in ISF1-X or in LT-X shaker machines (Kühner AG).

GC-MS analysis

The extraction and quantitation of geraniol in the plant cell cultures was performed as described earlier by Dong et al. [3] with small modifications. Briefly, plant suspension material was filtrated twice under vacuum and subsequently frozen at −20°C. Two-hundred mg plant cell material was taken, ground and resuspended homogenously in 1 ml citrate-phosphate buffer (pH 5.4), prepared by mixing 27.8 ml 0.2 M dibasic sodium phosphate and 22.2 ml 0.1 M citric acid and topping up to 100 ml with water [26]. The samples were ultra-sonificated for 15 min in a water bath at ambient temperature, followed by the addition of 0.5 ml "Viscozyme L" enzymatic mixture (Sigma-Aldrich, Germany). Next, 1 ml layer of heptane was added to the each sample, containing 10 μg/ml (*Z*)-nerolidol (Sigma-Aldrich) and incubated overnight at 37°C. Next day, each sample was centrifuged at 1750×*g* for 10 min at room temperature. The organic layer was removed and fresh 1 ml heptane was added to sample, vortexed and centrifuged at 1750×*g* for 10 min at room temperature. The last step was repeated once again and all three heptane extracts were pooled, passed through a glass Pasteur pipette provided with a small glass wool plug and ~1.5 cm anhydrous sodium sulfate (Sigma-Aldrich).

The heptane extract was concentrated under nitrogen gas and the geraniol content was measured quantitatively by a GC-MS machine coupled to a QP2010SE quadrupole mass spectrometer (Shimadzu, Japan). The separation was performed on a 30 m length ×0.25 mm internal diameter Zebron ZB-5 ms column (Phenomenex, USA) containing 0.25 μm stationary phase, preceded by a 5-m guard column. The injected 1 μl aliquot of each sample (split mode 1:10, injection port temperature 250°C) was subjected to the following temperature program: 45°C for 1 min followed by a gradient of 10°C per min until the temperature reached 300°C, which was held for a further 7 min. The helium inlet pressure was maintained at a constant column flow rate of 1.0 ml/min.

Geraniol was quantified using the base peak ($m/z = 69$) in SIM mode following electric ionization at 1 keV. The recorded mass-spectrum was compared with spectra of a reference standard from the NIST library.

Experimental design and statistical analysis

The experiment was accommodated in four shaker, satisfying the combinatorial states for the factors light intensity and shaking frequency. Temperature was fixed at 26°C based on our preliminary experiments (Table S1) in order to avoid the necessity for additional shaker machines.

We used an experimental design based on an orthogonal array with 72 runs created with the free open source R package DoE.base [27] (for details, see supplementary material). The design was optimized with respect to confounding of low order effects: all main effects are orthogonal to each other (orthogonal array), confounding between main effects and two-factor interactions was minimized, and in a second step, confounding among two-factor interactions was kept as low as possible. Figure 1 shows a mosaic plot for a triple of design factors that is one of the two triples with the worst-case confounding between main effects and two-factor interactions; even this worst case is reasonably balanced [28].

Results

Seven cultivation factors (four 2-level factors, two 3-level factors and one 4-level factor) were screened in a randomized experimental design with limited confounding: light, shaking frequency, inoculum size, filling culture volume, addition of conditioned medium, carbohydrate type (sucrose, glucose and D-mannitol) and cyclodextrin type (β-cyclodextrin, methyl-β-cyclodextrin, triacetyl-β-cyclodextrin and no cyclodextrin) (Table 1). Three responses were measured correspondingly: fresh weight, geraniol content and geraniol yield. The full experimental design with 72 runs in coded values and the measured responses are given in Table S2.

Geraniol yield model

The productivity of the plant cell platform depends on both the biomass production and the geraniol content per unit biomass. The multiplication of fresh weight by geraniol content in fresh weight stands for geraniol yield in our production system. The geraniol yield model ($p<0.05$) indicated that the strongest factors with effect on geraniol production are sucrose, filling volume and inoculum size, followed by light (Figure 2).

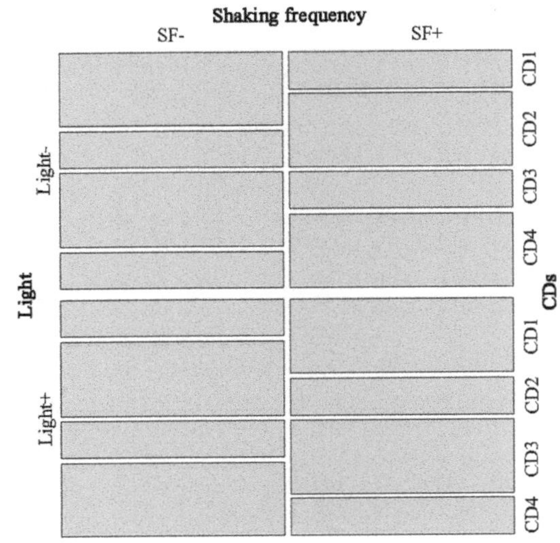

Figure 1. Mosaic plot for three selected factors (light, shaking frequency (SF) and cyclodextrins (CD1-no cyclodextrin; CD2-β-cyclodextrin; CD3-methyl-β-cyclodextrin and CD4-triacetyl-β-cyclodextrin)) showing that any level combination of factors CDs and SF does not determine completely the level of factor "light" which is indicative for limited confounding.

Biomass model

We created also biomass and geraniol content models ($p<0.05$) in order to demonstrate how plant cell growth and geraniol content influenced the geraniol yield. The biomass plot showed that the carbohydrate source from all screened factors had the strongest influence on biomass production of the transgenic plant cell cultures (Figure 3). The highest growth was achieved when sucrose was used as a carbon source in the cultivation medium. On the contrary, mannitol considerably reduced the plant cell growth when incorporated as a carbohydrate source into the medium. Filling volume and inoculum size exerted also a positive effect on biomass production but in a less profound extent as sucrose. Therefore, sucrose, filling volume and inoculum size affected biomass production in the same manner as geraniol yield. Shaking frequency, light and addition of conditioned medium and cyclodextrins have exhibited insignificant or negligible effect on the growth of plant cell suspensions.

Geraniol content model

The screening in terms of the geraniol content identified that the use of mannitol led to the highest geraniol levels accumulated in the tobacco transgenic cell cultures (Figure 4). Light had also a potent effect on geraniol biosynthesis. The rest of cultivation factors did not show a strong impact on geraniol production as seen from the geraniol content model ($p<0.05$).

The pattern of the most significant factors from the biomass model is very similar to the pattern of the strongest factors in the geraniol yield model. This serves as additional evidence that the growth of plant cells contributes to a greater extent to the geraniol yield formation in comparison to the factors acting mainly on the enhancement of geraniol content.

Significant interactions

We also identified several potential two-factor interactions concerning geraniol yield (Table S3). The most significant interactions ($p<0.01$) are filling volume:sugar (Figure 5A), sugar:-cyclodextrins (Figure 5B) and light:sugar (Figure 5C). Note that the figures do not show average geraniol yields at each level combination (as a naïve interaction plot would do) but estimated interaction effects from the model with all main effects and 2-factor interactions (obtained with R package effects by Fox 2003

Table 1. Levels of the screened factors.

Factor	Levels	Abbreviation
Light	11.50 (μmol/cm^2/s)	Lght-
	35.62 (μmol/cm^2/s)	Lght+
Shaking frequency	140 rpm	SF-
	180 rpm	SF+
Inoculum size	0.75 g	IS-
	1.40 g	IS+
Filling volume	15 ml	FV-
	20 ml	FV0
	25 ml	FV+
Conditioned medium	100% fresh medium	CM-
	1 ml conditioned medium	CM+
Carbohydrate	Sucrose	Suc
	Glucose	Gluc
	D-Mannitol	Mannit
Cyclodextrin (2 mM)	No cyclodextrin	CD1
	β-cyclodextrin	CD2
	Methyl-β-cyclodextrin	CD3
	Triacetyl-β-cyclodextrin	CD4

The screening experiment was conducted in 50 ml Erlenmeyer shaking flasks.

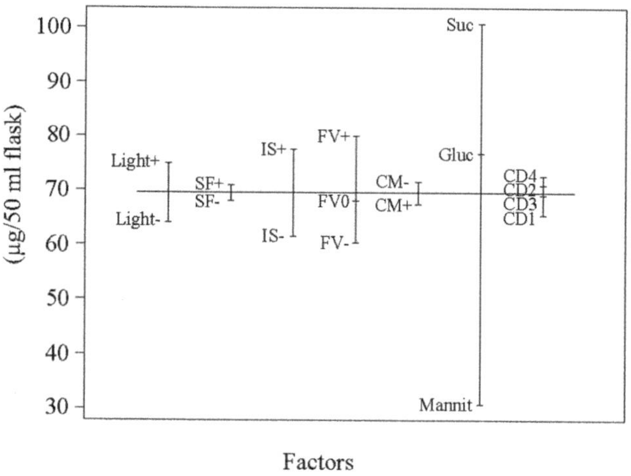

Figure 2. Geraniol yield model plot. The abbreviations are the same as in Table 1.

[29]). These are unbiased estimates in the presence of the partial confounding among 2-factor interactions that is present (and in some cases strong) in the experimental design. All three strongest interaction effects involve at least one factor with a strong main effect and are therefore plausible according to the weak effect heredity principle [30]. For two of the strong interactions, both factors have active main effects (strong effect heredity). For instance, Figure 5 shows that an increase of filling volume has a different effect in case mannitol was used as a source of carbohydrate than for the other two sugars. The effect of cyclodextrins on the geraniol yield goes into opposite directions for the sugars mannitol and sucrose, respectively. Light and sucrose potentiate reciprocally their actions on geraniol yield because the corresponding main effects are very strong. Remarkably, shaking frequency also appears to be involved in at least one, possibly several interactions, demonstrating its important role for maximizing geraniol production during plant cell cultivation (Table S3).

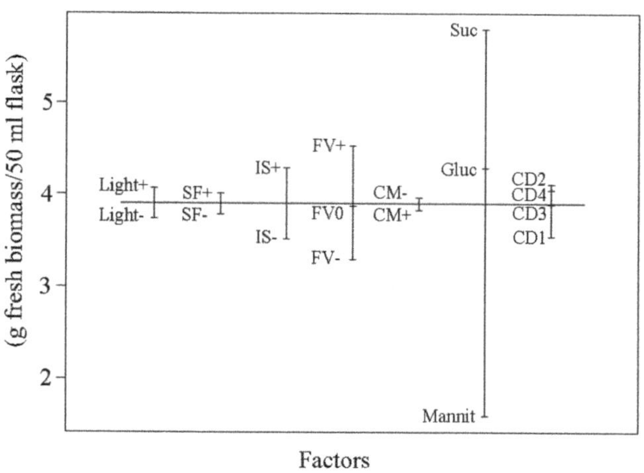

Figure 3. Biomass model plot. The abbreviations are the same as in Table 1.

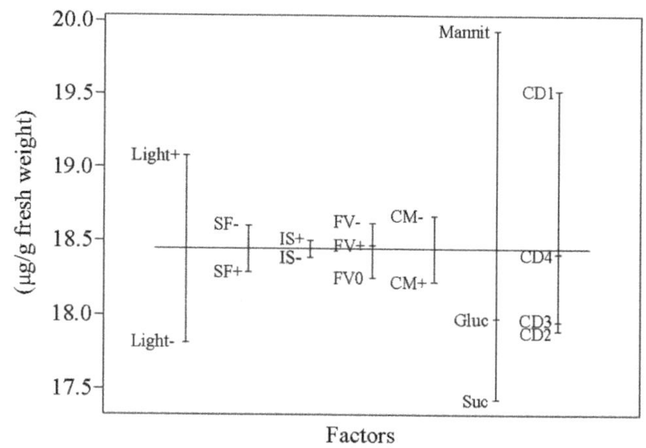

Figure 4. Geraniol content model plot. The abbreviations are the same as in Table 1.

Time-course dynamics

In the next step, the cultivation of the plant suspension was performed at the favourable levels of factors from the geraniol yield design in order to determine the time dynamics of plant cell productivity in larger shaking flasks (250 ml). We set the screened factors at the beneficial levels in Gamborg's B5-medium supplemented with vitamins and phytohormones in order to keep the corresponding proportions from the geraniol yield design. The following culture conditions were chosen: 20 g/l sucrose; 50% filling volume; 5.6% (w/v) inoculum size; 35.62 µmol/cm^2/s light; 2 mM triacetyl-β-cyclodextrin and 180 rpm shaking frequency. The suspension cultures grew steadily throughout the cultivation period (Figure 6A) and showed signs of cell death after 14 days post inoculation (dpi). The latter was due to the depletion of nutrients and sucrose in the plant cell cultivation medium which is also seen from the reduction of electroconductivity and osmolality (Figure 6B and 6C). Dissolved oxygen levels started to increase after its minimum at 5 dpi and reached its maximum at 12 dpi, which is indicative for the switch from heterotrophic to autotrophic growth (Figure 6D). The maximal geraniol yield in the plant cell cultures at the end of the cultivation period was 5.12 mg/l (Figure 6E) which is commensurate with the highest levels achieved in the screening design (runs 51 and 60 from Table S2) if these results are extrapolated to the same geraniol yield unit (mg/l).

Discussion

We initiated a multi-factorial experimental design based on orthogonal arrays to investigate the effects of different cultivation factors on biomass accumulation and geraniol production in transgenic tobacco cell cultures grown in shaking flasks. Such a preliminary screening step is ordinarily needed to define the basic conditions of the key cultivation factors prior the main medium optimization, focused on the sequential manipulation of nutrients and elicitors [25]. Furthermore, cultivation factors are known to demonstrate a variation in a species-dependent manner and thus should be adjusted on a case-by-case basis [18]. Finally, being holistic in nature the optimization based on experimental designs demonstrates improved reproducibility because it takes into account the interactive effects of screened factors [31].

Our study demonstrated that the geraniol yield is mainly dependent on the factors that regulate the growth of the plant cell

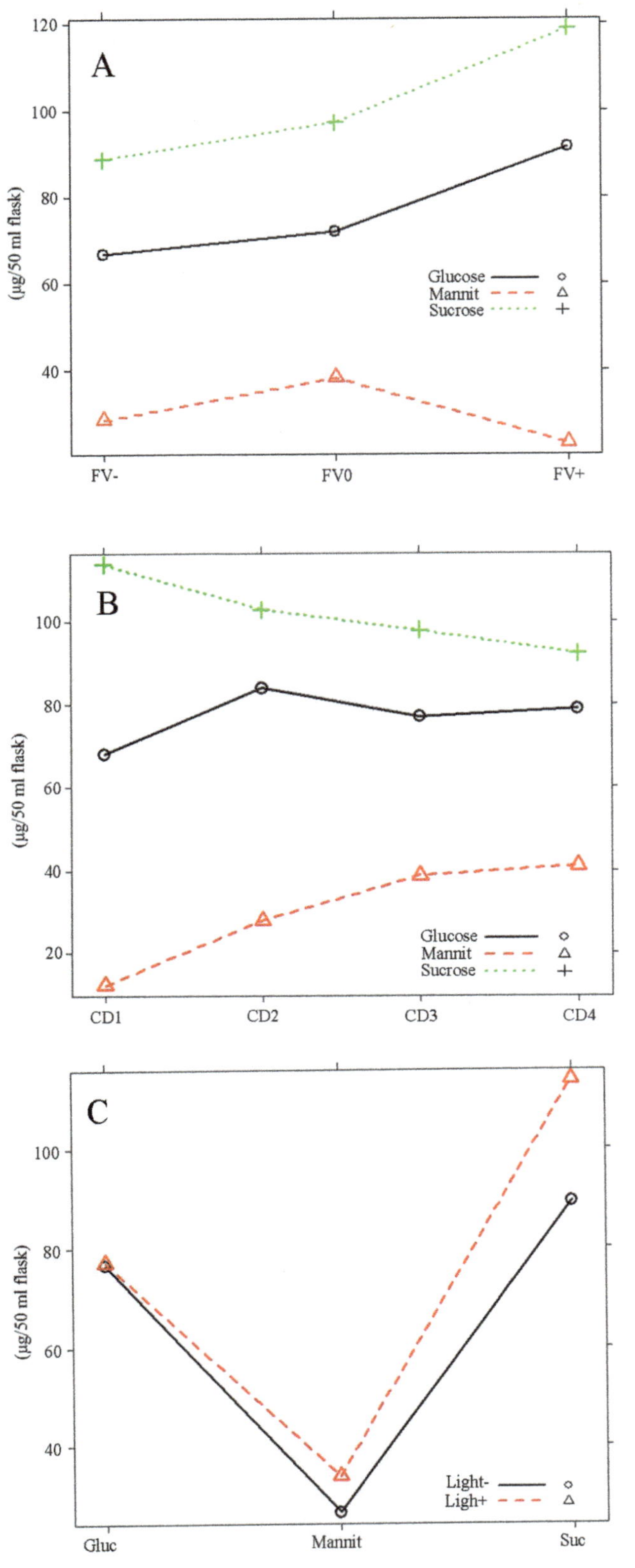

Figure 5. Significant two-factor interactions that affect geraniol yield, detected by the screening design. The abbreviations are the same as in Table 1. (A) Filling volume:sugar interaction. (B) Sugar:cyclodextrins interaction. (C) Light:sugar interaction.

cultures. The factors acting positively on biomass production had a prevailing influence over the factors that are active on geraniol content with regard to the final geraniol yield. Among the tested cultivation parameters, the carbohydrate source had the strongest effect on biomass formation in the transgenic tobacco cultures. Glucose was utilized in a lesser extent than sucrose and the supplementation of the plant cell cultures with mannitol inhibited strongly their growth. Sucrose is known to be the most universal sugar for the plant cell cultivation, followed by glucose, maltose and raffinose [19], whereas the addition of some sugar alcohols like mannitol may cause osmotic stress in the system [19,32]. Additionally, these sugar alcohols may serve not only as osmotic agents but they also regulate morphogenesis and metabolism [33]. We observed a similar metabolic effect of mannitol as it was the most potent factor stimulating the geraniol biosynthesis. However, the exact mechanism of induction of geraniol metabolism by mannitol remains to be clarified.

Filling volume is one of the operating parameters affecting the oxygen transfer rate in shaking flasks [34]. Filling volume had also an important effect on biomass production and respectively on geraniol yield of *VoGES* tobacco suspension cultures. Increasing the shaking speed and decreasing the medium volume improved biomass production to a certain extent in shake flask culture of *Atropa belladonna* hairy roots [16]. In our tobacco plant cell cultures, shaking speed did not play a significant role on geraniol yield. This indicates that the tobacco suspension cell cultures might not suffer from serious oxygen limitations at the applied shaking frequency and filling volume levels. Our study showed that shaking frequency may not function as a strong single factor but it is involved in several significant interactions, concerning biomass and geraniol yield mainly, and therefore, it should be also taken into consideration in the initial stage of optimization.

Our experiments identified that inoculum size had a moderate effect whilst the addition of conditioned medium had negligible impact on the biomass production of transgenic tobacco suspensions. Multi-factorial analyses using response surface experimental designs revealed that both inoculum size and conditioned medium had a positive effect on the biomass production in oil palm (*Elaeis guineensis*) cultures [18]. The inoculum size and the addition of conditioned medium had also a beneficial effect on alkaloid production in immobilized *Catharanthus roseus* cells [35]. Previous reports showed that the enhancement of inoculum density usually leads to higher biomass productivity but lower growth rate [13,15]. The high inoculum size can contribute for the elimination or even diminishment of the lag phase during the cultivation [11]. The experimental design in our investigation could also detect a positive influence of high inoculum levels on geraniol yield. This observation is in agreement with earlier reports in literature showing that several enzymes i.e. geraniol-10-hydroxylase, had increased their activity in high-density cultures of *Catharanthus roseus* [36].

Addition of conditioned medium can enhance the cell density and the accumulation of ginseng polysaccharide in *Panax notoginseng* cell cultures [37]. The lack of potent effect of the conditioned medium in our plant cell cultures may be due to various reasons like reduction of osmolarity, decreasing the sugar and nitrogen concentrations, or to a phytohormone alteration because of the conditioned medium addition, although in general conditioned medium may contain some favourable stimulating factors at the same time [18].

The geraniol content model identified light as an important factor for geraniol biosynthesis. Several studies showed previously that light exerts a strong effect on monoterpenoid metabolism by regulating transcript levels of monoterpenoid synthases, precursor

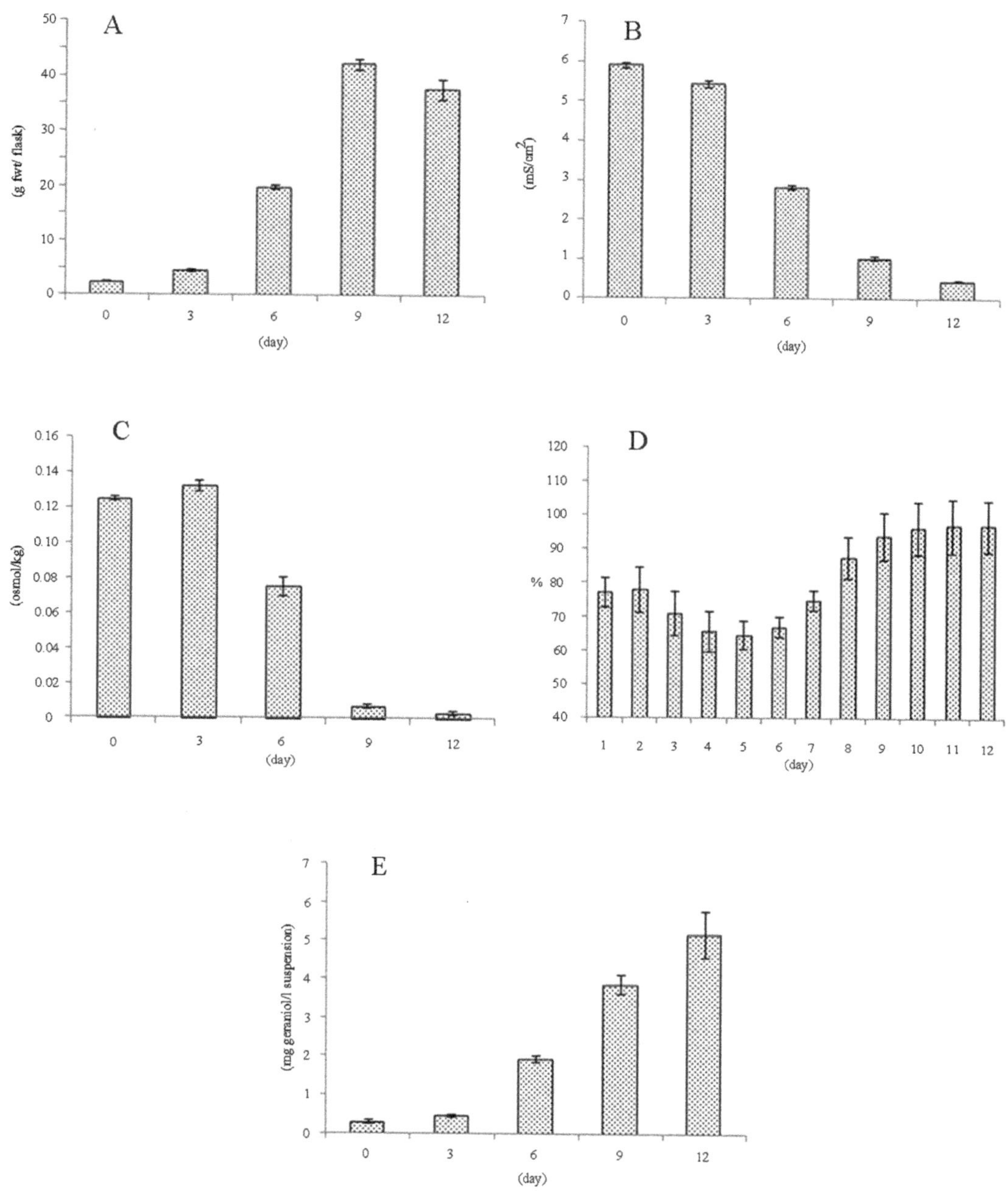

Figure 6. Time-course dynamics. Cultivation conditions: 250 ml shaking flasks; 20 g/l sucrose; 50% filling volume; 5.6% (w/v) inoculum size; 35.62 (μmol/cm^2/s) light; 2 mM triacetyl-β-cyclodextrin and 180 rpm shaking frequency. (A) Plant cell growth. (B) Electroconductivity. (C) Osmolality. (D) Dissolved oxygen. (E) Geraniol yield.

availability, constitutive promoter activity, etc. [20,21,38]. Furthermore, the interaction we found between light and sucrose might be also indicative for mixed contribution of both autotrophic and heterotrophic carbon metabolism in cultured tobacco cells [39].

We report herein that transgenic tobacco suspensions expressing *V. officinalis* GES could reach geraniol yield of 5.2 mg/l after 12 days cultivation. This accumulation level in non-fully-optimized conditions is close to the geraniol yield produced by yeast cells expressing the *O. basilium* GES [4]. Additional medium optimization may further improve the geraniol yield and turn transgenic plant cell cultures into an attractive production system of valuable natural products, matching the yields achieved by semi-synthesis and total chemical synthesis. Therefore, a combined, empirical study about optimization of plant suspensions productivity is needed, aiming at both physical and chemical factors and experimental statistical designs may serve as a useful tool.

Acknowledgments

We would like to thank Dr. Harro Bouwmeester (University of Wageningen, the Netherlands) for providing us with *N. tabacum* cv. Samsun NN transgenic seeds, expressing stably *VoGES*. We thank Dr. Richard Twyman for critical reading of the manuscript. None of the authors declares a conflict of interest.

Author Contributions

Conceived and designed the experiments: NV UG RF StS. Performed the experiments: NV CS. Analyzed the data: NV CS UG. Contributed to the writing of the manuscript: NV UG RF StS.

References

1. de Roode BM, Oliehoek L, van der Padt A, Franssen MC, Boom RM (2001) Downstream processing of enzymatically produced geranyl glucoside. Biotechnol Prog 17: 881–886.
2. Sathikge ND (2008) Process for the preparation of geraniol and nerol from crude sulphated turpentine stream. M. T. D. Thesis, Nelson Mandela Metropolitan University. Available: http://dspace.nmmu.ac.za:8080/jspui/bitstream/10948/734/1/Ndavheleseni%20David%20Sathikge.pdf. Accessed 2014 June 27.
3. Dong L, Miettinen K, Goedbloed M, Verstappen FW, Voster A, et al. (2013) Characterization of two geraniol synthases from *Valeriana officinalis* and *Lippia dulcis*: Similar activity but difference in subcellular localization. Metab Eng 20: 198–211.
4. Fischer MJ, Meyer S, Claudel P, Bergdoll M, Karst F (2011) Metabolic engineering of monoterpene synthesis in yeast. Biotechnol Bioeng 108: 1883–1892.
5. Fischer MJ, Meyer S, Claudel P, Perrin M, Ginglinger JF, et al. (2013) Specificity of *Ocimum basilicum* geraniol synthase modified by its expression in different heterologous systems. J Biotechnol 163: 24–29.
6. Gutensohn M, Orlova I, Nguyen TT, Davidovich-Rikanati R, Ferruzzi MG, et al. (2013) Cytosolic monoterpene biosynthesis is supported by plastid-generated geranyl diphosphate substrate in transgenic tomato fruits. Plant J 75: 351–363.
7. Iijima Y, Gang DR, Fridman E, Lewinsohn E, Pichersky E (2004) Characterization of geraniol synthase from the peltate glands of sweet basil. Plant Physiol 134: 370–379.
8. Ito M, Honda G (2007) Geraniol synthases from perilla and their taxonomical significance. Phytochemistry 68: 446–453.
9. Simkin AJ, Miettinen K, Claudel P, Burlat V, Guirimand G, et al. (2013) Characterization of the plastidial geraniol synthase from Madagascar periwinkle which initiates the monoterpenoid branch of the alkaloid pathway in internal phloem associated parenchyma. Phytochemistry 85: 36–43.
10. Yang T, Li J, Wang HX, Zeng Y (2005) A geraniol-synthase gene from *Cinnamomum tenuipilum*. Phytochemistry 66: 285–293.
11. Jeong GT, Park DH, Ryu HW, Hwang B, Woo JC (2004) Effects of inoculum conditions on growth of hairy roots of *Panax ginseng* C.A. Meyer. Appl Biochem Biotechnol 113–116: 1193–1203.
12. Smetanska I (2008) Production of secondary metabolites using plant cell cultures. Food Biotechnol 111: 187–228.
13. Cui XH, Chakrabarty D, Lee EJ, Paek KY (2010) Production of adventitious roots and secondary metabolites by *Hypericum perforatum* L. in a bioreactor. Bioresour Technol 101: 4708–4716.
14. Takahashi Y, Hitaka Y, Kino-oka M, Taya M, Tone S (2001) Evaluation of growth property of red beet hairy roots depending on condition of inocula and its application to culture control with fuzzy logic theory. Biochem Eng J 8: 121–127.
15. Zhang CH, Wu JY, He GY (2002) Effects of inoculum size and age on biomass growth and paclitaxel production of elicitor-treated *Taxus yunnanensis* cell cultures. Appl Microbiol Biotechnol 60: 396–402.
16. Kanokwaree K, Doran PM (1997) The extent to which external oxygen transfer limits growth in shake flask culture of hairy roots. Biotechnol Bioeng 55: 520–526.
17. Klockner W, Büchs J (2012) Advances in shaking technologies. Trends Biotechnol 30: 307–314.
18. Gorret N, bin Rosli AK, Oppenheim SF, Willis LB, Lessard PA, et al. (2004) Bioreactor culture of oil palm (*Elaeis guineensis*) and effects of nitrogen source, inoculum size, and conditioned medium on biomass production. J Biotech 108: 253–263.
19. George EF, Hall MA, De Klerk G-J (2008) Plant propagation by tissue culture. The background. Dordrecht, the Netherlands: Springer. 501 p.
20. Aharoni A, Giri AP, Deuerlein S, Griepink F, de Kogel WJ, et al. (2003) Terpenoid metabolism in wild-type and transgenic *Arabidopsis* plants. Plant Cell 15: 2866–2884.
21. Lu S, Xu R, Jia JW, Pang JH, Matsuda SPT, et al. (2002) Cloning and functional characterization of a beta-pinene synthase from *Artemisia annua* that shows a circadian pattern of expression. Plant Physiol 130: 477–486.
22. Brewster ME, Loftsson T (2007) Cyclodextrins as pharmaceutical solubilizers. Adv Drug Deliver Rev 59: 645–666.
23. Prakash G, Srivastava AK (2005) Statistical media optimization for cell growth and azadirachtin production in *Azadirachta indica* (A. Juss) suspension cultures. Process Biochem 40: 3795–3800.
24. Prakash G, Srivastava AK (2008) Statistical elicitor optimization studies for the enhancement of azadirachtin production in bioreactor *Azadirachta indica* cell cultivation. Biochem Eng J 40: 218–226.
25. Vasilev N, Gromping U, Lipperts A, Raven N, Fischer R, et al. (2013) Optimization of BY-2 cell suspension culture medium for the production of a human antibody using a combination of fractional factorial designs and the response surface method. Plant Biotechnol J 11: 867–874.
26. Pearse AGE (1980) Histochemistry, theoretical and applied: preparative and optical technology. Edinburgh: Churchill Livingstone. 728 p.
27. Grömping U, Amarov B, Xu H (2013) DoE.base: Full factorials, orthogonal arrays and base utilities for DoE packages. Package version 0.25-3. Available: http://CRAN.R-project.org/package = DoE.base. Accessed 26 June 2014.
28. Grömping U (2014) Mosaic plots are useful for visualizing low-order projections of factorial designs. Am Stat 68: 108–116.
29. Fox J (2003) Effect displays in R for generalised linear models. J Stat Softw 8: 1–27.
30. Yuan M, Joseph VR, Lin Y (2007) An efficient variable selection approach for analyzing designed experiments. Technometrics 49: 430–439.
31. Vaidya BK, Mutalik SR, Joshi RM, Nene SN, Kulkarni BD (2009) Enhanced production of amidase from *Rhodococcus erythropolis* MTCC 1526 by medium optimisation using a statistical experimental design. J Ind Microbiol Biotechnol 36: 671–678.
32. Lipavska H, Vreugdenhil D (1996) Uptake of mannitol from the media by in vitro grown plants. Plant Cell Tiss Org Cult 45: 103–107.
33. Steinitz B (1999) Sugar alcohols display nonosmotic roles in regulating morphogenesis and metabolism in plants that do not produce polyols as primary photosynthetic products. J Plant Physiol 155: 1–8.
34. Maier U, Büchs J (2001) Characterisation of the gas-liquid mass transfer in shaking bioreactors. Biochem Eng J 7: 99–106.
35. Lee CWT, Shuler ML (2000) The effect of inoculum density and conditioned medium on the production of ajmalicine and catharanthine from immobilized *Catharanthus roseus* cells. Biotechnol Bioeng 67: 61–71.
36. Moreno PR, Schlatmann JE, van der Heijden R, van Gulik WM, ten Hoopen HJ, et al. (1993) Induction of ajmalicine formation and related enzyme activities in *Catharanthus roseus* cells: effect of inoculum density. Appl Microbiol Biotechnol 39: 42–47.
37. Woragidbumrung KO, Sae-Tang P, Yao H, Han J, Chauvatcharin S, et al. (2001) Impact of conditioned medium on cell cultures of *Panax notoginseng* in an airlift bioreactor. Process Biochem 37: 209–213.
38. Schnurr JA, Guerra DJ (2000) The CaMV-35S promoter is sensitive to shortened photoperiod in transgenic tobacco. Plant Cell Rep 19: 279–282.
39. Wolf S, Kalman-Rotem N, Yakir D, Ziv M (1998) Autotrophic and heterotrophic carbon assimilation of in vitro grown potato (*Solanum tuberosum* L.) plants. J Plant Physiol 153: 574–580.

10

Icariin Ameliorates Neuropathological Changes, TGF-β1 Accumulation and Behavioral Deficits in a Mouse Model of Cerebral Amyloidosis

Zhi-Yuan Zhang*, Chaoyun Li, Caroline Zug, Hermann J. Schluesener

Division of Immunopathology of the Nervous System, Institute of Pathology and Neuropathology, University of Tuebingen, Tuebingen, Germany

Abstract

Icariin, a major constituent of flavonoids from the Chinese medicinal herb Epimedium brevicornum, exhibits multiple biological properties, including anti-inflammatory, neuroregulatory and neuroprotective activities. Therefore, Icariin might be applied in treatment of neurodegenerative disorders, including Alzheimer's disease (AD), which is neuropathologically characterized by β-amyloid aggregation, hyperphosphorylated tau and neuroinflammation. Potential therapeutic effects of Icariin were investigated in an animal model of cerebral amyloidosis for AD, transgenic APP/PS1 mouse. Icariin was suspended in carboxymethylcellulose and given orally to APP/PS1 mice. Therapeutic effects were monitored by behavioral tests, namely nesting assay, before and during the experimental treatment. Following an oral treatment of 10 days, Icariin significantly attenuated Aβ deposition, microglial activation and TGF-β1 immunoreactivity at amyloid plaques in cortex and hippocampus of transgenic mice 5 months of age, and restored impaired nesting ability. Our results suggest that Icariin might be considered a promising therapeutic option for human AD.

Editor: Jaya Padmanabhan, University of S. Florida College of Medicine, United States of America

Funding: Chaoyun Li is supported and financed by the Chinese government scholarship from the China Scholarship Council (CSC). The authors also acknowledge support by Deutsche Forschungsgemeinschaft (German Research Foundation) and Open Access Publishing Fund of Tuebingen University, for the open access publishing. The funders had no role in study design, data collection and analysis, decision to publish, or preparation of the manuscript.

Competing Interests: The authors have declared that no competing interests exist.

* Email: zhiyuan.zhang@medizin.uni-tuebingen.de

Introduction

Icariin is a natural flavonoid extracted from the Chinese tonic herb Epimedium and is considered the major pharmacologically active compound. Multi-functional Icariin possesses anti-tumor, anti-oxidant, vasorelaxant, anti-bacterial and anti-inflammatory activities [1] (Figure 1). Previous studies reported that Icariin ameliorated brain dysfunction induced by LPS [2], inhibited corticosterone-induced apoptosis in neurons [3], attenuated the ischemia/reperfusion damage to neurons [4], stimulated neurite growth [5] and thereby showed anti-neuroinflammatory and neuroprotective activities. Further, Icariin shows antidepressant-like activity [6]; protects against aluminium-induced learning and memory deficits due to its antioxidant activities; decreases Aβ1-40 content in the hippocampus of aluminium-intoxicated rodents [7]; and its metabolite Icaritin has neuroprotective effects on β amyloid-induced neurotoxicity to neuronal cells [8]. Therefore, Icariin is considered as a potential therapy against neurodegenerative diseases such as Alzheimer's disease (AD).

AD is the most common form of neurodegeneration and the major cause of dementia. AD is clinically defined by distinct behavioral and cognitive deficits, and the most characteristic neuropathological feature is extracellular deposit of aggregated β-amyloid (Aβ) peptide (amyloid plaques). It is proposed that Aβ peptides are toxic and causative in AD, contributing to memory loss, behavioral impairment and neurodegenerative pathology [9]. Beside the well known Aβ aggregation, neuroinflammation also plays an important role in the pathophysiology of this multifactorial disorder [10]. Neuroinflammation is characterized by release of numerous inflammatory mediators, microglial and astroglial activation, in particular around senile plaques [11,12]. Neuroinflammation may contribute to neural dysfunction and cell death, establishing a self-perpetuating vicious cycle by which inflammation induces further neurodegeneration [13]. Therefore, anti-inflammatory drugs may have beneficial effects on neurodegenerative disorders, including AD [13]. In addition, up-regulated TGF-β1 has been observed in the brain of AD patients and AD animal models [14,15]. As a potent immunomodulator and with a central role in the response to neuroinflammation, TGF-β1 is considered to be involved in pathological progression of AD.

Our aim was to study the potential therapeutic effect of Icariin in a transgenic mouse model of AD, focusing on its effect on Aβ deposition, neuroinflammation, TGF-β1 expression and behavioral deficits. We used the APP/PS1-21 double transgenic mouse model of cerebral amyloidosis, co-expressing the KM670/671NL mutated human amyloid precursor protein and the L166P mutated human presenilin 1 (APP/PS1-21 mice) under the neuron-specific Thy1 promoter element. In this transgenic line, amyloid deposition occurred as early as 2 months in the cortex and 4 months in the hippocampus, accompanied by inflammatory responses and robust impairment of cognitive function [16–18].

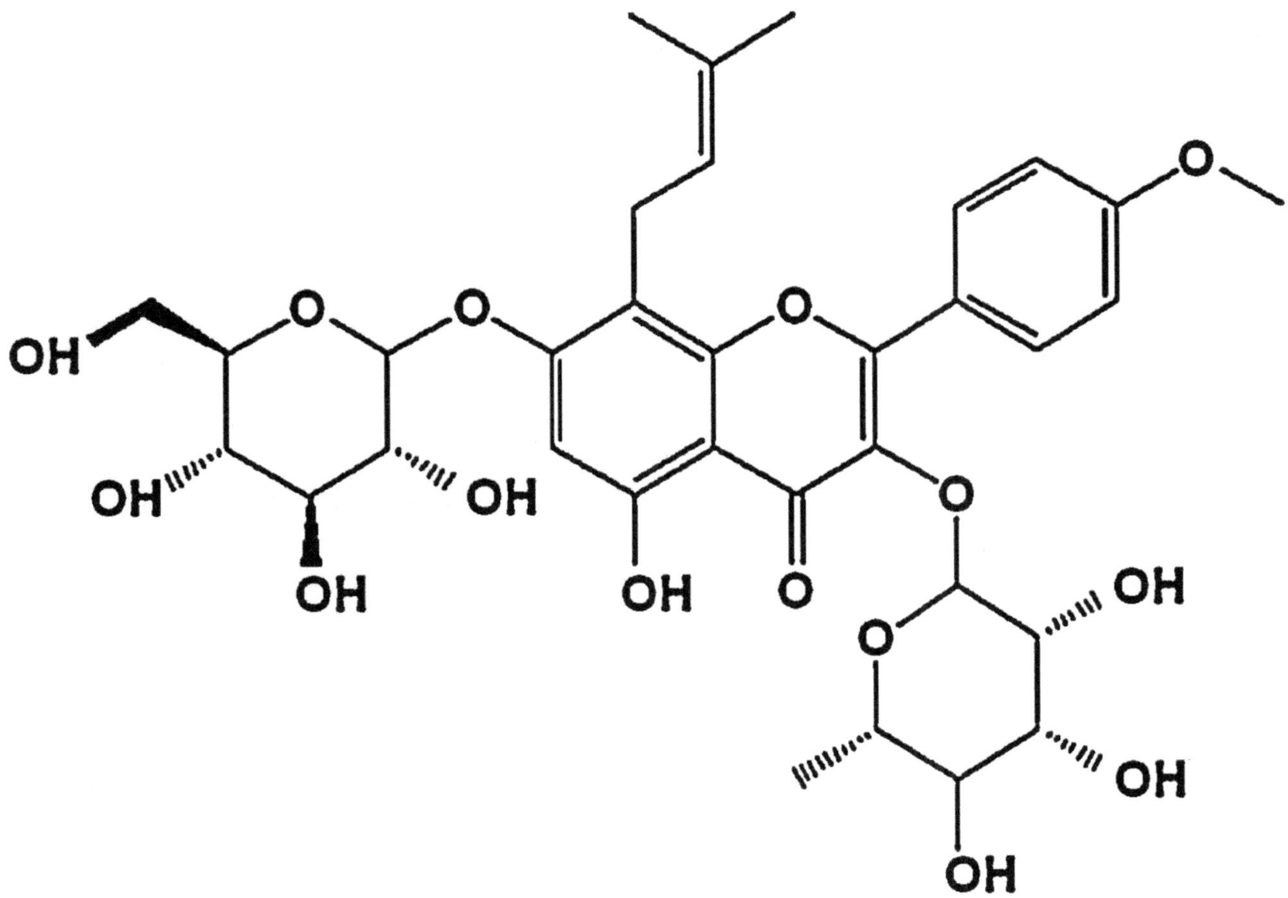

Figure 1. Molecular structure of Icariin.

Materials and Methods

Animals

Male APP/PS1-21 mice with a C57BL/6J background were obtained from Prof. M. Jucker. Heterozygous male APP/PS1-21 mice were bred with wild-type C57BL/6J females (Charles River Germany, Sulzfeld Germany). Offsprings were tail snipped and genotyped using PCR. Animals were housed under a 12 h light-12 h dark cycle with free access to food and water. Mouse diets were provided by SSNIFF Spezialdiaeten GmbH (Soest, Germany), diet number V1124-703 was for breeding pairs and diet number V2534-703 was for all the other mice. All experiments and protocols were licensed and approved by regional Administrative Council (Regierungspräsidium) Tuebingen according to The German Animal Welfare Act (TierSchG) of 2006.

Materials

Icariin (>98%) was purchased from MR Natural Product Co., Ltd. (Xi'an, China). For oral treatment, Icariin was suspended in 1% carboxymethylcellulose (CMC, Hercules-Aqualon, Düsseldorf, Germany) at a concentration of 12.5 mg/ml (Icariin/CMC solution).

Treatment with Icariin

14 transgenic APP/PS1-21 mice 5 months of age, 6 males and 8 females, were separated into two groups: group 1 received a 10-days Icariin treatment (100 mg/kg by daily gavage); group 2, as control, received the same volume of 1% CMC dissolved in water (approx. 200 ul).

Design and evaluation of nest construction assay

A nest construction assay [19] was modified to determine the deficits in affiliative/social behavior of APP/PS1 mice and potential changes following treatment.

Mice were individually housed for at least 24 hours in clean plastic cages with approximately 1 cm of wood chip bedding lining the floor and identification cards coded to render the experimenter blind to gender, age, and genotype of mice. Two hours prior to the onset of the dark phase of the light cycle, individual cages were supplied a 20×20 cm piece of paper towel torn into approximately 5×5 cm squared pieces. The next morning (approximately 16 hours later), cages were inspected for nest construction. Pictures were taken prior to evaluation for documentation. Paper towel nest construction was scored by a 3 point system: 1 = no biting or tears on the paper, 2 = moderate biting and/or tears on the paper but no coherent nest (not grouped into a corner of the cage) and 3 = the vast majority of paper torn into pieces and grouped into a corner of the cage [19].

Immunohistochemistry (IHC) and image evaluation/analysis

Icariin-treated and control mice were sacrificed after 10-days-treatment. According to our previous data, Aβ deposition starts as early as 2 months of age, but develops very fast (exponential phase) around the age of 5 months. We presumed that a potential treatment may affect Aβ deposition most effectively during the exponential phase; therefore mice at an age of 5 months were used. We balanced the treatment and control groups by gender-, age and body weight matched mice. Afterwards, we analysed possible difference in plaque burden and behavioural deficits between genders within a group and found no significant differences. But the brain sizes of female mice are generally smaller than that of males. Therefore, we introduced analysis of IR percentage to rule out interference of brain size differences between genders or single individuals of the same gender. Mice were deeply anesthetized with ether and perfused intracardially with 4°C, 4% paraformaldehyde in PBS. Brains were quickly removed and post-fixed in 4% paraformaldehyde overnight at 4°C. Post-fixed brains were cut into two hemispheres; hemispheres were embedded in paraffin, serially sectioned (3 μm) and mounted on silane-covered slides. Hemispheres sections were stained with HE or IHC as described previously [20]. The following antibodies were used: anti-β-amyloid (1:100; Abcam, Cambridge, UK) for Aβ deposition, anti-Iba-1 (1:200; Wako, Neuss, Germany) for activated microglia, anti-GFAP (1:500; Chemicon (Millipore), Billerica, MA, US) for astrocytes and anti-transforming growth factor beta 1 (TGF-β1) (1:50, Santa Cruz, Dallas, Texas, US). The rabbit polyclonal anti-β-amyloid antibody (ab2539) was generated against the synthetic peptide DAEFRHDSGYEVHH conjugated to KLH, corresponding to amino acids 1–14 of Human β-amyloid. Positive and negative controls were routinely performed in each staining experiment to validate the immunohistochemical staining quality and results. Negative controls were performed by deletion of primary antibody and no unspecific staining was observed.

After immunostaining, sections were examined by light microscopy (Nikon, Düsseldorf, Germany). Aβ deposition, Iba-1 and GFAP immunostaining were evaluated at cross-sections of hemispheres, especially focused on cortex and hippocampus. All sections were randomly numbered and analysed by two observers independently, who were not aware of the treatment and time points. Aβ plaques, Iba-1^+ and GFAP$^+$ cells in cortex and hippocampus were counted under a microscope with a 50-fold magnification, by clear deposition for plaques and clear counter staining of cellular nuclei for cells. Further, images of hemisphere cross-sections were captured using Nikon Cool-scope (Nikon, Düsseldorf, Germany) with fixed parameters; cortex and hippocampus of the images were outlined and analyzed using the software MetaMorph Offline 7.1 (Molecular Devices, Toronto, Canada). Area percentages of specific immunoreactivity (IR) of interesting regions were selected by color threshold segmentation and calculated. All parameters were fixed for all images of a particular staining. Results were given as arithmetic means of plaque/cell counts or area percentages of IR to interest areas on cross-sections and standard errors of means (SEM).

Statistical analysis

Difference of plaque/cell counts, area percentages of staining and scores of nest construction between treatments and controls were analysed by unpaired t-tests (Graph Pad Prism 5.0 software). For all statistical analyses, significance levels were set at $P<0.05$.

Results

Effect of Icariin treatment on behaviour impairments

Effect of Icariin treatment was monitored using a harmless behavioural test, the nesting assay. As a born instinct, nesting behaviour is important for small rodents in heat conservation, reproduction and shelter. As a baseline of this assay, an impaired nesting ability of these transgenic mice was confirmed in our previous study, compared to age- and gender matched naïve mice [21]. Nest construction with paper towel material was explored using a 3 point scaling system.

A significant difference between treatment and control groups of these age- and gender-matched APP/PS1 mice was already observed at Day 11, after 10 days of treatment (control = 1.5±0.1, Icariin = 2.1±0.2, $p<0.05$, $n=6$) (Figure 2A). In the Icariin treatment group, relatively immediate chewing and tearing of the paper towels were observed; paper towels were torn into pieces and grouped into a corner of the cage. In contrast, transgenic mice from the control group investigated and slightly chew but did not really destruct the paper towels; paper towels were found all over in the cage, not grouped or not in the corners. As a control, no significant difference between treatment and control groups could be observed right at the beginning of treatment, namely at Day 1 (control = 1.4±0.1, Icariin = 1.5±0.2, $p>0.5$, $n=6$) (Figure 2B).

After acquisition of these positive results, all mice were sacrificed and the effects of Icariin on neuropathological changes were further investigated.

Effects of Icariin on amyloid plaques

In brain of APP/PS1 transgenic mice from the control group, Aβ plaques were distributed throughout the whole cortex. Plaques were of different sizes, most small plaques had dense cores and larger plaques mainly consisted of a dense core surrounded by a large halo of diffuse amyloid (Figure 3A). In the hippocampus, plaque density was lower and most plaques were of smaller size (Figure 3E).

The transgenic mice received Icariin suspension or vehicle by gavage for 10 days. The treatment with Icariin attenuated neuropathological change, compared to the age- and gender-matched control mice. The Icariin treatment reduced the plaque counts significantly in cortex (control = 155.4±13.2, Icariin = 95.0±12.6, $p<0.05$, $n=6$) and hippocampus (control = 21.6±3.4, Icariin = 10.4±2.1, $p<0.05$; $n=6$) (Figure 4A and B). Further analysis of the micro photos showed highly significantly decreased Aβ IR area in both cortex and hippocampus from the Icariin treatment group (cortex: control = 0.72±0.06%, Icariin = 0.51±0.10%, $p<0.05$; hippocampus: control = 0.63±0.12%, Icariin = 0.33±0.09%, $p<0.05$; $n=6$) (Figure 4C and D). Notably, Aβ plaques had a smaller size and fewer branches (Figure 3A and B).

Effects of Icariin on microglial activation

In both cortex and hippocampus of non-transgenic mice, Iba-1 staining could be barely observed. In transgenic mice, however, amoeboid Iba1^+ microglia were observed clustered around amyloid deposits in both control and treatment groups (Figure 3C and G, Figure 5A with higher magnification), according to comparison with Aβ staining of serial sections. Numbers of Iba-1^+ cells in Icariin treatment group were significantly less than those of the control group. Notably, in the cortex; decreased Iba-1^+ cells were also less clustered around plaques (Figure 3D and H, Figure 5B with higher magnification). Further analysis showed that Icariin treatment significantly reduced the IR area of Iba-1 in cortex (control = 0.35±0.06%, Icariin = 0.21±0.04%, $p<0.05$, $n=6$), and hippocampus (control = 0.29±0.03%, Icariin = 0.16±0.05%, $p<0.05$, $n=6$) (Figure 4E and F). These results indicate reduced microglial activation.

Effects if Icariin on astrocyte GFAP expression

Numerous GFAP$^+$ cells (GFAP IR) were widely distributed throughout the hippocampus and cortex. They all showed typical

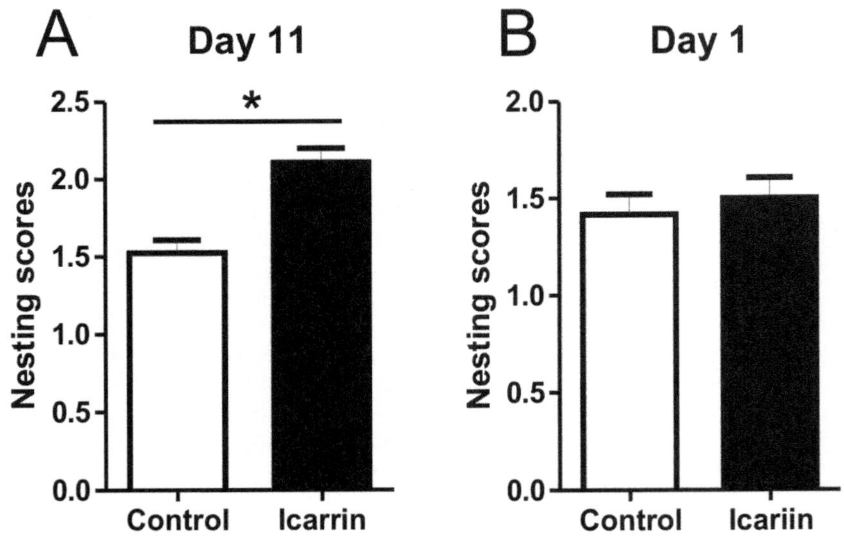

Figure 2. Effect of Icariin on impaired nesting ability. Therapeutic effects of Icariin were monitored using a nesting assay before and during the experimental treatment. Nest construction was explored with paper towel material using a 3 point scaling system in APP/PS1 mice. A: A significant difference between treatment and control group was observed after a 10-days treatment, namely at Day 11. B: No significant difference between the Icariin treatment and the control group could be observed right at the beginning of treatment, namely at Day 1.

morphology of astrocytes, including stellate shape and multiple branched processes; some of them had the typical morphology of perivascular astrocytes (Figure 5C and D). Double staining showed that all Aβ plaques were surrounded or covered by GFAP IR, but no massive aggregation of $GFAP^+$ cells was related to Aβ plaques. Further analysis showed that no significant changes in GFAP IR were observed between Icariin treatment and controls; GFAP-IR area was only slightly decreased by Icariin treatment: in cortex (control = 0.51±0.05%, Icariin = 0.45±0.04%, $p>0.05$, $n=6$), and in hippocampus (control = 0.38±0.04%, Icariin = 0.34±0.05%, $p>0.05$, $n=6$) (Figure 4G and H).

Expression pattern of TGF-β1 and effect of Icariin treatment on TGF-β1 IR

Until 3 months of age, TGF-β1 IR could not be seen in cortex or hippocampus of transgenic and naïve mice (data not shown). At age of 5 months, increased TGF-β1 IR was observed and mainly located on or around Aβ plaques, but could barely be seen on glial cells or neurons, according to the results of double staining and in comparison with serial sections of Aβ staining (Figure 6A and B). TGF-β1 IR therefore presented an Aβ plaque-like distribution pattern, but with much less intensity and smaller IR area compared to Aβ staining, as presented in the figure 6 (Figure 6A and B). TGF-β1 IR was more concentrated at the range of plaques and diffused branches, less in the center of plaques.

Following the 10 days-treatment with Icariin, TGF-β1 IR was obviously reduced in brains of transgenic mouse, especially in cortex (Figure 6C and D for control and Icariin group respectively). Numbers of TGF-β1 stained plaques in cortex were significantly reduced (control = 109.0±11.7, Icariin = 82.6±8.8, $p<0.05$, $n=6$) and areas of TGF-β1 IR in cortex were also significantly decreased (cortex: control = 0.19±0.02%, Icariin = 0.13±0.01%, $p<0.05$) (Figure 6E and F).

Discussion

In this work we report therapeutically beneficial effect of Icariin, a flavonoid from the Chinese medicinal herb Epimedium brevicornum (Herba epimedii), in a rodent APP/PS1 model of cerebral amyloidosis for AD. A 10-days oral treatment by Icariin significantly attenuated Aβ deposition, microglial activation and TGF-β1 IR in both cortex and hippocampus of APP/PS1 mice at an age of 5 months, and restored impaired nesting behavior as well.

As a tonic herbal in traditional Chinese medicine, Epimedium brevicornum has been used to treat various kinds of disorders, such as hypertension, coronary heart disease, osteoporosis, menopause syndrome, breast lump, rheumatism, arthritis or hypogonadism [1]. The most important active constituents of this herb are prenylated, isopentenyl flavonoids. Icariin is one of these isopentenyl flavonoids, a diglycoside with a glucose group at C-7, a rhamnose group at C-3 position, and a methoxyl group at C0–4, as shown in the (Figure 1) [22,23].

Recently, the structure-activity relationship (SAR) of flavonoids has been extensively investigated by theoretical calculations and experimental studies, including antitumor, antioxidant, vasorelaxant, antibacterial and anti-inflammatory activities [1]. As a multifunctional flavonoid, Icariin shows antioxidant [24], antidepressant [25], neuroprotective [7], anti-inflammatory activity/effects [26,27]. Icariin also has functions in regulating bone remodeling, including enhancing osteoblastic differentiation and mineralization, inhibiting bone resorption, and inducing apoptosis of osteoclasts [28,29]; and in controlling sexual, especially erectile dysfunction [30].

Icariin has shown efficient anti-neuroinflammatory activity against learning and memory deficits in animal models [7], through attenuating microglial activation by inhibiting NF-kappaB and p38 MAPK pathways [31]. This suggests Icariin's promising potential in aging-related neurodegenerative disease, since microglia are regarded as macrophage-like cells resident in the CNS and their activation has been implicated in many neurological disorders because of their inflammatory effects [32]. Therapeutic strategies controlling microglial activation and the excessive production of pro-inflammatory factors/molecules may be valuable to control neurodegeneration in dementia [33]. An important role of neuroinflammation involved in AD pathology has been

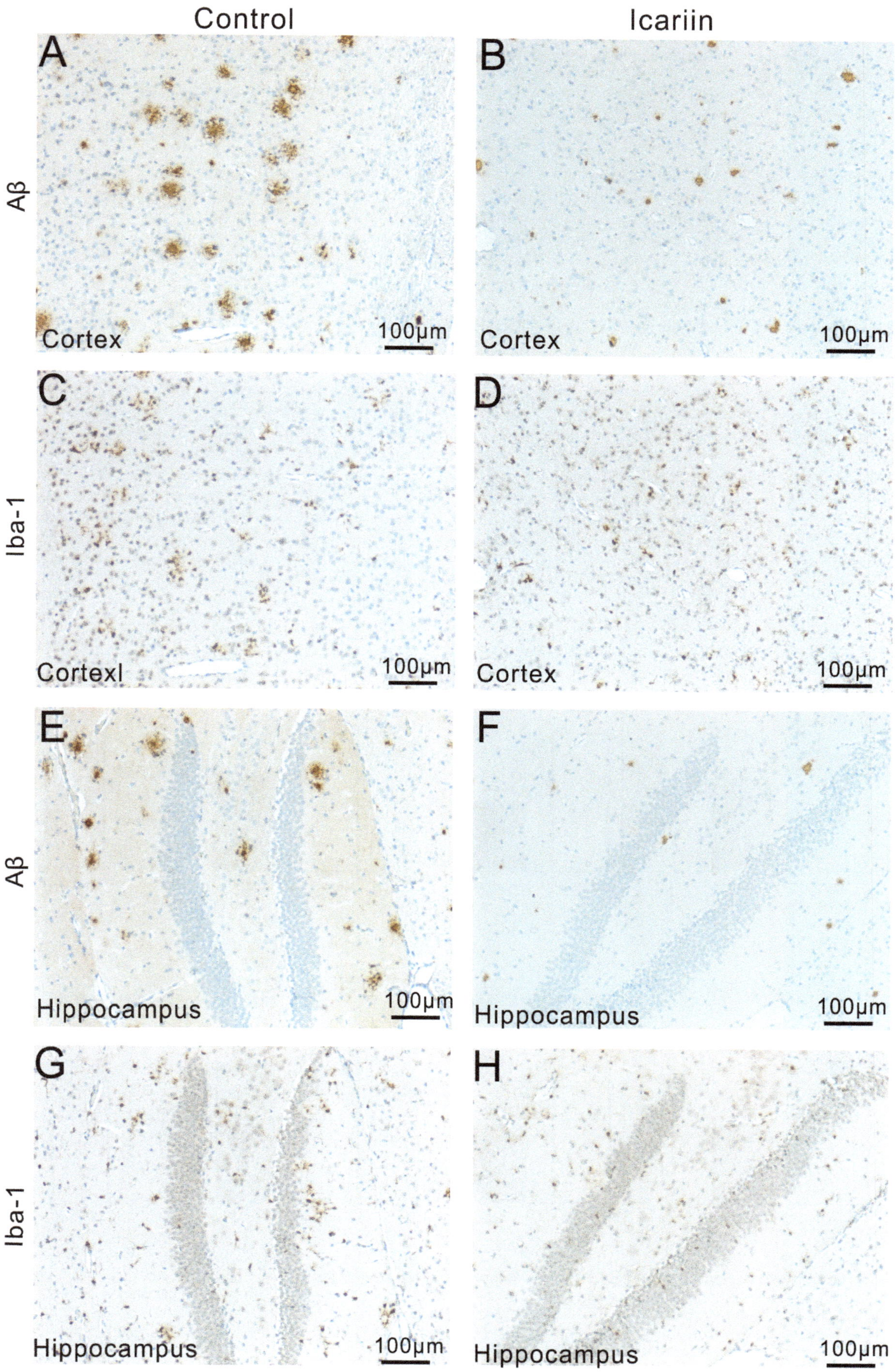
Control
Icariin
A
Aβ
Cortex
100μm
B
Cortex
100μm
C
Iba-1
Cortexl
100μm
D
Cortex
100μm
E
Aβ
Hippocampus
100μm
F
Hippocampus
100μm
G
Iba-1
Hippocampus
100μm
H
Hippocampus
100μm

Figure 3. Therapeutic effect of Icariin on Aβ deposition and microglial activation. Representative microimages show the changes in Aβ deposition and microglial activation in cortex and hippocampus following Icariin treatment. A–B and E–F: In both cortex (A–B) and hippocampus (E–F) of APP/PS1 mice from the Icariin group (B and F), reduced numbers of Aβ plaques with relatively smaller size were observed, compared to the control group (A and E). C–D (cortex) and G–H (hippocampus): According to serial sections of Aβ staining, most Iba-1$^+$ microglia accumulated at or surrounding Aβ plaques. In the treatment group (D and H) fewer numbers of Iba-1$^+$ cells and smaller IR area of Iba-1 could be seen, compared to the control group (C and G).

reported from rodent models and humans [13], and attenuated neuroinflammation has been proven to contribute to reduced hallmark features of AD pathology, including Aβ-plaque accumulation [34]. Icariin significantly attenuated microglial activation in cortex of APP/PS1 transgenic mice, suggesting an inhibitory effect of Icariin on neuroinflammation, which may contribute to the ameliorated pathology and improved behaviors.

Aβ deposition in brains of transgenic mice is also significantly reduced by Icariin, following a relatively short term treatment of 10-days. It may be attributed to direct down-regulated amyloid loading [35] and attenuated neuroinflammation, because it has been reported that Icariin administrated by gavage negatively regulated beta-amyloid peptide segment 25–35 production in a rat model [35] and attenuated neuroinflammation also contributed to reduced Aβ deposition during progression of AD pathology [34].

Further, Amelioration of behavioral deficits was observed following Icariin treatment. Distinct behavioral and cognitive deficits are the most characteristic clinical feature of AD. Not only cognitive impairment but also deficits in non-cognitive/non-mnemonic behaviors are found in most AD mouse models, they are therefore considered very valuable for modeling human AD [36]. Toxic Aβ peptides and amyloid precursor protein [37], inflammatory reaction and inflammatory cytokines/molecules are directly associated with these deficits in behaviors [38]. Nesting behavior is an affiliative, social behavior and deficit in this non-mnemonic behavior is a debilitating feature of neurodegenerative diseases, including AD. A previous study reported very early (3 months) occurrence of impaired nesting ability in transgenic APP and APPPS1 mice and suggested that social deficits precede other neuropsychiatric and cognitive AD-like symptoms and can be employed as early markers of AD pathology in transgenic mouse models [39]. Cognitive impairment, however, cannot be observed in the APP/PS1 mice until 8 months of age [17].

Amelioration in impaired nesting ability was observed following our Icariin treatment. The hippocampus and the prefrontal cortex damage in mice are demonstrated to lead to reduced nesting material consumption and nest quality, indicating that the impairment of nesting behavior in these transgenic mice might

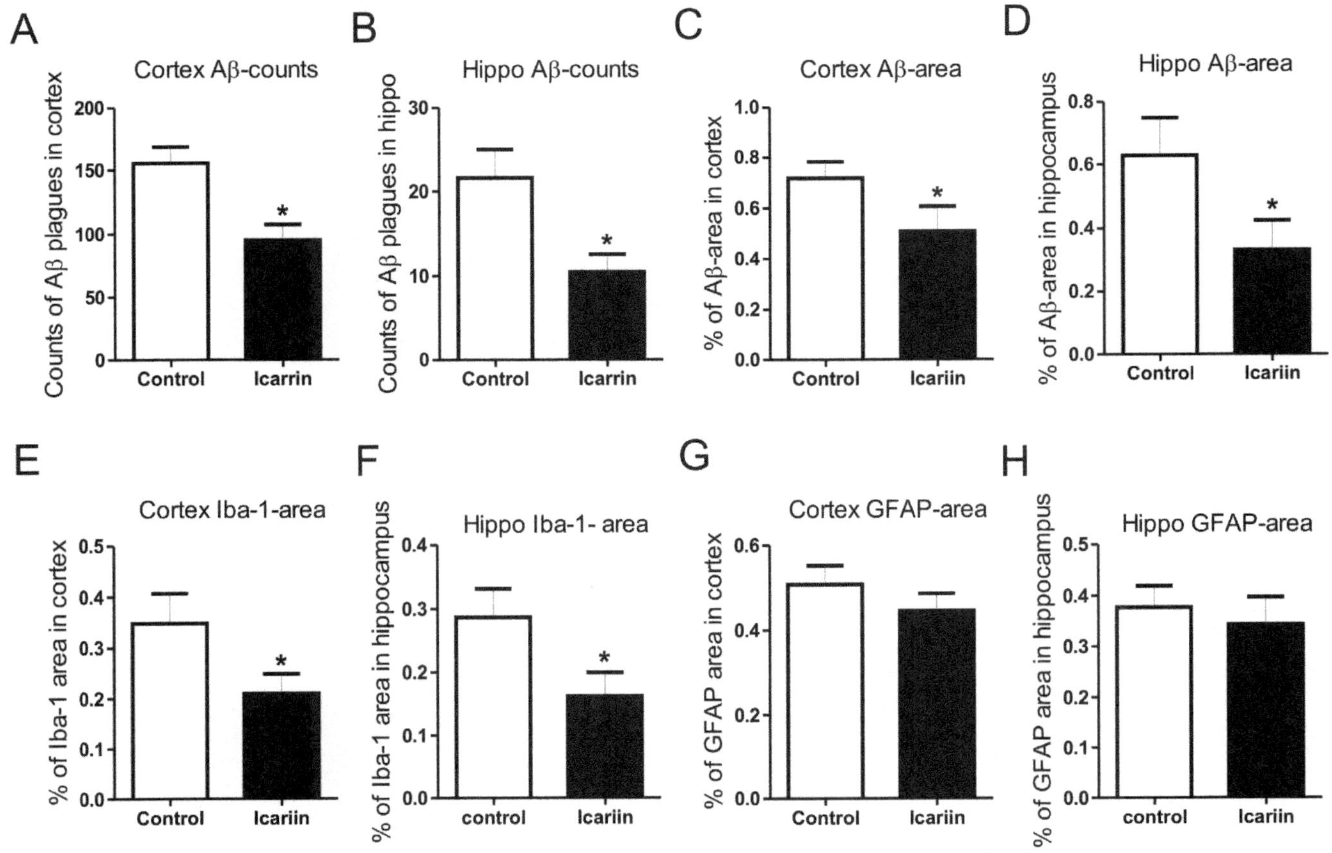

Figure 4. Icariin reduced β-amyloid (Aβ) deposition and microglial activation. Differences of Aβ plaque counts and Aβ$^+$/Iba-1$^+$/GFAP$^+$ area percentages between treatment and control were analysed by unpaired t-test and results are represented in the bar graphs. A and B: In cortex and hippocampus of transgenic mouse brains from the Icariin group, numbers of amyloid plaques were significantly reduced. C and D: IR area percentages of Aβ staining were highly significantly reduced. E and F: Following Icariin treatment, Iba-1 IR was significantly reduced in cortex and hippocampus. G and H: GFAP IR was not significantly changed by Icariin treatment.

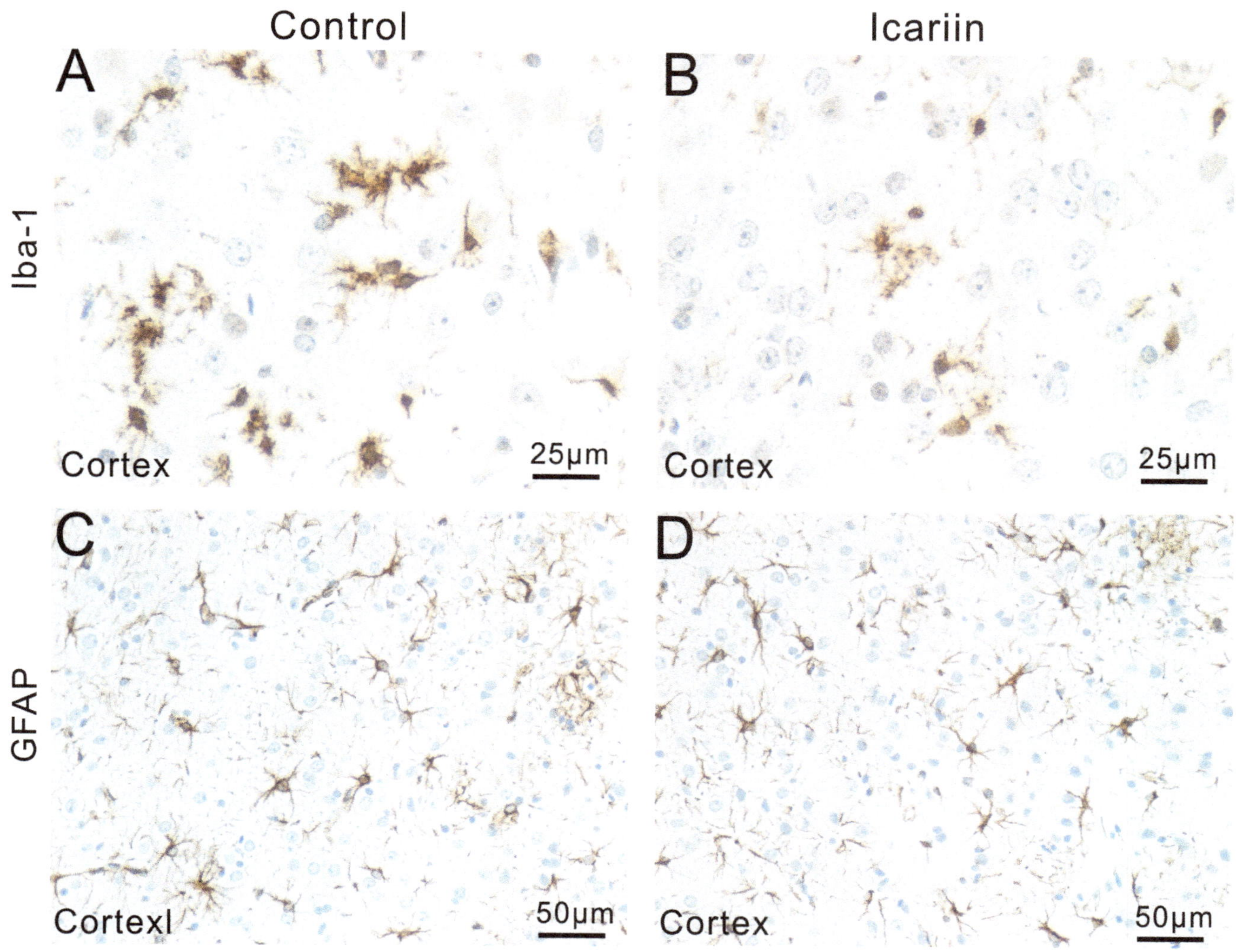

Figure 5. Therapeutic effect of Icariin on microglial and astrocytic activation. A and B: Representative microimages with higher magnification show the changes in microglial activation in cortex following Icariin treatment. C and D: Representative microimages show typical morphology of astrocytes, including stellate shape and multiple branched processes. No massive aggregation of $GFAP^+$ cells and no obvious changes of GFAP IR were observed.

be caused by toxic injury of Aβ and accompanying neuroinflammation in related brain areas [19]. In addition to reduced amyloid accumulation and decreased neuroinflammation, Icariin can also attenuate β-amyloid-induced neurotoxicity and neurite atrophy [31,40], all these may contribute to the improved affiliative/social behavior.

Moreover, Icariin also interferes with phosphodiesterases (PDEs), especially PDE4 and PDE5 [41,42]. This is of interest to AD research because inhibition of PDE4 is known to reverse Aβ-induced memory deficits [43]; PDE5 inhibition is known to ameliorate synaptic sprouting [44] and axonal remodelling [45] and PDE3 inhibition is known to enhance neurogenesis [46].

TGF-β1 is a potent immunomodulator and plays a central role in the response of the brain to inflammation and injury, but the specific role of TGF-β1 in AD pathogenesis still stays elusive. Some previous studies indicated a possibly protective effect of TGF-β1 in AD by its neuroproctive function [47] and by promoting Aβ clearance through activation of microglia cells [15]. But on the other hand, up-regulated expression of TGF-β1 has been reported from brains of AD patients and AD animal models [14,15]. Serveral studies further reported, that TGF-β1 potentiates/triggers increased production of APP and subsequent Aβ generation in murine and human astrocyte cultures [48,49] and transgenic mice overexpressing TGF-β1 in astrocytes elicit Aβ deposition [50]. Moreover, TGF-β1 can promote vascular abnormalities and Aβ deposition in cerebral blood vessels and cerebrovascular TGF-β may contribute to inflammation in AD brains [15,51,52]. We observed that TGF-β1 IR in cortex of transgenic mice was reduced following Icariin treatment. This may suggest a possible mechanism of Icariin's therapeutic effect through inhibition of local TGF-β deposition, which might be in part due to the reported inhibition of TGF-β expression [53]. In our study, TGF-β1 IR was observed in or around amyloid plaques and could barely be seen on glial cells or neurons, which is in accordance with previous studies of AD brains and animal models [15,51,54,55]. TGF-β1 is considered to be the most potent and ubiquitous profibrogenic cytokine and stimulates extracellular matrix (ECM) accumulation [56]. It has been reported, that Icariin protects tissue/organs from vascular pathological alterations by modulating expression of TGF-β1 [7,57]. ECMs such as collagen are important components of Aβ plaques [58], reduced TGF-β1 IR by Icariin may therefore, at least partialy, contribute

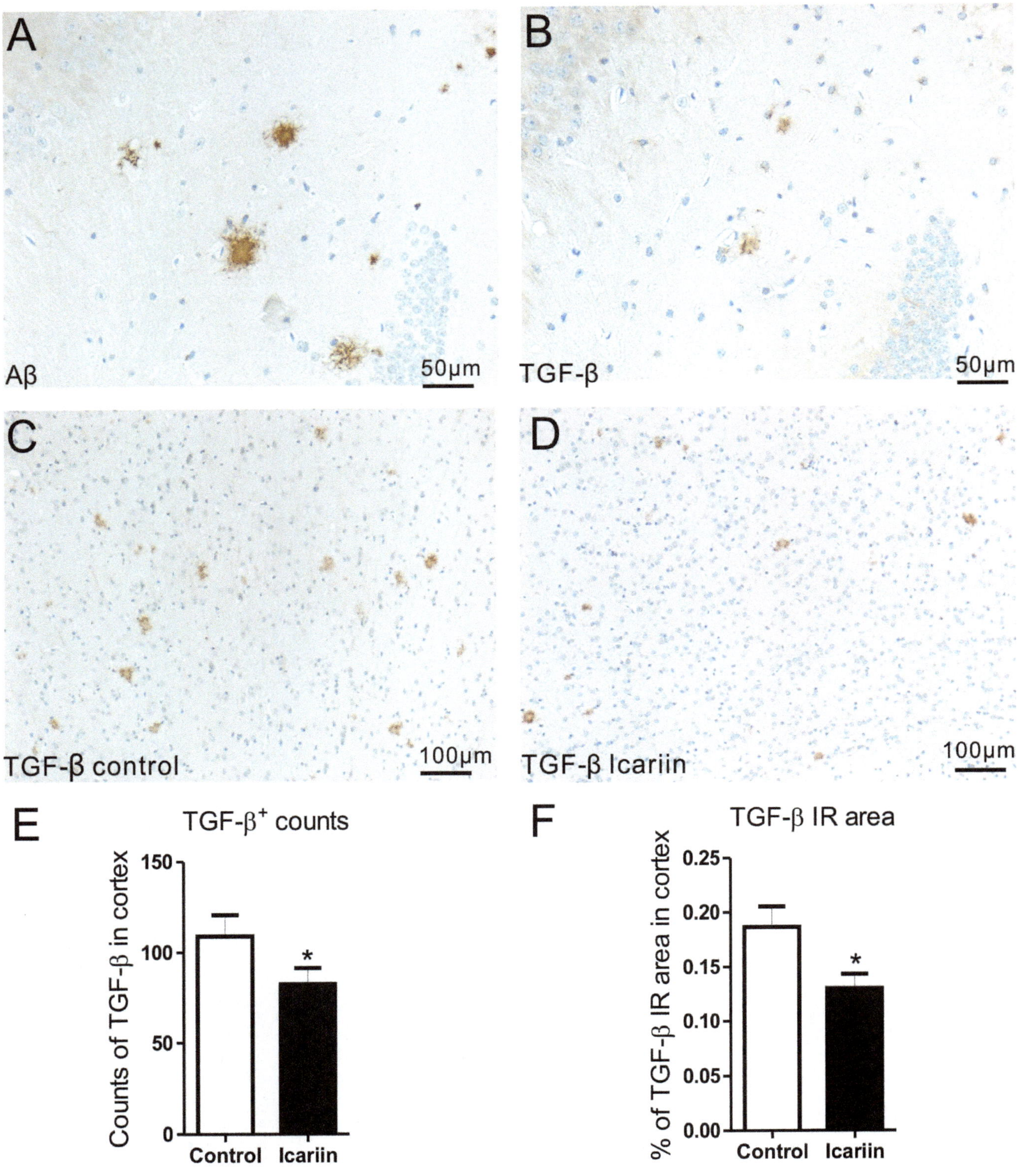

Figure 6. Effect of Icariin on TGF-β IR. At age of 5 months, increased TGF-β1 IR was observed in cortex of transgenic mice. A and B: In comparison of serial section of Aβ staining (A), TGF-β1 IR (B) had generally less intensity and smaller IR area, and was mainly located at or around Aβ plaques, but could be barely seen on glial cells or neurons. C and D: Much less intensity and smaller TGF-β1 IR area were seen in the Icariin treated group (D), compared to the control group (C). E and F: Statistical evaluation showed significant reductions in counts of TGF-β1^{+} plaques and area of TGF-β1 IR.

to reduction of Aβ plaques by decreasing the formation of the plaque's matrix.

Taken together, treatments with Icariin by gavage effectively ameliorated neuroinflammatory reaction and cerebral amyloidosis, reduced IR of TGF-β1 in cortex and hippocampus of transgenic APP/PS1 mice, and restored impaired nesting ability. All these results suggest that Icariin may be considered a promising therapeutic option for human AD.

Acknowledgments

The authors would like to thank Prof. M. Jucker for providing male transgenic APP/PS1-21 mice.

Author Contributions

Conceived and designed the experiments: ZZ HS. Performed the experiments: ZZ CL CZ. Analyzed the data: ZZ CL CZ. Contributed reagents/materials/analysis tools: ZZ CZ. Wrote the paper: ZZ.

References

1. Zhang D, Zhang J, Fong C, Yao X, Yang M (2012) Herba epimedii flavonoids suppress osteoclastic differentiation and bone resorption by inducing G2/M arrest and apoptosis. Biochimie 94: 2514–2522.
2. Guo J, Li F, Wu Q, Gong Q, Lu Y, et al. (2010) Protective effects of icariin on brain dysfunction induced by lipopolysaccharide in rats. Phytomedicine 17: 950–955.
3. Zhang H, Liu B, Wu J, Xu C, Tao J, et al. (2012) Icariin inhibits corticosterone-induced apoptosis in hypothalamic neurons via the PI3-K/Akt signaling pathway. Mol Med Rep 6: 967–972.
4. Li L, Zhou QX, Shi JS (2005) Protective effects of icariin on neurons injured by cerebral ischemia/reperfusion. Chin Med J (Engl) 118: 1637–1643.
5. Kuroda M, Mimaki Y, Sashida Y, Umegaki E, Yamazaki M, et al. (2000) Flavonol glycosides from Epimedium sagittatum and their neurite outgrowth activity on PC12h cells. Planta Med 66: 575–577.
6. Pan Y, Kong L, Xia X, Zhang W, Xia Z, et al. (2005) Antidepressant-like effect of icariin and its possible mechanism in mice. Pharmacol Biochem Behav 82: 686–694.
7. Luo Y, Nie J, Gong QH, Lu YF, Wu Q, et al. (2007) Protective effects of icariin against learning and memory deficits induced by aluminium in rats. Clin Exp Pharmacol Physiol 34: 792–795.
8. Wang Z, Zhang X, Wang H, Qi L, Lou Y (2007) Neuroprotective effects of icaritin against beta amyloid-induced neurotoxicity in primary cultured rat neuronal cells via estrogen-dependent pathway. Neuroscience 145: 911–922.
9. Selkoe DJ (2002) Alzheimer's disease is a synaptic failure. Science 298: 789–791.
10. Herrmann N, Chau SA, Kircanski I, Lanctot KL (2011) Current and Emerging Drug Treatment Options for Alzheimer's Disease: A Systematic Review. Drugs.
11. Akiyama H, Barger S, Barnum S, Bradt B, Bauer J, et al. (2000) Inflammation and Alzheimer's disease. Neurobiol Aging 21: 383–421.
12. Wyss-Coray T, Mucke L (2002) Inflammation in neurodegenerative disease–a double-edged sword. Neuron 35: 419–432.
13. Martin-Moreno AM, Brera B, Spuch C, Carro E, Garcia-Garcia L, et al. (2012) Prolonged oral cannabinoid administration prevents neuroinflammation, lowers beta-amyloid levels and improves cognitive performance in Tg APP 2576 mice. J Neuroinflammation 9: 8.
14. Lippa CF, Flanders KC, Kim ES, Croul S (1998) TGF-beta receptors-I and -II immunoexpression in Alzheimer's disease: a comparison with aging and progressive supranuclear palsy. Neurobiol Aging 19: 527–533.
15. Wyss-Coray T, Lin C, Yan F, Yu GQ, Rohde M, et al. (2001) TGF-beta1 promotes microglial amyloid-beta clearance and reduces plaque burden in transgenic mice. Nat Med 7: 612–618.
16. Gengler S, Hamilton A, Holscher C (2010) Synaptic plasticity in the hippocampus of a APP/PS1 mouse model of Alzheimer's disease is impaired in old but not young mice. PLoS One 5: e9764.
17. Radde R, Bolmont T, Kaeser SA, Coomaraswamy J, Lindau D, et al. (2006) Abeta42-driven cerebral amyloidosis in transgenic mice reveals early and robust pathology. EMBO Rep 7: 940–946.
18. Moehlmann T, Winkler E, Xia X, Edbauer D, Murrell J, et al. (2002) Presenilin-1 mutations of leucine 166 equally affect the generation of the Notch and APP intracellular domains independent of their effect on Abeta 42 production. Proc Natl Acad Sci U S A 99: 8025–8030.
19. Wesson DW, Wilson DA (2011) Age and gene overexpression interact to abolish nesting behavior in Tg2576 amyloid precursor protein (APP) mice. Behav Brain Res 216: 408–413.
20. Zhang Z, Zhang ZY, Fauser U, Schluesener HJ (2008) FTY720 ameliorates experimental autoimmune neuritis by inhibition of lymphocyte and monocyte infiltration into peripheral nerves. Exp Neurol 210: 681–690.
21. Zhang ZY, Schluesener HJ (2013) Oral administration of histone deacetylase inhibitor MS-275 ameliorates neuroinflammation and cerebral amyloidosis and improves behavior in a mouse model. J Neuropathol Exp Neurol 72: 178–185.
22. Chen Y, Wang J, Jia X, Tan X, Hu M (2011) Role of intestinal hydrolase in the absorption of prenylated flavonoids present in Yinyanghuo. Molecules 16: 1336–1348.
23. Sun P, Liu Y, Deng X, Yu C, Dai N, et al. (2013) An inhibitor of cathepsin K, icariin suppresses cartilage and bone degradation in mice of collagen-induced arthritis. Phytomedicine 20: 975–979.
24. Xie J, Sun W, Duan K, Zhang Y (2007) Chemical constituents of roots of Epimedium wushanense and evaluation of their biological activities. Nat Prod Res 21: 600–605.
25. Pan Y, Kong LD, Li YC, Xia X, Kung HF, et al. (2007) Icariin from Epimedium brevicornum attenuates chronic mild stress-induced behavioral and neuroendocrinological alterations in male Wistar rats. Pharmacol Biochem Behav 87: 130–140.
26. Chen SR, Xu XZ, Wang YH, Chen JW, Xu SW, et al. (2010) Icariin derivative inhibits inflammation through suppression of p38 mitogen-activated protein kinase and nuclear factor-kappaB pathways. Biol Pharm Bull 33: 1307–1313.
27. Wu J, Zhou J, Chen X, Fortenbery N, Eksioglu EA, et al. (2012) Attenuation of LPS-induced inflammation by ICT, a derivate of icariin, via inhibition of the CD14/TLR4 signaling pathway in human monocytes. Int Immunopharmacol 12: 74–79.
28. Ming LG, Chen KM, Xian CJ (2013) Functions and action mechanisms of flavonoids genistein and icariin in regulating bone remodeling. J Cell Physiol 228: 513–521.
29. Huang J, Yuan L, Wang X, Zhang TL, Wang K (2007) Icaritin and its glycosides enhance osteoblastic, but suppress osteoclastic, differentiation and activity in vitro. Life Sci 81: 832–840.
30. Ho CC, Tan HM (2011) Rise of herbal and traditional medicine in erectile dysfunction management. Curr Urol Rep 12: 470–478.
31. Zeng KW, Fu H, Liu GX, Wang XM (2010) Icariin attenuates lipopolysaccharide-induced microglial activation and resultant death of neurons by inhibiting TAK1/IKK/NF-kappaB and JNK/p38 MAPK pathways. Int Immunopharmacol 10: 668–678.
32. Xu Y, Xue Y, Wang Y, Feng D, Lin S, et al. (2009) Multiple-modulation effects of Oridonin on the production of proinflammatory cytokines and neurotrophic factors in LPS-activated microglia. Int Immunopharmacol 9: 360–365.
33. Agostinho P, Cunha RA, Oliveira C (2010) Neuroinflammation, oxidative stress and the pathogenesis of Alzheimer's disease. Curr Pharm Des 16: 2766–2778.
34. Tweedie D, Ferguson RA, Fishman K, Frankola KA, Van Praag H, et al. (2012) Tumor necrosis factor-alpha synthesis inhibitor 3,6′-dithiothalidomide attenuates markers of inflammation, Alzheimer pathology and behavioral deficits in animal models of neuroinflammation and Alzheimer's disease. J Neuroinflammation 9: 106.
35. Nie J, Luo Y, Huang XN, Gong QH, Wu Q, et al. (2010) Icariin inhibits beta-amyloid peptide segment 25–35 induced expression of beta-secretase in rat hippocampus. Eur J Pharmacol 626: 213–218.
36. Alexander G, Hanna A, Serna V, Younkin L, Younkin S, et al. (2011) Increased aggression in males in transgenic Tg2576 mouse model of Alzheimer's disease. Behav Brain Res 216: 77–83.
37. Rangasamy T, Cho CY, Thimmulappa RK, Zhen L, Srisuma SS, et al. (2004) Genetic ablation of Nrf2 enhances susceptibility to cigarette smoke-induced emphysema in mice. J Clin Invest 114: 1248–1259.
38. Ownby RL (2010) Neuroinflammation and cognitive aging. Curr Psychiatry Rep 12: 39–45.
39. Pietropaolo S, Delage P, Lebreton F, Crusio WE, Cho YH (2012) Early development of social deficits in APP and APP-PS1 mice. Neurobiol Aging 33: 1002 e1017–1027.
40. Sha D, Li L, Ye L, Liu R, Xu Y (2009) Icariin inhibits neurotoxicity of beta-amyloid by upregulating cocaine-regulated and amphetamine-regulated transcripts. Neuroreport 20: 1564–1567.
41. Xin ZC, Kim EK, Lin CS, Liu WJ, Tian L, et al. (2003) Effects of icariin on cGMP-specific PDE5 and cAMP-specific PDE4 activities. Asian J Androl 5: 15–18.
42. Ning H, Xin ZC, Lin G, Banie L, Lue TF, et al. (2006) Effects of icariin on phosphodiesterase-5 activity in vitro and cyclic guanosine monophosphate level in cavernous smooth muscle cells. Urology 68: 1350–1354.
43. Cheng YF, Wang C, Lin HB, Li YF, Huang Y, et al. (2010) Inhibition of phosphodiesterase-4 reverses memory deficits produced by Abeta25-35 or Abeta1-40 peptide in rats. Psychopharmacology (Berl) 212: 181–191.
44. Zhang R, Wang Y, Zhang L, Zhang Z, Tsang W, et al. (2002) Sildenafil (Viagra) induces neurogenesis and promotes functional recovery after stroke in rats. Stroke 33: 2675–2680.
45. Zhang L, Zhang RL, Wang Y, Zhang C, Zhang ZG, et al. (2005) Functional recovery in aged and young rats after embolic stroke: treatment with a phosphodiesterase type 5 inhibitor. Stroke 36: 847–852.
46. Zhao J, Harada N, Kurihara H, Nakagata N, Okajima K (2010) Cilostazol improves cognitive function in mice by increasing the production of insulin-like growth factor-I in the hippocampus. Neuropharmacology 58: 774–783.
47. Wyss-Coray T (2006) Tgf-Beta pathway as a potential target in neurodegeneration and Alzheimer's. Curr Alzheimer Res 3: 191–195.
48. Gray CW, Patel AJ (1993) Regulation of beta-amyloid precursor protein isoform mRNAs by transforming growth factor-beta 1 and interleukin-1 beta in astrocytes. Brain Res Mol Brain Res 19: 251–256.
49. Lesne S, Docagne F, Gabriel C, Liot G, Lahiri DK, et al. (2003) Transforming growth factor-beta 1 potentiates amyloid-beta generation in astrocytes and in transgenic mice. J Biol Chem 278: 18408–18418.
50. Wyss-Coray T, Masliah E, Mallory M, McConlogue L, Johnson-Wood K, et al. (1997) Amyloidogenic role of cytokine TGF-beta1 in transgenic mice and in Alzheimer's disease. Nature 389: 603–606.
51. Grammas P, Ovase R (2002) Cerebrovascular transforming growth factor-beta contributes to inflammation in the Alzheimer's disease brain. Am J Pathol 160: 1583–1587.

52. Wyss-Coray T, Lin C, Sanan DA, Mucke L, Masliah E (2000) Chronic overproduction of transforming growth factor-beta1 by astrocytes promotes Alzheimer's disease-like microvascular degeneration in transgenic mice. Am J Pathol 156: 139–150.
53. Li YC, Ding XS, Li HM, Zhang C (2013) Icariin attenuates high glucose-induced type IV collagen and fibronectin accumulation in glomerular mesangial cells by inhibiting transforming growth factor-beta production and signalling through G protein-coupled oestrogen receptor 1. Clin Exp Pharmacol Physiol 40: 635–643.
54. Peress NS, Perillo E (1995) Differential expression of TGF-beta 1, 2 and 3 isotypes in Alzheimer's disease: a comparative immunohistochemical study with cerebral infarction, aged human and mouse control brains. J Neuropathol Exp Neurol 54: 802–811.
55. van der Wal EA, Gomez-Pinilla F, Cotman CW (1993) Transforming growth factor-beta 1 is in plaques in Alzheimer and Down pathologies. Neuroreport 4: 69–72.
56. Liu RM, Gaston Pravia KA (2010) Oxidative stress and glutathione in TGF-beta-mediated fibrogenesis. Free Radic Biol Med 48: 1–15.
57. Qi MY, Kai C, Liu HR, Su YH, Yu SQ (2011) Protective effect of Icariin on the early stage of experimental diabetic nephropathy induced by streptozotocin via modulating transforming growth factor beta1 and type IV collagen expression in rats. J Ethnopharmacol 138: 731–736.
58. Brandan E, Inestrosa NC (1993) Extracellular matrix components and amyloid in neuritic plaques of Alzheimer's disease. Gen Pharmacol 24: 1063–1068.

11

Analysis of Functions of VIP1 and Its Close Homologs in Osmosensory Responses of *Arabidopsis thaliana*

Daisuke Tsugama[1,2¤], Shenkui Liu[3], Tetsuo Takano[1*]

1 Asian Natural Environmental Science Center (ANESC), The University of Tokyo, Nishi-tokyo-shi, Tokyo, Japan, **2** Laboratory of Plant Molecular Genetics, Graduate School of Agricultural and Life Sciences, The University of Tokyo, Bunkyo-ku, Tokyo, Japan, **3** Alkali Soil Natural Environmental Science Center (ASNESC), Northeast Forestry University, Harbin, China

Abstract

VIP1 is a bZIP protein in *Arabidopsis thaliana*. VIP1 accumulates in the nucleus under hypo-osmotic conditions and interacts with the promoters of hypo-osmolarity-responsive genes, *CYP707A1* and *CYP707A3* (*CYP707A1/3*), but neither overexpression of VIP1 nor truncation of its DNA-binding region affects the expression of *CYP707A3* in vivo, raising the possibility that VIP and other proteins are functionally redundant. Here we show further analyses on VIP1 and its close homologs, namely, Arabidopsis group I bZIP proteins. The patterns of the signals of the GFP-fused group I bZIP proteins were similar in onion and Arabidopsis cells, suggesting that they have similar subcellular localization. In a yeast one-hybrid assay, the group I bZIP proteins caused reporter gene activation in the yeast reporter strain. VIP1 and other group I bZIP proteins showed positive results in a yeast two-hybrid assay and a bimolecular fluorescence complementation assay, suggesting that they physically interact. These results support the idea that they have somewhat similar functions. By gel shift assays, VIP1-binding sequences in the *CYP707A1/3* promoters were confirmed to be AGCTGT/G. Their presence in the promoters of the genes that respond to hypo-osmotic conditions was evaluated using previously published microarray data. Interestingly, a significantly higher proportion of the promoters of the genes that were up-regulated by rehydration treatment and/or submergence treatment (treatment by a hypotonic solution) and a significantly lower proportion of the promoters of the genes that were down-regulated by such treatment shared AGCTGT/G. To further assess the physiological role of VIP1, constitutively nuclear-localized variants of VIP1 were generated. When overexpressed in Arabidopsis, some of them as well as VIP1 caused growth retardation under a mannitol-stressed condition, where VIP1 is localized mainly in the cytoplasm. This raises the possibility that the expression of VIP1 itself rather than its nuclear localization is responsible for regulating the mannitol responses.

Editor: Guangxiao Yang, Huazhong University of Science & Technology (HUST), China

Funding: This work was supported by JSPS Grant-in-aid for Scientific Research (22-2144 and 25-7247) to DT. The funder had no role in study design, data collection and analysis, decision to publish, or preparation of the manuscript.

Competing Interests: The authors have declared that no competing interests exist.

* Email: takano@anesc.u-tokyo.ac.jp

¤ Current address: Biology Department, The Pennsylvania State University, University Park, Pennsylvania, United States of America

Introduction

Water availability is a key factor determining the growth, productivity, and distribution of plants. When plants are dehydrated or rehydrated, expression of certain subsets of genes is changed, and this contributes to plant adaptation to dehydration or rehydration. For example, in *Arabidopsis thaliana*, dehydration induces the expression of *NCED3*, which plays a role in biosynthesis of a stress-related phytohormone, abscisic acid (ABA). This enhances the ABA level and thereby increases the stress tolerance of plants [1–4]. On the other hand, rehydration induces the expression of *CYP707A1* and *CYP707A3* (*CYP707A1/3*). CYP707A1/3 catalyze 8′-hydroxylation of ABA to decrease the ABA level [3–5].

Dehydration- and/or rehydration-responsive genes are thought to be under the control of osmosensory signaling, but upstream components of the plant osmosensory signaling are largely unknown. A transmembrane histidine kinase, Sln1, is thought to be an osmosensor in baker's yeast (*Saccharomyces cerevisiae*) ([6], for a review), and one of the Arabidopsis histidine kinases, ATHK1, has been suggested to be involved in dehydration responses. Knockout of *ATHK1* decreases drought tolerance and expression levels of ABA-responsive genes, while overexpression of *ATHK1* increases drought tolerance and the expression levels of the ABA-responsive genes [7–10]. However, it is still unclear how ATHK1 mediates transduction of dehydration signal.

VIP1 (VirE2-interacting protein 1) is an Arabidopsis bZIP protein that has roles in *Agrobacterium* responses [11–13], pathogen responses [14], [15] and low-sulfur responses [16]. In additions to these responses, we recently showed that VIP1 is involved in the Arabidopsis osmosensory responses. VIP1 is localized mainly in the cytoplasm both when extracellular osmolarity is stable and when cells are exposed to hyper-osmotic conditions. When cells are exposed to hypo-osmotic conditions, VIP1 accumulates in the nucleus and interacts with *CYP707A1/3* promoters. On the other hand, neither overexpression of VIP1 nor truncation of its C-terminal DNA-binding region affects the expression of *CYP707A3* in vivo [17], raising the possibility that VIP1 and other proteins redundantly regulate the expression hypo-osmolarity-responsive genes such as *CYP707A3*. Homologs of VIP1 are thought to be candidates for such factors. Arabidopsis bZIP proteins have been classified into 11 groups on the basis of similarities and features of their amino acid sequences. This classification somewhat reflects the functions of plant bZIP proteins. For example, group A bZIP proteins redundantly function in ABA signaling ([18], for a review). VIP1 belongs to Arabidopsis group I bZIP protein. The Arabidopsis group I contains 11 other members, but they have not been well characterized. Here we show that VIP1 and other group I bZIP proteins are redundant in the subcellular localization and the transcriptional activation function, and can interact with each other. We also further address the relationship between the group I bZIP proteins and Arabidopsis osmosensory responses.

Materials and Methods

Plant materials and growth conditions

Arabidopsis thaliana ecotype Columbia-0 (Col-0) was used throughout the experiments. Seeds of *vip1* (SALK_001014C) [13] were obtained from ABRC (Arabidopsis Biological Resource Center, https://abrc.osu.edu/). The T-DNA insertion was confirmed by genomic PCR analysis as previously described [13]. Seeds were surface sterilized and sown on 0.8% agar containing 0.5× MS salts (Wako), 1% (w/v) sucrose and 0.5 g/l MES, pH 5.8, with 0, 200 or 250 mM mannitol or 0.25 μM ABA, chilled at 4°C in the dark for 48 h (stratified), and germinated at 22°C. Plants were grown at 22°C under 16-hour light/8-hour dark condition (light intensity 120 $\mu mol \cdot m^{-2} \cdot s^{-1}$). Arabidopsis transformation was performed by the *Agrobacterium*-mediated floral-dip method [19], and T2 generation plants were used for analyses. GFP fluorescence in individual plants was checked immediately before measuring cotyledon lengths to confirm the expression of the transgenes.

Reverse transcription-PCR (RT-PCR)

To analyze stage- and tissue-specific expression of the group I bZIP protein genes, 10-day-old seedlings and roots, leaves, and inflorescences of mature wild-type plants were sampled. To analyze gene expression under a hypo-osmotic condition, 2-week-old wild-type seedlings were sampled 0, 15 and 90 minutes after being submerged in 20 mM Tris-HCl, pH 6.8. To analyze transgene expression, 10-day-old seedlings of each transgenic line were sampled. To analyze gene expression in transgenic plants, 18-day-old seedlings grown in the presence of 0 or 200 mM mannitol were sampled. In each condition, 10–20 plants were sampled and used for RNA extraction for one biological replicate. Total RNA was extracted as previously described [20] and cDNA was synthesized from 3 μg of the total RNA with PrimeScript Reverse Transcriptase (Takara Bio) using an oligo (dT) primer. Reaction mixtures were diluted 25 times with distilled water and used as templates for RT-PCR. To quantify the absolute transcript levels of the group I bZIP protein genes, cDNA clones were obtained from RIKEN BRC Experimental Plant Division [21] (see Table S1 for their clone names). Partial fragments of cDNA of *UNE4* (*AT2G12940*), *AtbZIP71* (*AT2G24340*), and *AtbZIP74* (*AT2G21235*) were amplified by RT-PCR using cDNA synthesized from total RNA from Arabidopsis seedlings as template, and cloned into pBluescript SK^{-}. Partial fragments of cDNA clones of *AtbZIP31* (*AT2G13150*) and *AtbZIP33* (*AT2G12900*), were obtained by PCR as described in Table S2. These plasmids and PCR products were used to make standard curves.

GoTaq Green Master Mix (Promega) was used for semi-quantitative RT-PCR, and GoTaq qPCR Master Mix (Promega) for quantitative RT-PCR. Quantitative PCR was run using a StepOne Real-Time PCR System (Applied Biosystems). The comparative C_T method was used to calculate the relative expression levels, and the standard curve method was used to calculate the absolute transcript levels. *UBQ5* was used as an internal control gene for normalization. Primers used for RT-PCR are listed in Table S2 (for the group I bZIP protein genes), Table S3 (for the transgenes) and Dataset S2 (for the genes that have AGCTGT/G in their promoters).

Yeast one-hybrid (Y1H) and two-hybrid (Y2H) assays

For Y1H and Y2H, the open reading frames (ORFs) of the group I bZIP protein genes and truncated versions of *VIP1* were amplified by PCR using the RIKEN cDNA clones as template, and cloned into pGADT7-Rec and pGBKT7 (Clontech). Primers and restriction sites used to generate the constructs are listed in Table S4. For Y1H, the *Saccharomyces cerevisiae* strain AH109 (Clontech) was transformed with the pGBK constructs. For Y2H, AH109 was co-transformed with pGBKT7 containing the ORF of the C-terminal region (amino acids 165–341) of VIP1 (pGBK-VIP1ΔN164) and the pGAD constructs. Reporter gene activation was checked by observing cell growth on SD (synthetic dextrose) media lacking adenine and histidine as previously described [17], [22].

Bimolecular fluorescence complementation (BiFC) and GFP localization studies

The ORFs of the group I bZIP protein genes and truncated versions of *VIP1* were amplified by PCR using the RIKEN cDNA clones as template, and inserted into pBS-35S-nYFP-2, pBS-35SMCS-cYFP [17], pBS-35SMCS-GFP [23] and/or pBI121-35SMCS-GFP [24]. Primers and restrictions sites used to generate the constructs are listed in Table S5. The ORFs of VIP1 variants that have point-mutations in the putative phosphorylation sites and the ORF of a VIP1 variant that has the SV40 NLS in its N-terminus were generated and inserted into pBS-35SMCS-GFP and/or pBI121-35SMCS-GFP as described in Table S6. The ORFs of the VIP1 variants in the generated constructs were sequenced to confirm the truncations, the mutations and the fusion of the SV40 NLS. The BiFC constructs (nYFP and cYFP constructs) and the pBS-35SMCS-GFP constructs were used for transient expression experiments using onion cells, while the pBI121-35SMCS-GFP constructs were used for Arabidopsis transformation. Onion epidermal cells were transformed by particle bombardment as previously described [25]. Arabidopsis transformation was performed as described above. Fluorescence was observed using an epifluorescence microscope (BX51, Olympus) in all the experiments.

Gel shift assay

To express the GST-fused C-terminal region (the amino acid position 292–553) of AtbZIP29 (GST-AtbZIP29ΔN291), the pGBK construct containing *AtbZIP29* (see above and Table S4) was digested by *Eco*RI and *Sal*I, and the resultant fragment corresponding to *AtbZIP29ΔN291* was inserted into the *Eco*RI-*Sal*I site of pGEX-6P-3 (GE Healthcare). The resultant construct was transformed into the *Escherichia coli* strain BL21 (DE3). To induce GST-AtbZIP29ΔN291, the transformed *E. coli* cells were grown in liquid LB medium at 37°C. When the OD_{600} reached

0.5, IPTG was added to the medium to a final concentration of 0.2 mM. The cells were further grown at 28°C for 3 h, collected by centrifugation (12000×g, 1 min), resuspended in 1× TBS (150 mM NaCl, 20 mM Tris-HCl, pH 7.5) containing 1 mg/ml lysozyme, frozen at −80°C, and thawed at room temperature. The freezing and thawing were repeated twice more, and 2 units recombinant DNase I was added to the solution. The solution was incubated for 20 min at room temperature, and centrifuged at 12000×g for 5 min at room temperature. GST-AtbZIP29ΔN291 in the supernatant was bound to Glutathione Sepharose 4 Fast Flow (GE Healthcare), washed five times by 1× TBS, eluted in 20 mM reduced glutathione in 50 mM Tris-HCl, pH 8.0, as manufacturer's instructions, and used as the purified protein sample.

Gel shift assays were performed using DIG-labeled DNA probes and the purified protein samples as previously described [17]. The sequences of the probes are provided in Fig. S3. To prepare the competitors for Fig. S4 and Fig. S5, two complementary oligonucleotides were mixed in equal concentrations, incubated at 90°C for 5 min, and gradually cooled to room temperature. The sequences of the competitors are provided in Fig. S5. The competitors were used at 0.5 μM unless otherwise stated.

Analysis of the VIP1-binding element in the promoters of the rehydration- and submergence-responsive genes

The lists of the rehydration- and submergence-responsive genes were obtained from the previous reports [26–29] (see Table 1 and Dataset S1 for the criteria of those genes). The list of the genes with promoters (regions either 500-bp or 1000-bp upstream of the start codon) containing the VIP1-binding element (AGCTGT or AGCTGG) was obtained with Patmatch in TAIR (https://www.arabidopsis.org/cgi-bin/patmatch/nph-patmatch.pl). These lists were used for one-sided binomial tests to evaluate the abundance of the VIP1-binding element in the promoters of the rehydration- and submergence-responsive genes. For the binomial tests, the numbers of the rehydration- and submergence-responsive genes were used as "the number of trials", and the numbers of the rehydration- and submergence-responsive genes with promoters containing the VIP1-binding element were used as "the numbers of success". For "the probabilities of success", the total number of the genes with promoters containing the VIP1-binding element was divided by the total number of genes in the Arabidopsis genome, and the resultant values was used. Those numbers used for the binomial tests are shown in Table 1 and Dataset S1.

Microarray

A custom microarray which contains approximately 1700 genes of Arabidopsis [30] was provided by Dr. Yoshihiro Narusaka (Research Institute for Biological Sciences, Okayama, Japan). For the microarray analysis, total RNA was extracted from 10-day-old seedlings of wild-type plants and transgenic plants expressing GFP-fused VIP1NLS using RNeasy Plant Mini Kit (Qiagen). cDNA synthesis and hybridization were performed using 3 DNA Array 50 Cy3/Cy5 Kit (Genisphere) following the manufacturer's instructions. Signal detection and normalization (global normalization) were performed at Hokkaido System Science, Japan. Experiments were performed in triplicate and data were averaged. The data are provided as Dataset S3. The genes that showed 2-fold or higher expression levels in the transgenic plants than in the wild type are indicated as up-regulated genes, and the genes that showed 0.5-fold or lower expression levels are indicated as down-regulated genes.

Results and Discussion

Functional redundancies among Arabidopsis group I bZIP proteins

The locus AT2G13130, which was listed as an Arabidopsis group I bZIP protein gene in an early study [18], was excluded from analyses because it matched a pseudogene in TAIR (The Arabidopsis Information Resource: http://www.arabidopsis.org/). The other 12 members of the Arabidopsis group I protein were further classified into subgroups 1 and 2, and others (AtbZIP71 and AtbZIP74) on the basis of a phylogenetic analysis using their amino acid sequences (Fig. 1A). The absolute transcript levels of the genes encoding these proteins were compared by quantitative RT-PCR (qRT-PCR). In all the tissues studied, the genes encoding the subgroup 1 proteins were more highly expressed than the genes encoding the proteins of the other groups. Among the subgroup 1 protein genes, *VIP1* was the most highly expressed in seedlings, leaves and roots, while *AtbZIP52* was the most highly expressed in flowers (Fig. 1B).

The nuclear-cytoplasmic shuttling of VIP1 is dependent on the nuclear localization signal (NLS) and the nuclear export signal (NES), which are present in the beginning and the end, respectively, of the C-terminal bZIP domain of VIP1 [12], [13], [17]. The NLS, the NES and the bZIP domain are highly conserved among the Arabidopsis group I bZIP proteins (Fig. S1, underlined region). Seven members of the Arabidopsis group I bZIP proteins (VIP1, PosF21, AtbZIP29, AtbZIP52, AtbZIP31, AtbZIP32 and AtbZIP33) were expressed as GFP-fused proteins in onion epidermal cells to examine their subcellular localization. GFP-fused AtbZIP31 showed intense signals in the nucleus, while the other six showed clear signals in the cytosol (Fig. 1C). The number of the amino acids following the putative NES is the smallest in AtbZIP31 among the Arabidopsis group I bZIP proteins (Fig. S1), and this might decrease the activity of the NES of AtbZIP31. Treatment with a hypotonic solution, not with a hypertonic solution, induces transient nuclear accumulation of VIP1 [17]. GFP-fused PosF21 (PosF21-GFP) and GFP-fused AtbZIP29 (AtbZIP29-GFP) were expressed in Arabidopsis (Fig. S2) to examine their subcellular localization under a hypo-osmotic condition. When plants were incubated in a hypotonic solution, the signals of both PosF21-GFP and AtbZIP29-GFP, as well as the signals of GFP-fused VIP1, were first detected mainly in the cytoplasm, then mainly in the nucleus, and then again mainly in the cytoplasm (Fig. 1D), suggesting that the transient nuclear accumulation under the hypotonic condition is somewhat conserved among the group I bZIP proteins. Besides the C-terminal bZIP domain, several conserved sites are present in the N-terminal regions of the group I bZIP proteins (Fig. S1). These conserved sites may be involved in regulating the subcellular localization of the group I bZIP proteins.

The N-terminal region of VIP1 shows a transcriptional activation function in yeast [17]. Transcriptional activation functions of other five members of the Arabidopsis group I bZIP proteins (AtbZIP52, PosF21, AtbZIP69, AtbZIP29, AtbZIP30) were examined by a yeast one-hybrid (Y1H) assay. The bZIP proteins were expressed as GAL4 DNA-binding domain (GAL4BD)-fused proteins in a yeast reporter strain which has a GAL4BD-binding site upstream of the auxotrophic marker genes *ADE2* and *HIS3*. All of the proteins, as well as VIP1, enabled yeast cells to grow on the selection medium lacking adenine and histidine (Fig. 1E), suggesting that these proteins have transcriptional activation functions in yeast.

The bZIP domain is responsible for the DNA-binding ability of bZIP proteins, and the VIP1 C-terminal region alone, which

Table 1. Proportions of the genes with promoters (500 bp upstream of the start codon) containing AGCTGT/G in the submergence- and/or rehydration-responsive gene sets detected in the previously performed microarray analyses.

*1Gene set	Proportion of genes with at least one AGCTGT/G	Cumulative binomial distribution	*4	Proportion of genes with at least two AGCTGT/G	Cumulative binomial distribution	*4
*1Total in the genome	0.185	0.504		0.033	0.508	
*2Up-regulated 30 min after submergence in root	0.263	0.986	+	0.068	0.983	+
*2Up-regulated 30 min after submergence in shoot	0.194	0.652		0.032	0.724	
*2Up-regulated 1 h after submergence in root	0.235	0.989	+	0.056	0.985	+
*2Up-regulated 1 h after submergence in shoot	0.176	0.5		0.029	0.604	
*2Up-regulated 3 h after submergence in root	0.198	0.839		0.069	~1	++
*2Up-regulated 3 h after submergence in shoot	0.147	0.103		0.021	0.239	
*2Up-regulated 6 h after submergence in root	0.194	0.792		0.034	0.58	
*2Up-regulated 6 h after submergence in shoot	0.166	0.159		0.028	0.32	
*2Up-regulated 12 h after submergence in root	0.196	0.849		0.037	0.776	
*2Up-regulated 12 h after submergence in shoot	0.175	0.244		0.028	0.247	
*3Up-regulated by 2-h rehydration and 2-h dehydration	0.293	0.999	++	0.102	~1	++
*2Down-regulated 30 min after submergence in root	0.23	0.908		0.027	0.478	
*2Down-regulated 30 min after submergence in shoot	0.143	0.618		0	0.789	
*2Down-regulated 1 h after submergence in root	0.193	0.683		0.017	0.037	*
*2Down-regulated 1 h after submergence in shoot	0.125	0.409		0	0.582	
*2Down-regulated 3 h after submergence in root	0.187	0.561		0.036	0.729	
*2Down-regulated 3 h after submergence in shoot	0.133	0.11		0.031	0.587	
*2Down-regulated 6 h after submergence in root	0.185	0.498		0.036	0.736	
*2Down-regulated 6 h after submergence in shoot	0.142	0.01	*	0.033	0.569	
*2Down-regulated 12 h after submergence in root	0.18	0.304		0.033	0.529	
*2Down-regulated 12 h after submergence in shoot	0.155	0.008	**	0.026	0.106	
*3Down-regulated by 2-h rehydration after 2-h dehydration	0.16	0.171		0.016	0.0786	

*1–3See Dataset S1 for the exact numbers of the genes and more gene sets, which include the gene sets detected in [26] and [28].
*2These gene sets were prepared using the data in [29]. In this study, the genes that showed at least 2-fold higher expression with *P*-values <0.05 under the submergence condition than the control condition (non-submergence condition) were defined as up-regulated genes, and the genes that showed at least 2-fold lower expression with *P*-values <0.05 as down-regulated genes.
*3This gene set was prepared using the data in [27]. In this study, the genes that showed at least 5-fold higher expression after the 2-hour rehydration treatment preceded by the 2-hour dehydration treatment were defined as up-regulated genes, and the genes that showed at least 2-fold lower expression as down-regulated genes.
*4++: $P>0.99$; +: $P>0.95$; **: $P<0.01$; *: $P<0.05$ in cumulative binomial.

includes the bZIP domain, can bind the *CYP707A1/3* promoters in vitro [17]. In a gel shift assay, the C-terminal region (the amino acid position 292–553) of AtbZIP29 caused a shift in the electrophoretic mobility of the *CYP707A1/3* promoter fragments, and this mobility shift was blocked when an excess amount of competitor DNA was present in the reaction mixture (Fig. 1F), suggesting that AtbZIP29 as well as VIP1 binds the *CYP707A1/3* promoters in vitro. A previous study predicted that plant group I bZIP proteins form homodimers and heterodimers via their bZIP domains [31]. Interactions between VIP1 and other Arabidopsis group I bZIP proteins were examined by a yeast two-hybrid (Y2H) assay and a bimolecular fluorescence complementation (BiFC) assay. In the Y2H assay, six Arabidopsis group I bZIP proteins (VIP1, PosF21, AtbZIP69, AtbZIP29, AtbZIP30 and AtbZIP52) enabled yeast cells to grow on selection medium when co-expressed with the C-terminal region (amino acids 165–341) of VIP1 (VIP1ΔN164), which contains the bZIP domain (Fig. 1G). In the BiFC assay, cYFP (the C-terminal region of yellow fluorescent protein)-fused VIP1, PosF21 and AtbZIP52 all recovered YFP fluorescence when co-expressed with nYFP (the N-terminal region of YFP)-fused VIP1 (Fig. 1H). These results support the above prediction of the dimerization between the bZIP domains of the group I bZIP proteins. bZIP domains are thought to determine the dimerization specificity and the DNA-binding specificity of the bZIP proteins [18], [31]. Together, it is possible that the dimers between VIP1, AtbZIP29 and other group I bZIP proteins have similar DNA-binding specificities, and share target genes.

Characterization of VIP1-binding sequence as hypo-osmolarity- and/or submergence-responsive element

In previous studies, VIP1 showed affinity to AGCNGT [15], [32], and a tobacco (*Nicotiana tabacum*) group I bZIP protein, RSG, bound to CAAGCTGG [33], [34]. The *CYP707A1* promoter has AGCTGT at the position −256 to −251 from the start codon, while the *CYP707A3* promoter has a part of the RSG-binding sequence, AGCTGG, at the corresponding position (Fig. S3). By gel shift assays, the VIP1-binding region in the *CYP707A1* promoter was narrowed to approximately 80 bp regions, and these regions contain AGCTGT (Fig. S4). The 21-bp

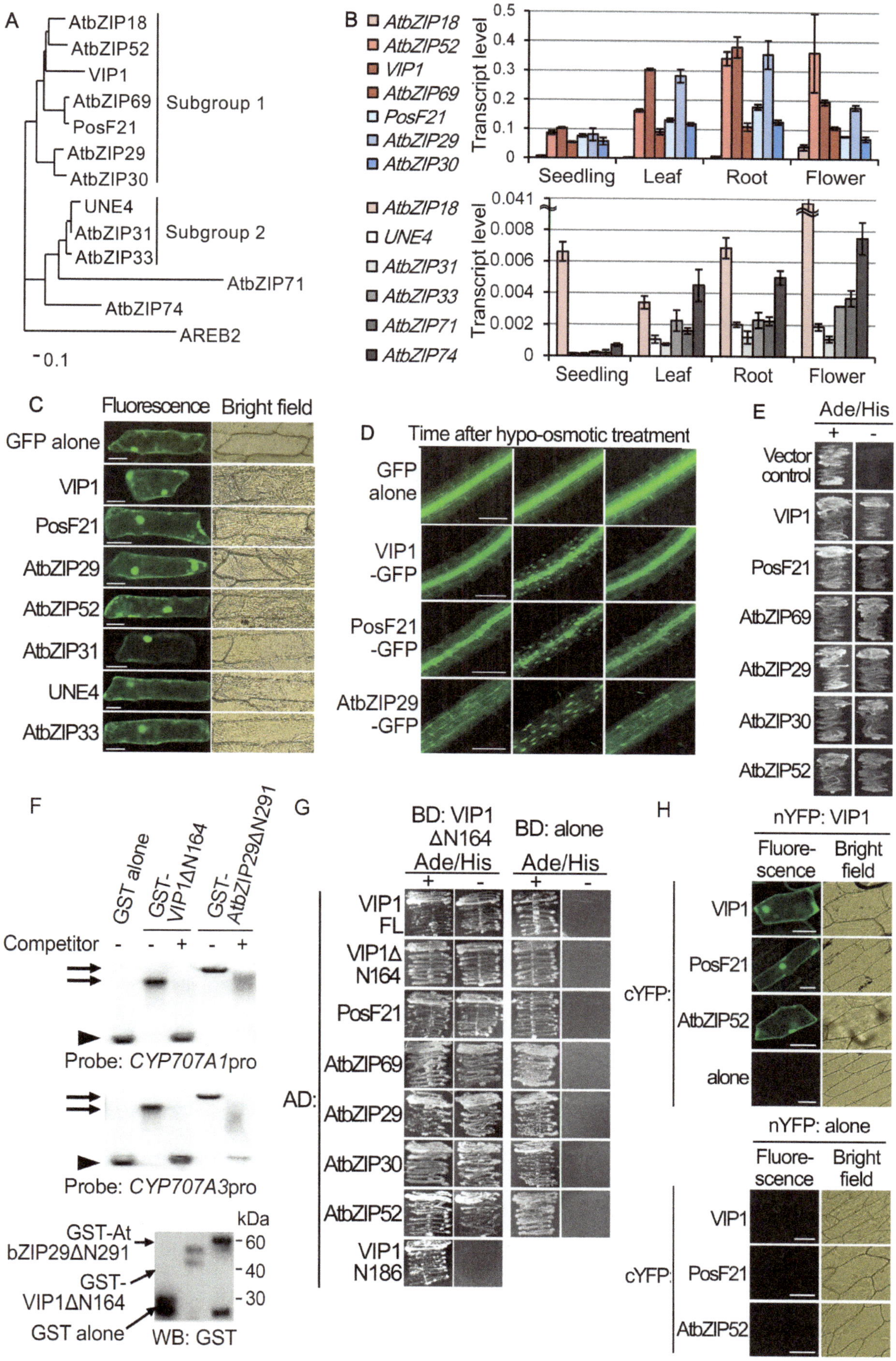

A
AtbZIP18
AtbZIP52
VIP1
AtbZIP69
PosF21
AtbZIP29
AtbZIP30
Subgroup 1
UNE4
AtbZIP31
AtbZIP33
Subgroup 2
AtbZIP71
AtbZIP74
AREB2
0.1
B
Transcript level
Seedling
Leaf
Root
Flower
C
Fluorescence
Bright field
GFP alone
D
Time after hypo-osmotic treatment
VIP1 -GFP
PosF21 -GFP
AtbZIP29 -GFP
E
Ade/His
Vector control
F
GST alone
GST-VIP1ΔN164
GST-AtbZIP29ΔN291
Competitor
Probe: CYP707A1pro
Probe: CYP707A3pro
kDa
WB: GST
G
BD: VIP1 ΔN164
BD: alone
AD:
VIP1 FL
VIP1 N186
H
nYFP: VIP1
nYFP: alone
cYFP:
Fluore-scence
Bright field
alone

Figure 1. Classification and functional redundancies of Arabidopsis group I bZIP proteins. (A) A rooted phylogenetic tree of the Arabidopsis group I bZIP proteins. The alignment of the amino acid sequences of the group I bZIP proteins (see Fig. S1) was used to generate the distance matrix and the neighbor-joining tree. AREB2 (a group A bZIP protein) was used as an outgroup. (B) Transcript levels of the Arabidopsis group I bZIP protein genes. The absolute amounts of cDNA of the group I bZIP protein genes in the template solution were quantified by qRT-PCR using the standard curve method. The values were normalized by the amount of cDNA of an internal control gene, *UBQ5*, and fold differences from the amount of *UBQ5* cDNA are shown as transcript levels. The expression levels of *AtbZIP18* in the lower panel are the same as those in the upper panel, and shown as control. Data are means of three biological replicates. Error bars indicate SD. (C) Subcellular localization of group I bZIP proteins in onion cells. The indicated proteins were expressed as GFP-fused proteins in onion epidermal cells. For each construct, more than 10 cells were observed, and a representative image is shown. Scale bars = 100 μm. (D) Subcellular localization of VIP1, PosF21 and AtbZIP29 under a hypo-osmotic condition. GFP-fused VIP1, PosF21 and AtbZIP29 (VIP1-GFP, PosF21-GFP and AtbZIP29-GFP, respectively) were expressed in Arabidopsis. Roots of the transgenic plants were submerged in 20 mM Tris-HCl, pH 6.8 (hypo-osmotic treatment), incubated for 0, 10, 60 minutes, and used for fluorescence microscopy. For each genotype, more than five plants were used for observation, and representative images in the same root are shown. Scale bars = 100 μm. (E) Transcriptional activation functions of group I bZIP proteins in yeast. The indicated proteins were expressed as GAL4 DNA-binding domain-fused proteins in the yeast reporter strain AH109. Transformed cells were grown on the medium with or without adenine and histidine (Ade/His + or -, respectively). For each construct, three individual colonies were assessed, and a representative result is shown. (F) Interaction between the C-terminal region of AtbZIP29 and *CYP707A1/3* promoters in vitro. The C-terminal regions of VIP1 and AtbZIP29 (the amino acid positions 165–341 for VIP1 and 292–553 for AtbZIP29) were expressed as GST-fused proteins (GST-VIP1Δ164 and GST-AtbZIP29Δ291, respectively) in *E. coli*, purified, and used for the gel shift assay. The presence of GST-VIP1Δ164, GST-AtbZIP29Δ291 and GST alone, which was used as control, in the purified protein solutions was confirmed by Western blotting using an anti-GST antibody (WB: GST, bottom panel). As probes for the gel shift assay, approximately 250-bp DIG-labeled fragments of the *CYP707A1* promoter (*CYP707A1*pro) and the *CYP707A3* promoter (*CYP707A3*pro) were used. As the competitors, the non-DIG-labeled DNA fragments that have the same sequences as the probes were used, and their presence and absence in the reaction mixtures are shown as + and -, respectively (top and middle panels). The experiments were performed three times, and representative results are shown. (G) Interactions between VIP1 and group I bZIP proteins in yeast. The VIP1 variants (VIP1FL: full-length VIP1; VIP1ΔN164: amino acids 165–341 of VIP1; VIP1N186: amino acids 1–186 of VIP1) and the group I bZIP proteins were expressed as GAL4 activation domain-fused proteins (shown as AD:) with GAL4 DNA-binding domain (BD)-fused VIP1ΔN164 (BD: VIP1Δ164) or BD alone (BD: alone) in AH109. VIP1N186 was used as a negative control. Transformed yeast cells were grown on the medium with or without adenine and histidine (Ade/His + or -, respectively). For each combination of the constructs, three individual colonies were assessed, and a representative result is shown. (H) BiFC between VIP1, PosF21 and AtbZIP52 in onion cells. The ORF of *VIP1* was cloned in-frame in front of the coding sequence of the N-terminal region of YFP (nYFP) to express nYFP-fused VIP1 (nYFP: VIP1). The ORFs of *VIP1*, *PosF21*, and *AtbZIP52* were cloned in frame in front of the coding sequence of the C-terminal region of YFP (cYFP) to express the cYFP-fused proteins (shown as cYFP:). These constructs were co-introduced into onion epidermal cells in the indicated combinations. Empty vectors which express nYFP alone and cYFP alone were used for negative controls (shown as nYFP: alone and cYFP: alone, respectively). For each combination, more than 10 cells were observed, and a representative image is shown. Scale bars = 100 μm.

CYP707A1 promoter fragment containing AGCTGT inhibited the VIP1-dependent band shift when used as competitor in gel shift assays. When AGCTGT was point-mutated, the resultant 21-bp fragment no longer inhibited the VIP1-dependent band shift. The corresponding 21-bp *CYP707A3* promoter fragment and the *CYP707A1* promoter fragment inhibited the VIP1-dependent band shift to the same extent (Fig. 2A). These results suggest that, in agreement with the previous studies, VIP1 has high affinity to AGCTGT or AGCTGG (AGCTGT/G), and that AGCTGT/G is largely responsible for the interactions between VIP1 and the *CYP707A1/3* promoters. RSG, a tomato group I bZIP protein, VSF-1, and rice group I bZIP proteins, Rf2a and Rf2b, can interact with DNA fragments that do not contain AGCTGT/G [33], [35–39]. However, those DNA fragments were less effective in inhibiting the VIP1-dependent band shift than either the *CYP707A1/3* promoter fragments or rbe (RSG-binding element), which contains AGCTGG, when used as competitors in gel shift assays (Fig. S5), suggesting that DNA containing AGCTGT/G has higher affinity to VIP1. Even though the *CYP707A1/3* promoter fragments and rbe share AGCTGT/G, the *CYP707A1/3* promoter fragments more effectively inhibited the VIP1-dependent band shift than rbe (Fig. S5, left 4 lanes), suggesting that sequences surrounding AGCTGT/G also affect the interaction between VIP1 and DNA.

Previously microarray analyses were performed to examine rehydration-responsive gene expression [26], [27] and submergence-responsive gene expression [28], [29] in Arabidopsis. In these studies, plants were exposed to hypotonic solutions as either submergence treatment or rehydration treatment. These treatments seemed similar to the treatment performed in our study, thus those microarray data were utilized to evaluate whether AGCTGT/G acts as a hypo-osmolarity-responsive element. Interestingly, when the regions within 500 bp upstream of the start codons were assessed, the proportions of the genes with promoters containing AGCTGT/G (or its complementary sequence C/ACAGCT) were significantly higher in the gene sets which were up-regulated in roots within 0.5–3 hours after the submergence treatment (using a hypotonic solution) and in the gene set which was up-regulated by 2-hour rehydration after 2-hour dehydration than would be expected from the abundance of AGCTGT/G in all the promoters in the whole Arabidopsis genome (Table 1, upper rows). In addition, the proportions of the genes with promoters (500 bp upstream of the start codons) containing AGCTGT/G were significantly lower in the gene sets which were down-regulated 1–12 hours after the root submergence than is expected (Table 1, lower rows). When the promoter regions assessed were extended to 1000 bp upstream of the start codons, the proportions of genes with promoters containing AGCTGT/G were not significantly higher in any gene sets which were up-regulated by the root submergence. However, the proportions of the genes with promoters containing AGCTGT/G were still significantly lower in the gene set which was down-regulated 1–12 hours after the submergence than is expected (Dataset S1). These results raise the possibility that AGCTGT/G within 500 bp upstream of the start codon acts as a cis-element which activates early submergence-responsive genes and that AGCTGT/G within the 1000 bp upstream of the start codon is necessary for maintaining the basal gene expression during the hypo-osmotic condition.

Expression of 26 genes with promoters (500 bp upstream of the start codons) containing at least one copy of AGCTGT/G were examined by qRT-PCR. Fourteen of those genes were more than 2-fold up-regulated either 15 or 90 minutes after the submergence treatment (Fig. 2B). No clear correlation was observed between the numbers of AGCTGT/G in the promoters and the

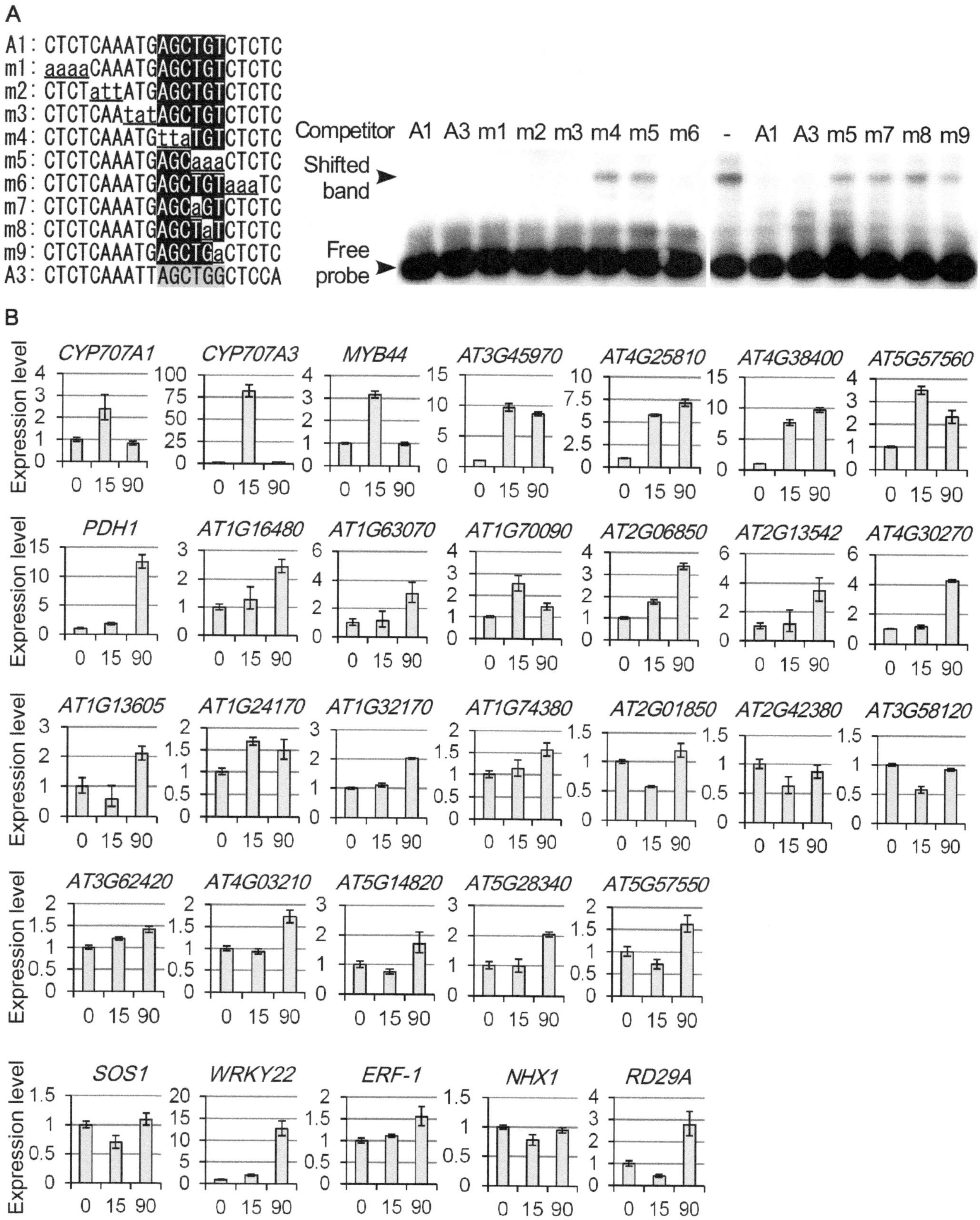

Figure 2. Interaction between VIP1 and AGCTGT/G and its relevance to hypo-osmolarity- and/or submergence-induced gene expression. (A) VIP1 binds AGCTGT/G. The gel shift assay was performed using purified GST-fused VIP1ΔN164 and a DIG-labeled *CYP707A1* promoter fragment as the probe. The 21-bp *CYP707A1/3* promoter fragments (A1 for the *CYP707A1* promoter and A3 for the *CYP707A3* promoter) and point-mutated versions of A1 (m1-9) were used as competitors. In the left panel, sequences mutated from A1 in m1-9 are in lower case and underlined. AGCTGT and AGCTGG in the competitors are indicated by black and gray boxes, respectively. "Competitor -" indicates the absence of the competitors in the reaction mixture. The results of A1 and A3 are shown in both the middle and the right panels for controls. The experiments were

performed three times, and a representative result is shown. (B) Quantitative RT-PCR analyses of expression of the genes with promoters containing at least one copy of AGCTGT/G (see Dataset S2 for the exact copy numbers of AGCTGT/G in the promoters of these genes) under a hypo-osmotic condition. Two-week-old wild-type seedlings were submerged in 20 mM Tris-HCl for 0, 15 and 90 minutes, and sampled for RNA extraction and cDNA synthesis. Numbers on x-axes indicate the time points of sampling. Genes up-regulated (>2 fold) either 15 or 90 minutes after the hypo-osmotic treatment are shown in the first two rows. Genes that were not up-regulated by the hypo-osmotic treatment are shown in the third and fourth rows. Genes in the bottom row (*SOS1, WRKY22, ERF-1, NHX1* and *RD29A*) do not have AGCTGT/G in the promoters and are shown as controls. Relative expression levels were calculated by the comparative cycle threshold (C_T) method using *UBQ5* as an internal control gene. Data are means of three biological replicates. Error bars indicate SD.

submergence-induced fold changes of the expression of those genes (Dataset S2).

Generation and characterization of constitutively nuclear-localized variants of VIP1

Because VIP1 variants that have different subcellular localization from the subcellular localization of wild-type VIP1 are thought to be useful to examine the functions of VIP1, several variants of VIP1 were generated and their subcellular localization was examined. First, VIP1 mutants with either Ser → Asp mutation at position 79 (VIP1S79D) or Ser → Ala mutation at position 115 (VIP1S115A) were tested. In previous studies, S79D enhanced the nuclear localization of VIP1 [14], and the mutation corresponding to S115A in tobacco RSG enhanced the nuclear localization of RSG [40], [41]. However, in our experiments, the patterns of the signals of GFP-fused VIP1S79D and GFP-fused VIP1S115A were both similar to the pattern of the signals of GFP-fused wild-type VIP1 (VIP1-GFP) in Arabidopsis treated with a hypotonic solution (Fig. S6), suggesting that neither S79D nor S115A single mutation affects the nuclear-cytoplasmic shuttling of VIP1 under the hypo-osmotic condition. VIP1 has many possible phosphorylation sites according to phosphorylation site-prediction programs (NetPhos: http://www.cbs.dtu.dk/services/NetPhos/, [42]; GPS: http://gps.biocuckoo.org/, [43]), thus multiple phosphorylation sites may be involved in regulating the subcellular localization of VIP1.

Next, N-terminal truncated versions of VIP1 were generated (Fig. 3A). When expressed as GFP-fused proteins in onion epidermal cells, the N-terminal 20 amino acid-truncated version of VIP1 (VIP1ΔN20, form b) showed clear cytoplasmic signals similar to the signals of VIP1-GFP (Fig. 3B, panels a and b), but the 50 amino acid- and 80 amino acid-truncated versions (VIP1ΔN50 and VIP1ΔN80, forms c and d, respectively) showed weaker cytoplasmic signals and intense nuclear signals (Fig. 3B, panels c and d). The signals for the GFP-fused 164 amino acid-truncated version (VIP1ΔN164, form e) were almost limited to the nucleus (Fig. 3B, panel e). These results suggest that the N-terminal truncations enhance the nuclear localization of VIP1.

SV40 (Simian virus 40) NLS functions as an active NLS in plants [44]. The signals of a VIP1 variant which has SV40 NLS, which functions in the N-terminus (VIP1NLS, Fig. 3A, form f) were detected only in the nucleus in onion cells when it was expressed as GFP-fused protein (Fig. 3B, panel f), suggesting that the N-terminal fusion of a functional NLS can convert VIP1 into a constitutively nuclear-localized protein.

In a Y1H assay, VIPΔN80 enabled yeast cells to grow on the selection medium when expressed as a GAL4BD-fused protein, while VIP1ΔN164 did not (Fig. S7), raising the possibility that these VIP1 variants have different effects on the expression of VIP1 target genes. VIP1ΔN80, VIP1ΔN164 and VIP1NLS were expressed as GFP-fused proteins (VIP1ΔN80-GFP, VIP1ΔN164-GFP and VIP1NLS-GFP, respectively) in Arabidopsis (Fig. S8) to further examine the physiological roles of VIP1 and group I bZIP proteins. In agreement with the results of the transient expression experiments using onion cells (Fig. 3B), in Arabidopsis, VIP1ΔN80-GFP, VIP1ΔN164-GFP and VIP1NLS-GFP all showed intense nuclear signals even when cells were not treated with a hypotonic solution (Fig. 3C). In a previous study, overexpression of VIP1-GFP did not affect the survival rate of plants that were dehydrated and then rehydrated, but it retarded plant growth in the presence of high concentrations of mannitol. In the same study, VIP1-GFP was localized mainly in the cytoplasm under the mannitol-stressed conditions [17]. In the presence of 200 mM mannitol, cotyledon growth was more severely inhibited in transgenic plants expressing VIP1-GFP, VIP1ΔN80-GFP, VIP1ΔN164-GFP and VIP1NLS-GFP than in the wild-type plants (WT) and the other control plants (Fig. 3D). The amount of VIP1 in the nucleus may be higher in VIP1-GFP-overexpressing plants than in WT, and this may be sufficient to cause the phenotype similar to the phenotypes caused by the expression of the constitutively nuclear-localized variants of VIP1. Alternatively, the expression of VIP1 itself rather than its nuclear localization might be responsible for causing the growth-retardation phenotype under the mannitol-stressed condition. Because VIP1ΔN80 and VIP1ΔN164, which showed different transcriptional activation functions in yeast (Fig. S8), caused the similar phenotype in the mannitol-stressed plants, the putative transcriptional activation domain of VIP1 may not be involved in regulating the mannitol responses. *CYP707A1/3*, which play roles in ABA catabolism [3–5], are two putative target genes of VIP1 [17], but no significant difference was observed between WT and the transgenic lines in ABA-induced inhibition of seed germination (Fig. S9).

In qRT-PCR, four out of the seven submergence-induced genes with promoters containing AGCTGT/G were more highly expressed in either the presence of 200 mM mannitol or its absence in the transgenic plants overexpressing VIP1-GFP and VIP1NLS-GFP than in WT. However, the expression levels of these genes were just slightly higher (less than 2 fold) in the transgenic plants overexpressing VIP1ΔN80-GFP, VIP1ΔN164-GFP than in WT (Fig. 3E, *CYP707A1*, *AT3G45970*, *AT4G25810* and *AT4G38400*). In addition, a larger-scale gene expression analysis using a custom microarray containing approximately 1700 Arabidopsis genes [30] detected only 10 genes down-regulated (more than 2-fold) in the VIP1NLS-GFP-expressing plants, and one gene up-regulated. No clear correlation was observed between these genes and either AGCTGT/G in the promoters or the submergence-responsive genes in the previously published microarray data (Dataset S3), suggesting that complicated mechanisms still underlie in regulating the hypo-osmolarity and/or submergence-responsive gene expression. VIP1 and other group I bZIP proteins can bind not only AGCTGT/G but also other DNA sequences (Fig. S5), and these non-AGCTGT/G sequences might also be involved in the hypo-osmolarity and/or submergence-responsive gene expression. It may also be possible that constitutively localizing VIP1 to the nucleus is not sufficient to fully regulate the expression of VIP1 target genes. For example, an Arabidopsis group A bZIP protein, AREB1, is constitutively nuclear-localized, but requires phosphorylation to fully activate its target genes [45]. In addition, Arabidopsis group C bZIP proteins,

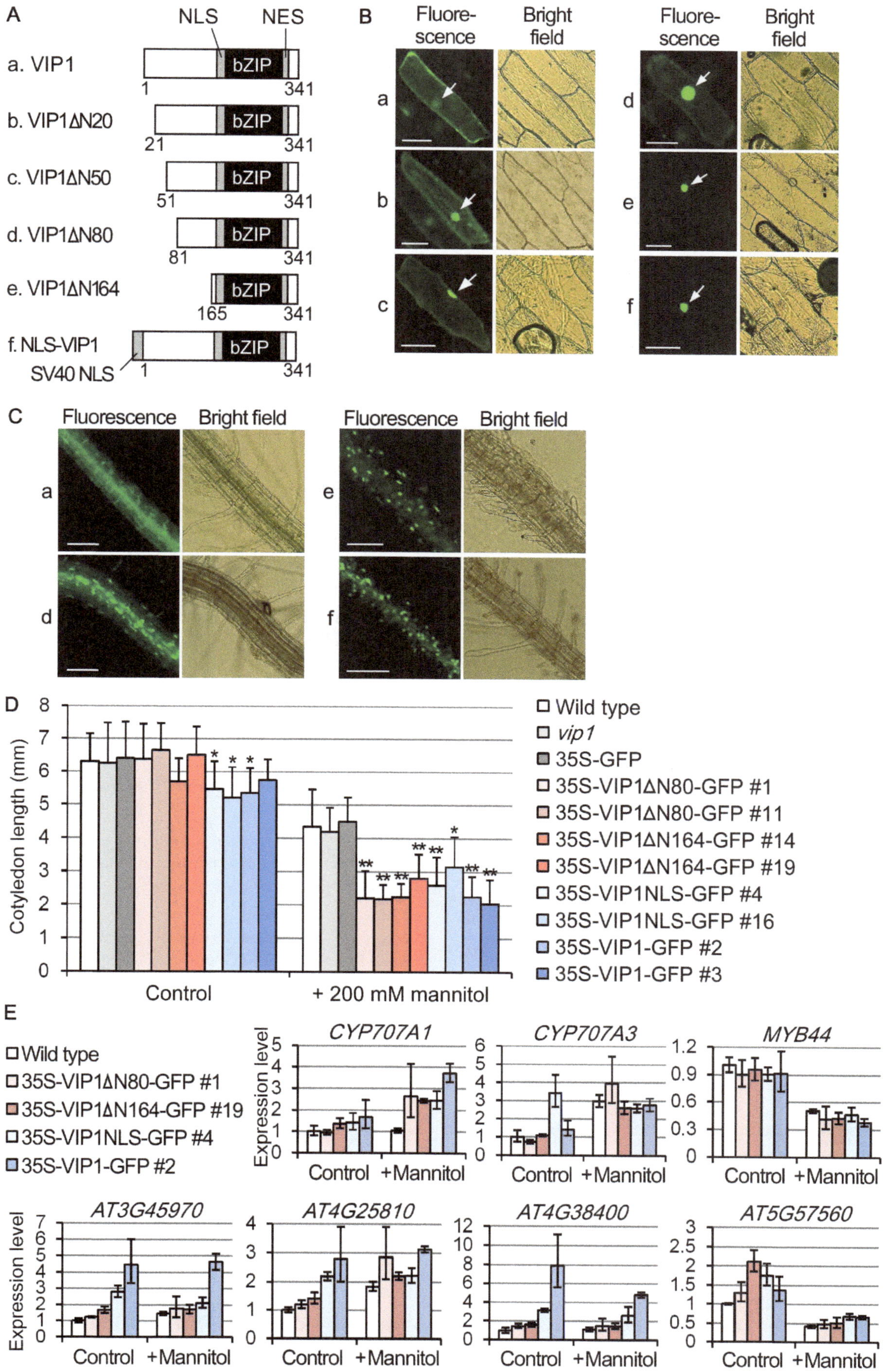
A
NLS
NES
a. VIP1
b. VIP1ΔN20
c. VIP1ΔN50
d. VIP1ΔN80
e. VIP1ΔN164
f. NLS-VIP1
SV40 NLS
bZIP
B
Fluore-scence
Bright field
C
Fluorescence
Bright field
D
Cotyledon length (mm)
Control
+ 200 mM mannitol
Wild type
vip1
35S-GFP
35S-VIP1ΔN80-GFP #1
35S-VIP1ΔN80-GFP #11
35S-VIP1ΔN164-GFP #14
35S-VIP1ΔN164-GFP #19
35S-VIP1NLS-GFP #4
35S-VIP1NLS-GFP #16
35S-VIP1-GFP #2
35S-VIP1-GFP #3
E
Wild type
35S-VIP1ΔN80-GFP #1
35S-VIP1ΔN164-GFP #19
35S-VIP1NLS-GFP #4
35S-VIP1-GFP #2
Expression level
CYP707A1
CYP707A3
MYB44
AT3G45970
AT4G25810
AT4G38400
AT5G57560
Control
+Mannitol

Figure 3. Generation and characterization of constitutively nuclear-localized variants of VIP1. (A) Schematic representations of the VIP1 variants. Numbers indicate amino acid positions of full-length VIP1. Lowercase letters (a–f) correspond to those in the panels B–D. NLS: nuclear localization signal; NES: nuclear export signal. (B) Subcellular localization of the VIP1 variants in onion cells. The indicated forms of VIP1 (a–f) were expressed as GFP-fused proteins in onion epidermal cells. For each construct, more than 10 cells were observed, and a representative image is shown. Scale bars = 100 μm. (C) Subcellular localization of the VIP1 variants in Arabidopsis. Roots of the transgenic Arabidopsis plants expressing GFP-fused VIP1 variants (forms a, d, e and f) were observed without being treated by a hypotonic solution. For each genotype, more than 10 plants were used for observation, and a representative image is shown. Scale bars = 100 μm. (D) Expression of the VIP1 variants retards the cotyledon growth under a mannitol-stressed condition. The transgenic plants expressing the GFP-fused VIP1 variants (35S-VIP1ΔN80-GFP, for example) were grown in the presence of 0 (Control) or 200 mM mannitol for 10 days, and their cotyledon lengths were measured. *vip1*, which lacks the 3′ region of the *VIP1* transcript because of a T-DNA insertion in the genomic region of *VIP1* [13], is shown as control. Values are means ± SD. (n = 15–20). *: $P<0.05$; **: $P<0.01$ vs. the wild type in Student's *t*-test. (E) Quantitative RT-PCR analyses of expression of the genes with promoters containing AGCTGT/G in the transgenic plants expressing the VIP1 variants. The transgenic plants were grown in the presence of 0 (Control) or 200 mM mannitol (+ Mannitol) for 18 days, and sampled for RNA extraction and cDNA synthesis. Relative expression levels were calculated by the comparative cycle threshold (C_T) method using *UBQ5* as an internal control gene. Data are means of three biological replicates. Error bars indicate SD.

AtbZIP10 and AtbZIP25, form heterodimers with a group S bZIP protein, AtbZIP53, and the heterodimers interact with a transcriptional coactivator, ABI3, to enhance the target gene activation [46]. As in these examples, even if localized in the nucleus, VIP1 may require post-translational modifications and/or protein-protein interactions to fully control the expression of its target genes.

In conclusion, our current data support the idea that VIP1 and the other group I bZIP proteins, especially those having higher expression levels, are similar in the subcellular localization, the transcriptional activation function and the DNA-binding specificity, and that the VIP1-binding sequence AGCTGT/G has a role in regulating the submergence-responsive gene expression. Further studies are required to elucidate the physiological roles as well as the above-described cytological and biochemical functions of the group I bZIP proteins.

Supporting Information

Figure S1 Alignment of the amino acid sequences of Arabidopsis group I bZIP proteins.

Figure S2 An RT-PCR analysis of transcripts of *GFP*-fused *PosF21* and *GFP*-fused *AtbZIP29* in Arabidopsis.

Figure S3 Alignment of the sequences of the *CYP707A1/3* promoters.

Figure S4 Gel shift assays for narrowing the VIP1-binding region in the *CYP707A1* promoter.

Figure S5 The *CYP707A1/3* promoter fragments most effectively inhibit the VIP1-dependent band shift.

Figure S6 Subcellular localization of VIP1 variants that have point mutations in the putative phosphorylation sites.

Figure S7 Transcriptional activation function of the truncated versions of VIP1 in yeast.

Figure S8 An RT-PCR analysis of expression of transcripts for GFP-fused VIP1 variants in Arabidopsis.

Figure S9 Expression of VIP1 variants do not affect ABA-induced inhibition of seed germination.

Table S1 cDNA clones obtained from RIKEN.

Table S2 Primers used for qRT-PCR analyses of expression of Arabidopsis group I bZIP protein genes.

Table S3 Primers used for semi-quantitative RT-PCR analyses of transgene expression.

Table S4 Primer pairs used to generate constructs for Y1H and Y2H assays.

Table S5 Primer pairs used to generate constructs for expression of nYFP-, cYFP- and GFP-fused proteins.

Table S6 Primers used to generate the point-mutated versions and the NLS-attached version of VIP1.

Dataset S1 The numbers and the proportions of the genes with promoters containing AGCTGT/G in the submergence-responsive genes.

Dataset S2 Complete list of the genes with promoters containing AGCTGT/G within 500 bp upstream of the start codons and summary of RT-PCR analyses using some of them.

Dataset S3 Data and summary of the microarray analysis. distributions using 0.185 and 0.033 (values in the first row) as the probabilities.

Acknowledgments

We are grateful to Dr. Yoshihiro Narusaka and Dr. Mari Narusaka (Research Institute for Biological Sciences, Japan) for providing the microarray and technical advices. We thank ABRC for providing the Arabidopsis mutant seeds, and Ms. Momoka Kobayashi (The University of Tokyo, Japan) for assistance in experiments.

Author Contributions

Conceived and designed the experiments: DT SL TT. Performed the experiments: DT SL TT. Analyzed the data: DT SL TT. Contributed reagents/materials/analysis tools: DT TT. Contributed to the writing of the manuscript: DT SL TT.

References

1. Iuchi S, Kobayashi M, Taji T, Naramoto M, Seki M, et al. (2001) Regulation of drought tolerance by gene manipulation of 9-cis-epoxycarotenoid dioxygenase, a key enzyme in abscisic acid biosynthesis in Arabidopsis. Plant J 27: 325–333.
2. Tan BC, Joseph LM, Deng WT, Liu L, Li QB, et al. (2003) Molecular characterization of the Arabidopsis 9-cis epoxycarotenoid dioxygenase gene family. Plant J 35: 44–56.
3. Kushiro T, Okamoto M, Nakabayashi K, Yamagishi K, Kitamura S, et al. (2004) The Arabidopsis cytochrome P450 CYP707A encodes ABA 8′-hydroxylases: key enzymes in ABA catabolism. EMBO J 23: 1647–1656.
4. Umezawa T, Okamoto M, Kushiro T, Nambara E, Oono Y, et al. (2006) CYP707A3, a major ABA 8′-hydroxylase involved in dehydration and rehydration response in *Arabidopsis thaliana*. Plant J 46: 171–182.
5. Okamoto M, Tanaka Y, Abrams SR, Kamiya Y, Seki M, et al. (2009) High humidity induces abscisic acid 8′-hydroxylase in stomata and vasculature to regulate local and systemic abscisic acid responses in Arabidopsis. Plant Physiol 149: 825–834.
6. O'Rourke SM, Herskowitz I, O'Shea EK (2002) Yeast go the whole HOG for the hyperosmotic response. Trends in Genet 18: 405–412.
7. Urao T, Yakubov B, Satoh R, Yamaguchi-Shinozaki K, Seki M, et al. (1999) A transmembrane hybrid-type histidine kinase in Arabidopsis functions as an osmosensor. Plant Cell 11: 1743–1754.
8. Tran LSP, Urao T, Qin F, Maruyama K, Kakimoto T, et al. (2007) Functional analysis of AHK1/ATHK1 and cytokinin receptor histidine kinases in response to abscisic acid, drought, and salt stress in Arabidopsis. Proc Natl Acad Sci USA 104: 20623–20628.
9. Wohlbach DJ, Quirino BF, Sussman MR (2008) Analysis of the Arabidopsis histidine kinase ATHK1 reveals a connection between vegetative osmotic stress sensing and seed maturation. Plant Cell 20: 1101–1117.
10. Kumar MN, Jane WN, Verslues PE (2013) Role of the putative osmosensor Arabidopsis histidine kinase1 in dehydration avoidance and low-water-potential response. Plant Physiol 161: 942–953.
11. Tzfira T, Vaidya M, Citovsky V (2001) VIP1, an Arabidopsis protein that interacts with Agrobacterium VirE2, is involved in VirE2 nuclear import and Agrobacterium infectivity. EMBO J 20: 3596–3607.
12. Tzfira T, Vaidya M, Citovsky V (2002) Increasing plant susceptibility to Agrobacterium infection by overexpression of the Arabidopsis nuclear protein VIP1. Proc Natl Acad Sci USA 99: 10435–10440.
13. Li J, Krichevsky A, Vaidya M, Tzfira T, Citovsky V (2005) Uncoupling of the functions of the Arabidopsis VIP1 protein in transient and stable plant genetic transformation by Agrobacterium. Proc Natl Acad Sci USA 102: 5733–5738.
14. Djamei A, Pitzschke A, Nakagami H, Rajh I, Hirt H (2007) Trojan horse strategy in Agrobacterium transformation: abusing MAPK defense signaling. Science 318: 453–456.
15. Pitzschke A, Djamei A, Teige M, Hirt H (2009) VIP1 response elements mediate mitogen-activated protein kinase 3-induced stress gene expression. Proc Natl Acad Sci USA 106: 18414–18419.
16. Wu Y, Zhao Q, Gao L, Yu XM, Fang P, et al. (2010) Isolation and characterization of low-sulphur-tolerant mutants of Arabidopsis. J Exp Bot 61: 3407–3422.
17. Tsugama D, Liu S, Takano T (2012) A bZIP protein, VIP1, is a regulator of osmosensory signaling in Arabidopsis. Plant Physiol 159: 144–155.
18. Jakoby M, Weisshaar B, Dröge-Laser W, Vicente-Carbajosa J, Tiedemann J, et al. (2002) bZIP transcription factors in Arabidopsis. Trends in Plant Sci 7: 106–111.
19. Clough SJ, Bent AF (1998) Floral dip: a simplified method for Agrobacterium-mediated transformation of *Arabidopsis thaliana*. Plant J 16: 735–743.
20. Chomczynski P, Sacchi N (1987) Single-step method of RNA isolation by acid guanidinium extraction. Anal Biochem 162: 156–159.
21. Seki M, Narusaka M, Kamiya A, Ishida J, Satou M, et al. (2002) Functional Annotation of a Full-Length Arabidopsis cDNA Collection. Science 296: 141–145.
22. Yamaji N, Huang CF, Nagao S, Yano M, Sato Y, et al. (2009) A zinc finger transcription factor ART1 regulates multiple genes implicated in aluminum tolerance in rice. Plant Cell 21: 3339–3349.
23. Tsugama D, Liu S, Takano T (2012) A putative myristoylated 2C-type protein phosphatase, PP2C74, interacts with SnRK1 in Arabidopsis. FEBS Lett 586: 693–698.
24. Tsugama D, Liu S, Takano T (2012) Drought-induced activation and rehydration-induced inactivation of MPK6 in Arabidopsis. Biochem Biophys Res Commun 426: 626–629.
25. Zhang CQ, Nishiuchi S, Liu S, Takano T (2008) Characterization of two plasma membrane protein 3 genes (*PutPMP3*) from the alkali grass, *Puccinellia tenuiflora*, and functional comparison of the rice homologues, *OsLti6a/b* from rice. BMB Rep 41: 448–454.
26. Seki M, Narusaka M, Ishida J, Nanjo T, Fujita M, et al. (2002) Monitoring the expression profiles of 7000 Arabidopsis genes under drought, cold and high-salinity stresses using a full-length cDNA microarray. Plant J 31: 279–292.
27. Oono Y, Seki M, Nanjo T, Narusaka M, Fujita M, et al. (2003) Monitoring expression profiles of Arabidopsis gene expression during rehydration process after dehydration using *ca.* 7000 full-length cDNA microarray. Plant J 34: 868–887.
28. Lee SC, Mustroph A, Sasidharan R, Vashisht D, Pedersen O, et al. (2011) Molecular characterization of the submergence response of the *Arabidopsis thaliana* ecotype Columbia. New Phytol 190: 457–471.
29. Hsu FC, Chou MY, Peng HP, Chou SJ, Shih MC (2011) Insights into hypoxic systemic responses based on analyses of transcriptional regulation in Arabidopsis. PLoS One 6: e28888.
30. Narusaka M, Abe H, Kobayashi M, Kubo Y, Kawai K, et al. (2006) A model system to screen for candidate plant activators using an immune-induction system in Arabidopsis. Plant Biotech 23: 321–327.
31. Deppmann CD, Acharya A, Rishi V, Wobbes B, Smeekens S, et al. (2004) Dimerization specificity of all 67 B-ZIP motifs in *Arabidopsis thaliana*: a comparison to *Homo sapiens* B-ZIP motifs. Nucleic Acids Res 32: 3435–3445.
32. Lacroix B, Citovsky V (2013) Characterization of VIP1 activity as a transcriptional regulator in vitro and *in planta*. Sci Rep 3: 2440.
33. Fukazawa J, Sakai T, Ishida S, Yamaguchi I, Kamiya Y, et al. (2000) Repression of shoot growth, a bZIP transcriptional activator, regulates cell elongation by controlling the level of gibberellins. Plant Cell 12: 901–915.
34. Fukazawa J, Nakata M, Ito T, Yamaguchi S, Takahashi Y (2010) The transcription factor RSG regulates negative feedback of *NtGA20ox1* encoding GA 20-oxidase. Plant J 62: 1035–1045.
35. Torres-Schumann S, Ringli C, Heierli D, Amrhein N, Keller B (1996) In vitro binding of the tomato bZIP transcriptional activator VSF-1 to a regulatory element that controls xylem-specific gene expression. Plant J 9: 283–296.
36. Yin Y, Zhu Q, Dai S, Lamb C, Beachy RN (1997) RF2a, a bZIP transcriptional activator of the phloem-specific rice tungro bacilliform virus promoter, functions in vascular development. EMBO J 16: 5247–5259.
37. Ringli C, Keller B (1998) Specific interaction of the tomato bZIP transcription factor VSF-1 with a non-palindromic DNA sequence that controls vascular gene expression. Plant Mol Biol 37: 977–988.
38. Dai S, Petruccelli S, Ordiz MI, Zhang Z, Chen S, et al. (2003) Functional analysis of RF2a, a rice transcription factor. J Biol Chem 278: 36396-36402.
39. Dai S, Zhang Z, Chen S, Beachy RN (2004) RF2b, a rice bZIP transcription activator, interacts with RF2a and is involved in symptom development of rice tungro disease. Proc Natl Acad Sci USA 101: 687–692.
40. Igarashi D, Ishida S, Fukazawa J, Takahashi Y (2001) 14-3-3 proteins regulate intracellular localization of the bZIP transcriptional activator RSG. Plant Cell 13: 2483–2497.
41. Ishida S, Fukazawa J, Yuasa T, Takahashi Y (2004) Involvement of 14-3-3 signaling protein binding in the functional regulation of the transcriptional activator REPRESSION OF SHOOT GROWTH by gibberellins. Plant Cell 16: 2641–2651.
42. Blom N, Gammeltoft S, Brunak S (1999) Sequence- and structure-based prediction of eukaryotic protein phosphorylation sites. J Mol Biol 294: 1351–1362.
43. Xue Y, Ren J, Gao X, Jin C, Wen L, et al. (2008) GPS 2.0, a tool to predict kinase-specific phosphorylation sites in hierarchy. Mol Cell Proteomics 7: 1598–1608.
44. Moes D, Himmelbach A, Korte A, Haberer G, Grill E (2008) Nuclear localization of the mutant protein phosphatase abi1 is required for insensitivity towards ABA responses in Arabidopsis. Plant J 54: 806–819.
45. Furihata T, Maruyama K, Fujita Y, Umezawa T, Yoshida R, et al. (2006) Abscisic acid-dependent multisite phosphorylation regulates the activity of a transcription activator AREB1. Proc Natl Acad Sci USA 103: 1988–1993.
46. Alonso R, Oñate-Sánchez L, Weltmeier F, Ehlert A, Diaz I, et al. (2009) A pivotal role of the basic leucine zipper transcription factor bZIP53 in the regulation of Arabidopsis seed maturation gene expression based on hetero-dimerization and protein complex formation. Plant Cell 21: 1747–1761.

12

Site-Directed Mutagenesis of IRX9, IRX9L and IRX14 Proteins Involved in Xylan Biosynthesis: Glycosyltransferase Activity Is Not Required for IRX9 Function in Arabidopsis

Yanfang Ren[1,2], Sara Fasmer Hansen[1], Berit Ebert[1,3], Jane Lau[1], Henrik Vibe Scheller[1,4]*

1 Joint Bioenergy Institute, Lawrence Berkeley National Laboratory, Berkeley, California, United States of America, **2** College of Agriculture, Guizhou University, Guiyang, China, **3** Department of Plant and Environmental Sciences, Faculty of Science, University of Copenhagen, Frederiksberg, Denmark, **4** Department of Plant and Microbial Biology, University of California, Berkeley, California, United States of America

Abstract

Xylans constitute the main non-cellulosic polysaccharide in the secondary cell walls of plants. Several genes predicted to encode glycosyltransferases are required for the synthesis of the xylan backbone even though it is a homopolymer consisting entirely of β-1,4-linked xylose residues. The putative glycosyltransferases IRX9, IRX14, and IRX10 (or the paralogs IRX9L, IRX14L, and IRX10L) are required for xylan backbone synthesis in Arabidopsis. To investigate the function of IRX9, IRX9L, and IRX14, we identified amino acid residues known to be essential for catalytic function in homologous mammalian proteins and generated modified cDNA clones encoding proteins where these residues would be mutated. The mutated gene constructs were used to transform wild-type Arabidopsis plants and the *irx9* and *irx14* mutants, which are deficient in xylan synthesis. The ability of the mutated proteins to complement the mutants was investigated by measuring growth, determining cell wall composition, and microscopic analysis of stem cross-sections of the transgenic plants. The six different mutated versions of IRX9 and IRX9-L were all able to complement the *irx9* mutant phenotype, indicating that residues known to be essential for glycosyltransferases function in homologous proteins are not essential for the biological function of IRX9/IRX9L. Two out of three mutated IRX14 complemented the *irx14* mutant, including a mutant in the predicted catalytic amino acid. A IRX14 protein mutated in the substrate-binding DxD motif did not complement the *irx14* mutant. Thus, substrate binding is important for IRX14 function but catalytic activity may not be essential for the function of the protein. The data indicate that IRX9/IRX9L have an essential structural function, most likely by interacting with the IRX10/IRX10L proteins, but do not have an essential catalytic function. Most likely IRX14 also has primarily a structural role, but it cannot be excluded that the protein has an important enzymatic activity.

Editor: Samuel P. Hazen, University of Massachusetts Amherst, United States of America

Funding: This work was funded by the U.S. Department of Energy, Office of Science (science.energy.gov), Office of Biological and Environmental Research, through contract DE-AC02-05CH11231 with Lawrence Berkeley National Laboratory. YR was supported by China Scholarship Council (www.csc.edu.cn) grant 2010852564. SFH was supported by a fellowship from The Carlsberg Foundation (www.carlsbergfondet.dk) contract 2010_01_0509. The funders had no role in study design, data collection and analysis, decision to publish, or preparation of the manuscript.

Competing Interests: The authors have declared that no competing interests exist.

* Email: hscheller@lbl.gov

Introduction

Xylan is the most abundant non-cellulosic polysaccharide in plants, comprising around 30% of the biomass. The structure of xylan is a β-1,4-linked backbone of D-xylose residues, decorated with various substituents. Glucuronic acid residues and arabinose substitutions are the most common monosaccharide substituents. The glucuronic acid residues are often 4-*O*-methylated and acetylation of the xylose residues at C-2 or C-3 positions is also a common modification [1]. Because of the importance of xylan, many studies have addressed genes and proteins involved in the biosynthesis, but it has turned out to be unexpectedly challenging and progress has been very slow. Identification of enzymes required for adding the substitutions has been more successful, and several glucuronosyltransferases, arabinosyltransferases, xylosyltransferases and actetyltransferases have now been reported [1–7].

The backbone synthesis has been more difficult to elucidate. Mutations in the Arabidopsis genes *Irregular Xylem* (*IRX*)*7* (and the homolog *IRX7L*), *IRX8* and *PARVUS* lead to decreased xylan content, but those plants still make xylan with extended backbones and it is suggested that this set of genes might be involved in synthesizing a primer or the reducing end oligosaccharide structure found in many plants [8]. In contrast, mutations in the genes *IRX9* and *IRX14*, which are members of Glycosyltransferase family (GT)43, or *IRX10* belonging to GT47 result in decreased xylan synthase activity, reduced xylan content, and xylan with short backbones [9–12].

In Arabidopsis all three genes have close homologs (*IRX9L, IRX14L, IRX10L*) that are partially functionally redundant with IRX9, IRX14 and IRX10, respectively [9,12]. The homologs IRX9/IRX9L, IRX14/IRX14L and IRX10/IRX10L appear to differ mostly in their organ specific expression patterns. Still, the proteins may not be completely redundant as e.g. overexpression lines of *IRX9L* in the *irx9* mutant background complemented the morphological phenotype but showed less xylosyltransferase activity than the wild type and varied in stem breaking strength [11–13]. However, it is difficult to conclude if this is due to too low expression in important target cells or to a difference in biochemical function of the two proteins. The double mutants *irx9/irx9L* and *irx14/irx14L* show an almost lethal phenotype with severe growth inhibition [12]. IRX9 and IRX14 are not redundant, as *irx9* mutants cannot be complemented by overexpression of IRX14 or IRX14L and vice versa [11,12]. That three glycosyltransferases are required to catalyze the formation of a homopolymer consisting entirely of β-1,4-linked xylose residues is surprising. Since no in vitro activity has been reported for the individual proteins it is unclear if one protein possesses the main activity with the other proteins having an accessory or structural role, or if all proteins are active at the same time. Overexpression of IRX9 leads to increased xylan synthase activity in Arabidopsis plants [11,13,14] and other reports have shown a synergistic effect of IRX9 and IRX14 overexpression on xylan synthase activity [15]. However, these studies do not prove that IRX9 and/or IRX14 are the actual glycosyltransferases responsible for the activity as their overexpression could also stimulate other proteins. For example one may hypothesize that IRX9, IRX14 and IRX10 function together in a complex and overexpression of IRX9 alone or together with IRX14 could stabilize the complex and lead to more activity of IRX10 and IRX10L. This outcome could be expected if IRX10/IRX10L were the catalytic subunits but the amount of IRX9 and IRX14 were limiting for formation of the complex. The idea that the proteins are somehow interacting in a complex has been proposed many times, but unambiguous evidence has been hard to obtain [1,16].

An essential anchoring rather than catalytic function of IRX9/IRX9L and IRX14/IRX14L is in agreement with a transcriptomic study of psyllium seeds (*Plantago ovata*), which are very rich in xylan [17]. Several *IRX10* homologs are expressed in the seeds, while *IRX9* and *IRX14* homologs have very low expression [18]. Interestingly, the most highly expressed *IRX10* homolog in psyllium (GenBank accession AGQ53953) encodes a protein with a strongly predicted N-terminal membrane anchor as in typical Type 2 membrane proteins. In contrast, the IRX10 and IRX10L proteins in Arabidopsis are not predicted to have any transmembrane helices, suggesting that they need to interact with other Golgi-resident proteins to be retained in the Golgi apparatus. Thus, these studies have led to the suggestion that IRX10 orthologs are the main or only catalytic proteins for xylan backbone biosynthesis, while IRX9 and IRX14 orthologs have accessory functions. Obviously, even if this is the case in psyllium seed mucilage formation, it need to be the case for other tissues in psyllium where *IRX9* and *IRX14* homologs are expressed or in other plant species.

We noted two peculiar properties of IRX9. Firstly, the *irx9-2* T-DNA mutant has a much milder phenotype than the *irx9-1* mutant. The *irx9-1* insertion is in the first intron and the mutant is likely a true knock-out. The *irx9-2* insertion is in the second exon in the first half of the protein, and if a protein is made in the mutant it would only contain the transmembrane domain and the linker region, whereas almost the entire domain predicted to contain catalytic activity would be missing. Secondly, IRX9 has an unusual amino acid sequence where the DxD motif that is found in most glycosyltransferases with a GT-A fold is not conserved. Instead in IRX9 the corresponding amino acid sequence is 'GLN' (Fig. S1). In contrast, in IRX9L and in IRX14/IRX14L the DxD motif is conserved (Fig. S1). These observations led us to suspect that IRX9 may not have an essential catalytic function and instead have a primary role in organizing and assembling a xylan synthase complex.

To probe the function of IRX9, IRX9L and IRX14 we mutagenized the proteins in specific amino acid residues known to be required for glycosyltransferase activity in mammalian GT43 homologs and determined if the mutated proteins could complement *irx9* and *irx14* mutants in Arabidopsis. Our results show that none of the mutations in regions that would be required for catalytic activity had any effect on the ability of IRX9 and IRX9L to complement the *irx9* mutant phenotype. In contrast, some but not all mutations in IRX14 did affect the complementation of *irx14* mutants. Thus, we conclude that IRX9 and IRX9L are not catalytically active and their main function is primarily or exclusively structural. For IRX14 our results demonstrate the importance of the DxD motif, which has been found to mainly interact with the phosphate groups of nucleotide sugar donors, but other residues presumed to be involved in catalysis are poorly conserved and seem not to be required for IRX14 function. Thus, the nucleotide sugar binding function of IRX14 is important but the protein is unlikely to have an essential catalytic role.

Materials and Methods

Plant Lines and Plant Growth Conditions

Arabidopsis thaliana accession Columbia-0 (Col-0) and T-DNA insertion lines (*irx9-2*, SALK_057033 and *irx14*, SALK_038212) were obtained from the Arabidopsis Biological Resource Center (ABRC, Ohio State University). Plants were grown on soil in a growth chamber under long-day conditions (16-h light/8-h dark cycle) at 22°C and 60% relative humidity after being stratified at 4°C for 4 days. Arabidopsis plants were transformed using *Agrobacterium tumefaciens* PGV 3805 via the floral dip method as described by Clough and Bent [19]. For BASTA selection seeds were germinated on soil as described above and 1-week-old seedlings were sprayed every 2 days for a total of five times with a glufosinate-ammonium (Crescent Chemical) solution (40 mg/mL). Resistant plants were transferred to new pots and grown as described above.

Bioinformatics and Sequence Identification

All GT43 proteins from *Arabidopsis thaliana* (At), rice (*Oryza sativa*, Os), poplar (*Populus trichocarpa*, Pt), spikemoss (*Selaginella moellendorffii*, Sm) and human (*Homo sapiens*, Sp) were selected for multiple sequence alignment. The protein sequences have the following Genbank accession numbers (locus numbers for Arabidopsis): AtIRX9 (AT2G37090), AtIRX9L (AT1G27600), AtIRX14 (AT4G36890), AtIRX14L (AT5G67230), Os01g06450 (AHW98784), OsIRX9L/Os01g48440 (NP_001043846), Os03g17850 (ABF95369), Os04g01280 (NP_001051997), Os04g55670 (AHW98789), Os05g03174 (AHW98781), Os05g48600 (NP_001056301), OsIRX14/Os06g47340 (AHW98790), OsIRX9/Os07g49370 (AHW98783), Os10g13810 (AAP52726), Pt002G107300 (XP_002301102), Pt005G141500 (XP_002306485), Pt006G131000 (XP_006381488), Pt006G240200 (XP_002309550), Pt007G047500 (XP_002310709), Pt016G086400 (XP_002323456), SmIRX9 (XP_002963107), SmIRX14 (XP_002976794), Hs_GlcAT-I (AAH71961), Hs_GlcAT-P (Q9P2W7), Hs_GlcAT-S (AAL57718).

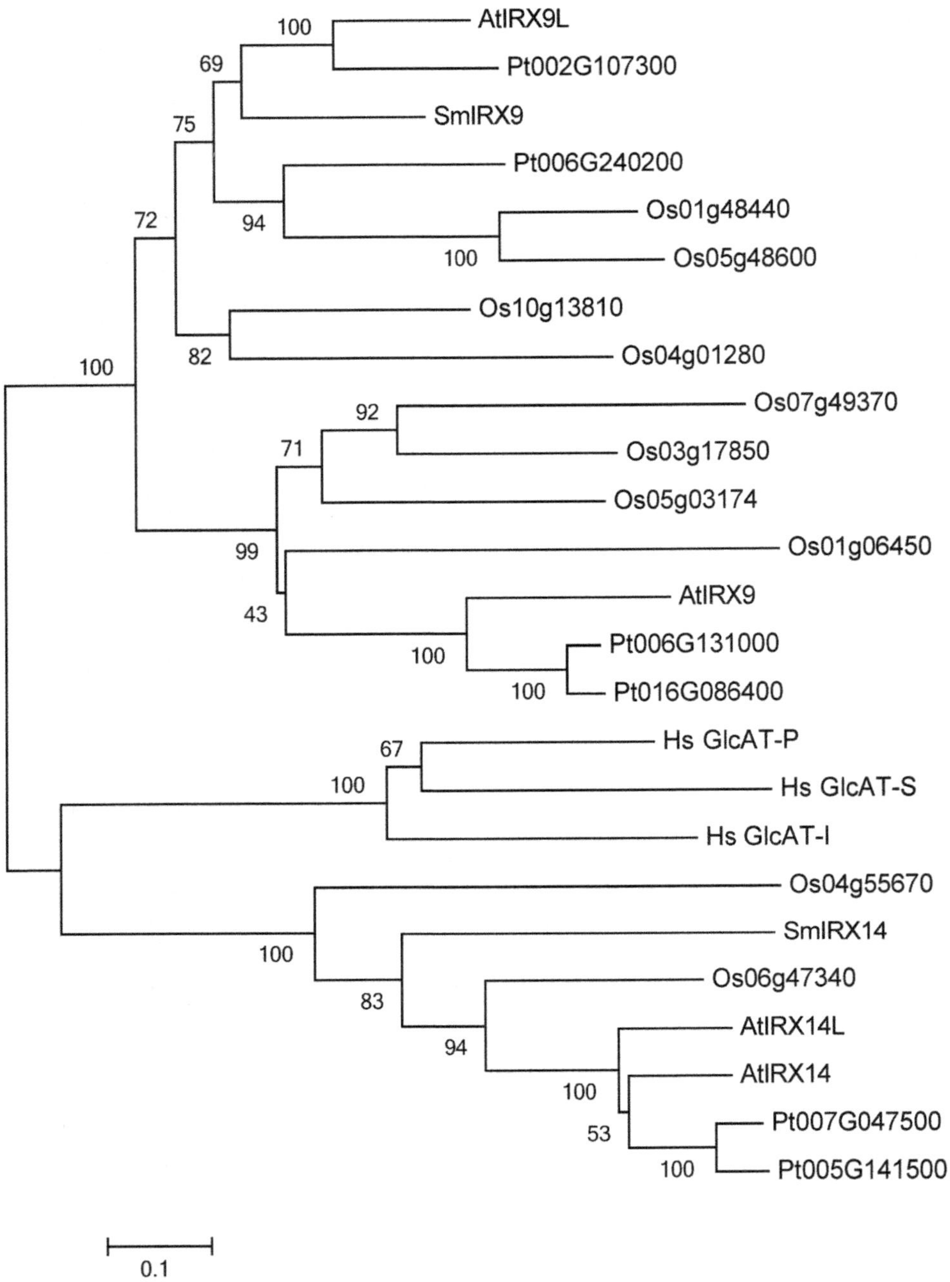

Figure 1. Phylogenetic tree of GT43 proteins from selected species. A multiple alignment of protein sequences from Arabidopsis (At), rice (*Oryza sativa*, Os), poplar (*Populus trichocarpa*, Pt), spikemoss (*Selaginella moellendorffii*, Sm) and human (*Homo sapiens*, Sp) was used to generate a neighbor-joining tree using MEGA5. Bootstrap values are indicated and the scale bar shows evolutionary distances in units of the number of amino acid substitutions per site. See Materials and Methods section for sequence IDs.

Protein sequence analysis was conducted with the software MEGA5 [20]. Multiple sequence alignments were created using the Muscle option with default settings and 500 bootstraps. Phylogenetic trees were constructed using the Neighbor-Joining method [21]. The evolutionary distances in units of the number of amino acid substitutions per site were computed using Poisson correction.

Site-Directed Mutagenesis

Wild-type sequences of *IRX9*, *IRX9L* and *IRX14* were amplified by PCR from Arabidopsis cDNA and cloned into pDONR223 (*IRX9* and *IRX14*) and pDONRZeo (*IRX9L*) as described previously [2]. Mutated versions of *IRX9, IRX9L* and *IRX14* were generated by site-directed mutagenesis using the QuikChange Kit from Stratagene according to the manufacturer's instructions and the pDONR clones as template. Oligonucleotide primers used to introduce the various mutations are listed in Table S1. The introduction of mutations was confirmed by sequencing, performed both before and after introduction of the mutated *IRX9, IRX9L* and *IRX14* coding sequences into pEarleyGate202 destination vector [22] using LR Clonase (Life Technologies). The pEarlyGate202 vector drives expression with the cauliflower mosaic virus 35S promoter. These constructs were introduced into homozygous *irx9-2 and irx14* mutant plants and wild type

Table 1. Nomenclature used for the mutated versions of IRX9, IRX9L and IRX14.

Designation	Mutation
IRX9-0	none
IRX9-1	W151A
IRX9-2	G215W
IRX9-3	G215D/L216D
IRX9-4	E252A
IRX9-5	C323A
IRX9L-0	none
IRX9L-1	D238A-D240A
IRX14-0	none
IRX14-1	D261A/D262A
IRX14-2	Q321A
IRX14-3	C415A

Col-0 plants via *Agrobacterium*-mediated transformation using the floral dip method.

Screening of Transformants for T-DNA Insertions and Transgenes

Transgenic plants were screened for Basta resistance and positive transformants were transferred to new pots and grown as described above. Positive transfomants were genotyped by PCR using gene-specific primers (Table S1) to verify the homozygous mutant background. To confirm the presence of the transgene in positive transformants, primer pairs annealing in the destination vector and the respective *IRX* gene were used (Table S1).

RT-PCR was used to confirm that the transgene was expressed in the selected plant lines. Total RNA was extracted from wild type and transgenic plants using the RNeasy Plant Mini Kit (QIAGEN). First-strand cDNA synthesis was performed using the SuperScript II RT Kit (Invitrogen) according to the manufacturer's protocol. RT-PCR was done with the OneStep RT-PCR Kit (QIAGEN) as described by the manufacturer using the primers listed in Table S1. *ACTIN* (*ACT2*) was used as an internal control.

Analysis of Cell Wall Monosaccharide Composition

Alcohol-insoluble residue (AIR) of inflorescence stems was prepared and destarched enzymatically as described previously [23]. AIR samples were subsequently hydrolyzed with 2 M triflouroacetic acid (TFA) for 1 h at 120°C. Samples were dried, resuspended in water, and centrifuged for 10 min at 13,000 g. The resulting supernatant was diluted and the monosaccharide composition was determined by high-performance anion exchange chromatography (HPAEC) on a ICS 3000 instrument (Dionex) using a CarboPac™ PA20 (3×150 mm, Dionex) anion exchange column as described by Obro et al. [24].

Xylan Immunolocalization and Microscopy

The basal regions of inflorescence stems (2.5 cm distal from the rosette) were collected from 6-week-old plants, and fixed in 4% paraformaldehyde in PEM buffer (50 mM PIPES, 5 mM EGTA, 5 mM $MgSO_4$, pH 6.9) overnight at 4°C. The stems were embedded in 7% agarose and cut into 60 µm thick sections using a Leica VT1000S vibratome. Sections were incubated with the LM10 antibody (1:10 dilution) [25], washed and subsequently incubated with rabbit anti-rat fluorescein isothiocyanate (FITC)-conjugated secondary antibodies (Sigma-Aldrich, 1:100 dilution). Finally, sections were analyzed using a fluorescence microscope (Leica DM4000B) and the Metamorph software.

Statistical Analysis

Quantitative growth and cell wall data were analyzed by ANOVA using SPSS v. 16.0 (IBM). Each value was expressed as means ± SD with 4–10 biological replicates. Averages were determined to be significantly different from each other ($P<0.05$) according to Tukey's multiple range test.

Results

Bioinformatic Analysis of GT43 Proteins

To identify targets for mutagenesis we first conducted a bioinformatics study of GT43 proteins. We especially focused on GlcAT-P, GlcAT-S and GlcAT-I, which are human β-1,3-glucuronosyltransferases. These proteins have been investigated in numerous site-directed mutagenesis studies and their three-dimensional structures have been resolved (see [26] for a recent review). Therefore, the key essential amino acid residues are known.

An alignment of GT43 members from plants and the three human GlcAT proteins is shown in Fig. S1. The phylogenetic tree shows that IRX14 and IRX14L in Arabidopsis result from a recent duplication, whereas IRX9 and IRX9L are products of a more ancient duplication in plants (Fig. 1) as also reported by Wu et al. [12]. The clade of the human GlcAT-P, GlcAT-S and GlcAT-I proteins is as closely related to IRX9 and IRX14 as the two plant clades are to each other (Fig. 1, Fig. S2).

Site-Directed Mutagenesis

Nine different constructs were created based on conserved sequence homology and known catalytic features among GT43 proteins [27–29] for site-directed mutagenesis of IRX9, IRX9L and IRX14. Our sequence alignment showed the residue Trp151 of IRX9, to be 100% conserved in all the proteins analyzed. Since tryptophan has unique features in terms of chemistry and size, the conserved position could indicate its function to be important. A construct replacing tryptophan with alanine (W151A) in the IRX9 sequence was made and designated 'IRX9-1' (Table 1). The DxD motif, shown to be involved in donor binding and direct

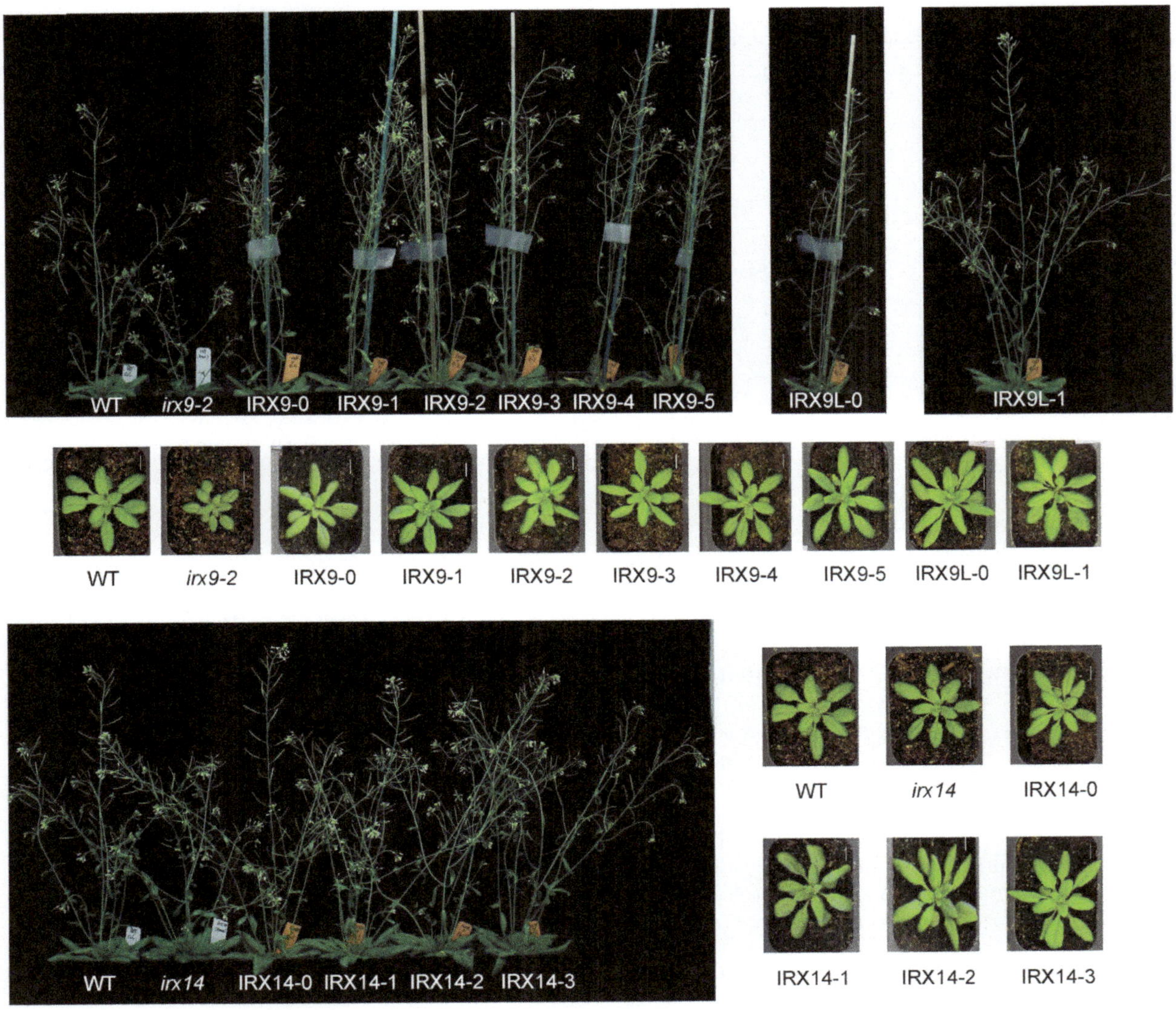

Figure 2. Representative photos of wild type, mutants and transformed plants. The nomenclature for the constructs used to transform the *irx9* and *irx14* plants is explained in Table 1. Rosettes were from 4-week-old plants and mature plants were 6-week-old.

interaction with the metal ion required for catalysis [28] and conserved in most glycosyltransferases with a GT-A fold [27] is also observed in the GT43 proteins. In the three mammalian GlcAT proteins and IRX9L of Arabidopsis as well as in IRX9L orthologs in rice and poplar, the motif 'DDD' can be found. A 'DDS' motif in IRX14 and IRX14L was found in all species aligned, except for one rice homolog comprising a 'DEN' instead. However the DxD motif is known to be able to vary to some degree, e.g., comprising a glutamic acid residue instead of the aspartic acid creating motifs as 'DxE' or 'ExD'. In *Selaginella*, which does not have an IRX9 ortholog, the motif is conserved as 'DDD' and 'DDS' as in the Arabidopsis IRX9L and IRX14 orthologs, respectively. Surprisingly, in the Arabidopsis and poplar IRX9 the alignment shows a conserved "GLx" motif instead, where x is N, S or A. In rice, which has four IRX9 orthologs, two share the "GLx" motif, whereas the other two proteins have 'AAS' and 'DAA' instead (Fig. S1). Site directed mutagenesis of the DxD motif was conducted for IRX9, IRX9L and IRX14. A construct replacing the small glycine with a bulky tryptophan residue (G215W) of the IRX9 sequence was designed and designated 'IRX9-2' (Table 1). Furthermore the 'GLN' motif of IRX9 was replaced with the 'DD(N)' motif similar to the corresponding sequence in IRX14 (G215D/L216D, designated 'IRX9-3'). In IRX9L the motif 'DDD' was changed to 'ADA' (D238A-D240A, designated 'IRX9L-1') and for IRX14 the 'DDS' was changed to 'AAS' (D261A/D262A, 'IRX14-1').

We next mutagenized a Glu residue that functions as the catalytic base [28,29]. Complete loss of catalytic activity of human GlcAT-I has been reported, when residue Glu227 was changed to Asp or Ala [29]. In the alignment all the species share a conserved Glu or Gln residue at this position (Fig. S1). Site directed mutations were made changing Glu252 to Ala in IRX9 and Gln321 to Ala in IRX14, generating 'IRX9-4' and 'IRX14-2', respectively (Table 1). Finally, the proteins have a conserved cysteine residue near the C-terminus that has proven essential for protein integrity and activity of mammalian GlcAT [30]. Changing Cys301 to Ala in GlcAT-I led to complete loss of activity. This cysteine residue is found conserved in all the plant proteins we looked at (Fig. S1), and mutations were made by changing the cysteines at this position in IRX9 and IRX14 to alanine, generating 'IRX9-5' and 'IRX14-3', respectively (Table 1).

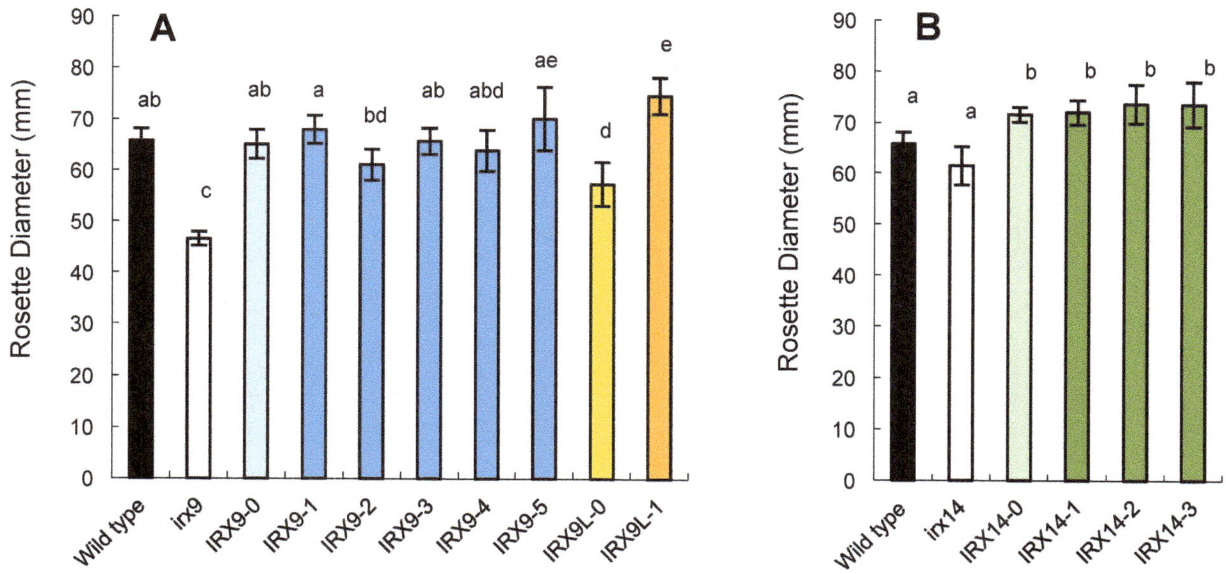

Figure 3. Rosette diameter of 4-week-old mutant plants transformed with the different constructs. The nomenclature for the constructs used to transform the *irx9* and *irx14* plants is explained in Table 1. The bars show average ± SD (n = 6). Averages that are not significantly different (ANOVA, Tukey's test, p>0.05) are indicated with the same letter.

The constructs with the mutated and wild-type versions of *IRX9* and *IRX9L* were used to transform wild-type and *irx9-2* plants, whereas the wild-type and mutated versions of *IRX14* were transformed into wild-type and *irx14* plants. Transgenic plants were identified and lines showing highest expression and maximum complementation of the mutant growth phenotype were selected for further analysis. The transformed wild-type plants served as controls to investigate if overexpression in the wild-type background would lead to increased xylan deposition or if mutated IRX proteins could have a dominant negative effect.

Analysis of Transformants

Several independent transformants (n = 2 − 10) were selected for each construct in the T1 generation. The most clearly complementing line was selected for each construct and further analyzed in T2. For the IRX14-1 construct, six independent T1 lines were obtained, none of which showed substantial complementation, despite expressing the transgene. Also for this construct, the line that showed most complementation was selected for analysis in T2. All further analysis reported below was done with the T2 plants.

Representative photos of the selected transgenic T2 plants are shown in Fig. 2. Generally, it was obvious that all the IRX9 and IRX9L constructs complemented the growth phenotype of *irx9-2*. The

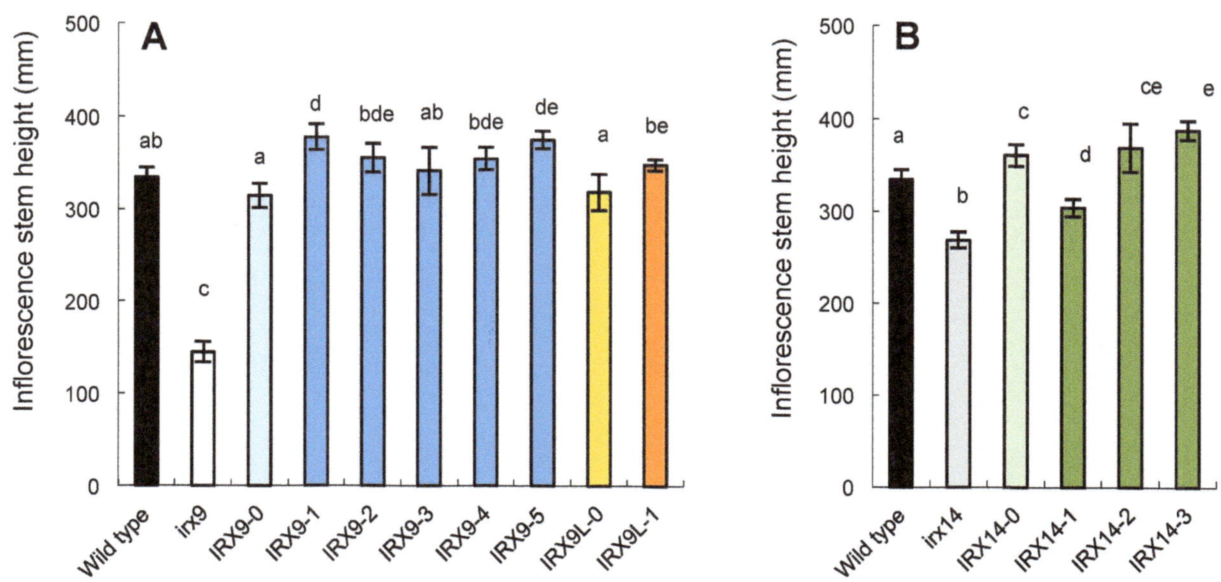

Figure 4. Stem height of 6-week-old mutant plants transformed with the different constructs. The nomenclature for the constructs used to transform the *irx9* and *irx14* plants is explained in Table 1. The bars show average ± SD (n = 6). Averages that are not significantly different (ANOVA, Tukey's test, p>0.05) are indicated with the same letter.

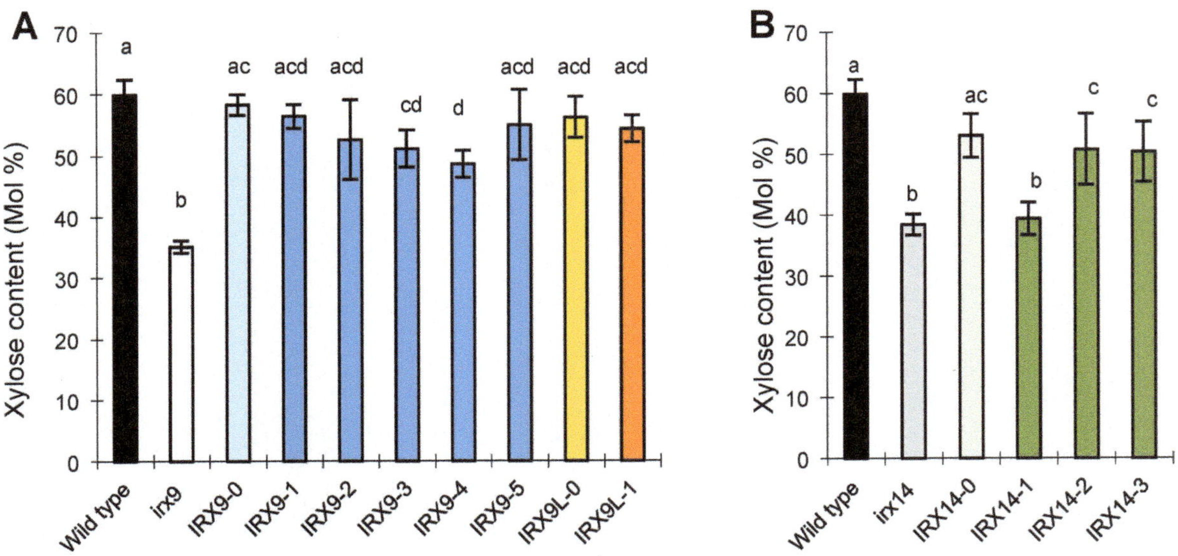

Figure 5. Xylose content in cell wall polysaccharides isolated from stems of 6-week-old mutant plants transformed with the different constructs. Cell walls were prepared from stems, hydrolyzed in TFA, and the xylose content determined by HPAEC. The nomenclature for the constructs used to transform the *irx9* and *irx14* plants is explained in Table 1. The bars show average ± SD (n = 4). Averages that are not significantly different (ANOVA, Tukey's test, p>0.05) are indicated with the same letter.

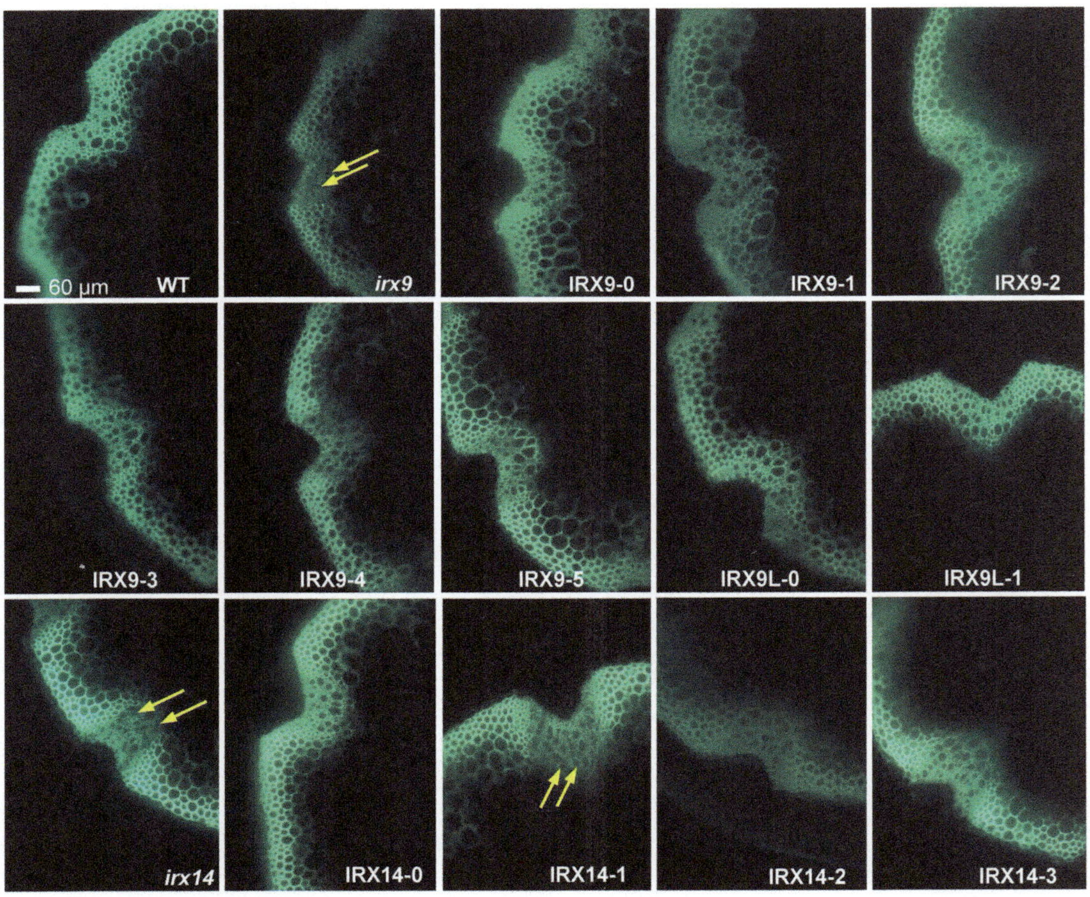

Figure 6. Microcopy analysis of stems from 6-week-old plants. The LM10 anti-xylan monoclonal antibody was used for immunodetection of xylan in transverse stem sections. The nomenclature for the constructs used to transform the *irx9* and *irx14* plants is explained in Table 1. Irregular xylem phenotype (indicated with arrows) was observed in the *irx9* and *irx14* mutants, as well as in the *irx14* plants transformed with the 35S:IRX14-1 construct. All the other transformants showed normal vessel phenotype.

irx14 mutant does not have a strong growth phenotype so it was not immediately obvious to what extent the IRX14 constructs complemented the phenotype. The level of complementation was quantified in several ways. Firstly, the rosette diameter (Fig. 3) and stem height were determined (Fig. 4). These results confirmed that the wild-type and all mutated versions of IRX9 and IRX9L complemented the *irx9* growth phenotype. The wild-type and mutated IRX14 proteins all complemented the *irx14* mutation, except for IRX14-1 where the 'DDS' motif was changed to 'AAS'. The IRX14-1 construct only partially complemented the growth phenotype as the stem height was intermediary between the *irx14* mutant and the wild type. None of the mutated proteins had a dominant negative effect in the wild-type background – this could have been expected with the IRX14-1 protein if the mutated protein would 'poison' a protein complex, but we did not observe that (Figs. S3–S5).

We next analyzed the sugar composition in the stems of T2 plants. Especially for IRX14 biochemical analysis of the phenotype is more accurate since the growth phenotype of the *irx14* mutant is mild. All wild-type and mutant versions of IRX9, IRX9L and IRX14 complemented the decreased xylose phenotypes of the *irx9* and *irx14* mutants, again with the notable exception of the IRX14-1 plants, for which the xylose content resembled mutant levels (Fig. 5). In some cases the complementation was not completely to the wild-type level, but the ability of the mutated proteins to complement also did not differ from the complementation seen with wild-type versions of the proteins – except for IRX14-1.

Finally, we investigated the irregular xylem phenotype in stem cross sections from T2 plants (Fig. 6). Irregular xylem (irx) phenotype was observed in the *irx9* and *irx14* as expected, and consistent with the other data, the irx phenotype was complemented in all the transformed plant lines, except for IRX14-1.

Discussion

This study shows that amino acid residues that have been shown to be essential in mammalian GT43 proteins are not important for the biological function of the Arabidopsis IRX9 and IRX9L proteins. All mutated versions of the proteins showed complementation of the *irx9* phenotype. This result is in agreement with the partial function of a truncated IRX9 protein in the *irx9-2* mutant, which is clearly too short to have catalytic glycosyltransferase function, but still plays a significant role as the growth phenotype is much milder than in *irx9* knock-out plants. Our observations and the phenotype of *irx9-2* strongly indicate that IRX9 and IRX9L have important non-catalytic functions. The absence of the DxD motif in IRX9 is surprising but also suggests that the protein is not catalytically active. Nevertheless, in the absence of an in vitro assay to test the activity of these proteins it is not possible to exclude that they possess glycosyltransferase activity. For IRX14 the results are more ambiguous since the DxD motif is clearly important for complementation, whereas the Gln residue in the predicted acceptor-binding and catalytic site is not. In the well-characterized human proteins the corresponding catalytic residue is actually Glu and not Gln. However, whether Gln and Glu could have the same function is not clear. Hence, we speculate that the UDP-xylose binding of IRX14 is important for xylan synthesis but that it does not essentially have to participate in xylosyl transfer. A conceivable scenario is that IRX14 passes the UDP-xylose substrate on to another protein, most likely IRX10 and IRX10L. However that raises the question, what is the role of IRX9 and IRX9L? We think our data indicate that IRX9 and IRX9L have a structural role, most likely helping to organize and/or assemble a protein complex, which is necessary for proper xylosyltransferase function in Arabidopsis. IRX10 and IRX10L are the likely catalytic subunits in such a complex. This is reminiscent of GAUT7, which is known to be catalytically inactive, despite being a member of the GAUT family of galacturonosyltransferases. The role of GAUT7 is instead the anchoring of GAUT1 in the Golgi, and thus GAUT7 is essential for the catalytical function of GAUT1 and ultimately for homogalacturonan synthesis despite not being a galacturonosyltransferase itself [31]. The interactions between IRX9, IRX14 and IRX10 have been difficult to show but can also not be excluded. Presumably interactions between these proteins are not very strong, which makes it difficult to prove them. Also protein tags that have been used to investigate interactions might interfere and thereby prohibit interactions. A study in wheat showed coimmunoprecipitation of IRX14 and IRX10, but IRX9 homologs were missing from the preparation and a direct interaction between the IRX14 and IRX10 homologs was not demonstrated [16].

Expression of the non-functional IRX14-1 protein in the wild-type background had no effect. In many cases, non-functional proteins that are part of a protein complex have a dominant negative effect on enzyme activity because the non-functional subunit interacts with the other subunits in the complex and prevent their normal function. However, we did not observe such an effect, perhaps because the normal IRX14 protein interacts more strongly or because the expression level of IRX14-1 was insufficient to bind a large proportion of the IRX10/IRX10L proteins. This could suggest that loss-of-function mutations in IRX10 or IRX10L would be more likely to have a dominant negative effect on xylan synthesis.

Supporting Information

Figure S1 Multiple sequence alignment of GT43 protein sequences. Multiple sequence alignment was created with the Muscle option in MEGA5. The N- and C-terminal regions that aligned less well are omitted in the figure. All GT43 proteins encoded in the genomes of Arabidopsis (At), rice (*Oryza sativa*, Os), poplar (*Populus trichocarpa*, Pt), spikemoss (*Selaginella moellendorffii*, Sm) and human (*Homo sapiens*, Sp) were included in the analysis. Conserved regions that were analyzed by site-directed mutagenesis are marked with blue, and the actual residues that were altered are marked with yellow. The changes made in the mutated proteins are listed in Table 1.

Figure S2 Identity matrix for protein sequences used to make the alignment is shown in Fig. S1 and the phylogenetic tree in Fig. 1.

Figure S3 Rosette diameter of 4-week-old wild-type plants transformed with the different constructs. The nomenclature for the constructs used to transform the plants is explained in Table 1. Wild type, *irx9*, and *irx14* were included for comparison. The bars show average ± SD (n = 10). Averages that are not significantly different (ANOVA, Tukey's test, $p>0.05$) are indicated with the same letter.

Figure S4 Stem height of 6-week-old wild-type plants transformed with the different constructs. The nomenclature for the constructs used to transform the plants is explained in Table 1. Wild type, *irx9*, and *irx14* were included for comparison. The bars show average ± SD (n = 10). Averages that are not significantly different (ANOVA, Tukey's test, $p>0.05$) are indicated with the same letter.

Figure S5 Xylose content in cell wall polysaccharides isolated from stems of 6-week-old wild-type plants transformed with the different constructs. Cell walls were prepared from stems, hydrolyzed in TFA, and the xylose content determined by HPAEC. The nomenclature for the constructs used to transform the plants is explained in Table 1. Wild type, *irx9*, and *irx14* were included for comparison. The bars show average ± SD (n = 4,except for WT+IRX9-4 transformants, where n = 3). Averages that are not significantly different (ANOVA, Tukey's test, p>0.05) are indicated with the same letter.

Table S1 Primers used for cloning, genotyping and RT-PCR.

Acknowledgments

Ms. Sherry Chan is thanked for her help with maintaining plants.

Author Contributions

Conceived and designed the experiments: YR SFH HVS. Performed the experiments: YR BE JL. Analyzed the data: YR BE JL HVS. Contributed to the writing of the manuscript: YR BE HVS.

References

1. Rennie EA, Scheller HV (2014) Xylan biosynthesis. Curr Opin Biotechnol 26: 100–107.
2. Rennie EA, Hansen SF, Baidoo EE, Hadi MZ, Keasling JD, et al. (2012) Three members of the Arabidopsis glycosyltransferase family 8 are xylan glucuronosyltransferases. Plant Physiol 159: 1408–1417.
3. Chiniquy D, Sharma V, Schultink A, Baidoo EE, Rautengarten C, et al. (2012) XAX1 from glycosyltransferase family 61 mediates xylosyltransfer to rice xylan. Proc Natl Acad Sci USA 109: 17117–17122.
4. Xiong G, Cheng K, Pauly M (2013) Xylan O-acetylation impacts xylem development and enzymatic recalcitrance as indicated by the Arabidopsis mutant *tbl29*. Mol Plant 6: 1373–1375.
5. Anders N, Wilkinson MD, Lovegrove A, Freeman J, Tryfona T, et al. (2012) Glycosyl transferases in family 61 mediate arabinofuranosyl transfer onto xylan in grasses. Proc Natl Acad Sci USA 109: 989–993.
6. Mortimer JC, Miles GP, Brown DM, Zhang ZN, Segura MP, et al. (2010) Absence of branches from xylan in Arabidopsis *gux* mutants reveals potential for simplification of lignocellulosic biomass. Proc Natl Acad Sci USA 107: 17409–17414.
7. Yuan Y, Teng Q, Zhong R, Ye ZH (2013) The Arabidopsis DUF231 domain-containing protein ESK1 mediates 2-O- and 3-O-acetylation of xylosyl residues in xylan. Plant Cell Physiol 54: 1186–1199.
8. York WS, O'Neill MA (2008) Biochemical control of xylan biosynthesis - which end is up? Curr Opin Plant Biol 11: 258–265.
9. Wu AM, Rihouey C, Seveno M, Hornblad E, Singh SK, et al. (2009) The Arabidopsis IRX10 and IRX10-LIKE glycosyltransferases are critical for glucuronoxylan biosynthesis during secondary cell wall formation. Plant J 57: 718–731.
10. Brown DM, Zhang ZN, Stephens E, Dupree P, Turner SR (2009) Characterization of IRX10 and IRX10-like reveals an essential role in glucuronoxylan biosynthesis in Arabidopsis. Plant J 57: 732–746.
11. Lee C, Teng Q, Huang W, Zhong R, Ye ZH (2010) The Arabidopsis family GT43 glycosyltransferases form two functionally nonredundant groups essential for the elongation of glucuronoxylan backbone. Plant Physiol 153: 526–541.
12. Wu AM, Hornblad E, Voxeur A, Gerber L, Rihouey C, et al. (2010) Analysis of the Arabidopsis *IRX9/IRX9-L* and *IRX14/IRX14-L* pairs of glycosyltransferase genes reveals critical contributions to biosynthesis of the hemicellulose glucuronoxylan. Plant Physiol 153: 542–554.
13. Chiniquy D, Varanasi P, Oh T, Harholt J, Katnelson J, et al. (2013) Three novel rice genes closely related to the Arabidopsis *IRX9*, *IRX9L*, and *IRX14* genes and their roles in xylan biosynthesis. Front Plant Sci 4: 83.
14. Lee CH, Teng QC, Zhong RQ, Ye ZH (2011) Molecular dissection of xylan biosynthesis during wood formation in poplar. Mol Plant 4: 730–747.
15. Lee C, Zhong R, Ye Z-H (2012) Arabidopsis family GT43 members are xylan xylosyltransferases required for the elongation of the xylan backbone. Plant Cell Physiol 53: 135–143.
16. Zeng W, Jiang N, Nadella R, Killen TL, Nadella V, et al. (2010) A glucurono(arabino)xylan synthase complex from wheat contains members of the GT43, GT47, and GT75 families and functions cooperatively. Plant Physiol 154: 78–97.
17. Jensen JK, Kim H, Cocuron JC, Orler R, Ralph J, et al. (2011) The DUF579 domain containing proteins IRX15 and IRX15-L affect xylan synthesis in Arabidopsis. The Plant journal: for cell and molecular biology 66: 387–400.
18. Jensen JK, Johnson N, Wilkerson CG (2013) Discovery of diversity in xylan biosynthetic genes by transcriptional profiling of a heteroxylan containing mucilaginous tissue. Front Plant Sci 4: 183.
19. Clough SJ, Bent AF (1998) Floral dip: a simplified method for *Agrobacterium*-mediated transformation of *Arabidopsis thaliana*. Plant J 16: 735–743.
20. Tamura K, Peterson D, Peterson N, Stecher G, Nei M, et al. (2011) MEGA5: molecular evolutionary genetics analysis using maximum likelihood, evolutionary distance, and maximum parsimony methods. Mol Biol Evol 28: 2731–2739.
21. Saitou N, Nei M (1987) The neighbor-joining method: A new method for reconstructing phylogenetic trees. Mol Biol Evol 4: 406–425.
22. Earley KW, Haag JR, Pontes O, Opper K, Juehne T, et al. (2006) Gateway-compatible vectors for plant functional genomics and proteomics. Plant J 45: 616–629.
23. Harholt J, Jensen JK, Sorensen SO, Orfila C, Pauly M, et al. (2006) ARABINAN DEFICIENT 1 is a putative arabinosyltransferase involved in biosynthesis of pectic arabinan in Arabidopsis. Plant Physiol 140: 49–58.
24. Obro J, Harholt J, Scheller HV, Orfila C (2004) Rhamnogalacturonan I in *Solanum tuberosum* tubers contains complex arabinogalactan structures. Phytochemistry 65: 1429–1438.
25. McCartney L, Marcus SE, Knox JP (2005) Monoclonal antibodies to plant cell wall xylans and arabinoxylans. J Histochem Cytochem 53: 543–546.
26. Anders N, Dupree P (2010) Glycosyltransferases of the GT43 Family. In: Ulvskov P, editor. Annual Plant Reviews: Plant Polysaccharides, Biosynthesis and Bioengineering. Hoboken: Wiley-Blackwell. Vol. 41, pp. 251–263.
27. Breton C, Snajdrova L, Jeanneau C, Koca J, Imberty A (2006) Structures and mechanisms of glycosyltransferases. Glycobiology 16: 29R–37R.
28. Pedersen LC, Darden TA, Negishi M (2002) Crystal structure of beta 1,3-glucuronyltransferase I in complex with active donor substrate UDP-GlcUA. J Biol Chem 277: 21869–21873.
29. Fondeur-Gelinotte M, Lattard V, Gulberti S, Oriol R, Mulliert G, et al. (2007) Molecular basis for acceptor substrate specificity of the human beta1,3-glucuronosyltransferases GlcAT-I and GlcAT-P involved in glycosaminoglycan and HNK-1 carbohydrate epitope biosynthesis, respectively. Glycobiology 17: 857–867.
30. Ouzzine M, Gulberti S, Netter P, Magdalou J, Fournel-Gigleux S (2000) Structure/function of the human Galbeta1,3-glucuronosyltransferase. Dimerization and functional activity are mediated by two crucial cysteine residues. J Biol Chem 275: 28254–28260.
31. Atmodjo MA, Sakuragi Y, Zhu X, Burrell AJ, Mohanty SS, et al. (2011) Galacturonosyltransferase (GAUT)1 and GAUT7 are the core of a plant cell wall pectin biosynthetic homogalacturonan:galacturonosyltransferase complex. Proc Natl Acad Sci USA 108: 20225–20230.

13

The Blast Resistance Gene *Pi54of* Cloned from *Oryza officinalis* Interacts with *Avr-Pi54* through Its Novel Non-LRR Domains

Navadagi B. Devanna, Joshitha Vijayan, Tilak R. Sharma*

National Research Centre on Plant Biotechnology, Indian Agricultural Research Institute, New Delhi, India

Abstract

The dominant rice blast resistance gene *Pi54* cloned by map-based cloning approach from indica rice cultivar Tetep confers broad spectrum resistance to *Magnaporthe oryzae*. In this investigation, an orthologue of *Pi54* designated as *Pi54of* was cloned from *Oryza officinalis* conferring high degree of resistance to *M. oryzae* and is functionally validated. We have also characterized the Pi54of protein and demonstrate its interaction with AVR-Pi54 protein. The *Pi54of* encoded ~43 kDa small and unique cytoplasmic LRR family of disease resistance protein having unique Zinc finger domain overlapped with the leucine rich repeat regions. *Pi54of* showed *Magnaporthe*-induced expression. The phylogenetic and western blot analysis confirmed orthologous nature of *Pi54* and *Pi54of* genes, whereas the identity of protein was confirmed through MALDI-TOF analysis. The *in silico* analysis showed that Pi54of is structurally more stable than other cloned Pi54 proteins. The molecular docking revealed that Pi54of protein interacts with AVR-Pi54 through novel non-LRR domains such as STI1 and RhoGEF. The STI1 and GEF domains which interact with AVR-Pi54 are also components of rice defensome complex. The Pi54of protein showed differential domain specificity while interacting with the AVR protein. Functional complementation revealed that *Pi54of* transferred in two rice lines belonging to indica and japonica background imparts enhanced resistance against three highly virulent strains of *M. oryzae*. In this study, for the first time, we demonstrated that a rice blast resistance gene *Pi54of* cloned from wild species of rice provides high degree of resistance to *M. oryzae* and might display different molecular mechanism involved in AVRPi54-Pi54of interaction.

Editor: Jauhar Ali, International Rice Research Institute, Philippines

Funding: T. R. Sharma is thankful to the Department of Biotechnology, Govt. of India (projects: BT/AB/FG-2 (PHII)/2009-3A and BT/HRD/34/04/2008) for financial assistance. BND is thankful to the Council of Scientific and Industrial Research, Govt. of India for senior research fellowship. The funders had no role in study design, data collection and analysis, decision to publish, or preparation of the manuscript.

Competing Interests: The authors have declared that no competing interests exist.

* Email: trsharma@nrcpb.org

Introduction

Rice is the staple food for larger part of the world population. The major biotic stress of rice which is considered as one of the most damaging diseases worldwide is rice blast caused by hemibiotrophic fungi *Magnaporthe oryzae*. In India alone, the total loss due to rice blast during 1960–1961 was 2, 65,000 T amounting to about 0.8% of total rice production. [1]. However, blast under severe epiphytic conditions may result between 70–90% loss in isolated fields/localities. Among the other control measures, use of resistance (*R*) -genes with overlapping resistance spectra is most viable approach [2]. *M. oryzae* is highly adaptive and infects rice plants at all developmental stages [3]. The adaptive nature of *M. oryzae* is mainly because of breakdown in host resistance due to mutations in the pathogen leading to instability of avirulence (*AVR*) genes [4]. The fungus can overcome even major *R* gene-mediated resistance with the mutation in the corresponding *AVR* gene limiting the life span of major *R*-genes [5,6]. This necessitates time bound release of resistant varieties for effective management of rice blast. In view of this, continued identification and cloning of new effective genes and their alleles are critical for *R*-gene mediated management of rice blast disease.

The wild species of rice harbour resistance genes that remain largely unexploited [7]. It is important to clone the *R*-gene (s) to establish their functional character for their effective and meaningful utilization. The resistance genes against bacterial blight from *O. longistaminata* (*Xa21*) and *O. officinalis*, brown plant hopper transmitting grassy stunt virus from *O. nivara* and white backed plant hopper from *O. officinalis* have been transferred in many varieties [8]. Wild species of rice such as *O. rufipogon* and *O. officinalis* have been reported to be a potential source for blast resistance genes [9]. In case of rice, many genes, such as *Pi9* from *O. minuta*, *Pi-40(t)* from *O. australiensis*, *Pirf2-1*(t) and *Pid3-A4* from *O. rufipogon*, *Pi54rh* from *O. rhizomatis* have been shown to be effective against blast disease [10]. Allele mining of the gene from the wild and cultivated species of rice not only aid in identification of genetic variation and to understand evolution, but also serves as a tool for identification of the better alleles. In rice, a strong rice blast resistance *Pi-ta(+)* allele has been isolated from wild rice *O. rufipogon* [11].

During the long co-evolution with pathogens, the plants have developed sophisticated mechanisms to perceive and defend against a pathogen's attack [12]. The plant innate immunity is classified into basal and *R*-gene-mediated resistance [13]. The first line of plant defense against invading pathogens is activated by membrane proteins called pattern recognition receptors (PRRs), which recognize the conserved pathogen associated molecular patterns (PAMPs). Some plant pathogens bypass this PAMP-triggered immunity (PTI) through effector molecules and trigger disease reaction by manipulating key components of the plant defense [14]. To overcome this infection, plants do have a second layer of defense called effector-triggered immunity (ETI) which relies on the specific recognition of certain pathogen-derived effector molecules called avirulence (AVR) proteins by the plant R- proteins. All the *R*-genes cloned till date from different plant species have been divided into five classes based on their protein structure [15]. The largest class of these *R*-genes encode NBS-LRR family of proteins. Pathogen AVR proteins are detected by R- proteins either through direct or indirect interaction. Direct R–AVR interaction has been reported for nine R- proteins: the first one being Pi-ta, followed by RRS1-R, N, L5/L6, M, RPP1, Pik/km/kp/ks/kh, RGA5, and Pb1 [16].

The recent developments in the field of molecular biology and genomic resources, besides the genetic tractability and economic significance of blast disease have made the *rice–Magnaporthe* interaction as a model system for the analysis of plant–pathogen relationships. Race specific *R*-gene-mediated resistance to *M. oryzae* closely follows the classical gene-for-gene relationship [17]. The isolation and characterization of *R*-genes will help to unravel varied molecular mechanisms underlying the interaction between host and the pathogen. Till-date, around 100 *R*-genes have been identified and many of them mapped on the rice genome and 23 of them (*Pib, Pita, Pi54, Pi-9, Pid2, Pi2, Piz-t, Pi-36, Pi-37, Pikm, Pi5, Pid3, pi21, Pit, Pb1, Pish, Pi-k, Pik-p, Pia, NLS1,Pi25, Pi54rh* and *Pid3-A4*) have been cloned and characterized [8]. Interestingly, 19 of them are NBS-LRR genes, while, *Pi54* and *Pi54rh* predicted to have rudimentary NBS domain and small LRR domains [18]; *Pi-d2* and *Pi21* code for non NBS-LRR proteins [16].

Among the cloned rice blast *R*-genes, only the functional characterization of protein encoded by *Pita* and its effector has been performed in detail [19]. The yeast two hybrid experiments showed that the rice blast resistance gene *RGA5* product interact directly with its AVR protein through non-LRR domain and *Pb1* gene product with WRKY45 through its CC domain [16,20]. However, details of molecular mechanisms linking R–AVR interaction with the activation of downstream disease resistance signalling pathways are largely unknown. These R- proteins encoded by rice blast resistance genes are very diverse in their size, AVR recognition specificity and resistance mechanism. So, *in vitro* expression systems are indispensable for expression, purification and structural and functional characterization of proteins involved in plant disease resistance [21].

Allele mining entails PCR-based amplification and sequencing of the different versions of the genes found in the germplasm. Variation in gene sequence is then correlated with the phenotypic trait or performance of the accession to enable the identification of favourable alleles for future experiments. The cloned dominant blast resistance gene *Pi54* has been effective against a selective set of rice blast pathogen isolates from different parts of India [22,23]. The wild rice line *O. officinalis* is highly resistant to virulent strains of *M. oryzae* like PLP-1, the most predominant isolate widely distributed in the North-western Himalayan region of India [24] and Mo ei-11 from Hazaribagh region of eastern India. Therefore, allele mining for *Pi54* gene in this wild rice might reveal more effective resistance orthologue of *Pi54* gene. We report in this study cloning of an orthologue of rice blast resistance gene *Pi54* from wild rice *O. officinalis* and its molecular and functional characterization.

Materials and Methods

Plant material, fungal strains and oligos

The wild rice *O. officinalis* (nrcpb004) is being maintained under controlled conditions at this institute. Seeds of indica rice lines IET16310 and HR12 were obtained from the Division of Genetics of the Indian Agricultural Research Institute, New Delhi. The japonica rice line Taipei 309 (TP309) was available with the authors. A highly virulent isolate of *M. oryzae*, Mo ei-11 was obtained from the Central Rainfed and Upland Rice Research Station, Hazaribagh, Jharkhand, India and maintained in the Centre's stock culture was used for phenotypic analysis of *O. officinalis*. The fungal strains RML21 and MG-79 were obtained from the Department of Agricultural Biotechnology, CSK Himachal Pradesh Agricultural University, Palampur, Himachal Pradesh, India and 6-Sikkim was obtained from the Division of Plant Pathology, Indian Agricultural Research Institute, New Delhi, India. The list of oligos used in this study is given in Table S1 in File S1.

Phenotyping with *Magnaporthe oryzae*

Phenotyping of fifteen-day-old young seedlings of rice variety HR12 and vegetatively propagated plantlets (two leaf stage) of wild rice *O. officinalis* were inoculated with *M. oryzae* spore suspension (1×105 spores/ml) as described by Sharma et al. [22]. The experiment was carried out under controlled growth conditions at 25±1°C and 90% RH for initial 24 h in dark and then shifted to 16/8-h light/dark regime. The disease reaction observed in each rice line was recorded after 7 days post inoculation using 0–5 disease rating scale [25].

Cloning of full length *Pi54of* gene

PCR amplification of *Pi54of* orthologue was performed with primers designed from the BAC clone sequence (AC145349.1) of the *O. sativa* cv. Nipponbare (IRGSP 2005) corresponding to the location of *Pi54* gene region. Amplified PCR product was gel purified and cloned in pGEMT-E vector (Promega) and sequenced (Xcelris Labs Ltd, India). To clone the complete functional *Pi54of* gene, inverse PCR (iPCR) was carried out using *Pi54of* specific primers following standard protocol [26]. For this, self ligated product of *Eco*RI digested *O. officinalis* genomic DNA was used as template. The iPCR product was cloned, sequenced and the sequence reads obtained were assembled using Phred/Pharp/Consed Software Package (http://www.phrap.org/). Further, 5′ Rapid Amplification of cDNA Ends (5′RACE) PCR was conducted by First Choice RLM-RACE kit (Ambion, Austin, Texas, USA), following the manufacturer's instructions. The total RNA isolated from *O. officinalis* plants at 72 hours post inoculation (hpi) with *M. oryzae* was used as template. The PCR product was cloned in pGEMT-Easy vector (Promega) and sequenced.

Real-time PCR analysis

Total RNA was isolated from HR12 and *O. officinalis* plantlets pre- and post- inoculation with the *M. oryzae* isolate Mo-ei-11 and mock (0.25% gelatine) control. Total RNA was extracted from leaf tissues after 12, 24, 48, 72 and 96 h of inoculation using the Spectrum Plant Total RNA Kit (Sigma). The qRT-PCR analysis was conducted using *Pi54of* ORF rice elongation factor-1 alpha (EF-1α) gene specific primers in a reaction mix of KAPA SYBR FAST qRT-PCR Master mix (Kapa Biosystems, Inc. USA) according to the manufacturer's protocol and reaction was run using Light Cycler 480 II PCR system (Roche).

Structural and comparative analysis of *Pi54of*

DNA sequence similarity search was done using BLASTn (blast.ncbi.nlm.nih.gov) and the nucleotide sequence comparison was done using BioEdit Ver. 7.0.9 (www.mbio.ncsu.edu). Structural annotation was done with FGENESH (http://linux1.softberry.com/) and PLACE (http://www.dna.affrc.go.jp/PLACE/) softwares. The amino acid sequence of *R*-gene products was downloaded from NCBI and was used for the phylogenetic tree construction based on the progressive multiple sequence alignment using the Neighbor-Joining method with 1000 replicates in the MEGA6 program (http://www.ncbi.nlm.nih.gov/) [27].

Analysis of physico-chemical properties of the proteins of *Pi54* orthologues

The physico-chemical properties of the protein sequences of *Pi54* orthologues were calculated by using Protparam software (http://web.expasy.org/protparam/). The secondary structures and number of bonds were calculated by using protein visualization software RasMol (http://www.openrasmol.org/doc/). The COILS server was used to predict coiled coil domain [28]. GlobPlot-2 analysis was performed to identify the presence of Globular domains and disorder regions (http://globplot.embl.de/). Various domains were predicted using SMART7 (http://smart.embl-heidelberg.de) and ExPASy software.

Subcellular localization of Pi54of protein

The deduced amino acid sequence of *Pi54of* was used for the *in silico* analysis using WoLF POSRT software [29]. The mGFP coding sequence was PCR amplified by the exon specific primers and cloned at *Xba*I site of pRT101.1 vector in frame with at N-terminus with *Pi54of* CDS already cloned under the control of CaMV 35S promoter. For localization study, the onion lower epidermis was used for bombardment by a pneumatic particle gun (PDS-1000/He, BIO-Rad, USA) and the cells were observed by a confocal microscope (TCS SP2, Leica, Wetzlar, Germany) following standard protocol [18].

In vitro expression and purification of *Pi54of* protein

The *Pi54of* ORF was PCR amplified and cloned into pET29a vector using gene specific primers with recognition sequences for *Kpn*I and *Sal*I restriction enzymes and transformed into BL21-DE3 bacterial expression systems (Novagen). Expression and protein purification were performed according to instructions provided in the pET System Manual (http://www.novagen.com).

Characterization of Pi54of protein

The protein concentration was determined by the bicinchoninic acid (BCA) assay [30]. Protein samples and 5X sample loading buffer (Thermo Scientific, USA) were mixed in a ratio of 4:1 (v/v) in an eppendorff tube, heated to 100°C for 2–5 min and were analysed by 12% SDS-PAGE. The SDS-PAGE resolved Pi54of and Pi54 proteins gel was used for western blot analysis using custom synthesized polyclonal antibodies for Pi54 protein (Sigma, USA) following standard protocol [31]. Further Pi54of was analyzed by MALDI-TOF/TOF with Autoflex, Mass spectrometer MALDI-TOF/TOF instrument (Bruker Daltonics, Billerica, MA, USA) as described by Condina et al. [32]. Peptides from mass spectra of in-gel digested samples were matched against protein databases such as Swissprot, NCBInr, and MSDB using Mascot search engine (Matrix Sciences, London, UK) for peptide mass fingerprinting (PMF).

Molecular docking of Pi54 and AVR-Pi54 proteins

The 3D models of Pi54 orthologues and AVR-Pi54 was generated by *ab initio* structure prediction method using I-TASSER web based server (zhanglab.ccmb.med.umich.edu/I-TASSER/). The structures were refined by ModRefiner server (http://zhanglab.ccmb.med.umich.edu/ModRefiner/). Energy calculation and minimization was performed using CHARMM force field for 2000 steps [33] (Table S2 in File S1). The potential energy, van der Waals energy and electrostatic energy of *Pi54* orthologues and AVR-Pi54 models was computed using Discovery Studio v 3.5 (http://accelrys.com/products/discovery-studio/). The dependability and quality of these predicted models were ensured by Ramachandran plot analysis (http://www.cgl.ucsf.edu/chimera/download.html). Molecular docking of Pi54 orthologues and AVR-Pi54 was performed using Z-Dock server of Discovery Studio v 3.5 by removing water molecules and adding hydrogen ions based on the CHARMm force field (Table S3 in File S1). The 3D structures and molecules involved in interactions were visualized at 4 Å distance using Accelrys DS 3.5 visualizer software. The relative binding efficiency of AVR-Pi54 with Pi54of, Pi54 and Pi54rh proteins was analyzed.

Transgenic development and complementation assay

A rice transformation vector (pRiTV) was constructed using pBlueScript SKII (Stratagene, USA) and full-length transcript of *Pi54of* was driven by CaMV35S and terminated at *nos* site. The full cassette was cloned in the *Sma*I site of the multiple cloning sites (MCS). The hygromycin gene (*hptII*) amplified from pCAMBIA 1305.1 was also cloned in the *Sac*II site of MCS. The embryogenic callus derived from mature seeds of rice cultivar IET16310 and TP309 were used for genetic transformation by biolistic approach using helium-driven Particle Delivery System (PDS 1000, BioRad, Washington DC, USA). The calli were regenerated into mature fertile plants following the standard protocol [34]. The putative transgenic rice plants were confirmed by PCR using *Pi54of*, marker gene and 35S-*nos* specific primers and also by Southern blot analysis using DIG labelled probes [31]. Phenotyping of T_1 generation transgenic IET16310 plants was performed with *M. oryzae* (RML21 and 6-Sikkim) spore suspension, as described earlier. Similarly for phenotyping of T_0 generation transgenic TP309 plants, detached leaf assay employing young leaves of grown up plants with *M. oryzae* (Sikkim) spore suspension was used [35]. The leaves were placed on the sterilized Whatman No.3 filter paper and 2 μl *M. oryzae* spore suspension (1×10^5 spores/ml) was placed on leaf surfaces. The whole set up was placed in a humid chamber (95%) at 28°C in dark for initial 24 h and then shifted to 16/8-h light/dark regime. The disease reaction was recorded 7 days after inoculation. Further for resistance spectrum comparison we performed phenotypic analysis of T_1 generation transgenics of TP309 and IET16310 plants along with independent transgenic TP309 plants containing *Pi54* and *Pi54rh* genes with spray inoculation of *M. oryzae* strain MG-79, which differentiates *Pi54* and *Pi54rh*. The non transgenics lines and highly susceptible rice cv. HR12 were used as controls in all the cases.

Development of allele specific functional CAPS marker

A functional Cleaved Amplified Polymorphic Site (CAPS) marker was designed using *Hind*III restriction enzyme which digests the ORF of *Pi54of* at 954 bp, but do not digest *Pi54tp* allele of TP309. The PCR fragments amplified using ORF specific primers from transgenic and non-transgenic TP309 plants were analyzed.

Results

Expression analysis and cloning of *Pi54of* gene

The phenotyping results showed that the wild rice *O. officinalis* was completely resistant when inoculated with *M. oryzae* isolate Mo ei-11, whereas the susceptible check rice line HR12 showed scale 4–5 susceptible reaction (Figure 1). A relatively higher cycle threshold (CT) value (33.22) was obtained at 0 hpi of the seedlings of *O. officinalis* indicating low abundance of transcript in the pre-inoculated plants. The relative expression revealed the increased upregulation of *Pi54of* from 24 hpi and reaches maximum expression of 23 folds at 72 hpi before decreasing drastically to 9.6 folds at 96 hpi (Figure 1). For the cloning of full length *Pi54of* gene, heterologous primers were used and a ~1.2 kb region was amplified. Further iPCR amplification could amplify ~2 kb flaking region. The 5′ RACE PCR results revealed the position of Transcription Start Site (TSS) at 76-bp upstream of the start codon (ATG) of the *Pi54of* transcript (Figure S1). The sequence reads were assembled to get a high quality (>Phred 30) consensus sequence for predicting full length *Pi54of* (Figure 1). The ORF sequence of the gene is available at EMBL Database (Accession no. HE589448).

Structural and comparative analysis of *Pi54of*

Among the cloned *Pi54* orthologues, *Pi54of* is the largest coding for 379 amino acids, whereas *Pi54* and *Pi54rh* code for 330 and 360 amino acids, respectively. The ORF region of *Pi54of* showed 80.7% and 84.1% identity at nucleotide and 77.3% and 84% identity at amino acid level with *Pi54* and *Pi54rh*, respectively. The *Pi54of* differed from *Pi54* at 45 transition and 28 transversion sites. Whereas, six transition and transversion sites differentiated *Pi54of* and *Pi54rh* orthologues.

Analysis of 1kb promoter region of *Pi54of* indicates presence of GATA (42 bp upstream of TSS), CAAT (73 bp upstream of TSS), BIHD (209 bp upstream of TSS) and three WRKY Box elements (362, 394 and 401 bp upstream of TSS). The structural features of *Pi54* allele from rice cv. Tetep and Nipponbare and the orhologues from *O. officinalis* and *O. rhizomatis* were compared. The susceptible *Pi54* orthologue in cv. Nipponbare has three insertions like 49 bp insertion in the promoter region, a 30 bp insertion in the open reading frame and 140 bp insertions in the Poly-A region. These insertions were not obtained in rest of the alleles. In the *Pi54* gene cloned from rice cv. Tetep, five regulatory elements, WUN-1 motif, MeJA responsive element, CAAT box, TATA box and GT-1 box were predicted in the promoter. In *O. rhizomatis*, four regulatory elements, G-box, BH-box, CAAT box and WRKY box were predicted in the promoter region (Figure 1). The phylogenetic analysis of cloned rice blast resistance gene products revealed divergent clusters. The three *Pi54* orthologues belong to a single sub-cluster and are closely related to two cloned *R*-genes, *Pi37* and *Pish-1* (Figure 2).

Physico-chemical properties of Pi54 proteins

We predicted that *Pi54of* codes for ~43 kDa cytoplasmic protein with pI of 5.16. It has functional motifs, such as N-glycosylation site (5 copies), Casein Kinase II phosphorylation site (12 copies), N-myristilation site (4 copies), Protein Kinase C phosphorylation site (3 copies) and Tyrosine Kinase phosphorylation site (1 copy). The multiple sequence alignment of the proteins of *Pi54* orthologues showed a large deletion corresponding to 58 amino acid residues in *Pi54tp* gene (Figure 2). The protein hydrophobicity analysis predicted that the Pi54 proteins of resistant alleles were slightly hydrophilic in nature compared to the Pi54tp protein of susceptible allele. The Pi54 was more hydrophilic followed by Pi54of, Pi54rh and Pi54tp proteins (Figure S2). Present analysis also revealed that Pi54 proteins are relatively thermostable, but overall these all resistance proteins are unstable in nature (Table S4 in File S1).

The SMART and ExPASy analysis of Pi54of protein predicted important domains such as, CC, LRR, Zinc finger domain, RhoGEF (Guanine nucleotide exchange factor for Rho/Rac/Cdc42-like GTPases) domain and STI1 domain (Heat shock co-chaperonin-binding motif) involved in the plant defense mechanism (Figure 3). The amino acid composition analysis of the four R- proteins as well as the their LRR domains showed the presence of amino acids Leucine (L), Glutamic acid (E) and Serine (S) in higher frequency (Figure 3). Among the Pi54 proteins, the CC domain was present in the resistant *Pi54* orthologues and absent in *Pi54tp* protein and significance of the prediction was more for Pi54of than Pi54 and Pi54rh.

In silico 3D structure prediction and molecular docking

The 3D structure of Pi54of and other Pi54 orthologues predicted typical horseshoe-shaped conformation (Figure 4). The global free energy of Pi54of protein was found to be much lower than Pi54, Pi54rh and Pi54tp proteins despite being larger in size (Table 1). Ramachandran plot analysis using UCSF-CHIMERA server showed that more than 98% amino acid residues of Pi54 proteins and 89.4% amino acid residues of AVR-Pi54 fell in the most favoured regions and allowed region of the plot (Figure S3). The molecular docking results showed that the binding of AVR-Pi54 with Pi54 was more robust followed by its binding with Pi54of and Pi54rh and no interaction was found with Pi54tp (Figure 4 and Table 1). Analysis of amino acid residues involved in the interactions showed that the Pi54 protein interacts with AVR-Pi54 through its CC, non-LRR and LRR regions. Whereas, Pi54rh interact through its non-LRR and LRR region and Pi54of interacts only through its non-LRR region (Figure 4, Table 2).

In vitro expression and characterization of Pi54of protein

Protein electrophoresis of IPTG induced and uninduced pET29a clones confirmed the induction of ~43 kDa and ~37 kDa protein products in *Pi54of* and *Pi54* clones, respectively (Figure 5). Western blot analysis revealed the hybridization of antibody to ~43 kDa Pi54of and ~37 kDa Pi54 proteins (Figure 5). The peptide mass fingerprinting analysis performed using MALDI-TOF mass spectrometry showed match of its spectra with the Pi54 proteins of rice lines resistant to the pathogen (Figure 5).

Subcellular localization of Pi54of protein

For subcellular localization, Pi54of protein was fused with N-terminal of the reporter gene *mGFP* under the control of constitutive CaMV 35S promoter and was expressed transiently in onion epidermal cells (Figure S4). The GFP fluorescence which was observed 24 h post bombardment revealed that Pi54of-mGFP was localized to the cytoplasm, whereas the GFP control was evenly distributed throughout the cell (Figure 6).

Functional validation of the gene

For functional validation, *Pi54of* gene was cloned in rice transformation vector (pRiTV) and transferred to the rice lines

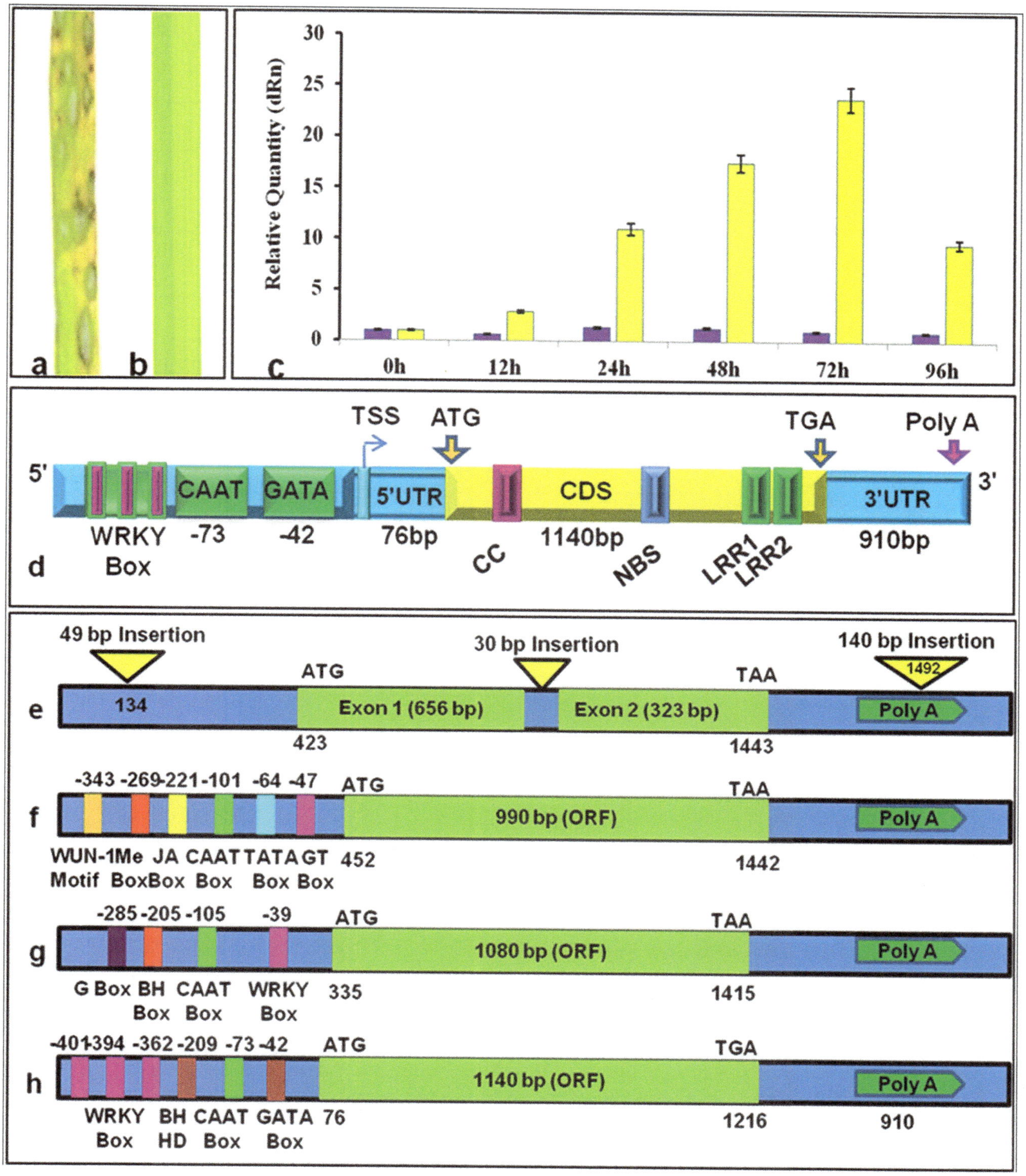

Figure 1. Phenotypic and expression analysis of *Pi54of* gene. Reaction of HR-12 (a) and *O. officinalis* (b) to with *M. oryzae*; c: Relative expression; pre-and-post inoculation of *O. officinalis* with *M. oryzae* using qRT-PCR at different intervals; d: Schematic representation of full length *Pi54of* gene; e: Structural analysis of *Pi54* orthologues in f: Nipponbare; g: Tetep (Sharma et al. 2005); h: *O. rhizomatis* (Das et al. 2012) and i: *O. officinalis*.

IET16310 and TP309 susceptible to *M. oryzae*. Molecular analysis of the putative transgenics of IET16310 (Figure S5) lines was performed. The PCR amplification showed the presence of 1, 140 bp *Pi54of*-specific and 500 bp *hptII*-specific amplicons in IET16310 transgenic plants (Figure S5). The Southern blot analysis of PCR positive transgenic plants revealed that the plant number IET01, IET02, IET04, IET06 and IET09 had single copy insertion, where as IET10 showed two copies of *Pi54of* integrated in its genome (Figure S5). Similarly, the putative transgenic TP309 plants (Figure S6) were confirmed by PCR to amplify 1,140 bp

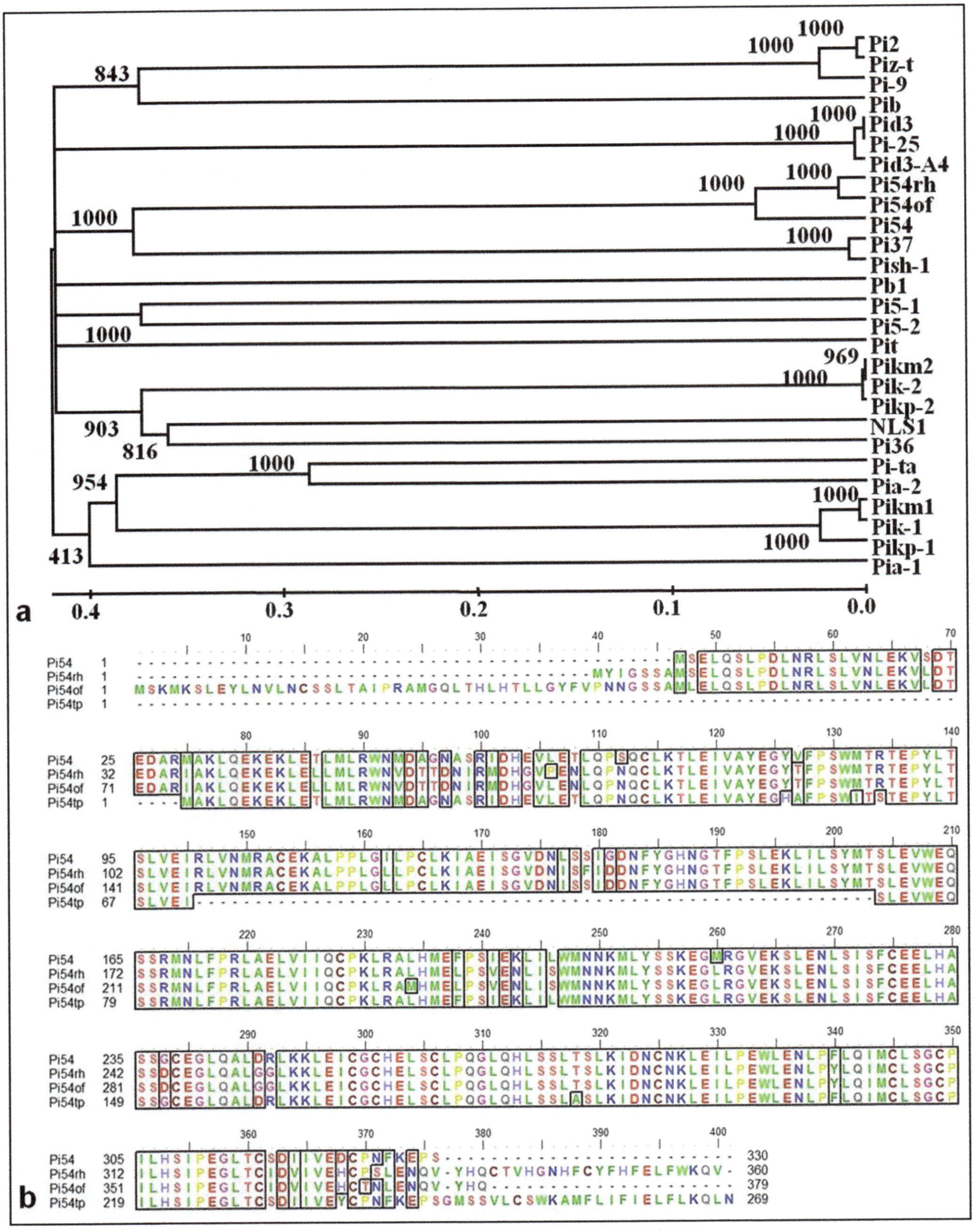

Figure 2. Phylogeny and alignment of amino acid residues from ORF of *Pi54, Pi54of* and *Pi54rh* orthologues. a: Phylogenetic relationship among the cloned rice blast resistance genes. Tree was generated using neighbor-joining method by MEGA.6 software with bootstrap value 1000 replications. The unit of branch length is 0.1 nucleotide substitutions per site, and is indicated by a bar at the bottom left corner of the tree. b: Aligned sequences of Pi54 protein orthologues using BioEdit tool.

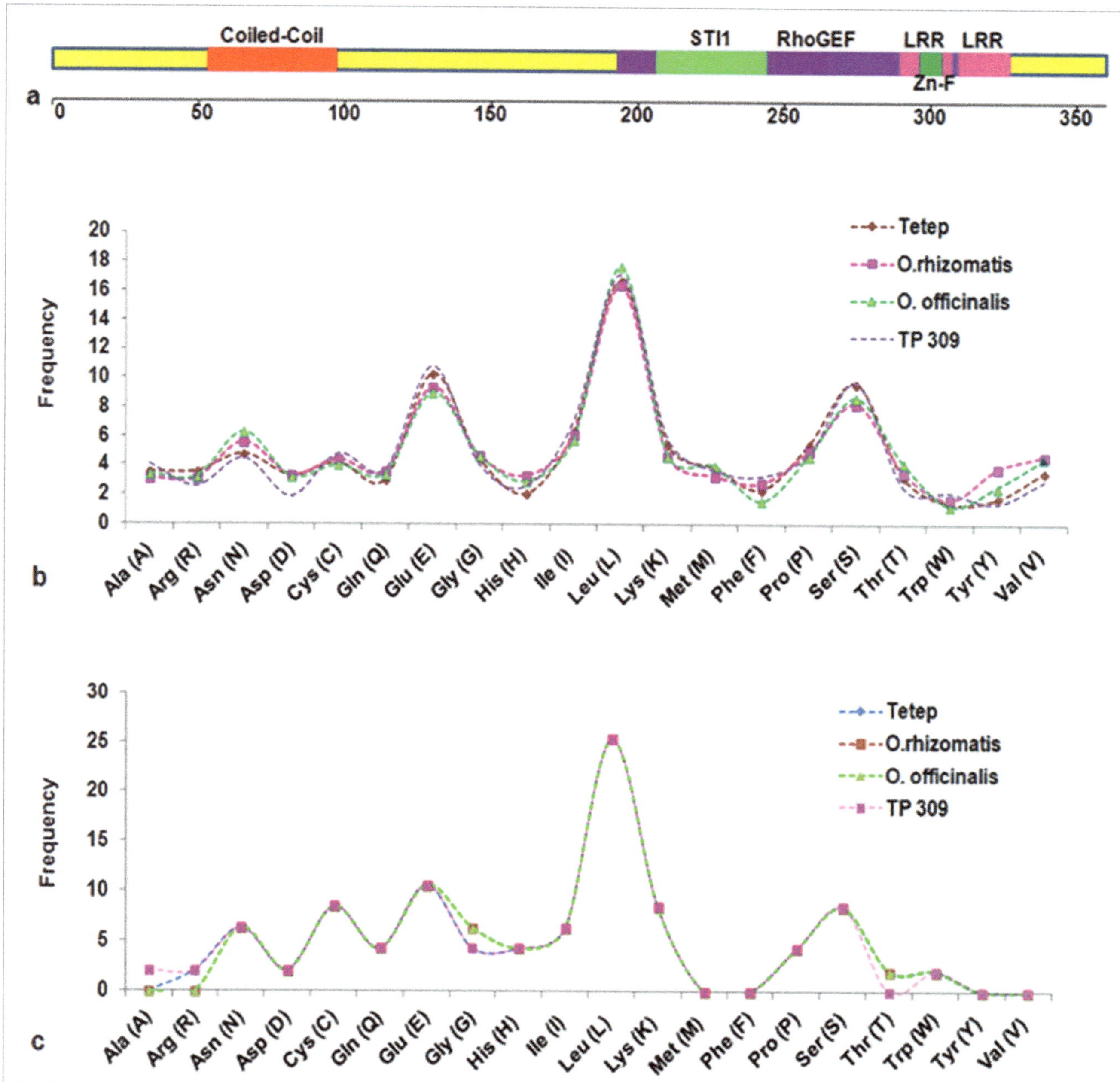

Figure 3. Schematic representation of predicted functional domains in Pi54of protein (a) and the frequency of amino acid residues found in four *Pi54* orthologue proteins; b: orthologue proteins; c: LRR region of all orthologue proteins.

Pi54of specific ORF and ~1, 851 bp fragment corresponding to *35S-Pi54of-nos* region. Further, the functional allele specific CAPS marker analysis also showed the presence of *Pi54of* gene in the five transgenics of TP309 events and its absence in non-transgenic plants (Figure S6). The phenotyping results of 6 transgenic IET16310 plants showed that, events IET01, IET02, IET04, IET06 and IET10 were resistant to *M. oryzae*, whereas IET09 showed susceptible reaction to the pathogen (Figure 7). Similarly, transgenic TP309 plants were challenged with *M. oryzae* isolate 6-Sikkim and the results showed that all five transgenic TP309 events (TPe1–TPe5) showed hypersensitive reaction (HR), where as non-transgenic control showed typical susceptible blast symptoms (Figure 7). The comparison of resistance response of transgenic TP309 lines containing *Pi54* orthologues revealed that TP-*Pi54* was susceptible and TP-*Pi54rh* and TP-*Pi54of* displayed resistance reaction against *M. oryzae* isolate MG-79. Further all the transgenic of IET series except IET09 harbouring *Pi54of* were resistant to isolate MG-79 (Figure S7).

Discussion

Over the years rice monocroping has lead to the increased susceptibility of many local rice cultivars to *M. oryzae*. Management of rice blast through host plant resistance is considered as viable and ecofriendly option. Since 1999 when first rice blast resistance gene *Pib* was cloned [36], efforts have been made to successfully clone many resistance genes from different sources. Currently 23 rice blast resistance genes have been cloned and

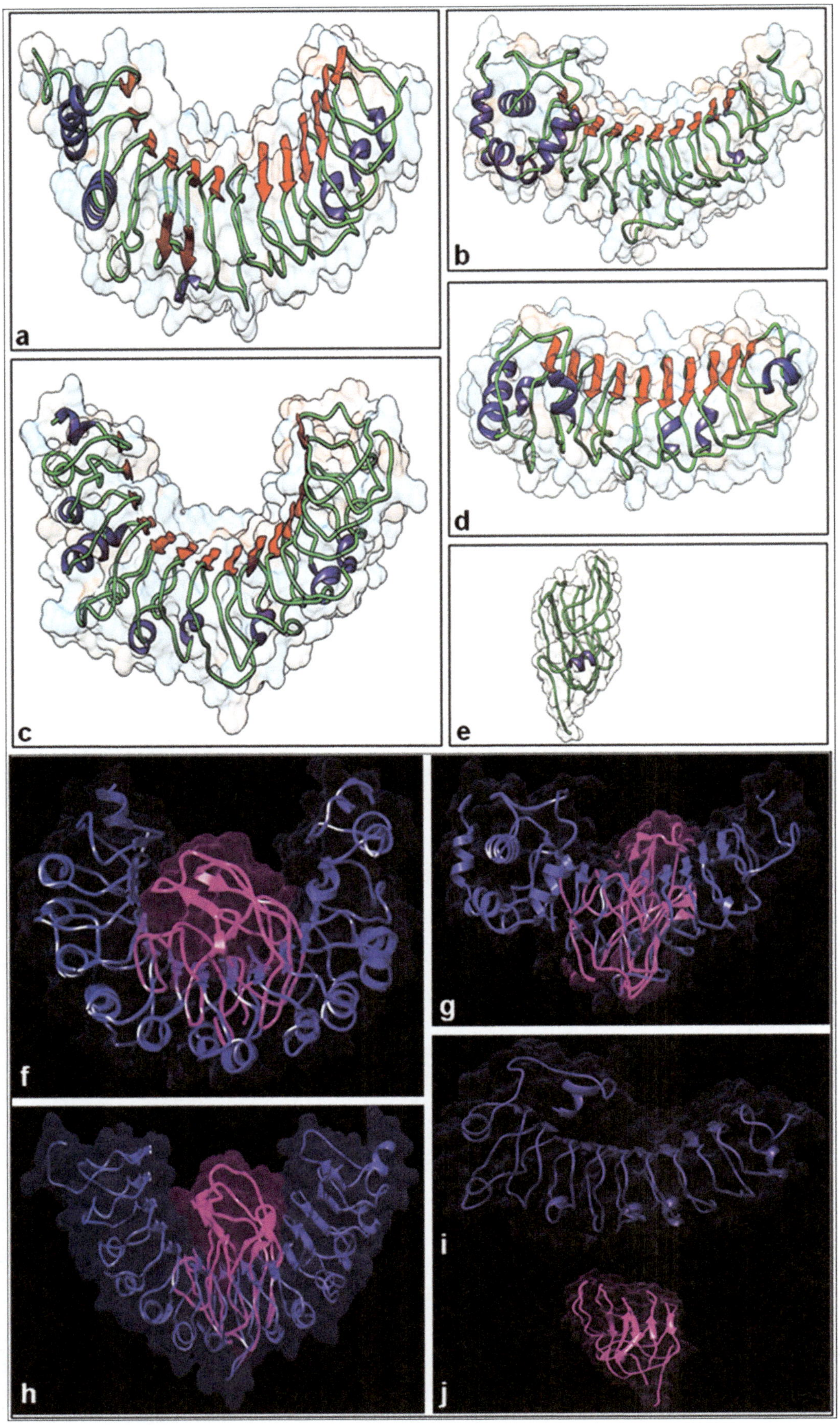
a
b
c
d
e
f
g
h
i
j

Figure 4. Molecular docking of Pi54 proteins. Predicted tertiary structures of Pi54 orthologue protein Pi54 (a); Pi54rh (b); Pi54of (c); Pi54tp (d) and AVR-Pi54 protein (e) generated by *ab initio* modeling using I-TASSER server. Various secondary structure; Alpha helices (blue), β-sheets (red), Coils (green) have been depicted. Docked AVR-Pi54 with protein Pi54 (f); Pi54rh (g); Pi54of (h) and Pi54tp (i) using Z-DOCK server in Discovery studio 3.5. In the interaction, AVR protein is depicted in pink color and R-protein in blue color.

characterized. Owing to high variability in *M. oryzae* pathogen population the average effective life of major rice blast resistance genes is reported to be very short. In view of this, continued identification and characterization of novel genes and its alleles is important for their use in stacking of multiple *R*-genes or transferring broad spectrum disease resistance genes to the cultivated varieties [37]. These cloned *R*-genes have untapped novel alleles in other rice backgrounds, providing useful source for efficient management of the rice blast disease. Some of the alleles of cloned *R*-genes have been reported to provide much stronger and broader resistance against rice blast disease [11,18]. The rice blast resistance gene *Pi54* provided broad spectrum resistance to *M. oryzae* population prevalent in different parts of India [22]. Allele mining for *Pi54* gene has resulted in the isolation and cloning of few better performing alleles and orthologues from various rice lines and wild rice species, respectively [18].

In the present investigation, an orthologue of *Pi54* designated as *Pi54of* (of-*officinalis*) was cloned from the wild rice species *O. officinalis*, characterized and functionally validated. The expression of *Pi54of* in the broad spectrum blast-resistant wild species of rice, *O.officinalis* pre-and-post challenge with *M. oryzae* found to be at the basal level before infection, but it showed highly induced expression at 72 hpi. This is similar to the expression pattern of *Pi54* gene, which is reported to be induced only upon pathogen challenge [21] and the expression of *Pi54rh* gene that was reported to be highly upregulated at 72 hpi [18]. The other blast resistance gene *Pib*, showed induction upon altered environmental conditions besides pathogen inoculation [36]. Similar upregulation was observed for rice bacterial blight resistance gene *Xa1* which showed induction upon pathogen challenge and wounding [38]. The late upregulation of the *Pi54of* gene might be attributed to its role in limiting the spread of pathogen to the cells adjacent to the site of infection and also secondary spread of fungal biomass at later stages [39] or by fortifying the membrane and plasmodesmata [40]. The rice blast resistance gene *Pi54* has also been reported to trigger the upregulation of five defence response genes (callose, chitinases, laccase, PAL and peroxidise) and defence related transcription factors (NAC6, Dof Zinc Finger, MAD Box, bZIP and WRKY) which results in blast resistance phenotype in rice [41].

The genetic variability of *Pi54of* at both the nucleotide and amino acid levels indicated its divergent nature from the *Pi54* orthologue. Among these orthologues, *Pi54of* is the largest and its single exon coding for 379 amino acids. The analysis of promoter region of the *Pi54of, Pi54* and *Pi54rh* shows the presence of CAAT Box in *Pi54of* and *Pi54rh* orthologues and TATA Box in *Pi54* gene. Besides, *Pi54of* has three WRKY71OS boxes, whereas *Pi54* and *Pi54rh* have one WRKY71OS box each. This cis-acting region is found in the Pathogenesis-Related Class10 (PR-10) genes [42]. The WRKY boxes are also identified in the promoter regions of the defence response genes like *NPR1* gene of *Arabidopsis*, Class I chitinase gene, *CHN50* in tobacco and parsely *PR-1* gene [18]. In case of rice, it has been reported that WRKY transcription factor acts as an activator of defence related genes after inoculation with *M. oryzae* [43] and rice OsWRKY76 expresses within 48 h after inoculation with *M. oryzae* [44]. BELL homeodomain (BH) box predicted in the *Pi54of* is specialized nucleotide sequences reported to be involved in the upregulation of defence response genes and is associated with the blast resistance response in rice [45]. The polyA tail predicted for *Pi54of* is longest of the three *Pi54* genes at 910 bp downstream to stop codon, which is important for nuclear export, stability and translation of mRNA.

The amino acid sequence analysis of Pi54of protein showed the presence of a CC domain, a Zinc finger domain, two LRR domains, apart from N-glycosylation, myristilation and phosphorylation sites. However no NBS domain was predicted in *Pi54of* protein. The previous studies involving *Pi54* and *Pi54rh* however predict a rudimentary NBS domain in both the orthologues using homology based approach and ExPASy tools, respectively [21,18]. Presence of a phosphorylation site suggests that the protein of this allele has important role in signal transduction and transmittance to the downstream genes. The CC domains predominant in *R*-genes from monocotyledons are the mediators in the recognition and activation of *R*-genes [46]. The LRR domain is the major determinant of recognition specificity of the pathogen avirulence factors. The leucine rich repeats are proven to be the mediators of interaction with Avr-proteins, in a typical protein-protein interaction leading to host resistance [47]. The Zinc finger domain found is unique among the other rice blast resistance gene and might play an important role in broad-spectrum resistance nature of *Pi54of* gene [48].

The evolutionary analysis performed in this study showed the structural relatedness and phylogenetic relationships among cloned *R*-genes. In the present study, a phylogenetic tree was constructed using cloned rice blast *R*-genes including *Pi54of*. The tree depicts the divergent origin of *R*-genes suggesting their

Table 1. Comparison of various energy values and docking features for each of the best of modelled complex of Pi54 orthologues and AVR-Pi54 using ZDOCKER in DS3.5.

Name	Forcefield	Potential Energy (kcal/mol)	Van der Waals Energy (kcal/mol)	Electrostatic Energy (kcal/mol)	Total Energy	Binding energy with AVR-PI54 (kcal/mol)
Pi54	CHARMm	−22362.03	−2303.26	−23069.97	−47735.26	**−2265.89**
Pi54rh	CHARMm	−22904.48	−2426.16	−23654.45	−48985.13	−1883.53
Pi54of	CHARMm	**−23888.54**	**−2579.13**	**−24675.68**	**−51143.35**	−1942.56
AVR-Pi54	CHARMm	−7388.96	−778.53	−7751.92	−15919.42	-

Binding energy = energy of complex − [energy (receptor) + energy (ligand)].

Table 2. Docking features (hydrogen bond forming residues with bond length <4Å) of modelled Pi54 rthologues and AVR-Pi54.

Gene	Amino acid residues involved in bond formation		
	Coiled Coil	**non-LRR**	**LRR**
Pi54	T-24, D-26	F-138, Y-139, G-140, H-141, N-142, G-143, F-145,, S-147, L-148, E-149, K-150, L-151, Y-155, M-156, T-157	R-246, K-248 and E-251
Pi54rh	–	F-145, Y-146, T-151, F-152, E-156, K-157, L-158, S-161, Y-162, M-163, T-164, S-165	G-253, L-254, L-256, L-257, E-258, I-259, C-260, C-262,S-276, S-277, K-282
Pi54of	–	F-191, S-193, L-194, E-195, I-198, L-199, S-200, Y-201 and M-202	–

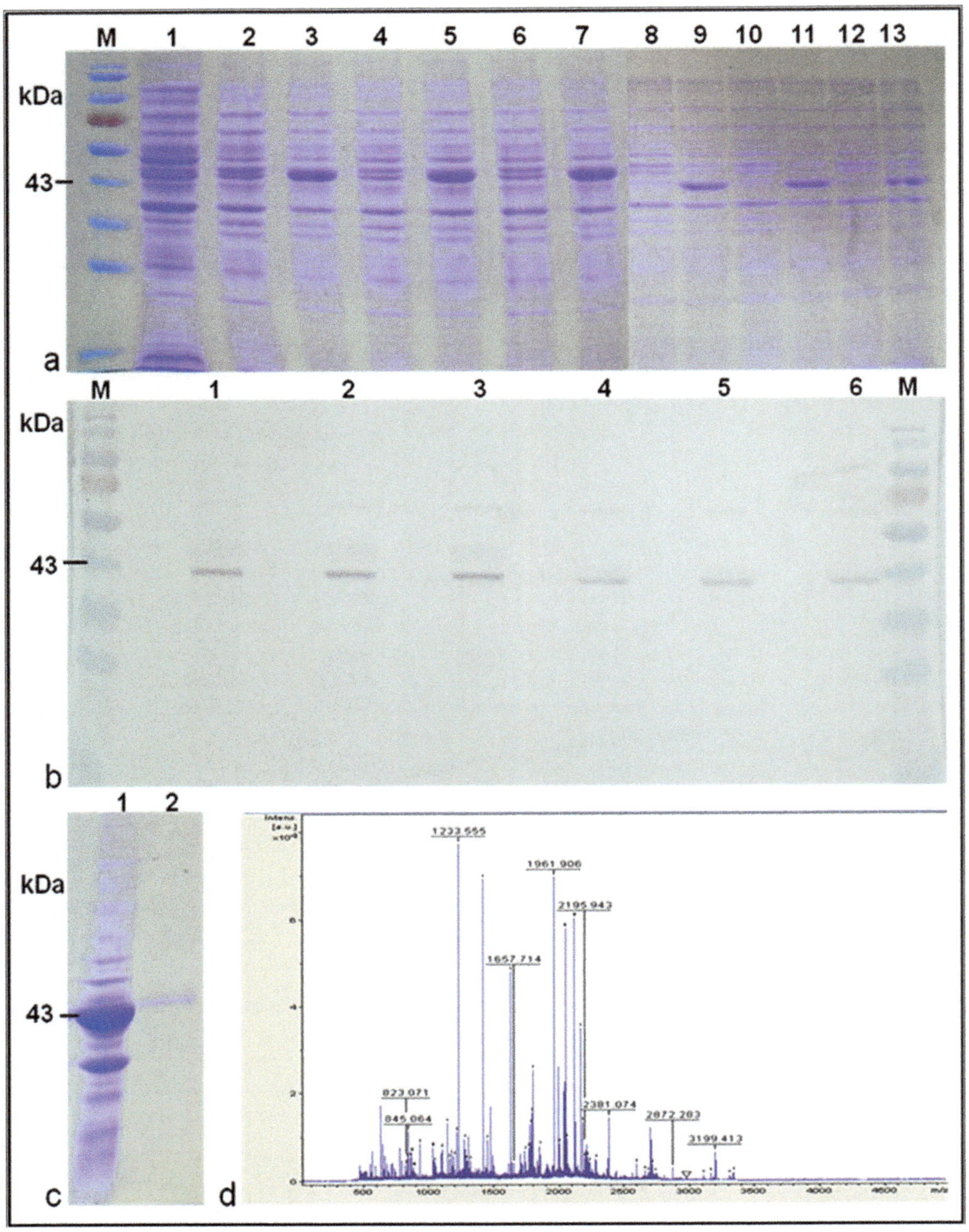

Figure 5. *In vitro* expression and characterization of Pi54of protein. a: SDS-PAGE analysis of IPTG induced BL21 clones. Lane 2–7: *Pi54of* clones; Lane 8–13: *Pi54* clones. b: Western blot of *Pi54of* (1–3) and *Pi54* (4–6) protein using polyclonal Ab developed for Pi54 protein. c: Purified Pi54of protein. d: MS mass spectra MALDI-TOF analyzed Pi54of protein; ID 1233.555 showed match with Pi54 protein.

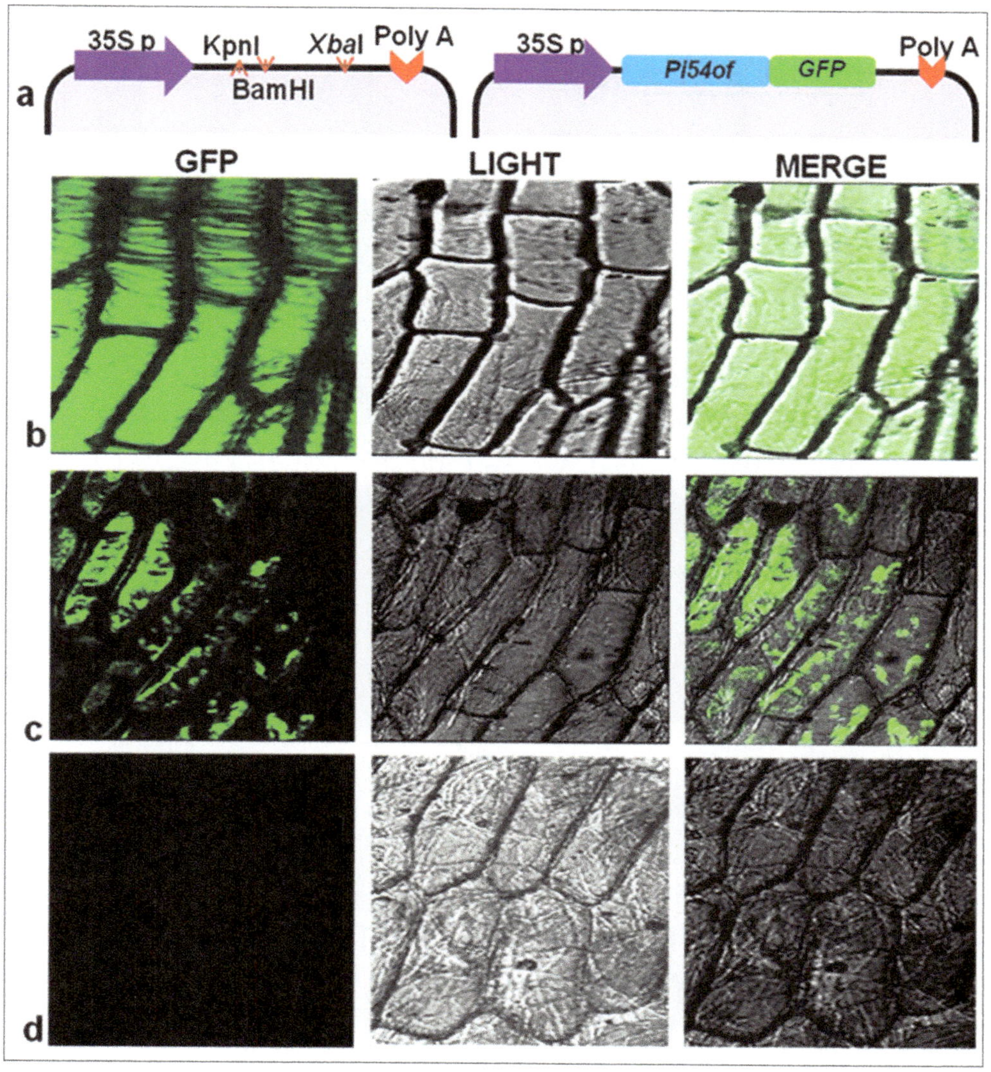

Figure 6. Sub-cellular localization analysis of Pi54of protein. Panel a: Schematic representation of construct used. Panel b: GFP alone under the control of *CaMV 35S* promoter; Panel c: GFP-Pi54of fusion protein under the control of *CaMV 35S*; Panel d- non bombarded onion cells.

evolutionary advantage for recognizing, binding and defending against a broad array of *M. oryzae* isolates [49]. The definitive clustering in the tree showed the common ancestral origin of these *R-* genes which have evolved by duplications and subsequent diversification leading to wide variation in their structure and specificity [50]. The clustering of *Pi54of* with *Pi54* and *Pi54rh* proves the orhologues nature of these three genes and clustering of *Pi54of* with other NBS-LRR type of blast resistance genes in sub-cluster III reveals that *Pi54of* is closely related to the NBS-LRR family of *R*-genes.

In vitro expression and characterization of the protein product of target gene is the first step in unravelling the molecular mechanism played by the corresponding gene in the living system. In rice, disease resistance related genes like *Pita-LRD, OsBIHD1, OsCDR1, OsPR-10, OsBBI1, Pb1* and *RGA5* [51,45,52,53,54,16;19] have been expressed *in vitro* and characterized. In the present study, *in vitro* expressed Pi54of protein was used for western blot hybridization and MALDI-TOF-MAS spectrometry. The above results confirmed that *Pi54of* is an orthologue of cloned and characterized *Pi54* gene.

Subcellular localization of R-proteins is an important factor which determines the site of interaction of R-proteins with the pathogen effector proteins in a typical gene-for-gene manner. Resistance proteins are localized to the cytoplasm, nucleus or both. Cytoplasmic localization of R-proteins enhances signaling mechanism related to HR response. Whereas, nuclear localization was the part of complete resistance reaction system [55]. In many cases, the R proteins are found to accumulate in the nucleus in response to the pathogen infection [56]. It has been reported that the nuclear localization is required for resistance response of R-proteins like, barley mildew A (*MLA*) 10, tobacco N, *Arabidopsis* RPS4 and RPS6 and rice blast resistance Pb1 [57,19]. The rice blast resistance protein Pb1 is found in both cytoplasm and nucleus, but nuclear localization of Pb1 was proved to be essential for blast resistance. In the present investigation, Pi54of protein was localized to the cytoplasm which is a subcellular localization site for most of the reported R-proteins. The protein product of the rice blast resistance gene *Pi37* which forms cluster with *Pi54of* was also found to be localized to the cytoplasmic region [58], whereas *Pi-d2* and *Pi54rh* showed their localization to the plasma membrane [18].

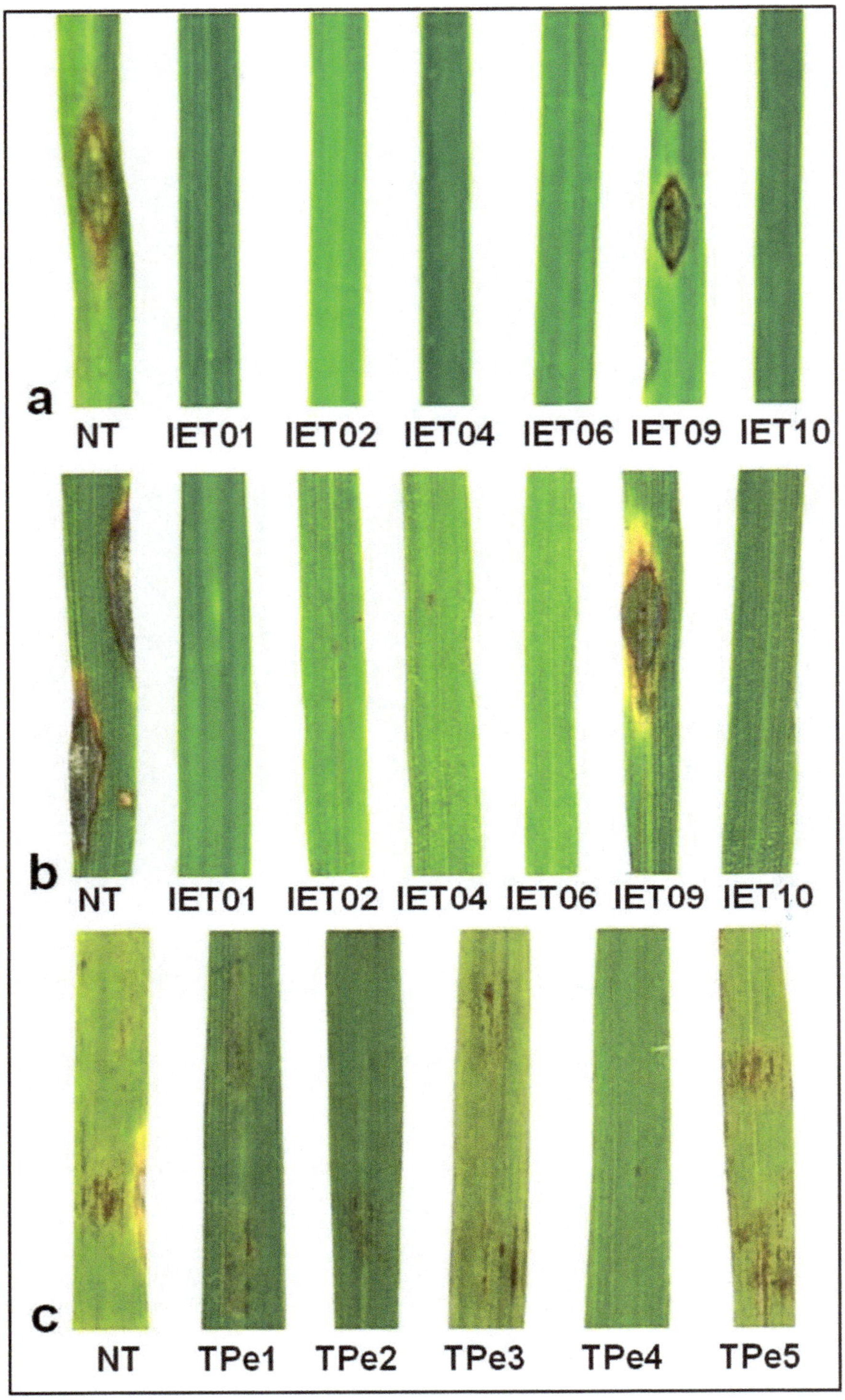

Figure 7. Genetic complementation of transgenic plants. Transgenic IET16310 plants challenged with *M. oryzae* stain RML21 (a) and 6-Sikkim (b). Detached leaves of transgenic TP309 plants challenged with *M. oryzae* (6-Sikkim) (c). NT (non transgenic) plants were used as control in all the experiments.

In this study, *in silico* comparative analysis of resistant Pi54, Pi54rh and Pi54of proteins along with susceptible protein Pi54tp was performed. Multiple sequence alignment of the four Pi54 proteins revealed the deletion of a large middle fragment between CC and LRR domain in *Pi54tp* which might be contributing to the susceptible nature of the rice line TP309 to *M. oryzae*. The amino acid composition analysis of the four R-proteins as well as their LRR domains revealed a high frequency of amino acids L, E and S. Leucine belongs to the aliphatic amino acids which would contribute to the hydrophobic interaction and to maintain the conformational stability in the interiors of proteins [59]. Hence, presence of higher L in Pi54of protein helps in maintaining 3D conformation and its stability. The charged amino acid, like E contribute to the electrostatic interaction, which is an important force for maintaining structural conformation and its stability in the outer part of the proteins [60]. The presence of this residue in the Pi54of protein makes it more stable in various stress conditions. The conformational stability of the LRR domain is important for interaction of R-protein with Avr proteins of the pathogen *M. oryzae* [5]. The aliphatic index which is >104 for these Pi54 orthologues further support the hypothesis that they are thermo-stable in nature. The other amino acid S, present in these proteins is the best known residue interacting with the water molecules surrounding the protein due to its hydrophilic nature [61]. Presence of higher S residues may help proteins to interact with the surface water molecules through hydrogen bond formation and helps for the stability of proteins at interacting surfaces [62]. Higher proportion of hydrophilic amino acids may facilitate the interaction of a protein with the water molecules and can form hydrogen bonds [63]. The physico-chemical analysis in the present study showed that resistant Pi54 proteins were comparatively more hydrophilic in nature than the susceptible Pi54tp protein. Hence, the interaction of the resistant Pi54 proteins with water molecules might be much better than the susceptible Pi54tp protein.

The plant NBS-LRR proteins form typical horseshoe like configuration *in vivo*. The crystal structures of more than 20 characterized LRR proteins have revealed that LRR domains typically contain a series of β- sheets, which form the inner surface of horseshoe like configuration [64]. *In silico* 3D conformation prediction Pi54, Pi54rh, Pi54of and Pi54tp proteins display typical horseshoe like structures. The parallel β-sheets formed the concave surface of the horseshoe structure, whereas the α-helices line the convex plain of these proteins. The authenticity of 3D structures was confirmed by Ramachandran plot analysis. It is known that, a good homology model should have more than 90% of its amino acid residues in the favourable region [65]. The Pi54 proteins of all the orthologues examined in this study have >98% of residues in the most allowed and allowed regions, while AVR-Pi54 proteins has 89.4% residues in the most allowed and allowed regions. Ordered regions are often termed globular and typically contain regular secondary structures packed into a compact globule [66] whereas the disordered regions are linear stretches of amino acids. Our GlobPlot-2 analysis showed the presence of three disorder regions at the N-terminal of Pi54rh and may be contributing to distorted horseshoe conformation of Pi54rh protein at the N-terminal end, which may be contributing to its higher global free energy and also comparatively more binding energy with AVR-PI54 protein.

AVR proteins are perceived by R-proteins through direct or indirect interactions. Direct protein-protein interaction has been reported for nine R-proteins with cognate effector molecules [16]. In case of rice-*Magnaporthe* pathosystem, experimental findings showed that recognition of AVR-Pita, AVR-Pia, AVR-Pik/km/kp, and AVR-CO39 occurs inside the host cells by their corresponding cytoplasmic R-proteins [16]. The molecular mechanism of interaction is reported for *AVR-Pita*, *AVR-Pik* and *AVR-CO39*. AVR-Pita interacts with C-terminal LRD domain of Pi-ta whereas, AVR-Pik interacts through the N-terminal domain including the CC domain of Pik-1, and sequences upstream of the NBS domain. The AVR-CO39 protein interacts with a novel non-LRR domain of RGA5A. Hence, the above examples demonstrate the cases of direct recognition wherein it seems to associate different R-protein domains with different resistance mechanisms. The molecular docking results of this study also revealed that the Pi54of interacted with AVR-Pi54 only through its novel non-LRR region corresponding to STI1 and GEF domains, whereas Pi54 and Pi54rh interacted through CC-non-LRR-LRR regions and non-LRR-LRR regions, respectively. The susceptible Pi54tp protein did not show interaction with AVR-Pi54. This differential preference of AVR protein for binding domains among the Pi54 proteins might be attributed to different resistance mechanisms against *M. oryzae*. The docking results of present study correlate well with the phenotypic reaction of all the rice genotypes containing Pi54 proteins with *M. oryzae* [21,18]. However, proteins coded by late blight resistance *RB* gene of potato and blast resistance genes *Pik* and RGA5A from rice showed that LRR regions need not be the site of interaction with AVR proteins [16].

In the present study two domains involved in the plant defensome complex were predicted. The one being STI1was predicted in Pi54of and the other one was Guanine Nucleotide Exchange Factor (GEF) found in both Pi54of and Pi54rh proteins. STI1 domain is a cochaperone STI1 binding domain. STI1 is a homologue of human cognate protein 70 (hsc70)/heat shock protein 90 (hsp90)-organising protein (Hop) and is a cochaperone for HSP70 and HSP90 [67]. There are two Sti1 homologues in the rice genome. The RhoGEF is also called Dbl-homologous (DH) domain. The presence of STI1 and RhoGEF domains together in *Pi54of* is of considerable significance as these two domains are part of the plant defensome complex [68]. The OsRac1 which is bound to Sti1 is initially in a GDP bound inactive state. The activation of OsRac1 needs a GEF molecule, which binds to RhoGEF domain. The pathogen derived PAMPs such as chitin, induce the activation of GEF and the functional GEF activates OsRac1. Further the receptor of PAMPs, a chitin receptor OsCERK1 transfer the signal to downstream target through Hop/Sti1a and Hsp90 complex which interact with OsCERK1 through its transmembrane domain [68]. The active OsRac1 then recruits RACK1A and RACK1A further interacts with N-terminus of RAR1 and SGT1 and also regulates ROS production by interacting directly with the N-terminus of Rboh, an NADPH oxidase [69]. Furthermore, the activation of rice MAPK6 requires OsRac1 for PAMP-induced regulation and it forms a complex with MAPK6 in rice cell extracts [70]. Considering the importance of stability and spatial arrangements of the components of plant defensome complex, Sti1 found in *Pi54of* might be stabilizing the interaction of kinases and HSP90 in MAPK cascade. The rice blast resistance gene *Pia* pathway was found to be affected by loss of Sti1 [67].

The genetic complementation study, which was performed to understand the functional role of *Pi54of* gene showed that the two susceptible rice lines IET16310 (indica) and TP309 (japonica) transformed with *Pi54of* gene could displays high level of resistance to highly virulent strains of *M. oryzae*. This analysis showed that *Pi54of* is a functional orthologue. The resistance spectrum comparison of independent transgenic TP309 plants harbouring *Pi54* and its two orthologues revealed that three orthologues have different resistance response to *M. oryzae* isolate MG-79 which might be due to their differential interaction with *AVR-Pi54* [18]. The functional Cleaved Amplified Polymorphic Site (CAPS) marker developed in this study, that differentiates the transgenic TP309 plants from non-transgenic lines as well as transgene integrated homozygous plants from heterozygous plants can be used for molecular breeding to analyze *Pi54of* integration and to differentiate homozygous and heterozygous TP309 plants. These findings showed that the cloned *Pi54of* gene provides resistance against three highly virulent strains of *M. oryzae* to two rice lines belonging to both indica and japonica genetic background.

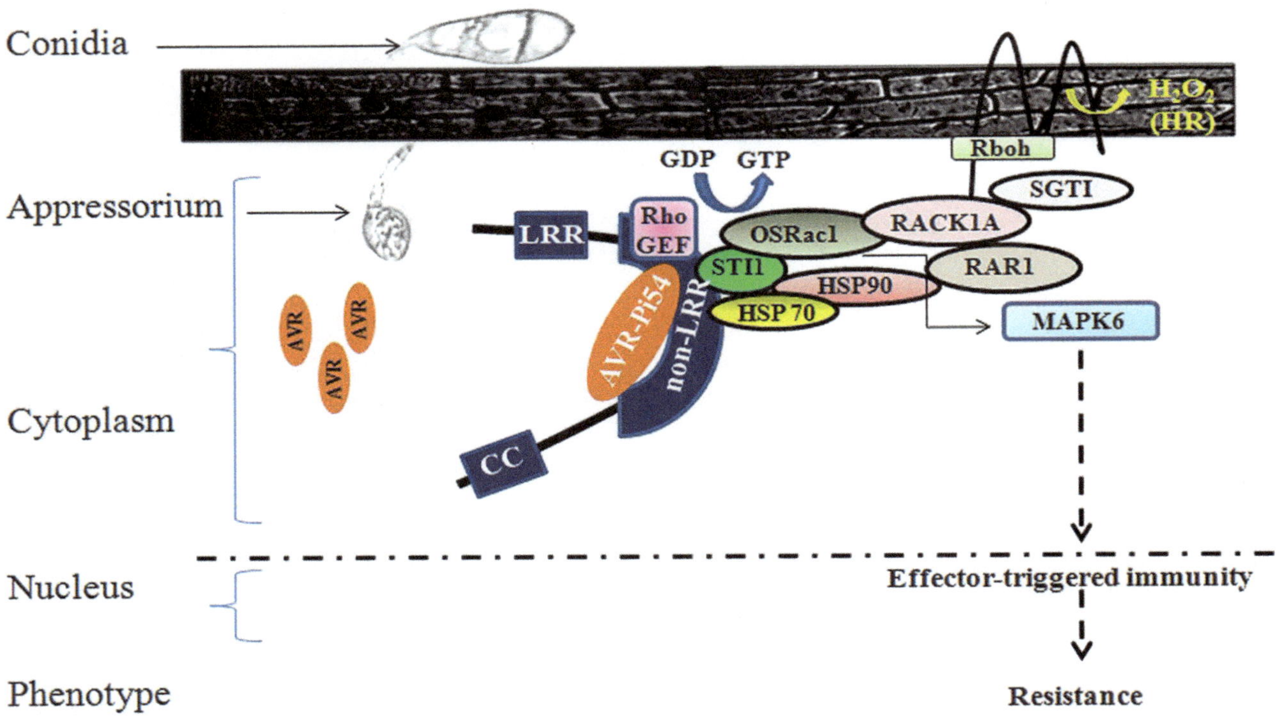

Figure 8. Possible molecular mechanism of *Pi54of* gene. Avr-Pi54 binds to Pi54of protein (blue) at the non-LRR region upstream to the LRR domain. Pi54of may perceive pathogen signals through STI1 which acts as an anchor for the so called defensome complex involving Os Rac1 (Rac/Rop GTPase), RACK1A (Receptor of Activated C Kinase), RAR (Required for Mla12 Resistance), SGT1 (Suppressor of the G2 allele of skp1), MAPK6 (a rice Mitogen-Activated Protein Kinase), Rboh (NADPH oxidases).

The present study provides a new blast resistance gene *Pi54of*, cloned and functionally characterized from wild rice species *O. officinalis*. This gene product is targeted to cytoplasm and is the longest among all the functionally validated *Pi54* genes. The resistance mechanism of the *Pi54of* might be different from other cloned *Pi54* orthologues due to its differential binding pattern with *AVR-Pi54*. It was also found that Pi54of protein interact with the AVR protein through its novel non-LRR region and might be involved in rice innate immunity by forming a so called defensome complex (Figure 8). This study is the first attempt in unravelling the actual molecular mechanism of *Pi54of* during its expression against virulent isolate of *M. oryzae* in two rice lines belonging to indica and japonica genetic background.

Supporting Information

Figure S1 Cloning of full length Pi54of orthologue from *O. officinalis*.

Figure S2 Composition of hydrophobic and hydrophilic amino acid residues of proteins of Pi54 orthologues.

Figure S3 Ramachandran plot validation of 3D structures of Pi54 orthologue proteins and AVR-Pi54 protein.

Figure S4 Construction of plant expression vector for sub-cellular localization of Pi54-of protein.

Figure S5 Genetic transformation of indica rice line IET16310 and molecular confirmation.

Figure S6 Genetic transformation of japonica rice line Taipei 309 and molecular confirmation.

Figure S7 Phenotypic reaction of *Pi54* orthologues in the TP309 genetic background challenged with *M. oryzae* isolate MG-79.

Acknowledgments

T R Sharma thanks all who provided seed material and fungal culture. We thank the in charge National Phytotron Facility at IARI, New Delhi for providing facilities to maintain wild species of rice.

Author Contributions

Conceived and designed the experiments: TRS. Performed the experiments: NBD. Analyzed the data: NBD JV. Contributed reagents/materials/analysis tools: TRS. Contributed to the writing of the manuscript: TRS NBD.

References

1. Padmanabhan SY (1965) Estimating losses from rice blast in India. In: The rice blast diseases. Johns Hopkins Press, Baltimore, Maryland, 203–221.
2. Wang JC, Wen JW, Liu WP, Yan SR, Wang JQ, et al. (2010) Interaction studies between rice and Pyricularia grisea in Jilin Province, P. R. China. In: Jia, Y. (Ed.), Proc. of the 5th Intern. Rice Blast Conf., USA. USDA-DBNRRC, 90.
3. Lee FN (1994) Rice breeding programs, blast epidemics and blast management in the United States. In: Zeigler, R.S., Leong, S.A., Teng, P.S. (Eds.), Rice Blast Disease. Oxon, CAB International, UK, 489e500.
4. Dai Y, Jia Y, Correll JC, Wang X, Wang Y (2010) Diversification and evolution of the avirulence gene *ARV-Pita1* in field isolates of *Magnaporthe oryzae*. Fungal Genet Biol 47: 973–80.
5. Flor HH (1971) Current status of the gene-for-gene concept. Annu Rev Phytopathol 9: 275–296.
6. Chen XW, Shang JJ, Chen DX, Lei CL, Zou Y (2006) A β-lectin receptor kinase gene conferring rice blast resistance. Plant J 46: 794–804.
7. Brar DS, Khush GS (1997) Alien introgression in rice. Plant Mol Biol 35: 35–47.
8. Sharma TR, Rai AK, Gupta SK, Vijayan J, Devanna BN, et al. (2012) Rice blast management through host-plant resistance: retrospect and prospects. Agric Res 1: 37–52.
9. Kumari A, Das A, Devanna BN, Thakur S, Singh PK, et al. (2013) Mining of rice blast resistance gene *Pi54* shows effect of single nucleotide polymorphisms on phenotypic expression of the alleles. Eur J Plant Pathol 137: 55–65.
10. Lv Q, Xu X, Shang J, Jiang G, Pang Z, et al. (2013) Functional Analysis of Pid3-A4, an Ortholog of Rice Blast Resistance Gene Pid3 Revealed by Allele Mining in Common Wild Rice. Phytopathol 103: 594–599.
11. Yang MZ, Cheng ZQ, Chen SN, Qian J, Xu LL, et al. (2007) A rice blast-resistance genetic resource from wild rice in Yunnan, China. Zhi Wu Sheng Li Yu Fen Zi Sheng Wu Xue Xue Bao 33: 589–595.
12. Dangl JL, Jones JD (2001) Plant pathogens and integrated defence responses to infection. Nature 411: 826–833.
13. Abramovitch RB, Anderson JC, Martin GB (2006) Bacterial elicitation and evasion of plant innate immunity. Nat Rev Mol Cell Biol 7: 601–611.
14. Jones JDG, Dangl JL (2006) The plant immune system. Nature 444: 323–329.
15. Hammond-Kosack KE, Jones JDG (1996) Resistance gene–dependent plant defense responses. Plant Cell 8: 1773–1791.
16. Cesari S, Thilliez G, Ribot C, Chalvon V, Michel C, et al. (2013) The rice resistance protein pair RGA4/RGA5 recognizes the *Magnaporthe oryzae* effectors *AVR-Pia* and *AVR1-CO39* by direct binding. Plant Cell 25: 1463–1481.
17. Silue D, Notteghem JL, Tharreau D (1992) Evidence for a gene-for-gene relationship in the Oryza sativa-Magnaporthe grisea pathosystem. Phytopathol 82: 577–582.
18. Das A, Soubam D, Singh PK, Thakur S, Singh NK, et al. (2012) A novel blast resistance gene, Pi54rh cloned from wild species of rice, Oryza rhizomatis confers broad spectrum resistance to Magnaporthe oryzae. Funct Integr Genomics 12: 215–228.
19. Jia Y, Martin R (2008) Identification of a new locus, Ptr(t), required for rice blast resistance gene Pi-ta-mediated resistance. Mol Plant Microbe Interact 21(4): 396–403.
20. Inoue H, Hayashi N, Matsushita A, Xinqiong L, Nakayama A, et al. (2013) Blast resistance of CC-NB-LRR protein Pb1 is mediated by WRKY45 through protein–protein interaction. Proc Natl Acad Sci USA 110: 9577–82.
21. Kato M, Kezuka Y, Kobayashi C, Ishibashi K, Nonaka T, et al. (2013) Crystallization and preliminary X-ray crystallographic analysis of the inhibitory domain of the tomato mosaic virus resistance protein Tm-1. Acta Cryst F69: 1411–1414.
22. Sharma TR, Madhav MS, Singh BK, Shanker P, Jana TK, et al. (2005) High resolution mapping, cloning and molecular characterization of the *Pi-kh* gene of rice, which confers resistance to *Magnaporthe grisea*. Mol Genet Genomics 274: 569–578.
23. Rai AK, Kumar SP, Gupta SK, Gautam N, Singh NK, et al. (2011) Functional complementation of rice blast resistance gene *Pi-kh* (*Pi54*) conferring resistance to diverse strains of *Magnaporthe oryzae*. J Plant Biochem Biot 20: 55–65.
24. Sharma TR, Chauhan RS, Singh BM, Paul R, Sagar V, et al. (2002) RAPD and pathotype analyses of Magnaporthe grisea populations from the north-western Himalayan Region of India. J Phytopath 50: 649–656.
25. Bonman JM, Vergel F, Khin MM (1986) Physiological specialization of Pyricularia oryzae in the Philippines. Annu Rev Phytopathol 30: 508–520.
26. Ochman H, Gerber AS, Hartl DL (1988) Genetic applications of an inverse polymerase chain reaction. Genetics 120: 621–625.
27. Tamura K, Stecher G, Peterson D, Filipski A, Kumar S (2013) MEGA6: Molecular Evolutionary Genetics Analysis Version 6.0. Mol Biol Evol 30: 2725–2729.
28. Lupas A, Van Dyke M, Stock J (1991) Predicting coiled coils from protein sequences. Science 252 (5010): 1162–1164.
29. Horton P, Park KJ, Obayashi T, Fujita N, Harada H, et al. (2007) WoLF PSORT: protein localization predictor. Nucleic Acids Res 35: W585–W587.
30. Smith PK, Krohn RI, Hermanson GT, Mallia AK, Gartner FH, et al. (1985) Measurement of protein using bicinchoninic acid. Anal Biochem 150: 76–85.
31. Sambrook J, Rusell DW (2007) Molecular Cloning: A Laboratory Manual, 3rd ed. Cold Spring Harbor Laboratory Press, Cold Spring Harbor, New York, USA.
32. Condina MR, Guthridge MA, McColl SR, Hoffmann P (2009) A sensitive magnetic bead method for the detection and identification of tyrosine phosphorylation in proteins by MALDI-TOF/TOF MS. Proteomics 9: 3047–3057.
33. Brooks BR, Bruccoleri RE, Olafson BD, States DJ, Swaminathan S, et al. (1983) J Comput Chem 4: 187–217.
34. Hiei Y, Ohta S, Komari T, Kumashiro T (1994) Efficient transformation of rice (*Oryza sativa* L.) mediated by *Agrobacterium* and sequence analysis of the boundaries of T-DNA. Plant J 6: 271–282.
35. Jia Y, Valent B, Lee FN (2003) Determination of host responses to Magnaporthe grisea on detached rice leaves using a spot inoculation method. Plant Dis 87: 129–133.
36. Wang ZX, Yano M, Yamanouchi U, Iwamoto M, Monna L, et al. (1999) The Pib gene for rice blast resistance belongs to the nucleotide binding and leucine rich repeat class of plant disease resistance genes. Plant J19: 55–64.
37. Liu X, Lin F, Wang L, Pan Q (2007) The *in Silico* Map-Based cloning of *Pi36*, a rice Coiled-Coil-Nucleotide-Binding Site-Leucine-Rich repeat gene that confers race-specific resistance to the blast fungus. Genetics 176: 2541–2549.
38. Yoshimura S, Yamanouchi U, Katayose Y, Toki S, Wang ZX, et al. (1998) Expression of Xa1, a bacterial blight resistance gene in rice, is induced by bacterial inoculation. Proc Natl Acad Sci USA 95: 1663–1668.
39. Parker D, Beckmann M, Enot DP, Overy DP, Rios ZC, et al. (2008) Rice blast infection of Brachypodium *distachyon* as a model system to study dynamic host/pathogen interactions. Nature Protoc 3: 435–445.
40. Kankanala P, Czymmek K, Valent B (2007) Roles of rice membrane dynamics and plasmodesmata during biotrophic invasion by the blast fungus. Plant Cell 19: 706–724.
41. Gupta SK, Rai AK, Kanwar SS, Chand D, Singh NK, et al. (2012) The single functional blast resistance gene Pi54 activates a complex defence mechanism in rice. J Exp Bot 63(2): 757–72.
42. Eulgem T, Rushton PJ, Schmelzer E, Hahlbrock K, Somssich IE (1999) Early nuclear events in plant defence signalling: rapid gene activation by WRKY transcription factors. EMBO J 18: 4689–4699.
43. Chujo T, Takai R, Akimoto-Tomiyama C, Ando S, Minami E, et al. (2007) Involvement of the elicitor-induced gene OsWRKY53 in the expression of defense-related genes in rice. Biochim Biophys Acta 1769: 497–505.
44. Yokotani N, Sato Y, Tanabe S, Chujo T, Shimizu T, et al. (2013) WRKY76 is a rice transcriptional repressor playing opposite roles in blast disease resistance and cold stress tolerance. J Exp Bot 64 (16): 5085–97.
45. Luo H, Song F, Goodman RM, Zheng Z (2005) Up-regulation of OsBIHD1, a rice gene encoding BELL homeodomain transcriptional factor, in disease resistance responses. Plant Biol 7(5): 459–68.
46. Hwang C, Williamson VM (2003) Leucine-rich-repeat mediated intramolecular interactions in nematode recognition and cell death signalling by the tomato resistance protein Mi. Plant J 34: 585–593.
47. Marone D, Russo MA, Laido G, De Leonardis AM, Mastrangelo AM (2013) Plant Nucleotide Binding Site-Leucine-Rich Repeat (NBS-LRR) Genes: Active Guardians in Host Defense Responses. Int J Mol Sci 14(4): 7302–26.
48. Gupta SK, Rai AK, Kanwar SS, Sharma TR (2012) Comparative Analysis of Zinc Finger Proteins Involved in Plant Disease Resistance. PLoS ONE 7(8): e42578. doi:10.1371/journal.pone.0042578.
49. Lehman P (2002) Structure and evolution of plant disease resistance genes. J Appl Genet 43: 403–413.
50. Martin GB, Bogdanove AJ, Sessa G (2003) Understanding the functions of plant disease resistance proteins. Annu Rev Plant Biol 54: 23–61.
51. Jia Y, McAdams SA, Bryan GT, Hershey HP, Valent B (2000) Direct interaction of resistance gene and avirulence gene products confers rice blast resistance. EMBO J 19: 4004–4014.
52. Prasad BD, Creissen G, Lamb C, Chattoo BB (2010) Heterologous expression and characterization of recombinant OsCDR1, a rice aspartic proteinase involved in disease resistance. Protein Expr Purif 72(2): 169–74.
53. Kim ST, Kang YH, Wang Y, Wu J, Park ZY, et al. (2009) Secretome analysis of differentially induced proteins in rice suspension-cultured cells triggered by rice blast fungus and elicitor. Proteomics 9(5): 1302–13.
54. Li W, Zhong S, Li G, Li Q, Mao B, et al. (2011) Rice RING protein OsBBI1 with E3 ligase activity confers broad spectrum resistance against *Magnaporthe oryzae* by modifying the cell wall defence. Cell Res 21: 835–48.
55. Bai S, Liu J, Chang C, Zhang L, Maekawa T, et al. (2012) Structure-function analysis of barley NLR immune receptor MLA10 reveals its cell compartment specific activity in cell death and disease resistance. PLoS Pathog doi: 10.1371/journal.
56. Meier I, Somers DE (2011) Regulation of nucleocytoplasmic trafficking in plants. Curr Opin Plant Biol 14: 538–546.
57. Bhattacharjee S, Garner CM, Gassmann W (2013) Newclues in the nucleus: transcriptional reprogramming in effector-triggered immunity. Front Plant Sci. doi:10.3389/fpls.2013.00364.
58. Lin F, Chen S, Que Z, Wang L, Liu X, et al. (2007) The Blast Resistance Gene Pi37 Encodes a Nucleotide Binding Site–Leucine-Rich Repeat Protein and Is a

Member of a Resistance Gene Cluster on Rice Chromosome 1. Genetics 177: 1871–1880.
59. Pack SP, Yoo YJ (2004) Protein thermostability: Structure-based difference of amino acid between thermophilic and mesophilic proteins. J Biotechnol 111: 269–277.
60. Dill KA (1990) Dominant forces in protein folding. Biochemistry 29: 7133–7155.
61. Mattos C (2002) Protein-water interactions in a dynamic world. Trends Biochem Sci 27: 203–208.
62. Xu J, Baase WA, Quillin ML, Baldwin EP, Matthews BW (2001) Structural and thermodynamic analysis of the binding of solvent at internal sites in T4 lysozyme. Protein Science 10: 1067–1078.
63. Yu H, Li J, Zhang D, Yang Y, Jiang W, et al. (2008) Improving the thermostability of N-carbamyl-d-amino acid amidohydrolase by error-prone PCR. Appl Microbiol Biotechnol 82: 279–285.
64. Enkhbayar P, Kamiya M, Osaki M, Matsumoto T, Matsumisha M (2003) Structural principles of Leucine Rich Repeat proteins. Proteins 54: 393–403.
65. Dehury B, Sahu M, Patra MC, Sarma K, Sahu J, et al. (2013) Insights into the structure-function relationship of disease resistance protein HCTR in maize (*Zea mays* L.): a computational structural biology approach. J Mol Graph Model 45: 50–64.
66. Wright P, Dyson H (1999) Intrinsically unstructured proteins: re-assessing the protein structure-function paradigm. J Mol Biol 293: 321–331.
67. Thao NP, Chen L, Nakashima A, Hara S, Umemura K, et al. (2007) RAR1 and HSP90 form a complex with Rac/Rop GTPase and function in innate-immune responses in rice. Plant Cell 19: 4035–45.
68. Chen L, Hamada S, Fujiwara M, Zhu T, Thao NP, et al. (2010) The Hop/Sti1-Hsp90 chaperone complex facilitates the maturation and transport of a PAMP receptor in rice innate immunity. Cell Host Microbe 7(3): 185–96.
69. Nakashima A, Chen L, Thao NP, Fujiwara M, Wong HL, et al. (2008) RACK1 functions in rice innate immunity by interacting with the Rac1 immune complex. Plant Cell 20: 2265–2279.
70. Lieberherr D, Thao NP, Nakashima A, Umemura K, Kawasaki T, et al. (2005) A sphingolipid elicitor-inducible mitogen-activated protein kinase is regulated by the small GTPase OsRac1 and heterotrimeric G-protein in rice. Plant Physiol 138(3): 1644–52.

14

Characterization of *MORE AXILLARY GROWTH* Genes in *Populus*

Olaf Czarnecki[1], Jun Yang[1,2], Xiaoping Wang[1,3], Shucai Wang[3], Wellington Muchero[1], Gerald A. Tuskan[1], Jin-Gui Chen[1]*

1 Biosciences Division, Oak Ridge National Laboratory, Oak Ridge, Tennessee, United States of America, **2** National Key Laboratory of Plant Molecular Genetics, Institute of Plant Physiology and Ecology, Shanghai Institutes for Biological Sciences, Chinese Academy of Sciences, Shanghai, China, **3** Key Laboratory of Molecular Epigenetics of MOE, Institute of Genetics and Cytology, Northeast Normal University, Changchun, China

Abstract

Background: Strigolactones are a new class of plant hormones that play a key role in regulating shoot branching. Studies of branching mutants in Arabidopsis, pea, rice and petunia have identified several key genes involved in strigolactone biosynthesis or signaling pathway. In the model plant Arabidopsis, *MORE AXILLARY GROWTH1* (*MAX1*), *MAX2*, *MAX3* and *MAX4* are four founding members of strigolactone pathway genes. However, little is known about the strigolactone pathway genes in the woody perennial plants.

Methodology/Principal Finding: Here we report the identification of MAX homologues in the woody model plant *Populus trichocarpa*. We identified the sequence homologues for each MAX protein in *P. trichocarpa*. Gene expression analysis revealed that *Populus MAX* paralogous genes are differentially expressed across various tissues and organs. Furthermore, we showed that *Populus MAX* genes could complement or partially complement the shoot branching phenotypes of the corresponding Arabidopsis *max* mutants.

Conclusion/Significance: This study provides genetic evidence that strigolactone pathway genes are likely conserved in the woody perennial plants and lays a foundation for further characterization of strigolactone pathway and its functions in the woody perennial plants.

Editor: Jin-Song Zhang, Institute of Genetics and Developmental Biology, Chinese Academy of Sciences, China

Funding: This work was supported by the Plant-Microbe Interfaces Scientific Focus Area (http://pmi.ornl.gov:8080/pmi/index.jsp) in the Genomic Science Program, United States Department of Energy, Office of Science, Biological and Environmental Research (http://science.energy.gov/ber/). Oak Ridge National Laboratory is managed by UT-Battelle, LLC, for the United States Department of Energy under contract DE-AC05-00OR22725. The early phase of this work was supported by the Laboratory Directed Research and Development Program (Seed Money Fund) of Oak Ridge National Laboratory (http://www.ornl.gov/). J.Y. was partially supported by a visiting scholarship from the Chinese Academy of Sciences (http://english.cas.cn/) (grant number: 201019). X.W. was partially supported by a visiting scholarship from the China Scholarship Council (http://en.csc.edu.cn/). The funders had no role in study design, data collection and analysis, decision to publish, or preparation of the manuscript.

Competing Interests: The authors have declared that no competing interests exist.

* Email: chenj@ornl.gov

Introduction

Plant architecture plays a major role in determining photosynthetic light use efficiency and biomass yield. Optimal plant architecture is critical for achieving maximum carbon capture per unit land area especially when plants are grown in high-density stands. One recent breakthrough in the field of plant biology was the discovery of strigolactones (SLs) as a new class of plant hormones controlling shoot branching [1,2]. More importantly, the synthesis of SLs in plants is regulated by the nutrient availability, particularly Pi deficiency, in soil and SLs exuded by roots serve as host recognition signals for symbiotic fungi [2–14]. Therefore, SLs are viewed as integrative signaling molecules that can couple nutrient availability and microbial symbiosis to the control of plant architecture and productivity. This breakthrough offers a new opportunity to dissect the interaction between plants and their environmental systems and to link this interaction to the control of plant architecture and productivity.

Substantial progress has been made in the last decade to understand the biosynthesis, signal transduction and physiological functions of SLs (reviewed by [12,15–28]). *MORE AXILLARY GROWTH1* (*MAX1*), *MAX2*, *MAX3* and *MAX4* are four founding members of strigolactone pathway genes in the model plant Arabidopsis [29–32]. While *MAX1*, *MAX3* and *MAX4* are key genes involved in SL biosynthesis, *MAX2* is a key gene involved in SL signaling. *MAX1* encodes a cytochrome P450 monooxygenase [30]. *MAX3* and *MAX4* encode two carotenoid cleavage dioxygenases (CCD), CCD7 and CCD8, respectively [29,31]. *MAX2* encodes an F-box leucine-rich protein [32]. Loss-of-function mutations in each of these four *MAX* genes resulted in increased shoot branching [29–32]. Orthologues for *MAX* genes have been identified and corresponding mutants have been characterized in pea (*ramosus, rms*), rice (*dwarf, d; high tillering dwarf, htd*) and petunia (*decreased apical dominance, dad*) (reviewed by [15–17,19–21,23]). In addition to *MAX* genes, several other genes involving in SL biosynthesis or signaling have been identified. For example, *D27* encoding a novel iron-containing protein is involved in SL biosynthesis [33,34]. *PLEIOTROPIC DRUG RESISTANCE1* (*PDR1*) encoding an ATP-binding cassette transporter functions as a cellular SL exporter [35]. *D14* encoding a protein of the α/β-fold hydrolase superfamily is involved in SL signaling and likely functions as a receptor for SLs [36–40]. D53, a member of class I Clp ATPase protein family, is a substrate of the SCF^{D3} ubiquitination complex and the degradation of D53 protein is promoted by SL and is dependent on D14 and D3 (a rice ortholog of MAX2) [41,42]. The degradation of BES1, a positive regulator in brassinosteroid signaling pathway, is also dependent on MAX2 [43].

However, despite this important discovery, little is known about this new class of plant hormones in perennial woody plants. So far, among perennial woody plants, strigolactone pathway genes have only been recently studied in willow [44,45]. Because differences have been observed between monocots and dicots, between herbaceous and woody plants, and between annual and perennial plants, it has remained elusive whether the SL pathways are conserved in the perennial woody plants, such as *Populus*. For example, studies in herbaceous plants (such as Arabidopsis and pea) suggested that SLs act as long-distance signaling molecules that can be transported from roots to shoots to exert functional control on shoot branching. However, the woody perennial plants, such as *Populus*, are typically several orders of magnitude taller than annual herbaceous plants. This raises the question of whether SL can function similarly in *Populus*.

As a first step towards exploring SL pathways in the model woody plant *Populus*, we conducted a genome-wide search of sequence homologues of strigolactone pathway genes in the sequenced genome of *Populus trichocarpa* [46]. By using Arabidopsis strigolactone pathway genes as templates, we identified *Populus* sequence homologues for each of those four founding members of strigolactone pathway genes, namely *MAX1*, *MAX2*, *MAX3* and *MAX4*. We found that *Populus MAX* paralogous genes are differentially expressed across various tissues and organs. We showed that *Populus MAX* genes could complement or partially complement the shoot branching phenotypes of corresponding Arabidopsis *max* mutants. These findings provide genetic evidence that SL pathways are likely conserved in the woody perennial plants.

Materials and Methods

Plant Materials and Growth Conditions

Arabidopsis wild type Columbia-0 (Col-0) and mutant *max1-4* (SAIL_25_A05, ABRC stock #: CS862413) and *max2-4* (SALK_028336) were obtained from the Arabidopsis Biological Resources Center (Columbus, Ohio). Arabidopsis *max3-12* mutant has been previously described [47]. Seeds were surface sterilized by serial washing with 96% (v/v) ethanol, 20% (v/v) household bleach supplemented with 0.05% (v/v) Tween-20, and water, and placed at 4°C for 2 days. Seeds were subsequently plated on ½ Murashige and Skoog (MS) medium [48] supplemented with 1% (w/v) sucrose and 0.8% (w/v) agar, and germinated in 12 h/12 h photoperiod at 23°C, approximately 90 µmol photons $m^{-2} s^{-1}$. Seven day-old Arabidopsis seedlings were transferred from ½ MS medium to soil and grown in growth chambers at 23°C, approximately 125 µmol photons $m^{-2} s^{-1}$ with 14 h/10 h (long-day conditions) or 10 h/14 h (short-day conditions) photoperiods.

Cloning of *Populus MAX* Homologous Genes

The full-length open reading frame (ORF) of each *PtrMAX* gene was determined according to the sequence information available at Phytozome [46,49]. Gene-specific primers were designed to amplify the full-length ORF of each *PtrMAX* gene from cDNA library derived from RNA isolated from leaves and roots of *Populus trichocarpa* plants. Subsequently, the full-length ORF of each *PtrMAX* gene was introduced into the pENTR vector by using pENTR/D-TOPO Cloning Kit (Life Technologies). The cloned RT-PCR products were validated by sequencing and then transferred into plant Gateway destination vector pGWB502Ω (2×CaMV35SΩ) [50] using LR clonase (Life Technologies). All primers used for cloning are listed in Table S1.

Genetic Complementation

pGWB502Ω (2×CaMV35SΩ) binary vectors containing *35S:PtrMAX* plasmid were transformed into Arabidopsis *max* mutants via *Agrobacterium tumefaciens* strain GV3101 mediated flower dipping transformation [51,52]. T1 transformants were selected using 20 µg/L hygromycin B. A minimum of 20 independent transgenic lines were selected for each transgene. Two independent transgenic lines were used for further studies.

When plants reached maturity, the number of primary rosette-leaf branches was counted. A minimum of 10 individual plants per genotype were used.

RT-PCR

To examine the absence or presence of Arabidopsis *MAX* transcript in the *max* mutant, total RNA was extracted from 7-day-old seedlings using the Invisorb Spin Plant Mini Kit (Stratec Molecular). Two µg of total RNA were reversely transcribed to cDNA using Fermentas RevertAid Reverse Transcriptase (Thermo Scientific). For semi-quantitative RT-PCR, Arabidopsis *MAX*-specific primers spanning the full-length ORF of *MAX* gene were used. PCR amplification of *AtACTIN8 (AtACT8)* served as a control.

For the examination of expression of *Populus MAX* genes in Arabidopsis transgenic lines, *PtrMAX* gene-specific primers were used. All primers used for semi-quantitative RT-PCR are listed in Table S1.

Quantitative RT-PCR (qRT)

qRT was conducted to examine the transcript level of each *PtrMAX* genes across various tissues and organs using a StepOnePlus (Applied Biosystems), Maxima SYBR Green/ROX qPCR Master Mix (Thermo Scientific) and cDNA corresponding to 80 ng RNA in a total volume of 25 µl. Three biological replicates were used for qRT analysis. The following cycling conditions were used for PCR: 10 min at 95°C, 35 cycles of 15 s at 95°C and 60 s at 60°C. The transcript level was normalized against that of *PtrACTIN5 (PtrACT5)*. Gene-specific qRT primers were designed using QuantPrime [53]. All primers used for qRT analysis are listed in Table S1.

Results

Sequence Homologues of MAX Proteins in *Populus*

MAX1, *MAX2*, *MAX3* and *MAX4* are four founding members of SL pathway genes in the model plant Arabidopsis. In order to identify sequence homologs of *MAX* genes in *Populus trichocarpa* (hereafter referred to as *Populus*), we used amino acid sequences of

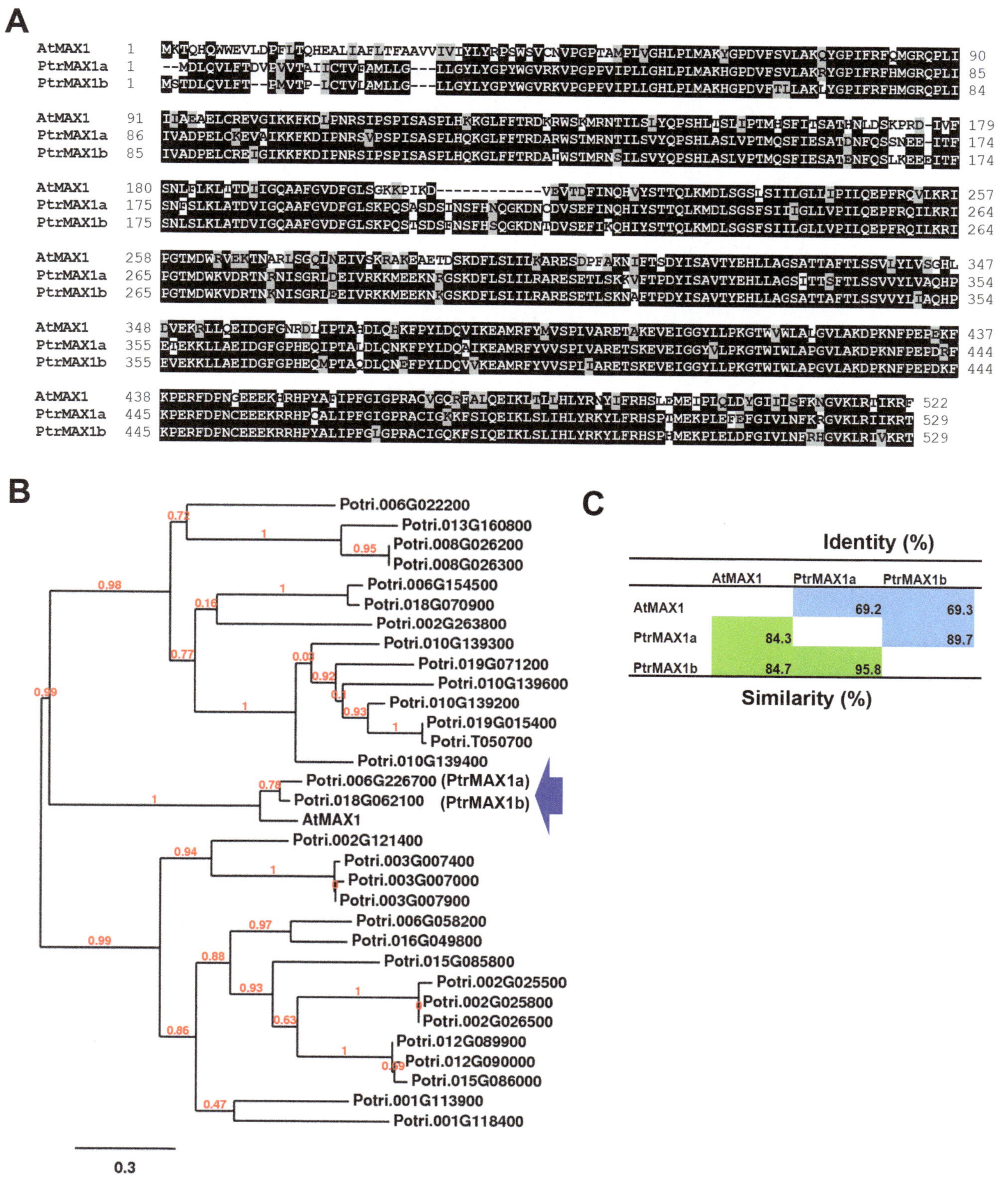

Figure 1. Bioinformatic analysis of MAX1 proteins from *Populus* and *Arabidopsis*. (A) Amino acid sequence alignment. **(B)** Phylogenetic analysis of *Populus* sequence homologues of *Arabidopsis* MAX1 protein. **(C)** Amino acid sequence similarity and identity.

Arabidopsis MAX proteins as templates to perform protein homology search in the fully-sequenced *Populus* genome.

Based on the current annotation by TAIR (http://www.arabidopsis.org) [54] and Phytozome (www.phytozome.net) [49], Arabidopsis *MAX1* (*AtMAX1*) has two transcript variants. The primary variant encoded by gene locus At2g26170.1 was used as a template for searching protein sequence homologues encoded by the *Populus* genome using the "Protein Homologs" search tool at Phytozome. The search identified two close sequence homologues encoded by loci Potri.006G226700 and Potri.018G062100, designated as PtrMAX1a and PtrMAX1b, respectively (Figure 1). Based on the current annotation by Phytozome, *PtrMAX1a* has

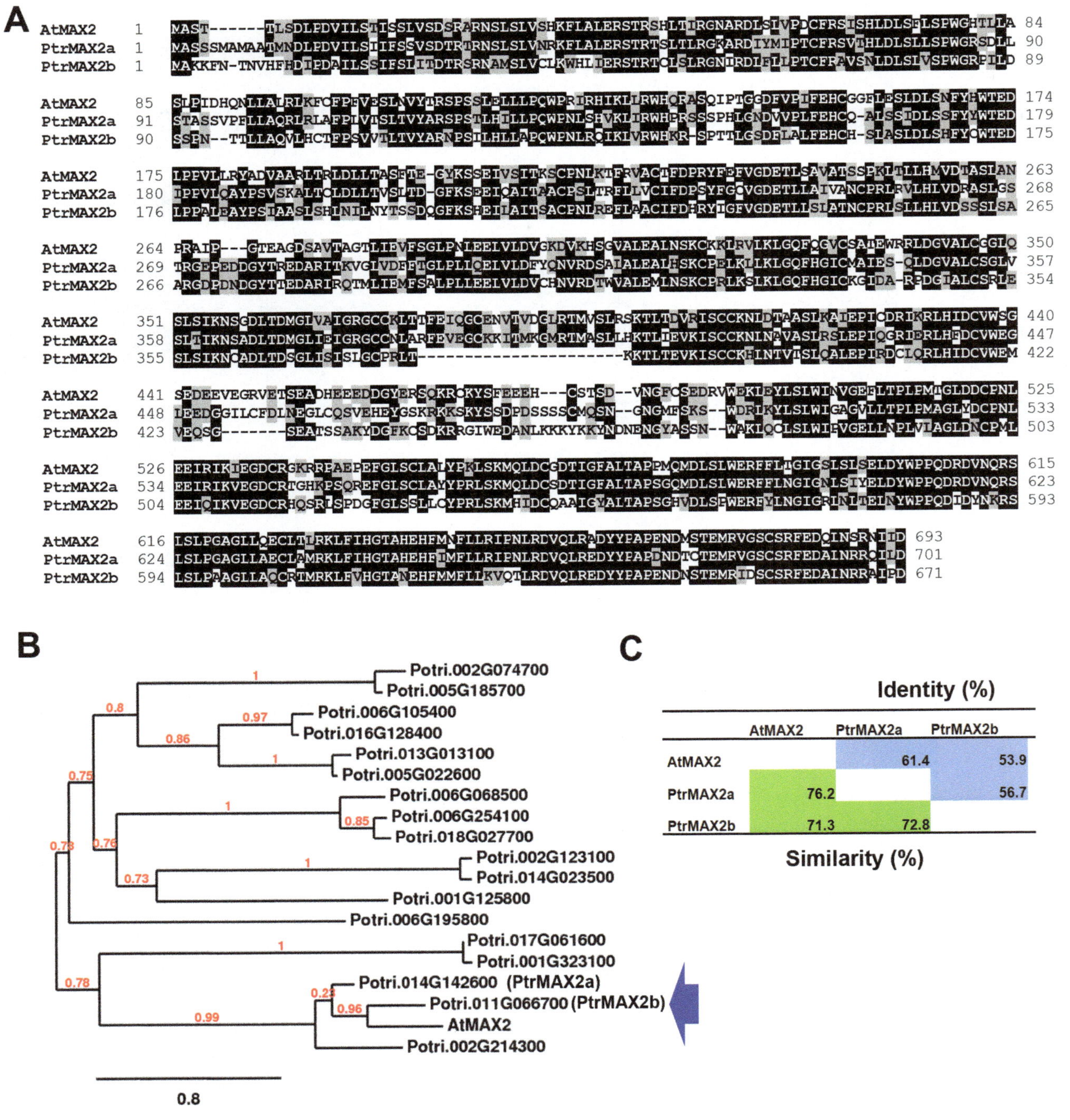

Figure 2. Bioinformatic analysis of MAX2 proteins from *Populus* and *Arabidopsis*. (**A**) Amino acid sequence alignment. (**B**) Phylogenetic analysis of *Populus* sequence homologues of *Arabidopsis* MAX2 protein. (**C**) Amino acid sequence similarity and identity.

three transcript variants and *PtrMAX1b* has four transcript variants. The primary variants (e.g., Potri.006G226700.1 and Potri.018G062100.1) were used for subsequent analysis. PtrMAX1a showed approximately 84% similarity and 69% identity with AtMAX1 at the amino acid level. PtrMAX1b showed approximately 85% similarity and 69% identity with AtMAX1 at the amino acid level. PtrMAX1a and PtrMAX1b each other shared approximately 96% similarity and 90% identity at the amino acid level. The number of amino acids of PtrMAX1a (529 aa) and PtrMAX1b (529 aa) are comparable to that of AtMAX1 (522 aa). No other *Populus* proteins showed more than 43% similarity with AtMAX1 at the amino acid level.

The two closest *Populus* sequence homologues of Arabidopsis MAX2 (AtMAX2) were encoded by loci Potri.014G142600 and Potri.011G066700, designated as *PtrMAX2a* and *PtrMAX2b*, respectively (Figure 2). A third homolog, Potri.002G214300 consisting of 316 aa is only half the size of AtMAX2 (693 aa). Thus, it was not included in our further analysis. PtrMAX2a showed approximately 76% similarity and 61% identity with AtMAX2 at the amino acid level. PtrMAX2b showed approximately 71% similarity and 54% identity with AtMAX2 at the

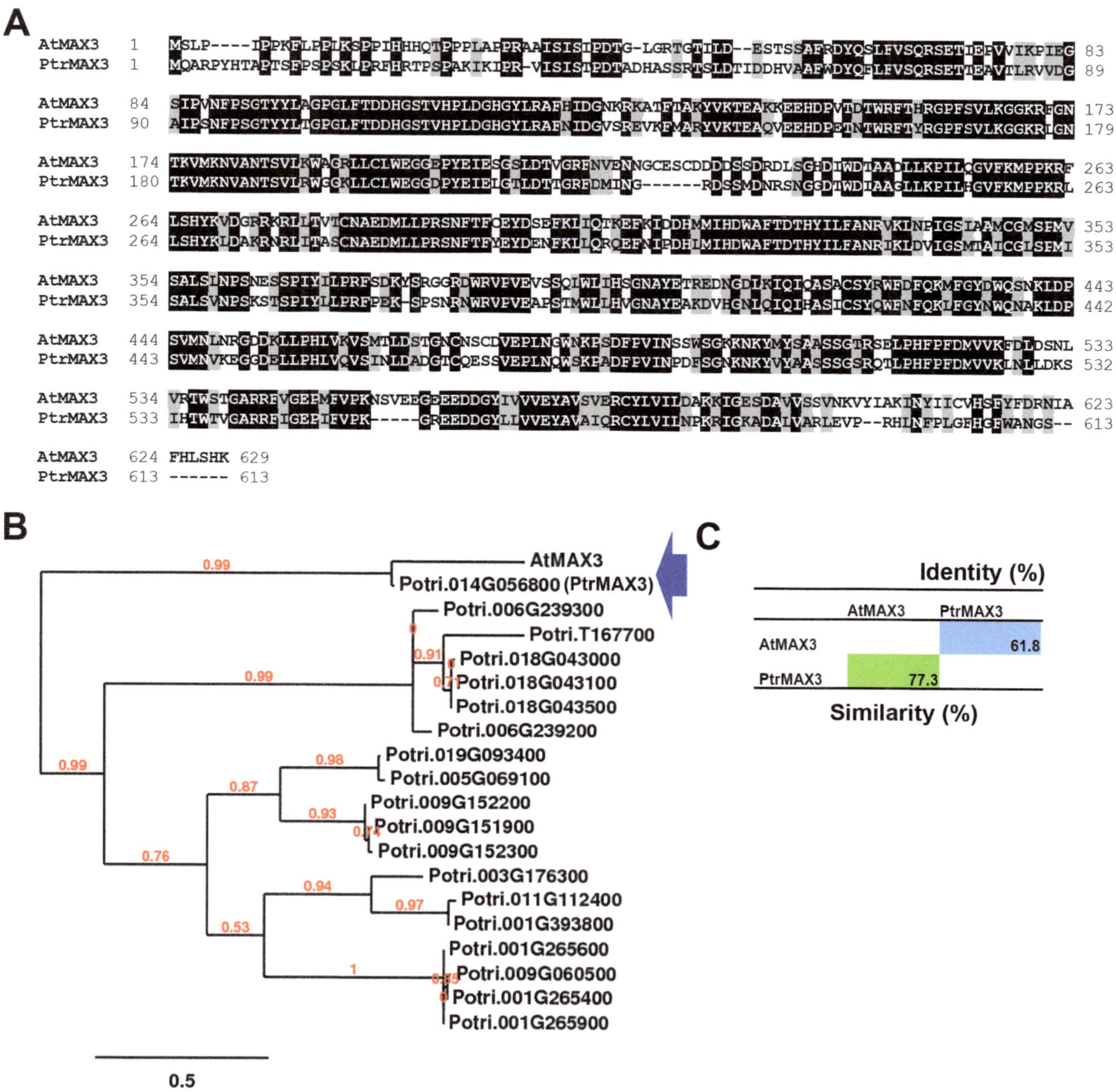

Figure 3. Bioinformatic analysis of MAX3 proteins from *Populus* and *Arabidopsis*. **(A)** Amino acid sequence alignment. **(B)** Phylogenetic analysis of *Populus* sequence homologues of *Arabidopsis* MAX3 protein. **(C)** Amino acid sequence similarity and identity.

amino acid level. PtrMAX2a and PtrMAX2b each other shared approximately 73% similarity and 57% identity at the amino acid level. The number of amino acids of PtrMAX2a (701 aa) and PtrMAX2b (671 aa) are comparable to that of AtMAX2 (693 aa). No other *Populus* proteins showed more than 32% similarity with AtMAX2 at the amino acid level.

Arabidopsis MAX3 (AtMAX3) has only one close sequence homologue in *Populus* encoded by loci Ptri.014G056800, designated as *PtrMAX3* (Figure 3). PtrMAX3 showed approximately 77% similarity and 62% identity with AtMAX3 at the amino acid level. The number of amino acids of PtrMAX3 (613 aa) is comparable to that of AtMAX3 (629 aa). No other *Populus* proteins showed more than 35% similarity to AtMAX3 at the amino acid level.

Arabidopsis MAX4 (AtMAX4) has two close sequence homologues in *Populus* encoded by loci Potri.018G044100 and Potri.006G238500, designated as *PtrMAX4a* and *PtrMAX4b*, respectively (Figure 4). Both *PtrMAX4a* and *PtrMAX4b* have two transcript variants based on the current annotation by Phytozome. The primary variant (e.g., Potri.018G044100.1 and Potri.006G238500.1) were used for subsequent analysis. PtrMAX4a showed approximately 78% similarity and 65% identity with AtMAX4 at the amino acid level. PtrMAX4b showed approximately 78% similarity and 64% identity with AtMAX4 at the amino acid level. PtrMAX4a and PtrMAX4b each other shared approximately 97% similarity and 93% identity at the amino acid level. The number of amino acids of PtrMAX4a (557 aa) and PtrMAX4b (557 aa) are comparable to that of AtMAX4 (570 aa).

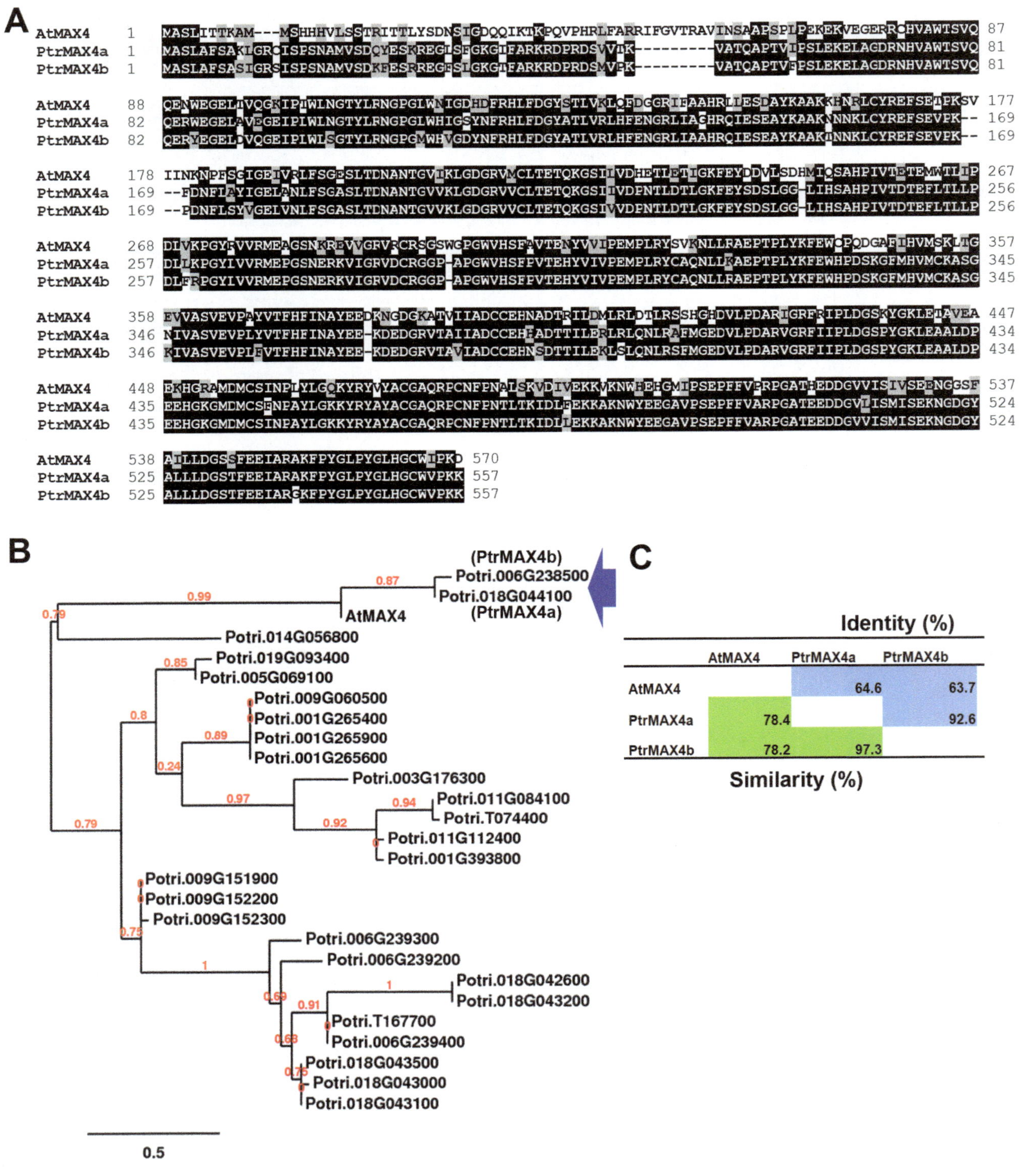

Figure 4. Bioinformatic analysis of MAX4 proteins from *Populus* and *Arabidopsis*. (**A**) Amino acid sequence alignment. (**B**) Phylogenetic analysis of *Populus* sequence homologues of *Arabidopsis* MAX4 protein. (**C**) Amino acid sequence similarity and identity.

No other *Populus* proteins showed more than 42% similarity to AtMAX4 at the amino acid level.

In total, we identified two *Populus* sequence homologues each for Arabidopsis MAX1, MAX2 and MAX4, and one *Populus* sequence homologues for Arabidopsis MAX3.

Tissue and Organ Expression Patterns of *Populus MAX* Genes

Subsequently, we examined the expression patterns of each *PtrMAX* gene across various tissues and organs. We used qRT to examine the transcript level of each *PtrMAX* gene and normalized their transcript levels against the transcript level of *PtrACT5*. We made three interesting observations. Firstly, each *PtrMAX* gene

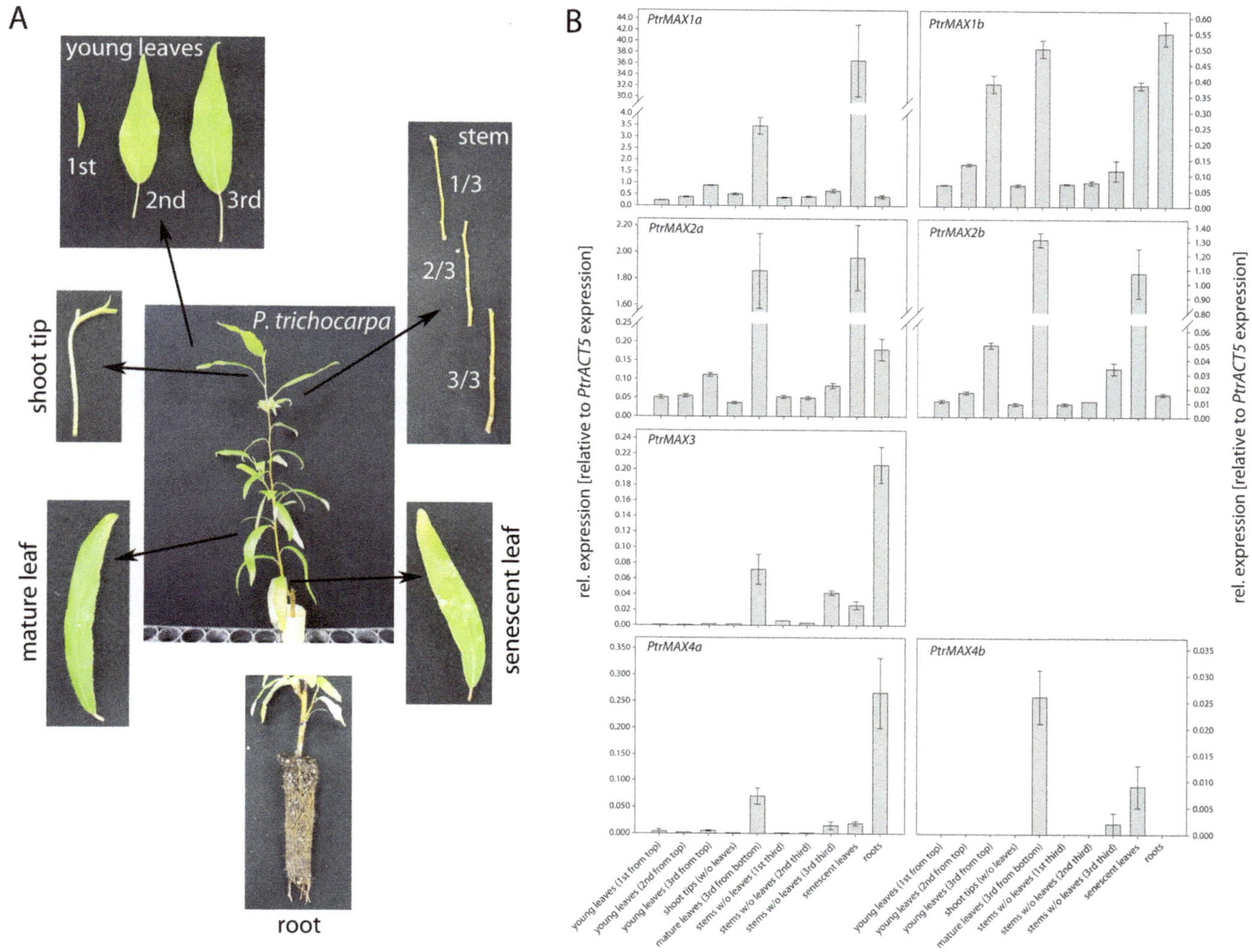

Figure 5. Expression of *Populus MAX* homologous genes across various tissues and organs. **(A)** Illustration of tissues and organs used for expression analysis. **(B)** Quantitative RT-PCR data. Shown are means ± S.E. of three biological replicates.

was expressed at different levels with *PtrMAX1a* at the highest level and *PtrMAX4b* at the lowest level in most tissues and organs (Figure 5). Secondly, differential expression in terms of transcript level was observed between *PtrMAX* paralogous genes. For example, *PtrMAX1a* was expressed over 10-fold higher than *PtrMAX1b* across most tissues and organs. Similarly, *PtrMAX4a* was expressed over 10-fold higher than *PtrMAX4b* across most tissues and organs. In contrast, *PtrMAX2a* and *PtrMAX2b* were expressed at similar levels in most tissues and organs. Finally, differences in tissue and organ expression patterns were observed between *PtrMAX* paralogous genes. For example, *PtrMAX4a* was highly expressed in roots but the transcript of *PtrMAX4b* was undetectable in roots (Figure 5).

Genetic Complementation of Arabidopsis *max* Mutants with *Populus MAX* Genes

In order to further test whether *PtrMAX* genes may function similarly as *AtMAX* genes, we conducted genetic complementation studies by heterologously expressing each *PtrMAX* gene in corresponding Arabidopsis *max* mutants. The expression of each *PtrMAX* gene was driven by the constitutive *35S* promoter [55]. At least 20 independent transgenic lines were selected from each transformation and two independent transgenic lines were selected for further analysis. We used the number of primary rosette-leaf branches as a phenotype to determine whether heterologous expression of each *PtrMAX* gene could rescue the corresponding Arabidopsis *max* mutants.

For the test of *PtrMAX1* genes, we used a homozygous T-DNA insertional *max1* mutant allele designated as *max1-4* as the transformation background. In this mutant allele, a T-DNA is insertion in the 1st exon, 74 bp downstream of the start codon of MAX1 (Figure S1). RT-PCR analysis indicated that the full-length transcript of *MAX1* was absent in *max1-4* mutant, suggesting that it is likely a loss-of-function allele (Figure S1). Consistent with the RT-PCR result, *max1-4* mutants displayed high number of primary rosette-leaf branches (Figure S1). For the complementation, the full-length ORF of *PtrMAX1a* and *PtrMAX1b* was each cloned into the binary plant expression vector pGWB502Ω (2×CaMV35SΩ) [50] and transformed into *max1-4* mutant background via *Agrobacterium*-mediated transformation. RT-PCR analysis indicated that in the transgenic lines (lines #7 and #8 for *PtrMAX1a*, and lines #2 and #6 for *PtrMAX1b*), *PtrMAX1a* or *PtrMAX1b* transgene was expressed in the Arabidopsis *max1-4* mutant background (Figure 6A, B). We counted the number of primary rosette-leaf branches and compared it with that of Columbia-0 (Col-0) wild type and

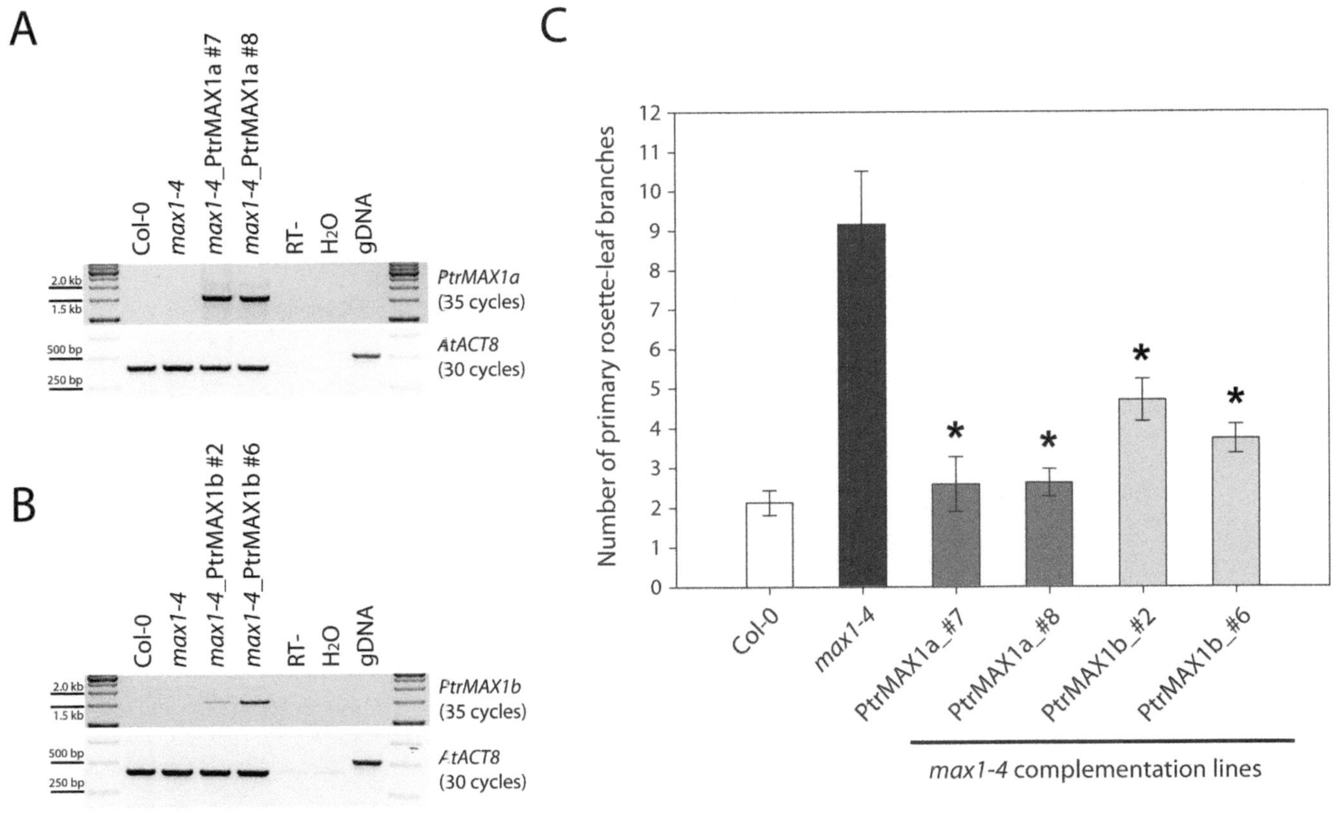

Figure 6. Genetic complementation of *Arabidopsis max1* mutants with *Populus MAX1* genes. (**A**) RT-PCR analysis of *35S:PtrMAX1a* transgenic lines. (**B**) RT-PCR analysis of *35S:PtrMAX1b* transgenic lines. (**C**) Number of primary rosette-leaf branches. Shown are average numbers of primary rosette-leaf branches from at least 10 individual plants ± S.E. *, significant difference from *max1-4*, $p<0.05$.

max1-4 mutants which were grown side-by-side with transgenic lines under identical conditions. We found that the number of primary rosette-leaf branches in the transgenic lines was largely reverted to a similar number as in the wild type. Under our growth conditions, Col-0 produced about two primary rosette-leaf branches whereas *max1-4* mutants produced about nine primary rosette-leaf branches. The *PtrMAX1a* transgenic lines #7 and #8 produced approximately two primary rosette-leaf branches and *PtrMAX1b* transgenic lines #2 and #6 produced about four primary rosette-leaf branches (Figure 6C). These results indicated that both *PtrMAX1a* and *PtrMAX1b* were able to complement or partially complement the shoot branching phenotypes of Arabidopsis *max1-4* mutant.

For the test of *PtrMAX2* genes, we used a T-DNA insertional *max2* mutant allele, *max2-4*, as the transformation background. The *max2-4* allele had been previously described by Umehara et al. [2] and Ha et al. [56] but the genotyping and branching phenotype were not reported. These data are provided here as supplemental materials (Figure S2). In this mutant allele, a T-DNA insertion site was found within the 1st exon. RT-PCR analysis indicated that the full-length transcript of *MAX2* was absent in *max2-4* mutant, suggesting that it is likely a loss-of-function allele. Consistent with the RT-PCR result, *max2-4* mutants displayed more shoot branches (Figure S2). For the complementation, the full-length ORF of *PtrMAX2a* and *PtrMAX2b* was each cloned into the binary plant expression vector pGWB502Ω and transformed into *max2-4* mutant background. RT-PCR analysis indicated that in the transgenic lines (lines #4 and #8 for *PtrMAX2a*, and lines #3 and #6 for *PtrMAX2b*), *PtrMAX2a* or *PtrMAX2b* transgene was expressed in the Arabidopsis *max2-4* mutant background (Figure 7). Compared with *max2-4* mutants, the number of primary rosette-leaf branches in the transgenic lines was significantly reduced and in line #4 of *PtrMAX2a*, the number was reverted to a similar number in the wild type (Figure 7C). These results indicated that both *PtrMAX2a* and *PtrMAX2b* were able to complement or partially complement the shoot branching phenotypes of Arabidopsis *max2-4* mutant.

For the test of *PtrMAX3* gene, we used a T-DNA insertional *max3* mutant allele, *max3-12*, as the transformation background. *max3-12* was previously described by Umehara et al. [2] and the genotypic and phenotypic data had been reported by Li et al. [47]. The full-length ORF of *PtrMAX3* was cloned into the binary plant expression vector pGWB502Ω and transformed into *max3-12* mutant background. RT-PCR analysis indicated that in the transgenic lines (lines #2 and #4), *PtrMAX3* transgene was expressed in the Arabidopsis *max3-12* mutant background (Figure 8). Compared with *max3-12* mutants, the number of primary rosette-leaf branches in the transgenic lines was significantly reduced but was not completely reverted to wild-type branch number (Figure 8C). These results indicated that *PtrMAX3* was able to partially complement the shoot branching phenotypes of Arabidopsis *max3-12* mutant.

For the test of *PtrMAX4* genes, *35S:PtrMAX4a* and *35S:PtrMAX4b* transgenic lines were generated using Arabidopsis *max4-1* mutant [31] as background. RT-PCR analysis indicated that in the *35S:PtrMAX4* transgenic lines, *PtrMAX4a* (lines #4 and #5) or *PtrMAX4b* (lines #3 and #9) transgene was expressed (Figure 9). Compared with *max4-1* mutants, the numbers of primary rosette-leaf branches were significantly reduced in these transgenic lines but were not completely reverted to wild-type

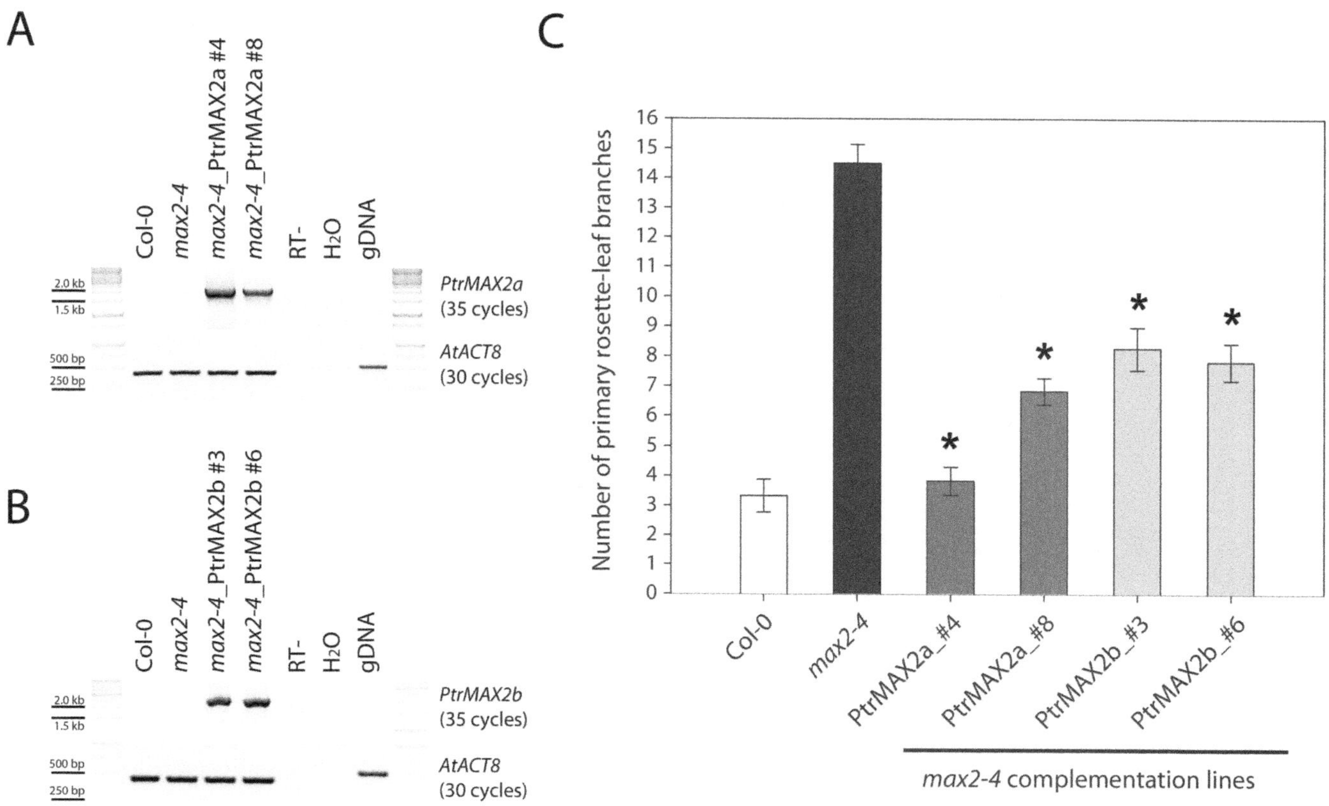

Figure 7. Genetic complementation of *Arabidopsis max2* mutants with *Populus MAX2* genes. (**A**) RT-PCR analysis of *35S:PtrMAX2a* transgenic lines. (**B**) RT-PCR analysis of *35S:PtrMAX2b* transgenic lines. (**C**) Number of primary rosette-leaf branches. Shown are average numbers of primary rosette-leaf branches from at least 10 individual plants ± S.E. *, significant difference from *max2-4*, $p<0.05$.

branch number (Figure 9C). These results indicated that both *PtrMAX4a* and *PtrMAX4b* were able to partially complement the shoot branching phenotype of Arabidopsis *max4* mutants.

Discussion

SLs are a new class of plant hormones controlling shoot branching. Due to the importance of shoot branching in the determination of photosynthetic light use efficiency and biomass yield in woody perennial trees, we wanted to examine whether the strigolactone pathway is conserved in the woody model plant *Populus*. We report the identification of sequence homologues of four founding members of SL pathway genes, namely *MAX1*, *MAX2*, *MAX3* and *MAX4*. Furthermore, we provide genetic evidence that these *Populus MAX* genes likely function similarly in controlling shoot branching.

Due to recent genome duplication, in general, each Arabidopsis gene can have two paralogous genes in *Populus*. As such, we found two *Populus* paralogous genes for *MAX1*, *MAX2* and *MAX4* (Figures 1, 2 and 4). The exception here is *MAX3*; only one *MAX3* homologous gene was identified in the *Populus* genome (Figure 3). Our gene expression analysis indicated that *Populus MAX* paralogous genes are differentially expressed across various tissues and organs, in terms of expression level and expression patterns. Over 10-fold differences in transcript level were observed between *PtrMAX1a* and *PtrMAX1b*, and between *PtrMAX4a* and *PtrMAX4b* (Figure 5). It remains unclear whether differences in expression level or pattern of *Populus MAX* genes may result in differences in their functioning. When their expression was driven by the constitutive *35S* promoter, *PtrMAX1* and *PtrMAX4* paralogous genes could rescue the shoot branching phenotype of the corresponding Arabidopsis *max* mutants (Figure 6, Figure 9), suggesting that both paralogous genes could function similarly. The expression analysis also indicated that *PtrMAX3* and *PtrMAX4a* were expressed at the highest level in roots (Figure 5). These results are consistent with the view that roots represent one of the major biosynthesis sites for SLs (reviewed by [25]). This has been demonstrated in a grafting experiment where grafting wild-type Arabidopsis rootstocks to *max1*, *max3*, or *max4* scions were able to restore a wild-type branching pattern to the mutant shoots [30].

For the complementation tests, we found that all *PtrMAX* genes were able to rescue or partially rescue the shoot branching phenotypes of corresponding Arabidopsis *max* mutants (Figures 6, 7, 8 and 9). For *PtrMAX1* and *PtrMAX2*, we could identify transgenic lines in which the numbers of primary rosette-leaf branches were completely reverted to wild-type level (Figure 6, Figure 7), suggesting that *PtrMAX1* and *PtrMAX2* may function equivalently to Arabidopsis *MAX1* and *MAX2*, respectively. On the other hand, although the numbers of primary rosette-leaf branches in *PtrMAX3* and *PtrMAX4* transgenic lines were significantly reduced when compared with Arabidopsis *max3* and *max4* mutants, respectively, the numbers were not completely reverted to wild-type level. There are several possible explanations. Firstly, the expression level of *PtrMAX3* and *PtrMAX4* in the transgenic lines may not be high enough to render complete rescue. Secondly, PtrMAX3 and PtrMAX4 may function similarly, but not equivalently, as Arabidopsis MAX3 and MAX4, respectively. PtrMAX3 showed approximately 77% similarity and 62% identity with MAX3 (Figure 3) and PtrMAX4 showed

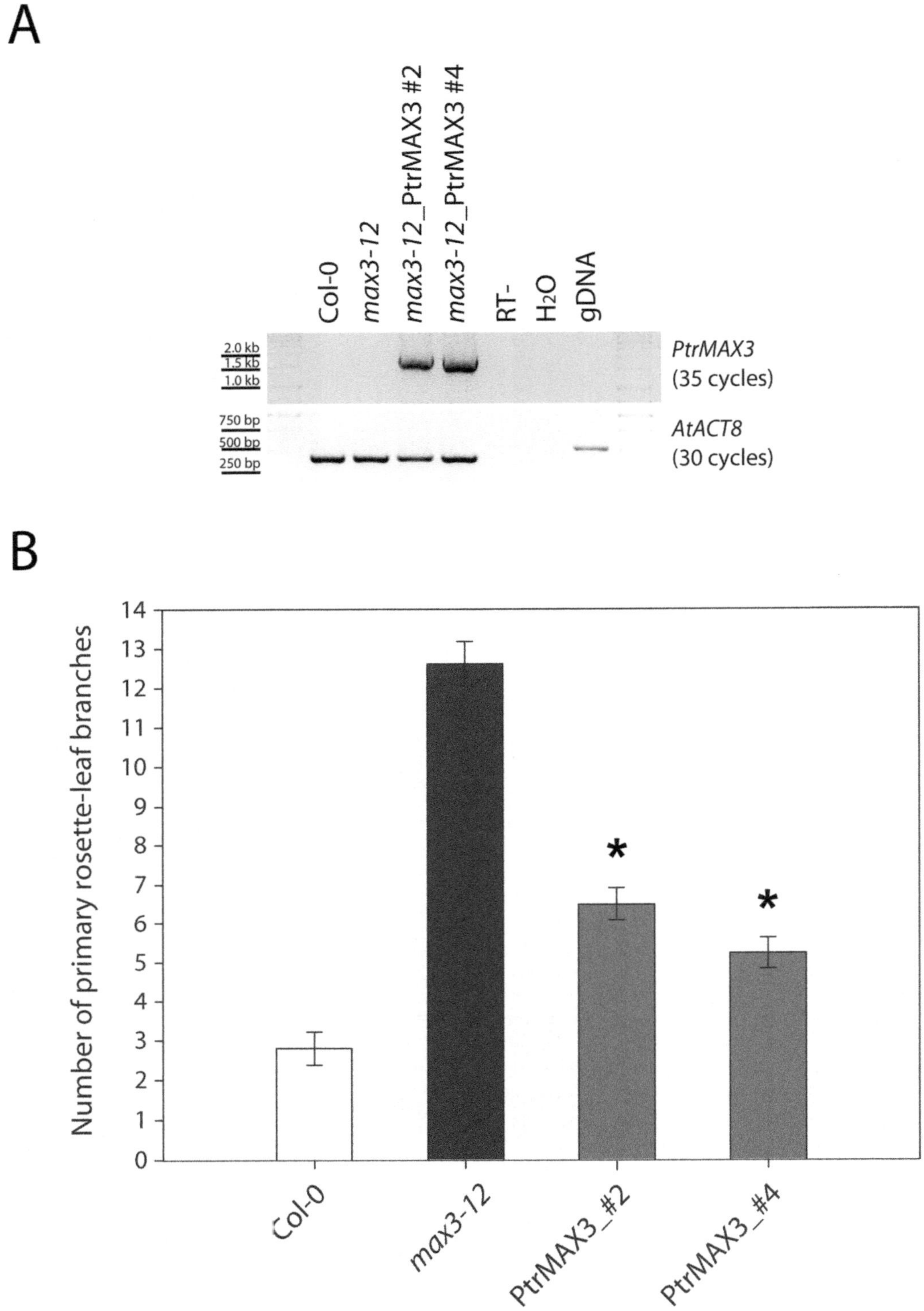

Figure 8. Genetic complementation of *Arabidopsis max3* mutants with *Populus MAX3* gene. (**A**) RT-PCR analysis of *35S:PtrMAX3* transgenic lines. (**B**) Number of primary rosette-leaf branches. Shown are average numbers of primary rosette-leaf branches from at least 10 individual plants ± S.E. *, significant difference from *max3-12*, $p<0.05$.

approximately 78% similarity and 65% identity with AtMAX4 at the amino acid level (Figure 4). It is unclear whether differences in amino acid sequence between *Populus* and *Arabidopsis* MAX3 and MAX4 may have contributed to such partial rescue phenotypes in the transgenic lines. Finally, it is also possible that there may be other MAX3 and MAX4 homologues in *Populus* though no other *Populus* proteins showed more than 35% similarity with AtMAX3 and no other *Populus* proteins showed more than 42% similarity with AtMAX4 at the amino acid level. This deserves further investigation.

In addition to *MAX* genes, several other genes are involved in SL biosynthesis or signaling such as D27 [33,34] and D14 [36,38–

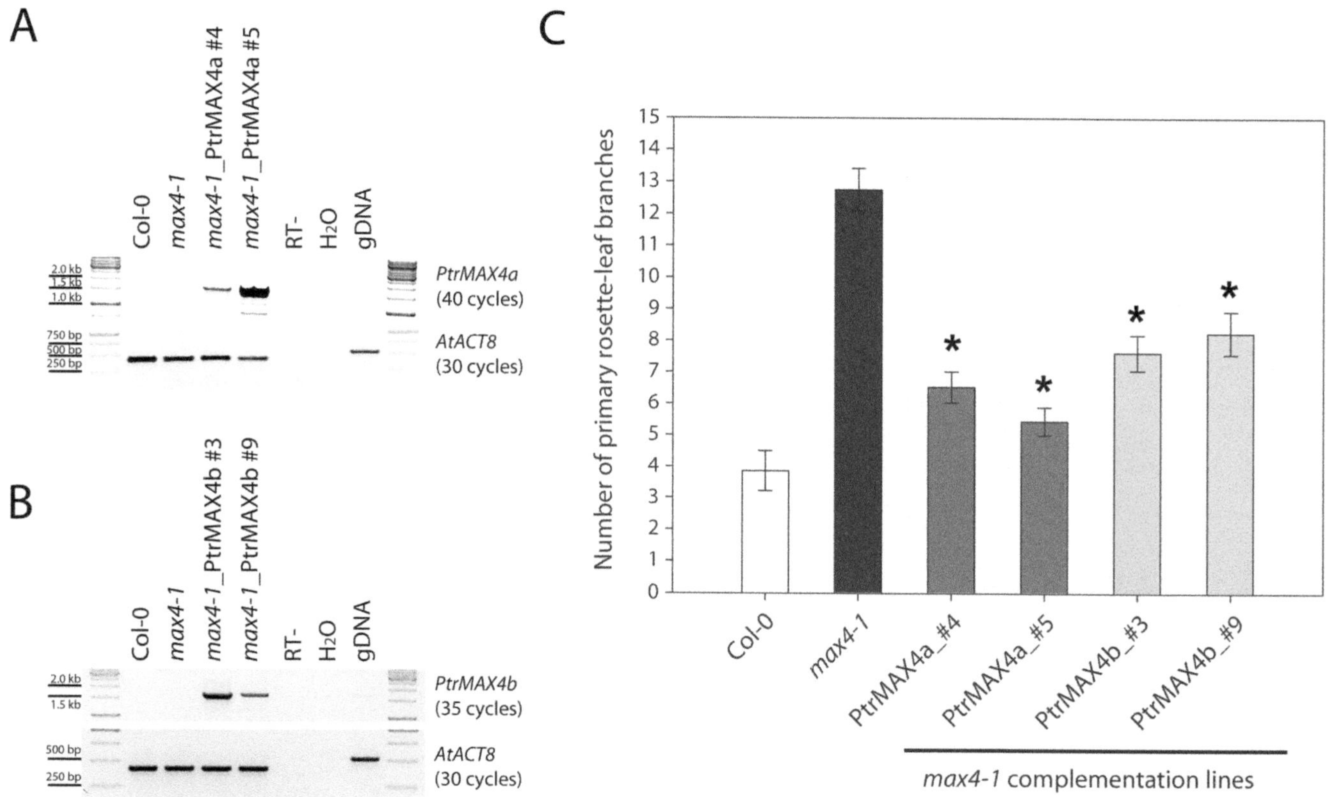

Figure 9. Genetic complementation of *Arabidopsis max4* mutants with *Populus MAX4* genes. (**A**) RT-PCR analysis of *35S:PtrMAX4a* transgenic lines. (**B**) RT-PCR analysis of *35S:PtrMAX4b* transgenic lines. (**C**) Number of primary rosette-leaf branches. Shown are average numbers of primary rosette-leaf branches from at least 10 individual plants ± S.E. *, significant difference from *max4-1*, $p<0.05$.

40]. It would be helpful to identify sequence homologs of each of those genes in *Populus* and conduct genetic complementation studies similar to what have been described in this study in order to strengthen the view that strigolactone pathway exists and operates in the woody perennial plant *Populus*. Our preliminary analysis indicated sequence homologues of AtD14 (encoded by locus At3g03990) are present in the genome of *Populus*. For example, two proteins encoded by locus Potri.002G118900 and Potri.014G016500 showed about 89% and 88% similarity with AtD14 at the amino acid level. Our preliminary analysis indicated that sequence homologues of AtD27 (encoded by locus At1g03055) are also present in the genome of *Populus*. For example, a protein encoded by locus Potri.005G216400 showed about 59% similarity with AtD27 at the amino acid level. In order to fully assign SL pathway in *Populus*, future studies should focus on the functional characterization of these SL pathway genes (e.g. generating *Populus* transgenic lines), the determination of SL molecules, biosynthetic intermediates and derivatives, and the determination of the physiological roles of SLs in *Populus*.

In summary, this study provided a preliminary characterization of strigolactone pathway genes in the model woody plant *Populus*. Sequence homologues of those four founding members of Arabidopsis SL pathway genes have been identified in *Populus*. Gene expression analysis indicated that *Populus MAX* paralogous genes are differentially expressed across various tissues and organs. Genetic complementation studies indicated *Populus MAX* genes are able to complement or partially complement the shoot branching phenotypes of corresponding Arabidopsis *max* mutants. These findings lay a foundation for further characterization of SL pathway and its functions in woody perennial plants.

Supporting Information

Figure S1 Arabidopsis *max1-4* mutant. (**A**) T-DNA insertion site. (**B**) PCR genotyping. (**C**) RT-PCR analysis. (**D**) Shoot branching phenotypes. (**E**) Number of primary rosette-leaf branches. Shown are average numbers of primary rosette-leaf branches from at least 10 individual plants ± S.E. *, significant difference from Col-0, $p<0.05$.

Figure S2 Arabidopsis *max2-4* mutant. (**A**) T-DNA insertion site. (**B**) PCR genotyping. (**C**) RT-PCR analysis. (**D**) Shoot branching phenotypes. (**E**) Number of primary rosette-leaf branches. Shown are average numbers of primary rosette-leaf branches from at least 10 individual plants ± S.E. *, significant difference from Col-0, $p<0.05$.

Acknowledgments

Disclaimer: This manuscript has been authored by UT-Battelle, LLC, under Contract No. DE-AC05-00OR22725 with the U.S. Department of Energy. The United States Government retains and the publisher, by accepting the article for publication, acknowledges that the United States Government retains a non-exclusive, paid-up, irrevocable, world-wide license to publish or reproduce the published form of this manuscript, or allow others to do so, for United States Government purposes.

Author Contributions

Conceived and designed the experiments: OC J-GC. Performed the experiments: OC JY XW. Analyzed the data: OC JY XW J-GC. Contributed to the writing of the manuscript: OC JY XW SW WM GAT J-GC.

References

1. Gomez-Roldan V, Fermas S, Brewer PB, Puech-Pages V, Dun EA, et al. (2008) Strigolactone inhibition of shoot branching. Nature 455: 189–194.
2. Umehara M, Hanada A, Yoshida S, Akiyama K, Arite T, et al. (2008) Inhibition of shoot branching by new terpenoid plant hormones. Nature 455: 195–200.
3. Akiyama K, Matsuzaki K, Hayashi H (2005) Plant sesquiterpenes induce hyphal branching in arbuscular mycorrhizal fungi. Nature 435: 824–827.
4. Besserer A, Puech-Pages V, Kiefer P, Gomez-Roldan V, Jauneau A, et al. (2006) Strigolactones stimulate arbuscular mycorrhizal fungi by activating mitochondria. PLoS Biol 4: 1239–1247.
5. Yoneyama K, Xie XN, Kusumoto D, Sekimoto H, Sugimoto Y, et al. (2007) Nitrogen deficiency as well as phosphorus deficiency in sorghum promotes the production and exudation of 5-deoxystrigol, the host recognition signal for arbuscular mycorrhizal fungi and root parasites. Planta 227: 125–132.
6. Yoneyama K, Yoneyama K, Takeuchi Y, Sekimoto H (2007) Phosphorus deficiency in red clover promotes exudation of orobanchol, the signal for mycorrhizal symbionts and germination stimulant for root parasites. Planta 225: 1031–1038.
7. Lopez-Raez JA, Charnikhova T, Gomez-Roldan V, Matusova R, Kohlen W, et al. (2008) Tomato strigolactones are derived from carotenoids and their biosynthesis is promoted by phosphate starvation. New Phytol 178: 863–874.
8. Umehara M, Hanada A, Magome H, Takeda-Kamiya N, Yamaguchi S (2010) Contribution of strigolactones to the inhibition of tiller bud outgrowth under phosphate deficiency in rice. Plant Cell Physiol 51: 1118–1126.
9. Kohlen W, Charnikhova T, Liu Q, Bours R, Domagalska MA, et al. (2011) Strigolactones are transported through the xylem and play a key role in shoot architectural response to phosphate deficiency in nonarbuscular mycorrhizal host Arabidopsis. Plant Physiol 155: 974–987.
10. Mayzlish-Gati E, De-Cuyper C, Goormachtig S, Beeckman T, Vuylsteke M, et al. (2012) Strigolactones are involved in root response to low phosphate conditions in Arabidopsis. Plant Physiol 160: 1329–1341.
11. Yoneyama K, Xie XN, Kim HI, Kisugi T, Nomura T, et al. (2012) How do nitrogen and phosphorus deficiencies affect strigolactone production and exudation? Planta 235: 1197–1207.
12. Brewer PB, Koltai H, Beveridge CA (2013) Diverse roles of strigolactones in plant development. Mol Plant 6: 18–28.
13. Czarnecki O, Yang J, Weston DJ, Tuskan GA, Chen JG (2013) A dual role of strigolactones in phosphate acquisition and utilization in plants. Int J Mol Sci 14: 7681–7701.
14. Foo E, Yoneyama K, Hugill CJ, Quittenden LJ, Reid JB (2013) Strigolactones and the regulation of pea symbioses in response to nitrate and phosphate deficiency. Mol Plant 6: 76–87.
15. Beveridge CA, Kyozuka J (2010) New genes in the strigolactone-related shoot branching pathway. Curr Opin Plant Biol 13: 34–39.
16. Xie XN, Yoneyama K, Yoneyama K (2010) The strigolactone story. Annu Rev Phytopathol 48: 93–117.
17. Domagalska MA, Leyser O (2011) Signal integration in the control of shoot branching. Nat Rev Mol Cell Biol 12: 211–221.
18. Koltai H (2011) Strigolactones are regulators of root development. New Phytol 190: 545–549.
19. Wang YH, Li JY (2011) Branching in rice. Curr Opin Plant Biol 14: 94–99.
20. Seto Y, Kameoka H, Yamaguchi S, Kyozuka J (2012) Recent advances in strigolactone research: chemical and biological aspects. Plant Cell Physiol 53: 1843–1853.
21. Tsuchiya Y, McCourt P (2012) Strigolactones as small molecule communicators. Mol Biosyst 8: 464–469.
22. Mason MG (2013) Emerging trends in strigolactone research. New Phytol 198: 975–977.
23. de Saint Germain A, Bonhomme S, Boyer FD, Rameau C (2013) Novel insights into strigolactone distribution and signalling. Curr Opin Plant Biol 16: 583–589.
24. Rasmussen A, Depuydt S, Goormachtig S, Geelen D (2013) Strigolactones fine-tune the root system. Planta 238: 615–626.
25. Ruyter-Spira C, Al-Babili S, van der Krol S, Bouwmeester H (2013) The biology of strigolactones. Trends Plant Sci 18: 72–83.
26. Janssen BJ, Drummond RS, Snowden KC (2014) Regulation of axillary shoot development. Curr Opin Plant Biol 17: 28–35.
27. Waldie T, McCulloch H, Leyser O (2014) Strigolactones and the control of plant development: lessons from shoot branching. Plant J. 2014 Feb 25. doi:10.1111/tpj.12488 [Epub ahead of print].
28. Zheng Z, Germain Ade S, Chory J (2014) Unfolding the mysteries of strigolactone signaling. Mol Plant 7: 934–936.
29. Booker J, Auldridge M, Wills S, McCarty D, Klee H, et al. (2004) MAX3/CCD7 is a carotenoid cleavage dioxygenase required for the synthesis of a novel plant signaling molecule. Curr Biol 14: 1232–1238.
30. Booker J, Sieberer T, Wright W, Williamson L, Willett B, et al. (2005) MAX1 encodes a cytochrome P450 family member that acts downstream of MAX3/4 to produce a carotenoid-derived branch-inhibiting hormone. Dev Cell 8: 443–449.
31. Sorefan K, Booker J, Haurogne K, Goussot M, Bainbridge K, et al. (2003) MAX4 and RMS1 are orthologous dioxygenase-like genes that regulate shoot branching in Arabidopsis and pea. Genes Dev 17: 1469–1474.
32. Stirnberg P, van de Sande K, Leyser HMO (2002) MAX1 and MAX2 control shoot lateral branching in Arabidopsis. Development 129: 1131–1141.
33. Lin H, Wang RX, Qian Q, Yan MX, Meng XB, et al. (2009) DWARF27, an iron-containing protein required for the biosynthesis of strigolactones, regulates rice tiller bud outgrowth. Plant Cell 21: 1512–1525.
34. Waters MT, Brewer PB, Bussell JD, Smith SM, Beveridge CA (2012) The Arabidopsis ortholog of rice DWARF27 acts upstream of MAX1 in the control of plant development by strigolactones. Plant Physiol 159: 1073–1085.
35. Kretzschmar T, Kohlen W, Sasse J, Borghi L, Schlegel M, et al. (2012) A petunia ABC protein controls strigolactone-dependent symbiotic signalling and branching. Nature 483: 341–344.
36. Arite T, Umehara M, Ishikawa S, Hanada A, Maekawa M, et al. (2009) d14, a strigolactone-insensitive mutant of rice, shows an accelerated outgrowth of tillers. Plant Cell Physiol 50: 1416–1424.
37. Gaiji N, Cardinale F, Prandi C, Bonfante P, Ranghino G (2012) The computational-based structure of Dwarf14 provides evidence for its role as potential strigolactone receptor in plants. BMC Res Notes 5: 307.
38. Waters MT, Nelson DC, Scaffidi A, Flematti GR, Sun YKM, et al. (2012) Specialisation within the DWARF14 protein family confers distinct responses to karrikins and strigolactones in Arabidopsis. Development 139: 1285–1295.
39. Hamiaux C, Drummond RSM, Janssen BJ, Ledger SE, Cooney JM, et al. (2012) DAD2 is an alpha/beta hydrolase likely to be involved in the perception of the plant branching hormone, strigolactone. Curr Biol 22: 2032–2036.
40. Nakamura H, Xue YL, Miyakawa T, Hou F, Qin HM, et al. (2013) Molecular mechanism of strigolactone perception by DWARF14. Nat Commun 4: 2613.
41. Jiang L, Liu X, Xiong G, Liu H, Chen F, et al. (2013) DWARF 53 acts as a repressor of strigolactone signalling in rice. Nature 504: 401–405.
42. Zhou F, Lin Q, Zhu L, Ren Y, Zhou K, et al. (2013) D14-SCF(D3)-dependent degradation of D53 regulates strigolactone signalling. Nature 504: 406–410.
43. Wang Y, Sun S, Zhu W, Jia K, Yang H, et al. (2013) Strigolactone/MAX2-induced degradation of brassinosteroid transcriptional effector BES1 regulates shoot branching. Dev Cell 27: 681–688.
44. Ward SP, Salmon J, Hanley SJ, Karp A, Leyser O (2013) Using Arabidopsis to study shoot branching in biomass willow. Plant Physiol 162: 800–811.
45. Salmon J, Ward SP, Hanley SJ, Leyser O, Karp A (2014) Functional screening of willow alleles in Arabidopsis combined with QTL mapping in willow (Salix) identifies SxMAX4 as a coppicing response gene. Plant Biotechnol J 12: 480–491.
46. Tuskan GA, Difazio S, Jansson S, Bohlmann J, Grigoriev I, et al. (2006) The genome of black cottonwood, Populus trichocarpa (Torr. & Gray). Science 313: 1596–1604.
47. Li Z, Czarnecki O, Chourey K, Yang J, Tuskan GA, et al. (2014) Strigolactone-regulated proteins revealed by iTRAQ-based quantitative proteomics in Arabidopsis. J Proteome Res 13: 1359–1372.
48. Murashige T, Skoog F (1962) A revised medium for rapid growth and bio assays with tobacco tissue cultures. Physiol Plant 15: 473–497.
49. Goodstein DM, Shu S, Howson R, Neupane R, Hayes RD, et al. (2012) Phytozome: a comparative platform for green plant genomics. Nucleic Acids Res 40: D1178–1186.
50. Nakagawa T, Suzuki T, Murata S, Nakamura S, Hino T, et al. (2007) Improved gateway binary vectors: High-performance vectors for creation of fusion constructs in Transgenic analysis of plants. Biosci, Biotechnol, Biochem 71: 2095–2100.
51. Holsters M, Silva B, Van Vliet F, Genetello C, De Block M, et al. (1980) The functional organization of the nopaline A. tumefaciens plasmid pTiC58. Plasmid 3: 212–230.
52. Clough SJ, Bent AF (1998) Floral dip: a simplified method for Agrobacterium-mediated transformation of Arabidopsis thaliana. Plant J 16: 735–743.
53. Arvidsson S, Kwasniewski M, Riano-Pachon DM, Mueller-Roeber B (2008) QuantPrime - a flexible tool for reliable high-throughput primer design for quantitative PCR. BMC Bioinformatics 9: 465.
54. Lamesch P, Berardini TZ, Li D, Swarbreck D, Wilks C, et al. (2012) The Arabidopsis Information Resource (TAIR): improved gene annotation and new tools. Nucleic Acids Res 40: D1202–1210.
55. Benfey PN, Chua NH (1990) The cauliflower mosaic virus 35S promoter: Combinatorial regulation of transcription in plants. Science 250: 959–966.
56. Ha CV, Leyva-Gonzalez MA, Osakabe Y, Tran UT, Nishiyama R, et al. (2014) Positive regulatory role of strigolactone in plant responses to drought and salt stress. Proc Natl Acad Sci U S A 111: 851–856.

15

Comparison of the Rhizosphere Bacterial Communities of Zigongdongdou Soybean and a High-Methionine Transgenic Line of This Cultivar

Jingang Liang[1]☯, Shi Sun[2]☯, Jun Ji[1], Haiying Wu[3], Fang Meng[1], Mingrong Zhang[3], Xiaobo Zheng[1], Cunxiang Wu[2]*, Zhengguang Zhang[1]*

1 Department of Plant Pathology, College of Plant Protection, Nanjing Agricultural University, and Key Laboratory of Integrated Management of Crop Diseases and Pests, Ministry of Education, Nanjing, China, 2 The National Key Facility for Crop Gene Resources and Genetic Improvement (NFCRI), MOA Key Laboratory of Soybean Biology (Beijing), Institute of Crop Science, The Chinese Academy of Agricultural Sciences, Beijing, China, 3 Nanchong Academy of Agricultural Science, Nanchong, China

Abstract

Previous studies have shown that methionine from root exudates affects the rhizosphere bacterial population involved in soil nitrogen fixation. A transgenic line of Zigongdongdou soybean cultivar (ZD91) that expresses *Arabidopsis* cystathionine γ-synthase resulting in an increased methionine production was examined for its influence to the rhizosphere bacterial population. Using 16S rRNA gene-based pyrosequencing analysis of the V4 region and DNA extracted from bacterial consortia collected from the rhizosphere of soybean plants grown in an agricultural field at the pod-setting stage, we characterized the populational structure of the bacterial community involved. In total, 87,267 sequences (approximately 10,908 per sample) were analyzed. We found that Acidobacteria, Proteobacteria, Bacteroidetes, Actinobacteria, Chloroflexi, Planctomycetes, Gemmatimonadetes, Firmicutes, and Verrucomicrobia constitute the dominant taxonomic groups in either the ZD91 transgenic line or parental cultivar ZD, and that there was no statistically significant difference in the rhizosphere bacterial community structure between the two cultivars.

Editor: Raffaella Balestrini, Institute for Sustainable Plant Protection, C.N.R., Italy

Funding: This research was supported by Genetically Modified Organisms Breeding Major Projects of China (Grant No. 2014ZX08011-003) and Jiangsu Provincial Coordination Panel for Recruitment and Selection of Specially-Appointed Professors. The funders had no role in study design, data collection and analysis, decision to publish, or preparation of the manuscript.

* Email: zhgzhang@njau.edu.cn (ZZ); wucx@mail.caas.net.cn (CW)

☯ These authors contributed equally to this work.

Introduction

The global commercial cultivation of transgenic crops has increased from 1.7 million hectares in 1996 to 170.3 million hectares in 2012 [1]. Modern agricultural biotechnology and genetic engineering have allowed the development of crops with improved properties, such as the transgenic high-methionine soybean line Zigongdongdou 91 (ZD91) [2]. Most studies to date have suggested that the release of transgenic plants results in only minor changes to the microbial community structure and that these minor changes are often transient and short lived [3–15].

Soil microorganisms are affected by soil characteristics, environmental conditions, and crop management strategies [16–19]. Organic compounds in root exudates and rhizosphere microbial communities perform fundamental processes that contribute to nutrient cycling and plant health and root growth, which are governed by complex interactions driven by various factors such as soil types and plant species [3,20–23]. The type and amount of root exudation may be an inherent property of a plant, and the rhizosphere microbial community is also expected to be unique [3,24–26]. Thus, an altered exudate composition may result in a corresponding shift in the rhizosphere microorganism community [27–30]. According to Faragova et al. [31], genetically modified plants release root exudates different from those of non-transgenic counterparts, therefore affecting the microbial community of the rhizosphere. This occurs even if the genetically modified plant is of different cultivars of the same plant species [32–34]. It was reported that methionine has the effect of inhibiting nitrification in soil, and transgenic alfalfa that produces high levels of methionine has a profound effect on bacteria population involved in the nitrogen cycle of the soil [31,35]. In a recent comparative study that employed Polymerase Chain Reaction-Denaturing Gradient Gel Electrophoresis (PCR-DGGE) technology to examine the bacterial communities in the rhizosphere of a transgenic high-methionine soybean, we found that the growth stage was the main factor affecting the community structure of the soil bacteria [36]. This finding is consistent with that of van Overbeek and van Elsas [25], Jin et al. [37] and Inceoglu et al. [38], in which bacterial community structures vary per growth stage changes. In addition, a study of rape (*Brassica napus*) indicated that the rhizosphere bacterial composition of a transgenic cultivar could be distinguished from that of its non-transgenic cultivar [39]. Conversely, in a study of potatoes, the microbial community structure in the rhizosphere was not significantly affected by genetic engineering [40,41].

The biological nitrogen cycle is one of the most important nutrient cycles in the terrestrial ecosystem. It includes four major processes: nitrogen fixation, mineralization (decay), nitrification and denitrification. Studies have shown that there are three functional genes: *nifH*, *amoA*, and *nosZ*, which encode the key enzymes involved in nitrogen fixation, ammonia oxidization and complete denitrification, respectively. Microorganisms capable of nitrogen fixation are called diazotrophs whose composition is determined by associated plant species. In addition, agricultural practices can also influence soil diazotrophs [42–45]. There are two types of diazotrophs: free-living and symbiotic [46,47]. The Rhizobia, *Frankia*, and Cyanobacteria are three important symbiotic diazotrophs. Rhizobia make up a paraphyletic group that falls into two classes in the phylum Proteobacteria: Alphaproteobacteria and Betaproteobacteria Nitrification, which converts ammonium to nitrate, includes two steps: ammonia oxidation to nitrite, and nitrite oxidation to nitrate. The oxidation of ammonia to nitrite is performed by two groups of organisms, ammonia-oxidizing bacteria (AOB) and ammonia-oxidizing archaea (AOA) [48]. AOB is found among the Betaproteobacteria and Gammaproteobacteria [49,50]. Denitrification is a microbially-facilitated process of nitrate reduction. So far, more than 60 genera of denitrifying microorganisms have been identified [51].

High-throughput TAG-encoded FLX amplicon pyrosequencing (using 16S rRNA genes) is a great technical advance for microbial ecology studies [52]. This technique can produce ~400,000–1,000,000 reads in a single 4 h run, with an accuracy of approaching 99%, and the tag sequence contains adequate information for taxonomic assignment [53]. Several recent studies have used this technique with the 16S rRNA gene to determine the microbial diversity in various environmental samples [3,52–60]. Given the wealth of information available due to the deep sequencing of environmental DNA, it is possible to determine whether these lineages are ecologically coherent [3]. It also has been proposed that some taxa are extremely coherent ecologically. Such taxa can be said to follow an r- or K-type life strategy [61–63].

In the present study, we used 16S rRNA gene-based pyrosequencing to compare the community dynamics of bacteria inhabiting the rhizosphere of field-grown transgenic cultivar ZD91 during the pod-setting stage to those of the corresponding non-transgenic ZD plants. Our study showed a wide variety of rhizosphere bacterial population associated with these cultivars but there was no significant difference in the bacterial population between transgenic cultivar ZD91 and parental cultivar ZD.

Results

Quantification of *nifH*, *amoA*, and *nosZ* genes, and carbon, nitrogen concentrations in the rhizospheres

To examine the effect of transgenic soybean on the abundance of diazotrophs, ammonia-oxidizing bacteria, and denitrifiers, we used quantitative real-time PCR (qPCR) assays to quantify *nifH*, *amoA*, and *nosZ* genes to inform nitrogen-cycling related bacteria collected from the rhizosphere soil. The abundance of *nifH*, *amoA*, and *nosZ* genes between ZD and ZD91 showed no significant differences (Table 1).

We also estimated whether there are any differences in carbon and nitrogen levels in the rhizosphere of ZD and ZD91 and found no significant differences (Table 1).

Amino acids in root exudates

Amino acids included aspartic acid, threonine, serine, glutamic acid, glycine, alanine, valine, methionine, isoleucine, leucine, tyrosine, phenylalanine, lysine and histidine in root exudates were determined. However, there were no differences in amino acids contents between ZD and ZD91 (Table 2).

Pyrosequencing analysis of the bacterial communities of the rhizospheres

We performed 16S rRNA gene-based pyrosequencing analysis of the V4 region to characterize the bacterial community composition in the rhizospheres surrounding ZD and ZD91 cultivars. A total of 87,267 high-quality sequences was obtained, with an average of 10,908 sequences per sample. The results were summarized in Table 3 with detailed sample characteristics shown in Table S1. We used the grouping of cultivars based on the number of replicates (ZD, four replicates: ZD_1, ZD_2, ZD_3, and ZD_4; ZD91, four replicates: ZD91_1, ZD91_2, ZD91_3, and ZD91_4).

The rarefaction curves tended towards the saturation plateau while Good's coverage estimations revealed that 78–84% of the species were from eight samples, indicating that the sequence coverage was sufficient to capture the diversity of the bacterial communities in the samples (Fig. 1). The estimators of community richness (Chao and ACE) and diversity (Shannon and Simpson) were shown in Table 3, and there were no significant differences in these indices between ZD and ZD91. The null hypothesis was tested through AMOVA analysis (Table 4) and it did show that there were no significant differences between ZD and ZD91.

Taxonomic composition

Although probably present, this study did not attempt to characterize any Archaea. All sequences were classified from phylum to genus according to the RDP classifier using the default settings. 22 different phyla, including 1 archaeal phylum, were identified from the samples. The overall microbiota structure for ZD and ZD91 at the phylum level was shown in Figure 2 and Table S2. Acidobacteria, Proteobacteria, Bacteroidetes, Actinobacteria, Chloroflexi, Planctomycetes, Gemmatimonadetes, Firmicutes, and Verrucomicrobia accounted for >90% of the reads. Again, there were no observable differences among these categories between ZD and ZD91 at this taxonomic level (Fig. 3).

A total of 13,299 operational taxonomic units was found in the complete data set (Table S3). The ten most abundant OTUs within each sample were identified (Table S4). There were 12 dominant OTUs in total, including Gp4, Gp6, *Terrimonas, Sphingosinicella, Sphingomonas, Flavisolibacter, Sphingomonadaceae, Lysobacter, Chryseobacterium, Acinetobacter, Levilinea and Opitutus.* A total of 2047 common OTUs was determined in rhizosphere soil and there was no significant difference between ZD and ZD91 ($P<0.01$) (Fig. 4). The ten most abundant genera from each of the eight samples were classified as a percentage of the total sequences per sample (Table S5). This analysis showed that there were 17 dominant genera in total, including four members of the Acidobacteria, five of the Bacteroidetes, three Alphaproteobacteria, two Gammaproteobacteria, and one each of the Actinobacteria, Gemmatimonadetes and Verrucomicrobia.

To analyze whether the distribution of low-abundance genera differs between samples, sequences with abundances below 0.1% were analyzed. A total of 481 such low-abundance genera was found. Of the 481 genera, 214 (70.63%), 203 (68.35%), 200 (61.54%), and 219 (64.60%) were observed in ZD_1, ZD_2,

Table 1. Changes in abundance of *nifH*, *amoA*, and *nosZ* genes (Log_{10} copies/g soil), and carbon (C), nitrogen (N) contents in rhizosphere soil ($P<0.01$).

	ZD	ZD91
nifH	9.05±0.02 A	9.06±0.12 A
amoA	8.03±0.02 A	8.13±0.01 A
nosZ	7.70±0.08 A	7.90±0.02 A
Total C (%)	1.68±0.02 A	1.65±0.03 A
Total N (%)	0.06±0.01 A	0.07±0.01 A

ZD_3, and ZD_4, respectively. In comparison, 208 (67.97%), 205 (61.75%), 219 (68.65%), and 206 (65.19%) were detected in ZD91_1, ZD91_2, ZD91_3, and ZD91_4, respectively. There were no genera unique to cultivar ZD or cultivar ZD91. Rare genera (abundances below 0.1%) accounted for over 60% in all samples, and there was no significant difference between cultivar ZD (66.28±4.02%) and cultivar ZD91 (65.89±3.14%).

Root-colonizing pseudomonads are known to be involved in plant growth and health. These bacteria increase plant growth either directly through the production of phytohormones or other stimulants and increasing the bioavailability of nutrients in the soil, or indirectly by the suppression of plant diseases and the induction of systemic resistance in plants [10]. A total of 6 pseudomonad species was found in our samples and there was no significant difference between ZD (0.07±0.03%) and ZD91 (0.05±0.04%).

Nitrogen-fixing Bacteria

In the present study, *Frankia* was not found in any samples. Sequences related to Cyanobacteria, Alphaproteobacteria, and Betaproteobacteria were not significantly different between cultivar ZD (0.22±0.07%, 14.04±5.07%, and 1.23±0.30%) and cultivar ZD91 (0.14±0.08%, 13.56±4.88%, and 1.11±0.25%) samples.

Nitrifying bacteria

In the present study, there were 35 Gammaproteobacteria genera detected. There was no significant difference between the sequences for cultivar ZD (3.47±0.17%) and ZD91 (4.49±1.38%). Nitrifying bacteria, which oxidize nitrite, include *Nitrobacter*, *Nitrospina*, *Nitrococcus*, and *Nitrospira* [65], but only *Nitrospira* was detected in this study. In addition, there was no significant difference between the sequences for cultivar ZD (0.36±0.16%) and cultivar ZD91 (0.44±0.16%).

Denitrifying bacteria

In the present study, 36 such bacterial genera were found. The sequences related to denitrifying bacteria were not significantly different between cultivar ZD (3.13±1.31%) and cultivar ZD91 (2.80±0.28%).

Shared OTUs in ZD and ZD91 libraries

The bacterial OTUs in the cultivar ZD and cultivar ZD91 libraries were further investigated for shared OTUs. As shown in Table S6, ZD_1, ZD_2, ZD_3, and ZD_4 had 700 OTUs in common. A statistical analysis revealed that the OTUs common to the four libraries comprised 58.96%, 59.62%, 52.19%, and 46.82% of the reads in the ZD_1, ZD_2, ZD_3, and ZD_4 libraries, respectively. Acidobacteria, Proteobacteria, and Bacteroidetes comprised 531 of the shared OTUs (75.86% in propor-

Table 2. Amino acids secreted by soybean roots in the root box at pod-setting stage (μmol/ml) ($P<0.01$).

	ZD	ZD91
Asp	3.44±1.56 A	1.13±0.48 A
Thr	2.85±1.28 A	0.99±0.41 A
Ser	12.13±6.05 A	4.11±1.82 A
Glu	2.31±0.59 A	1.30±0.23 A
Gly	4.83±2.43 A	1.60±0.86 A
Ala	3.63±1.72 A	1.32±0.51 A
Val	2.14±0.83 A	1.01±0.25 A
Met	0.32±0.28 A	0.31±0.27 A
Ile	1.04±0.41 A	0.40±0.17 A
Leu	1.54±0.55 A	0.77±0.24 A
Tyr	2.03±0.58 A	1.33±0.20 A
Phe	1.22±0.47 A	0.65±0.17 A
Lys	1.03±0.57 A	0.33±0.20 A
His	2.12±1.36 A	0.68±0.46 A

Table 3. Pyrosequencing data summary.

Sample	Sequences	OTUs	ACE	Chao	Shannon	Simpson	Coverage
ZD	10534±1353A	3391±505A	8900±1019A	6381±780A	7.34±0.21A	0.002±0.001A	0.82±0.02A
ZD91	11283±1818A	3649±370A	9637±795A	6911±688A	7.47±0.17A	0.001±0.000A	0.82±0.03A

Values are the mean±standard deviation of four replicates. Within each vertical column, values followed by the same letter are not statistically different from each other ($P<0.01$). The number of OTUs, richness estimators Chao and ACE, diversity estimators Shannon and Simpson, and Good's coverage were calculated at 3% distance.

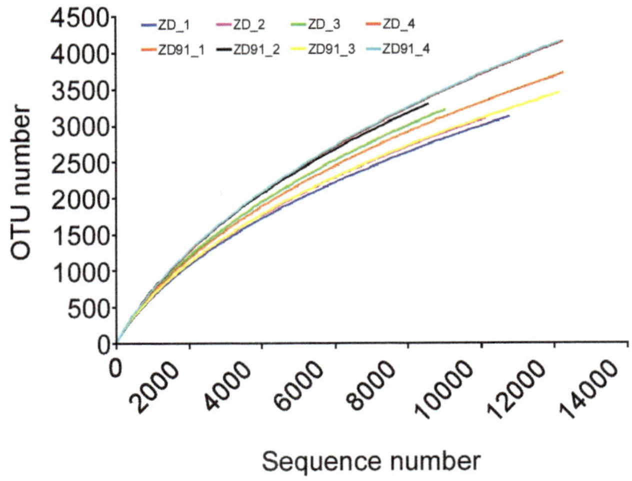

Figure 1. Rarefaction analysis. Rarefaction curves of OTUs clustered at 97% sequence identity across the samples. The sample labeled with ZD_1, ZD_2, ZD_3 and ZD_4 correspond to four replicates of cultivar ZD; ZD91_1, ZD91_2, ZD91_3 and ZD91_4 represent four replicates of transgenic cultivar ZD91.

tion). Further, the OTUs common to the four libraries of the transgenic cultivar comprised 54.04%, 49.01%, 52.17%, and 42.50% of the reads in the ZD91_1, ZD91_2, ZD91_3, and ZD91_4 libraries, respectively. Acidobacteria, Proteobacteria, and Bacteroidetes included 534 of the shared OTUs (76.29% in proportion).

Bacterial community dynamics

To examine the effect of cultivar on the total distribution of phyla and genera, we performed principal coordinate analysis (PCoA) of our data. The PCoA analysis of weighted UniFrac distances was used to calculate pairwise distances between the rhizosphere bacterial communities of the eight samples [64]. The first two axes of the PCoA represented 40.52% and 19.06% of the total variation, respectively (Fig. 5). AMOVA analysis (Table S7) was used to find out whether the separation of cultivar ZD and cultivar ZD91 in the PCoA was statistically significant. The result showed p-value was 0.859 indicating no significantly different between cultivar ZD and cultivar ZD91. Furthermore, we calculated ADONIS differences between cultivar ZD and cultivar ZD91 (Table 5), and found no significant difference between these two cultivars. Thus, the bacterial communities in the cultivars were not different from each other.

Discussion

In this study, we monitored the quantitative changes of the key genes involved in N-cycling as well as elements carbon, nitrogen concentrations in rhizosphere soil of ZD and ZD91, and analyzed the amino acids contents in the root exudates of different cultivars. We found that there were no significant differences between cultivar ZD and cultivar ZD91 in nitrogen-cycling genes and carbon, nitrogen concentrations. The results showed that the amino acids contents in the root exudates of ZD91 were lower than ZD, but they were not statistically significant.

We assessed the dynamics of the relative soil abundance of bacteria and the structure of soil bacterial communities as a function of cultivar type using the transgenic soybean cultivar ZD91 and its parental non-transgenic soybean cultivar ZD in a single experimental field. According to the pyrosequencing results,

Table 4. AMOVA analysis between ZD and ZD91.

	ZD91-ZD	Among	Within	Total
Based on OTU (Jclass distance coefficient)	SS	0.261484	1.603340	1.86483
	df	1.000000	6.000000	7.00000
	MS	0.261484	0.267224	
	Fs: 0.978522			
	p-value: 0.745			
Based on OTU (Thetayc distance coefficient)	SS	0.0321963	0.290100	0.322296
	df	1.0000000	6.000000	7.000000
	MS	0.0321963	0.048350	
	Fs: 0.665901			
	p-value: 0.565			
Based on weighted UniFrac distances	SS	0.0172751	0.1477470	0.165022
	df	1.0000000	6.0000000	7.000000
	MS	0.0172751	0.0246245	
	Fs: 0.701539			
	p-value: 0.861			

Experiment-wise error rate: 0.05.

genetic modification via the insertion and expression of *AtD-CGS* did not affect the composition of the bacterial community in the rhizosphere of soybean. However, the profiles of independent replicates differed in terms of the abundance of various phyla (Fig. 5 and Table S2), but this variation did not correlate with the experimental variables. All plots were on the same field, but due to the large size of the field, the replicates were located an average of 20 meters apart from each other. Thus, field heterogeneity was not considered when evaluating the impact of transgenic soybean on rhizosphere bacterial community. Furthermore, we analyzed the OTUs of high abundances and repeated well (occurred in more than three samples, defined as the common OTUs) in the soil. Since the singletons and low abundance OTUs were detected randomly, and there was no significant difference within the treatment (among the 4 replicates) and between the treatments (Fig. 4).

Due to logistic and technical (e.g., low sequence numbers) constraints, studies of soil microbial diversity and function have focused mainly on the most dominant species [3]. Previous studies have demonstrated weaknesses in traditional molecular methods

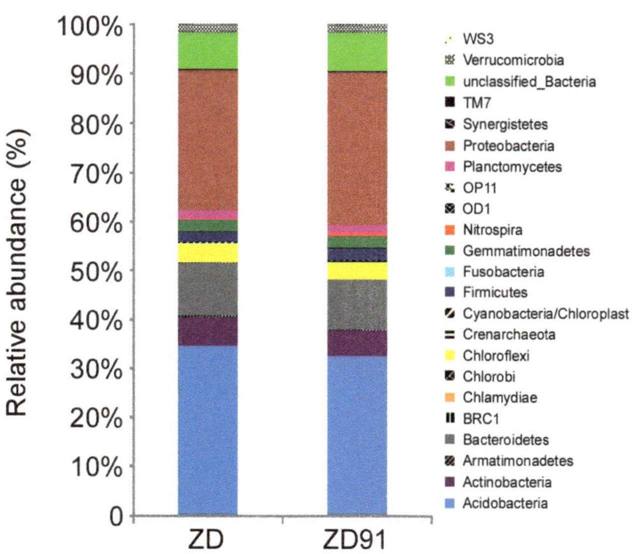

Figure 2. Bacterial composition at the phylum level. Relative read abundance of bacterial phyla within the communities. Sequences that could not be classified into any known group were assigned as unclassified_Bacteria.

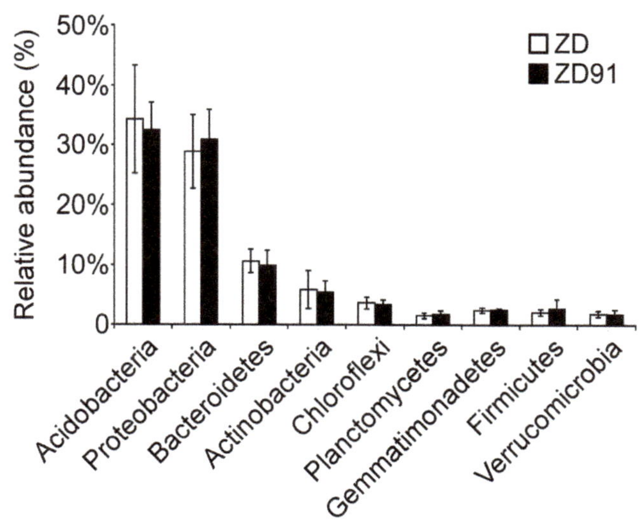

Value less than ten:

	Bacteroidetes	Actinobacteria	Chloroflexi	Planctomycetes	Gemmatimonadetes	Firmicutes	Verrucomicrobia
ZD		5.83%	3.63%	1.54%	2.43%	2.09%	1.83%
ZD91	9.92%	5.43%	3.35%	1.70%	2.51%	2.69%	1.73%

Figure 3. Relative abundance of main phyla in the rhizosphere of soybean cultivars. Error bars indicate standard errors. The *p*-values of the nine groups of bacteria, in order, were 0.734, 0.633, 0.662, 0.834, 0.687, 0.690, 0.751, 0.493, and 0.840 when testing for differences between cultivar ZD and transgenic cultivar ZD91.

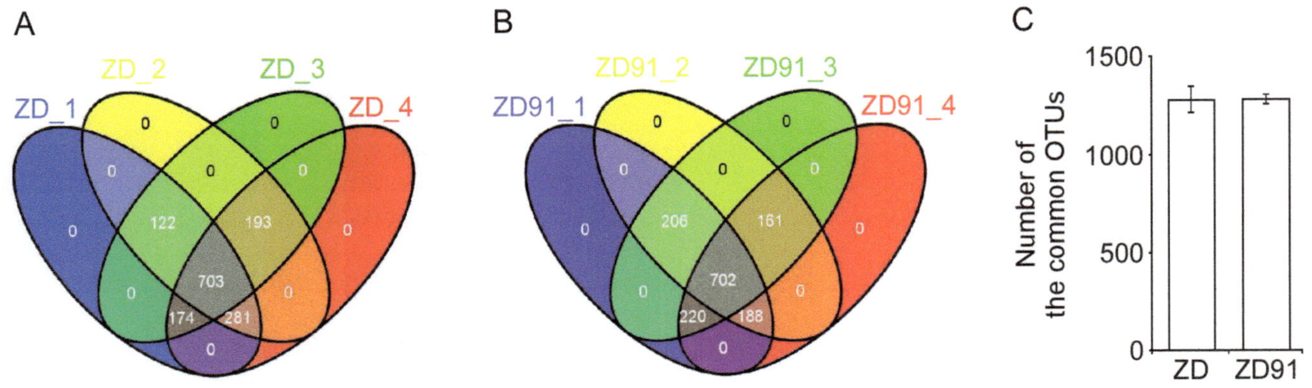

Figure 4. Common OTUs recovered from soil. (A) Venn diagram showing variable overlap between four replicates of ZD. (B) Venn diagram showing variable overlap between four replicates of ZD91. (C) Number of the common OTUs between ZD and ZD91.

(e.g., DGGE, RFLP, ARDRA, and SSCP) in terms of resolution; specifically, it is difficult to detect subtle changes in banding patterns or changes in closely related species, leading to the underestimation of bacterial diversity by one to two orders of magnitude [4,54,57,66]. Moreover, some rare species may have large effects on soil function, in spite of their low total biomass or rarity [67,68]. In this study, pyrosequencing provided a promising, fast, economical, and unique opportunity to access the less abundant bacterial taxa surrounding soybean plants, as well as the chance to compare their presence across cultivars in the pod-setting stage [54,55,57]. Similar to the results of Liu et al. [20] and Xu et al. [21,69], our sequencing results revealed that numerous bacterial phyla (e.g., Acidobacteria, Proteobacteria, Bacteroidetes, Actinobacteria, Chloroflexi, Planctomycetes, Gemmatimonadetes, Firmicutes, Nitrospirae, Verrucomicrobia, BRC1, Chlamydiae, Cyanobacteria, Fusobacteria, Chlorobi, OD1, OP11, Synergistetes, TM7, Armatimonadetes, and WS3) commonly inhabit the soybean rhizosphere.

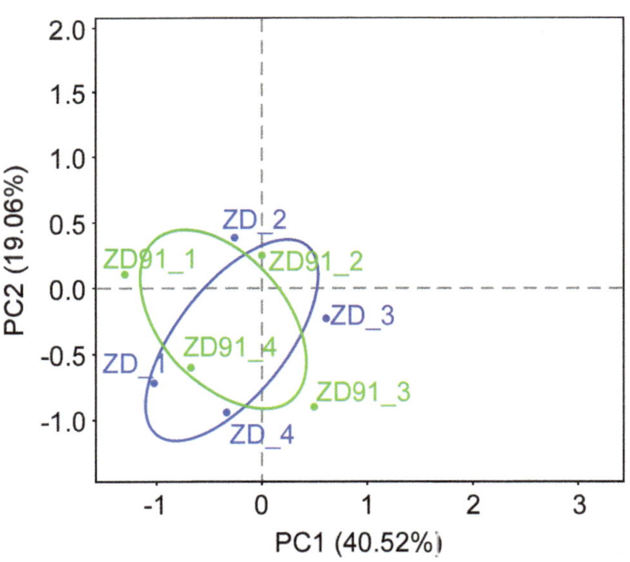

Figure 5. Principal coordinate analysis (PCoA) plot of samples using the weighted UniFrac distance metric. The variance explained by each principal coordinate axis is shown in parentheses. Datasets were subsampled to equal depth prior to the UniFrac distance computation. A circle was drawn with R around the samples of the same type (blue represents cultivar ZD, and green represents transgenic cultivar ZD91).

Moreover, dominant members of particular phyla or classes were used as indicators to reveal the prevailing ecological pattern (e.g., along the copiotroph/oligotroph scale) [3]. Members of the Betaproteobacteria and Bacteroidetes, as well as species belonging to the genus *Pseudomonas*, tend to be favored in r-selection/copiotrophic soils, which have higher carbon availability, while Acidobacteria are favored under K-selection/oligotrophic conditions [61–63]. Consequently, such groups may have higher abundances in soils with higher versus lower levels of easily available carbon, respectively. Furthermore, the ratio of Proteobacteria to Acidobacteria can be used as an indicator of the soil trophic level [70]. Although it is rather implausible that a whole phylum would have shared ecological characteristics [3], our indicators revealed variability (based on the ratio of Proteobacteria to Acidobacteria) occurs in the response of Proteobacteria versus Acidobacteria to shifting ecological conditions in the soil. The ratio is 0.16 in oligotrophic soil [71,72], 0.34 in low-input agricultural soil [73], 0.46 in low-nutrient systems [62], and 0.87 in high-input agricultural systems [74]. The values obtained in this study, 0.93 ± 0.46 in cultivar ZD, and 0.98 ± 0.31 in cultivar ZD91, indicating a high-nutrient system, but again there was no significant difference between the two cultivars.

The pyrosequencing protocol used in previous investigations and here carries certain limitations [3,52–60]. Any PCR primer set for the amplification of 16S rDNA genes may miss a considerable amount of the extant microbial diversity; thus, at present, no clear answer can be given with respect to the real extant community make-up [3]. Because PCR biases mostly affect the detection of rare sequences in a sample, it is important to view such data on the rare biosphere with proper cautions.

Of the ten most abundant bacterial OTUs in the samples, several were related to Acidobacteria. Of the ten most abundant bacterial genera, several were related to both Acidobacteria and Proteobacteria. The Acidobacteria are physiologically diverse and ubiquitous, especially in soil, but are underrepresented in culture [75–77]. Because most Acidobacteria have not been cultured, their ecology and metabolism are not well understood [76]. However, these bacteria may be important contributors to ecosystems because they are particularly abundant in soil [78]. The Proteobacteria are a major group of bacteria, and include many of the species responsible for nitrogen fixation.

We also monitored structural and quantitative changes in the bacterial communities of each soybean regarding nitrogen cycling.

Table 5. ADONIS differences between ZD and ZD91.

	Df	SumsOfSqs	MeanSqs	F. Model	R^2	Pr(>F)
qiime.data$map [[opts$category]]	1	0.017275	0.017275	0.70124	0.10464	0.8492
Residuals	6	0.147814	0.024636		0.89536	
Total	7	0.165089			1.00000	

ADONIS differences is a non-parametric method to measure statistical significance of sample grouping. It was calculated based on weighted UniFrac distance metric.

The abundance of nitrogen-transforming bacteria were not significantly different between cultivars ZD and ZD91.

In conclusion, we found that 454-pyrosequencing can be a useful tool for risk assessment studies to analyze the immediate impact of plants on the diversity of soil bacteria in rhizospheres. Of all DNA reads, 7.56% belonged to unclassified bacteria, containing a suite of diverse taxa. To understand the ecological impact of unclassified sequences, they must be identified and ideally an organism should be cultured. This once again stresses that culture-independent techniques should go hand-in-hand with culture-dependent ones [3]. For ecological risk identification, however, the observation of shifts during one stage of plant development and in a one-year period is insufficient. The positive identification of risk requires that changes be demonstrated at all stages of soybean development and over an extended time frame. The fact that no effect between transgenic and non-transgenic plants was detected suggests only minor physiological changes caused by the insertion of *AtD-CGS* into the soybean genome.

Materials and Methods

Soybean Cultivars

Transgenic soybean cultivar (ZD91) contains the *Arabidopsis* cystathionine γ-synthase (*AtD-CGS*) gene which has been introduced artificially into the soybean cultivar Zigongdongdou (ZD) using *Agrobacterium*-mediated transformation, and which exhibits a high content of methionine in the seeds.

Field tests setup and sampling

An experimental field - Nanchong (30°48′N, 106°04′E), Sichuan Province, China, in which a completely randomized block design was set out in 2011, was used. The basic properties of the soil were 46.98 g/kg organic matter, 0.51 g/kg total nitrogen, 11.73 mg/kg available phosphorus, 220.40 mg/kg available potassium, and 52.81 mg/kg alkali-hydrolyzed nitrogen. The studied region was 2.8 acres. For each soybean cultivar, four replicate plots, which were randomly distributed over the field, were used. The field was under standard agricultural practice. The samples were collected at the pod-setting stage (80 days after seedling emergence) of the two soybean cultivars. Rhizosphere soil samples were collected as described previously [38]. Briefly, five plants were carefully sampled using five-point sampling method and the plants with adhering soil were immediately taken to the laboratory. The loosely adhering soil on the roots was shaken off, and the resulting roots (containing rhizosphere soil) were pooled per plot. Using the pooled sample, soil tightly adhering to the roots was brushed off and collected (constituting rhizosphere soil).

Ethics statements

This study was approved by the Ministry of Agriculture of the People's Republic of China and the genetically modified organisms safety team of Nanjing Agricultural University, China. The field studies did not involve endangered or protected species. The land was not privately owned or protected in any way.

Soil DNA extraction, PCR amplification and pyrosequencing

Soil DNA was extracted from rhizosphere soil samples by employing the FastDNA SPIN Kit for Soil (MP Biomedicals, USA) as recommended by the manufacturer. DNA concentrations were quantified by using a NanoDrop 1000 Spectrophotometer (Thermo Scientific, USA) according to the manufacturer's protocol. The 16S rDNA gene (V4 region) was amplified using the barcode primers 515F [79] and 926R [80]. The primers were designed as 5′-adapter+barcode+gene specific primer-3′ (Table S8). The PCR reaction mixture (20 μL) contained 2 μL 10-fold buffer, 1.6 μL dNTP mix (2.5 mM each), 1 μL BSA (20 mg/ml), 0.25 μL rTaq polymerase (2.5 U/μL), 0.2 μL of each of the primers (10 μM), 0.5–5 μL (1–50 ng) of isolated DNA as template. The mixtures were placed in a TaKaRa PCR Thermal Cycler Dice (TaKaRa, Japan) and thermal cycling was performed as follows: initial denaturation consisting of 5 min at 95°C; followed by 30 cycles consisting of 30 sec at 94°C, 30 sec at 57°C and 30 sec at 72°C; and final extension for 10 min at 72°C. Analysis of PCR products on 1% agarose gel revealed bands of the corresponding size. These bands were cut and purified using the QIAquick Gel Extraction Kit as recommended by the manufacturer (Qiagen, Netherlands), and then purified by Agencourt AMPure XP (Beckman Coulter, USA). Quantification of the purified PCR products was performed using the Quant-iT PicoGreen dsDNA Assay Kit (Invitrogen, Netherlands) and a TBS-380 Fluorometer (Turner Biosystems, USA) as recommended by the manufacturer. The PCR products from different samples were then mixed in equal ratios (7.2 ng per sample) for pyrosequencing with a Roche Genome Sequencer GS-FLX Titanium platform [56].

Real-time PCR

The abundance of *nifH*, *amoA*, and *nosZ* genes in all samples were quantified using real-time PCR. Quantitative real-time PCR was performed according to the methods from previous studies: *nifH* (as a measure of N-fixing bacteria) used primers nifH-F (5′-AAA GGY GGW ATC GGY AAR TCC ACC AC-3′) and nifH-R (5′-TTG TTS GCS GCR TAC ATS GCC ATC AT-3′) [81]; *amoA* (as a measure of ammonia-oxidizing bacteria) used primers amoA-1F (5′-GGG GTT TCT ACT GGT GGT-3′) and amoA-2R (5′-CCC CTC KGS AAA GCC TTC TTC-3′) [82]; and *nosZ* (as a measure of denitrification bacteria) used primers nosZ-F (5′-CGY TGT TCM TCG ACA GCC AG-3′) and nosZ 1622R (5′-CGS ACC TTS TTG CCS TYG CG-3′) [83]. Purified PCR products from a common DNA mixture (equal amounts of DNA from all samples) were used to prepare sample-derived quantification standards as previously described [84]. In comparison to using a clone (plasmid) as standard, this method avoids the difference of PCR amplification efficiency between standards and samples caused by the different sequence composition in the PCR templates (single sequence in a plasmid for the standard vs. mixture of thousands of sequences in a soil sample) [83]. The copy number of gene in each sample was calculated by http://cels.uri.edu/gsc/cndna.html and the data were Log_{10} transformed [84]. The 20 μL reaction mixture contained 10 μL SYBR *Premix Ex Taq* (Tli RNaseH Plus) (2×) (TaKaRa, Japan), 0.4 μL of each primer (10 μM), 0.4 μL ROX Reference Dye II (50×) (TaKaRa, Japan), 2 μL template, and 6.8 μL dH_2O (sterile distilled water). Real-time amplification was performed in an Applied Biosystems 7500 Fast Real-Time PCR System using the following program: 95°C for 30 s; 40 cycles of 95°C for 5 s, (58°C for *nifH* gene; 57°C for *amoA* gene; 64°C for *nosZ* gene) for 31 s, 72°C for 34 s. A dissociation step was added at the end of the qPCR to assess amplification quality. The R^2 of all these standard curves were higher than 0.99. All quantitative PCR reactions were performed in triplicate.

Root box design and determination of amino acids secreted by roots

The root box used in this study was designed as reported previously [85]. There were three replicates of root box for each genotype. The lower chamber of the root box was filled with 500 mL of sterile ultrapure water, which was used to collect the amino acids exuded by the roots during a 72-h period at the pod-setting stage. Samples of root exudate were collected, filtered through a 0.45 μm membrane, freeze-dried, and redissolved in 1 mL of deionized water. Amino acids contents in the root exudates were analyzed by L-8900 Amino Acid Analyzer (Hitachi, Japan).

Chemical analysis of carbon and nitrogen contents in rhizosphere soil

The concentrations of carbon and nitrogen in rhizosphere soil were determined by Vario MICRO cube (Elementar, Germany). The detection was based on JY/T 017–1996 General rules for elemental analyzer which was approved by Ministry of Education of the People's Republic of China.

Bioinformatic analysis

All sequences generated in this study can be downloaded from NCBI Sequence Read Archive, accession number: SRP026381. On the basis of several previous reports describing sources of errors in 454 sequencing runs, the valid reads should comply with the following rules: the sequences with low quality (length< 200 bp, with ambiguous base 'N', and average base quality score< 20) were removed; the sequences that matched the primer and one of the used barcode sequences and had at least an 80% match to a previously determined 16S rRNA gene sequence were retained [54,56,83,86]. The sequences were used for chimera check using UCHIME method in Mothur [87], and chimeric sequences were removed.

The high-quality sequences were assigned to samples according to barcodes. Sequences were aligned in accordance with RDP (Ribosomal Database Project) alignment [88] and clustered into operational taxonomic units (OTUs). OTUs were defined using a threshold of 97% identity, within and between groups, which was a criterion for species level delineation in previous studies [56]. OTUs that reached 97% similarity level were used for diversity (Shannon and Simpson), richness (Chao and ACE), Good's coverage, and Rarefaction curve analysis by using Mothur [87]. The OTUs which occurred in more than three samples from either ZD or ZD91 were defined as the common OTUs [89]. Taxonomic assignments of OTUs exhibiting 97% similarity were performed by using Mothur in accordance with RDP Classifier at 80% confidence threshold. The null hypothesis was tested using OTU and weighted UniFrac distances. The study objective was to test whether the transgenic soybean had an effect on rhizosphere bacterial communities. AMOVA analysis was done by using Mothur. ADONIS differences were calculated by using R.

To compare bacterial community structures across all samples based on the relative abundance of bacterial phyla, Principal coordinate analysis (PCoA) was performed by using Qiime [90]. AMOVA analysis was used to find out whether the separation of ZD and ZD91 in the PCoA was statistically significant.

Statistical analysis

One-way analysis of variance and Duncan pair-wise comparisons ($P<0.01$) were used to determine the minimum significant difference between soybean cultivars (rhizosphere soils) by employing SPSS version 17.0 for Windows.

Supporting Information

Table S1 Pyrosequencing data and estimator index of each sample. The number of OTUs, richness estimator Chao and ACE, diversity estimator Shannon and Simpson, and Good's coverage were calculated at 3% distance.

Table S2 Bacterial composition of the communities at the phylum level. The number in the table (except the first line and column) represent pyrosequencing reads.

Table S3 Different OTUs for each sample.

Table S4 Classification of the 10 most abundant bacterial OTUs in the eight samples. Relative abundance (%, a percentage of the total sequences per sample) of each OTU is under the genus name/group number. OTUs were identified using 97% cutoffs. All Gp4 OTU classify in the same class 'Acidobacteria_Gp4' and all Gp6 OTU classify in the same class 'Acidobacteria_Gp6'.

Table S5 The ten most abundant genera found in the eight samples. Relative abundance (%, a percentage of the total sequences per sample) of each genus is included in parentheses. All Gp6 classify in the same class 'Acidobacteria_Gp6', all Gp4 classify in the same class 'Acidobacteria_Gp4' and all Gp3 classify in the same class 'Acidobacteria_Gp3'.

Table S6 Shared OTUs in the samples. (A) Shared OTUs in the ZD libraries. (B) Shared OTUs in the ZD91 libraries.

Table S7 AMOVA analysis between ZD and ZD91 in PCoA. AMOVA analysis in PCoA (classfily method: UPGMA). AMOVA analysis was used to find out whether the separation of ZD and ZD91 in the PCoA is statistically significant.

Table S8 Primers with tags and adapters used in pyrosequencing.

Acknowledgments

We are grateful to Ping Wang of LSUHSC (USA), Ruifu Zhang and Zhongli Cui of Nanjing Agricultural University (China) for their helpful suggestions.

Author Contributions

Conceived and designed the experiments: ZZ CW SS XZ. Performed the experiments: JL JJ FM. Analyzed the data: JL. Contributed reagents/materials/analysis tools: CW SS HW MZ. Wrote the paper: JL ZZ.

References

1. James C (2012) Global Status of commercialized biotech/GM crops: 2012. Ithaca: The International Service for the Acquisition of Agri-biotech Applications. 1 p.
2. Song S, Hou W, Godo I, Wu C, Yu Y, et al. (2013) Soybean seeds expressing feedback-insensitive cystathionine gamma-synthase exhibit a higher content of methionine. J Exp Bot 64: 1917–1926.
3. Inceoglu O, Al-Soud WA, Salles JF, Semenov AV, van Elsas JD (2011) Comparative analysis of bacterial communities in a potato field as determined by pyrosequencing. PLoS One 6: e23321.
4. Liu B, Zeng Q, Yan FM, Xu HG, Xu CR (2005) Effects of transgenic plants on soil microorganisms. Plant Soil 271: 1–13.
5. Milling A, Smalla K, Maidl FX, Schloter M, Munch JC (2004) Effects of transgenic potatoes with an altered starch composition on the diversity of soil and rhizosphere bacteria and fungi. Plant Soil 266: 23–39.
6. Li XG, Zhang TL, Wang XX, Hua K, Zhao L, et al. (2013) The composition of root exudates from two different resistant peanut cultivars and their effects on the growth of soil-borne pathogen. Inter J Biol Sci 9: 164–173.
7. Hur M, Kim Y, Song HR, Kim JM, Choi YI, et al. (2011) Effect of genetically modified poplars on soil microbial communities during the phytoremediation of waste mine tailings. Appl Environ Microbiol 77: 7611–7619.
8. Hannula SE, de Boer W, van Veen J (2012) A 3-year study reveals that plant growth stage, season and field site affect soil fungal communities while cultivar and GM-trait have minor effects. PLoS One 7: e33819.
9. Duc C, Nentwig W, Lindfeld A (2011) No adverse effect of genetically modified antifungal wheat on decomposition dynamics and the soil fauna community - A Field Study. PLoS One 6: e25014.
10. Meyer JB, Song-Wilson Y, Foetzki A, Luginbuhl C, Winzeler M, et al. (2013) Does wheat genetically modified for disease resistance affect root-colonizing Pseudomonads and Arbuscular Mycorrhizal Fungi? PLoS One 8: e53825.
11. Di Giovanni GD, Watrud LS, Seidler RJ, Widmer F (1999) Comparison of parental and transgenic alfalfa rhizosphere bacterial communities using Biolog GN metabolic fingerprinting and enterobacterial repetitive intergenic consensus sequence-PCR (ERIC-PCR). Microb Ecol 37: 129–139.
12. Rasche F, Hodl V, Poll C, Kandeler E, Gerzabek MH, et al. (2006) Rhizosphere bacteria affected by transgenic potatoes with antibacterial activities compared with the effects of soil, wild-type potatoes, vegetation stage and pathogen exposure. FEMS Microbiol Ecol 56: 219–235.
13. Heuer H, Kroppenstedt RM, Lottmann J, Berg G, Smalla K (2002) Effects of T4 lysozyme release from transgenic potato roots on bacterial rhizosphere communities are negligible relative to natural factors. Appl Environ Microbiol 68: 1325–1335.
14. Duke SO, Lydon J, Koskinen WC, Moorman TB, Chaney RL, et al. (2012) Glyphosate effects on plant mineral nutrition, crop rhizosphere microbiota, and plant disease in glyphosate-resistant crops. J Agric Food Chem 60: 10375–10397.
15. Demaneche S, Sanguin H, Pote J, Navarro E, Bernillon D, et al. (2008) Antibiotic-resistant soil bacteria in transgenic plant fields. Proc Natl Acad Sci U S A 105: 3957–3962.
16. Govaerts B, Mezzalama M, Unno Y, Sayre KD, Luna-Guido M, et al. (2007) Influence of tillage, residue management, and crop rotation on soil microbial biomass and catabolic diversity. Appl Soil Ecol 37: 18–30.
17. McLaughlin A, Mineau P (1995) The impact of agricultural practices on biodiversity. Agric, Ecosyst & Environt 55: 201–212.
18. Garbeva P, Postma J, Van Veen J, Van Elsas J (2006) Effect of above-ground plant species on soil microbial community structure and its impact on suppression of *Rhizoctonia solani* AG3. Environ Microbiol 8: 233–246.
19. Salles J, Van Elsas J, Van Veen J (2006) Effect of agricultural management regime on Burkholderia community structure in soil. Micro Ecol 52: 267–279.
20. Liu J, Wang G, Jin J, Liu J, Liu X (2011) Effects of different concentrations of phosphorus on microbial communities in soybean rhizosphere grown in two types of soils. Ann Microbiol 61: 525–534.
21. Xu Y, Wang G, Jin J, Liu J, Zhang Q, et al. (2009) Bacterial communities in soybean rhizosphere in response to soil type, soybean genotype, and their growth stage. Soil Biol Biochem 41: 919–925.
22. Wang G-H, Jin J, Xu M-N, Pan X-W, Tang C (2007) Inoculation with phosphate-solubilizing fungi diversifies the bacterial community in rhizospheres of maize and soybean. Pedosphere 17: 191–199.
23. Albareda M, Dardanelli MS, Sousa C, Megías M, Temprano F, et al. (2006) Factors affecting the attachment of rhizospheric bacteria to bean and soybean roots. FEMS Microbiol Lett 259: 67–73.
24. Garbeva P, van Veen JA, van Elsas JD (2004) Microbial diversity in soil: selection microbial populations by plant and soil type and implications for disease suppressiveness. Annu Rev Phytopathol 42: 243–270.
25. van Overbeek L, van Elsas JD (2008) Effects of plant genotype and growth stage on the structure of bacterial communities associated with potato (Solanum tuberosum L.). FEMS Microbiol Ecol 64: 283–296.
26. Germida JJ, Siciliano SD (2001) Taxonomic diversity of bacteria associated with the roots of modern, recent and ancient wheat cultivars. Biol Fertility Soils 33: 410–415.
27. Duineveld BM, Rosado AS, van Elsas JD, van Veen JA (1998) Analysis of the dynamics of bacterial communities in the rhizosphere of the chrysanthemum via denaturing gradient gel electrophoresis and substrate utilization patterns. Appl Environ Microbiol 64: 4950–4957.
28. Gelsomino A, Keijzer-Wolters AC, Cacco G, van Elsas JD (1999) Assessment of bacterial community structure in soil by polymerase chain reaction and denaturing gradient gel electrophoresis. J Microbiol Methods 38: 1–15.
29. Schmalenberger A, Tebbe CC (2002) Bacterial community composition in the rhizosphere of a transgenic, herbicide-resistant maize (Zea mays) and comparison to its non-transgenic cultivar Bosphore. FEMS Microbiol Ecol 40: 29–37.
30. Nannipieri P, Ascher J, Ceccherini M, Landi L, Pietramellara G, et al. (2008) Effects of root exudates in microbial diversity and activity in rhizosphere soils. Molecular mechanisms of plant and microbe coexistence: Springer. 339–365.
31. Faragova N, Faragó J, Drabekova J (2005) Evaluation of abundance of aerobic bacteria in the rhizosphere of transgenic and non-transgenic alfalfa lines. Folia Microbiol 50: 509–514.
32. da Silva KRA, Salles JF, Seldin L, van Elsas JD (2003) Application of a novel Paenibacillus-specific PCR-DGGE method and sequence analysis to assess the diversity of Paenibacillus spp. in the maize rhizosphere. J Microbiol Methods 54: 213–231.
33. Miller HJ, Henken G, Vanveen JA (1989) Variation and composition of bacterial-populations in the rhizospheres of maize, wheat, and grass cultivars. Can J Microbiol 35: 656–660.
34. Rengel Z, Ross G, Hirsch P (1998) Plant genotype and micronutrient status influence colonization of wheat roots by soil bacteria. J Plant Nutrition 21: 99–113.
35. Vasantharajan V, Bhat J (1968) Interrelations of micro-organisms and mulberry. Plant Soil 29: 156–169.
36. Pu CL, Liang JG, Gao JY, Wu CX, Zhang MR, et al. (2012) Effects of high producing methionine soy-bean transferred cystathionine γ-synthase gene on community structure of bacteria in soil. J Nanjing Agric Univ 35: 8–14.
37. Jin J, Wang GH, Liu XB, Liu JD, Chen XL, et al. (2009) Temporal and spatial dynamics of bacterial community in the rhizosphere of soybean genotypes grown in a black Soil. Pedosphere 19: 808–816.
38. Inceoglu O, Salles JF, van Overbeek L, van Elsas JD (2010) Effects of plant genotype and growth stage on the Betaproteobacterial communities associated with different potato cultivars in two fields. Applied and Environ Microbiol 76: 3675–3684.
39. Siciliano SD, Germida JJ (1999) Taxonomic diversity of bacteria associated with the roots of field-grown transgenic Brassica napus cv. Quest, compared to the non-transgenic B-napus cv. Excel and B. rapa cv. Parkland. FEMS Microbiol Ecol 29: 263–272.
40. Lukow T, Dunfield PF, Liesack W (2000) Use of the T-RFLP technique to assess spatial and temporal changes in the bacterial community structure within an agricultural soil planted with transgenic and non-transgenic potato plants. FEMS Microbiol Ecol 32: 241–247.
41. Lottmann J, Berg G (2001) Phenotypic and genotypic characterization of antagonistic bacteria associated with roots of transgenic and non-transgenic potato plants. Microbiol Res 156: 75–82.
42. Tan XY, Hurek T, Reinhold-Hurek B (2003) Effect of N-fertilization, plant genotype and environmental conditions on nifH gene pools in roots of rice. Environ Microbiol 5: 1009–1015.
43. Shen JP, Zhang LM, Zhu YG, Zhang JB, He JZ (2008) Abundance and composition of ammonia-oxidizing bacteria and ammonia-oxidizing archaea communities of an alkaline sandy loam. Environ Microbiol 10: 1601–1611.
44. Briones AM, Okabe S, Umemiya Y, Ramsing NB, Reichardt W, et al. (2002) Influence of different cultivars on populations of ammonia-oxidizing bacteria in the root environment of rice. Appl Environ Microbiol 68: 3067–3075.
45. Wang YA, Ke XB, Wu LQ, Lu YH (2009) Community composition of ammonia-oxidizing bacteria and archaea in rice field soil as affected by nitrogen fertilization. Syst Appl Microbiol 32: 27–36.
46. Cleveland CC, Townsend AR, Schimel DS, Fisher H, Howarth RW, et al. (1999) Global patterns of terrestrial biological nitrogen (N-2) fixation in natural ecosystems. Global Biogeochem Cycles 13: 623–645.
47. Hsu SF, Buckley DH (2009) Evidence for the functional significance of diazotroph community structure in soil. Isme J 3: 124–136.
48. Treusch AH, Leininger S, Kletzin A, Schuster SC, Klenk HP, et al. (2005) Novel genes for nitrite reductase and Amo-related proteins indicate a role of

uncultivated mesophilic crenarchaeota in nitrogen cycling. Environ Microbiol 7: 1985–1995.
49. Purkhold U, Pommerening-Roser A, Juretschko S, Schmid MC, Koops HP, et al. (2000) Phylogeny of all recognized species of ammonia oxidizers based on comparative 16S rRNA and amoA sequence analysis: Implications for molecular diversity surveys. Appl Environ Microbiol 66: 5368–5382.
50. Kowalchuk GA, Stephen JR (2001) Ammonia-oxidizing bacteria: A model for molecular microbial ecology. Annu Rev Microbiol 55: 485–529.
51. Philippot L, Hallin S, Schloter M (2007) Ecology of denitrifying prokaryotes in agricultural soil. Adv Agronomy 96: 249–305.
52. Zhou HW, Li DF, Tam NFY, Jiang XT, Zhang H, et al. (2011) BIPES, a cost-effective high-throughput method for assessing microbial diversity. Isme J 5: 741–749.
53. Qiu MH, Zhang RF, Xue C, Zhang SS, Li SQ, et al. (2012) Application of bio-organic fertilizer can control Fusarium wilt of cucumber plants by regulating microbial community of rhizosphere soil. Biol Fertility Soils 48: 807–816.
54. Nacke H, Thurmer A, Wollherr A, Will C, Hodac L, et al. (2011) Pyrosequencing-based assessment of bacterial community structure along different management types in German forest and grassland soils. PLoS One 6: e17000.
55. Hirsch J, Strohmeier S, Pfannkuchen M, Reineke A (2012) Assessment of bacterial endosymbiont diversity in Otiorhynchus spp. (Coleoptera: Curculionidae) larvae using a multitag 454 pyrosequencing approach. Bmc Microbiol 12: S6.
56. Zhang CH, Zhang MH, Wang SY, Han RJ, Cao YF, et al. (2010) Interactions between gut microbiota, host genetics and diet relevant to development of metabolic syndromes in mice (vol 4, pg 232, 2010). Isme J 4: 312–313.
57. Lee OO, Wang Y, Yang JK, Lafi FF, Al-Suwailem A, et al. (2011) Pyrosequencing reveals highly diverse and species-specific microbial communities in sponges from the Red Sea. Isme J 5: 650–664.
58. Yu K, Zhang T (2012) Metagenomic and metatranscriptomic analysis of microbial community dtructure and gene expression of activated sludge. PLoS One 7: e38183.
59. Zhang T, Zhang XX, Ye L (2011) Plasmid metagenome reveals high levels of antibiotic resistance genes and mobile genetic elements in activated sludge. PLoS One 6: e26041.
60. Chen WG, Liu FL, Ling ZX, Tong XJ, Xiang C (2012) Human intestinal lumen and mucosa-associated microbiota in patients with colorectal cancer. PLoS One 7: e39743.
61. Fierer N, Bradford MA, Jackson RB (2007) Toward an ecological classification of soil bacteria. Ecology 88: 1354–1364.
62. Smit E, Leeflang P, Gommans S, van den Broek J, van Mil S, et al. (2001) Diversity and seasonal fluctuations of the dominant members of the bacterial soil community in a wheat field as determined by cultivation and molecular methods. Appl Environ Microbiol 67: 2284–2291.
63. Castro HF, Classen AT, Austin EE, Norby RJ, Schadt CW (2010) Soil microbial community responses to multiple experimental climate change drivers. Appl Environ Microbiol 76: 999–1007.
64. Yashiro E, McManus PS (2012) Effect of streptomycin treatment on bacterial community structure in the apple phyllosphere. PLoS One 7: e37131.
65. Gerardi MH (2002) Nitrification and denitrification in the activated sludge process. New York: John Wiley and Sons, Inc. 43–55.
66. Sogin ML, Morrison HG, Huber JA, Welch DM, Huse SM, et al. (2006) Microbial diversity in the deep sea and the underexplored "rare biosphere". Proc Natl Acad Sci U S A 103: 12115–12120.
67. Power ME, Tilman D, Estes JA, Menge BA, Bond WJ, et al. (1996) Challenges in the quest for keystones. BioSci 46: 609–620.
68. Hooper D, Chapin Iii F, Ewel J, Hector A, Inchausti P, et al. (2005) Effects of biodiversity on ecosystem functioning: a consensus of current knowledge. Ecol Monogr 75: 3–35.
69. Xu YX, Wang GH, Jin J, Liu JD, Zhang QY, et al. (2007) Comparison of bacterial community in rhizosphere of soybean by anaylysis of 16S rDNA obtained directly from soil and culture plate. Soybean Sci 26: 907–913.
70. Hartman WH, Richardson CJ, Vilgalys R, Bruland GL (2008) Environmental and anthropogenic controls over bacterial communities in wetland soils. Proc Natl Acad Sci U S A 105: 17842–17847.
71. Dunbar J, Takala S, Barns SM, Davis JA, Kuske CR (1999) Levels of bacterial community diversity in four arid soils compared by cultivation and 16S rRNA gene cloning. Appl Environ Microbiol 65: 1662–1669.
72. Kuske CR, Barns SM, Busch JD (1997) Diverse uncultivated bacterial groups from soils of the arid southwestern United States that are present in many geographic regions. Appl Environ Microbiol 63: 3614–3621.
73. Borneman J, Skroch PW, O'Sullivan KM, Palus JA, Rumjanek NG, et al. (1996) Molecular microbial diversity of an agricultural soil in Wisconsin. Appl Environ Microbiol 62: 1935–1943.
74. McCaig AE, Glover LA, Prosser JI (1999) Molecular analysis of bacterial community structure and diversity in unimproved and improved upland grass pastures. Appl Environ Microbiol 65: 1721–1730.
75. Barns SM, Cain EC, Sommerville L, Kuske CR (2007) Acidobacteria phylum sequences in uranium-contaminated subsurface sediments greatly expand the known diversity within the phylum. Appl Environ Microbiol 73: 3113–3116.
76. Quaiser A, Ochsenreiter T, Lanz C, Schuster SC, Treusch AH, et al. (2003) Acidobacteria form a coherent but highly diverse group within the bacterial domain: evidence from environmental genomics. Mol Microbiol 50: 563–575.
77. Rappé MS, Giovannoni SJ (2003) The uncultured microbial majority. Annu Rev Microbiol 57: 369–394.
78. Eichorst SA, Breznak JA, Schmidt TM (2007) Isolation and characterization of soil bacteria that define Terriglobus gen. nov., in the phylum Acidobacteria. Appl Environ Microbiol 73: 2708–2717.
79. Turner S, Pryer KM, Miao VP, Palmer JD (1999) Investigating deep phylogenetic relationships among Cyanobacteria and plastids by small subunit rRNA sequence analysis1. J Eukaryot Microbiol 46: 327–338.
80. Liu Z, Lozupone C, Hamady M, Bushman FD, Knight R (2007) Short pyrosequencing reads suffice for accurate microbial community analysis. Nucleic Acids Res 35: e120.
81. Rösch C, Mergel A, Bothe H (2002) Biodiversity of denitrifying and dinitrogen-fixing bacteria in an acid forest soil. Appl Environ Microbiol 68: 3818–3829.
82. Rotthauwe J-H, Witzel K-P, Liesack W (1997) The ammonia monooxygenase structural gene amoA as a functional marker: molecular fine-scale analysis of natural ammonia-oxidizing populations. Appl Environ Microbiol 63: 4704–4712.
83. Mao Y, Yannarell AC, Mackie RI (2011) Changes in N-transforming archaea and bacteria in soil during the establishment of bioenergy crops. PLoS One 6: e24750.
84. Chen J, Yu Z, Michel FC, Wittum T, Morrison M (2007) Development and application of real-time PCR assays for quantification of erm genes conferring resistance to macrolides-lincosamides-streptogramin B in livestock manure and manure management systems. Appl Environ Microbiol 73: 4407–4416.
85. Yang T, Liu G, Li Y, Zhu S, Zou A, et al. (2012) Rhizosphere microbial communities and organic acids secreted by aluminum-tolerant and aluminum-sensitive soybean in acid soil. Biol Fertility Soils 48: 97–108.
86. Wu S, Wang G, Angert ER, Wang W, Li W, et al. (2012) Composition, diversity, and origin of the bacterial community in grass carp intestine. PLoS One 7: e30440.
87. Schloss PD, Westcott SL, Ryabin T, Hall JR, Hartmann M, et al. (2009) Introducing mothur: open-source, platform-independent, community-supported software for describing and comparing microbial communities. Appl Environ Microbiol 75: 7537–7541.
88. Wang Q, Garrity GM, Tiedje JM, Cole JR (2007) Naive Bayesian classifier for rapid assignment of rRNA sequences into the new bacterial taxonomy. Appl Environ Microbiol 73: 5261–5267.
89. Yang W, Zheng Y, Gao C, He X, Ding Q, et al. (2013) The arbuscular mycorrhizal fungal community response to warming and grazing differs between soil and roots on the Qinghai-Tibetan Plateau. PLoS One 8: e76447.
90. Caporaso JG, Kuczynski J, Stombaugh J, Bittinger K, Bushman FD, et al. (2010) QIIME allows analysis of high-throughput community sequencing data. Nat methods 7: 335–336.

16

Cloning and Characterization of a *Phragmites australis* Phytochelatin Synthase (*PaPCS*) and Achieving Cd Tolerance in Tall Fescue

Cuizhu Zhao[1,2], Jin Xu[3], Qiang Li[1], Shuo Li[1], Peng Wang[1], Fengning Xiang[1]*

1 The Key Laboratory of Plant Cell Engineering and Germplasm Innovation, School of Life Sciences, Shandong University, Jinan, Shandong, China, **2** College of Agronomy, Northwest A&F University, Yangling, Shanxi, China, **3** Key Laboratory of Agricultural Water Resources, Center for Agricultural Resources Research, Institute of Genetics and Developmental Biology, Chinese Academy of Sciences, Shijiazhuang, China

Abstract

The production of phytochelatins (PCs) provides an important means for plants to achieve tolerance to cadmium (Cd) toxicity. A reed gene encoding PC synthase (*PaPCS*) was isolated and its function tested through its heterologous expression in a strain of yeast sensitive to Cd. Subsequently, the Cd sensitive and high biomass accumulating species tall fescue was transformed either with *PaPCS* or *PaGCS* (a glutamyl cysteine synthetase gene of reed) on their own (single transformants), or with both genes together in the same transgene cassette (double transformant). The single and double transformants showed greater Cd tolerance and accumulated more Cd and PC than wild type plants, and their Cd leaf/root ratio content was higher. The ranking in terms of Cd and PC content for the various transgenic lines was double transformants>*PaGCS* single transformants>*PaPCS* single transformants>wild type. Thus *PaGCS* appears to exert a greater influence than *PaPCS* over PC synthesis and Cd tolerance/accumulation. The double transformant has interesting potential for phytoremediation.

Editor: Manuel Reigosa, University of Vigo, Spain

Funding: This research was supported by Key Projects in the National Science & Technology Pillar Program in the 12th Five year Plan of China (grant no. 2011BAI06B01), the National Natural Science Foundation of China (grant no. 30970243), International technology cooperation project of Shandong Province of China (grant no. 2011176), Science & Technology Plan of Shandong Province (grant 2009GG10002001) and the Chinese Natural Education Ministry Doctor Station Foundation Fellowship (grant no. 913111006). The funders had no role in study design, data collection and analysis, decision to publish, or preparation of the manuscript.

Competing Interests: The authors have declared that no competing interests exist.

* Email: xfn0990@sdu.edu.cn

Introduction

Heavy metal pollution caused by a combination of natural leaching and anthropogenic activity is becoming a significant environmental problem. Thousands of hectares of arable land have been contaminated in China, representing a significant health hazard to the population [1–3]. Phytoremediation is seen as a favorable strategy to remove the contamination. The optimal phytoremediating plant needs to be highly productive in terms of biomass and efficient in terms of accumulation of heavy metals [4,5]. One such species is the reed *Phragmites australis* (Cav.) Trin. ex Steud, a most effective accumulator of Cd, Pb and Zn, and has been widely exploited as a sewage treatment wetland species [6]. However, the most efficient accumulators appear to be poor in terms of biomass production [7–9]. As a result, it has been proposed that an effective approach could be based on the engineering of heavy metal accumulation and tolerance into a species already recognized as being an effective biomass producer [10–12]. Tall fescue (*Festuca arundinacea* Schreb), a particularly vigorous grass species with broad climatic adaptation, could be an attractive recipient species, since turf grasses tolerate regular mowing, thus allowing for the ready disposal of the heavy metals translocated into their foliage. A number of genes involved in the uptake of and tolerance to heavy metals have been identified, and some have already been successfully transferred into plants [13–16], and it has been demonstrated that the heterologous expression of some of these improves the levels of heavy metal tolerance and accumulation.

The presence of heavy metals tends to induce the production of phytochelatins (PCs) in plants [17,18]. The PCs form a family of oligopeptides able to neutralize the toxicity of heavy metals by chelation [19,20]. Two key rate-limiting enzymes in the synthesis of PCs are phytochelatin synthase (PCS) and γ-glutamyl cysteine synthetase (γ-GCS) [21]. Their involvement in the accumulation of heavy metals has been experimentally confirmed in both *Arabidopsis thaliana* and *Brassica juncea* [22,23] but it is unclear which of these two enzymes is the most important [24,25]. We have previously shown that the γ-GCS glutamyl cysteine synthetase gene of *P. australis* (*PaGCS*) is a key component of the species' heavy metal tolerance, and that when this gene is expressed in the fast-growing species *Agrostis palustris*, tolerance to Cd toxicity is measurably enhanced [26]. In the present study we further explored the potential of *P. australis* as a donor of phytoremediation genes by testing the effect of expressing *PaPCS* and/or *PaGCS* in tall fescue.

Materials and Methods

Cloning of *Phragmites australis* phytochelatin synthase gene

Total RNA extracted from frozen *P. australis* leaf tissue using the TRIzol reagent (Invitrogen, Carlsbad, CA, USA) formed the template for the M-MLV reverse transcriptase driven synthesis of cDNA (TaKaRa Bio Group, Otsu, Japan). The resulting cDNA was amplified using the primer pair 5′-CTTCCAG(A/T)CTCA(G/A)TCGGAGC and 5′-ATTGC(G/C)ACTCCT(T/A)GACAGCA to obtain an initial *PaPCS* sequence. After sequencing this amplicon, further primers were derived to perform 3′-RACE and 5′-RACE (TaKaRa 3′- and 5′-Full RACE Core set), according to the manufacturer's protocols.

Yeast complementation assay

SacI/XbaI PaPCS cDNA fragments were sub-cloned into the pYES2 vector and transformed into *Saccharomyces cerevisiae* mutant strain YK44 (ura3-52 his3-200, *ΔZRCΔCot1*, mating type α) using the lithium acetate method [27]. Yeast cells harboring either an empty pYES2 vector or pYES2-PaPCS were cultured on SD-Ura medium (Difco BRL, USA) at 30°C until the OD_{600} reached 0.5. The cells were pelleted by centrifugation from a 500 μL aliquot of the culture and re-suspended in deionized water in 10^{-1}, 10^{-2}, and 10^{-3} concentrations, and then inoculated into 40 μL YPGAL (1% w/v yeast extract, 2% w/v peptone, 2% w/v galactose, Difco) medium in the presence of 100 μM $CdCl_2$. All experiments were done in quadruplicate.

Construction of plant expression vectors

The *XbaI/SacI PaPCS* cDNA fragment was cloned into the pROKII vector (Fig. S1A), and the resulting pROK/PaPCS fusion was transformed into *E. coli* DH10B by thermal shock. The *35S::PaPCS-Tnos* fragment released from pROK/PaPCS and the equivalent for *PaGCS* [26] were fused and inserted into the plant gene expression vector pCAMBIA3301 based on NPTII as the selectable marker (Fig. S1B). The recombinant vector (p3301/PG) was then transformed into *E. coli* DH10B.

Transformation of tall fescue

The binary plasmids pROK/PaGCS, pROK/PaPCS and p3301/PG were separately introduced into the *Agrobacterium tumefaciens* strain AGL1. Putative transformants were selected by including 50 mg/L kanamycin in the culture medium. Subsequent tissue culture and agroinfection of the hypocotyledonary axis were performed as reported by Fu et al (2007). Prior to agroinfection, *F. arundinacea* embryogenic calli were grown for one week on MS medium [28] containing 2 mg/L 2,4-dichlorophenoxyacetic acid (2,4-D). Following the agroinfection process, the embryogenic calli were exposed to 2 mg/L and 50 mg/L kanamycin for two weeks, after which the surviving calli were transferred to a differentiation medium (MS medium containing 1 mg/L 6-BA, 1 mg/L IAA and 50 mg/L kanamycin) for one month. Rooted seedlings were hardened on differentiation medium with the addition of 2 mg/L clobutrazol and 50 mg/L kanamycin for two weeks before being transplanted into soil in a greenhouse.

Plant material and stress treatment

Transgenic and wild type (WT) *F. arundinacea* rooted seedlings were transplanted into soil in a greenhouse (22±3°C under a 16 h photoperiod with a photosynthetic photon flux density of 45 μmol m^{-2} s^{-1}). The seedlings were at tillering stage after being transplanted into the soil for 30 days. Then, WT and transgenic tiller clones were transferred into Hoagland's solution, and exposed for five days to 150 μM $CdCl_2$.

Genetic analysis of putative transformants

Genomic DNA was isolated from the leaf of candidate transformants using the CTAB method [29]. A fragment beginning within the 35S promoter and ending within the *PaGCS* coding region was amplified as previously described [26] for the detection of *PaGCS* transformants. To detect candidate *PaPCS* transformants, a primer pair was designed to target a similar fragment of the *PaPCS* transgene (primer sequences were 5′-CTTCCAGACTCAGTCGGAGC and 5′-ATTGC-GACTCCTTGACAGCA). Each 20 mL PCR reaction contained 0.4 μM of each primer, 400 μM dNTP, 1.5 mM $MgCl_2$, 1.25 U rTaq DNA polymerase (Takara, Japan) and either 300 ng plant DNA or 15 pg plasmid. Each PCR consisted of a denaturation step (95°C/5 min), followed by 35 cycles of 95°C/40 s, 55°C/60 s 72°C/60 s, ending with an extension step of 72°C/10 min. PCR products were separated by 1% agarose gel electrophoresis. The presence of the transgene was also tested by Southern hybridization. For this purpose, 15 μg *Eco*RI- or *Sac*I digested genomic DNA was electrophoresed through a 0.8% agarose gel and electrophoretically transferred onto a Hybond N^+ membrane (Amersham, Piscataway, NJ, USA). The membranes were hybridized with the *PaGCS* or *PaPCS* amplicons labeled with digoxigenin using the random primer method (DIG High Prime DNA Labeling and Detection Starter Kit II, Roche, Basel, Switzerland). Hybridization, washing and signal detection were performed according to the manufacturer's instructions.

Biomass assay and the assessment of Cd^{2+} uptake

The biomass of WT and transgenic tiller clones exposed to 150 μM $CdCl_2$ for five days was recorded. Each plant was thoroughly washed with distilled water, divided into leaves, shoots, and roots. Root samples were washed for 10 min at 4°C in 2 mM $CaSO_4$ and 10 mM EDTA, and rinsed three times in water, and then placed into oven dried to a constant weight. A sample of 30 mg dry matter was digested in HNO3 and HClO4 (4:1 v/v). Then it was evaporated to dryness and dissolved in 2 mol/L HNO_3. The concentration of Cd^{2+} in the solvent was determined as described by Vitória, Lea & Azevedo [30], using an inductively coupled plasma mass spectrometer (ICP-MS). All experiments were done in quadruplicate.

MDA content

Top most expanded leaves of transgenic and wild-type tall fescue with or without Cd treatment were collected. Malondialdehyde (MDA) content was determined by measuring the concentration of thiobarbituric acid-reacting material present, using thiobarbituric acid (TBA) [31]. Fresh leaves were homogenized with 2 mL of TBA reagent (15% w/v ti-ichloroacetic acid and 0.25 M HCl), and treated in a boiling water bath for 15 minutes. After cooling, the suspension was centrifuged at 10,000× g for 10 minutes. The supernatant was then separated. The absorbance was measured at 535 nm. The MDA concentration was determined by the specific absorbance coefficient (156 mM^{-1} cm^{-1}). The MDA concentrations were measured after a 0 h, 6 h, 12 h, 24 h, 48 h, 72 h and 120 h exposure to 150 μM $CdCl_2$. All treatments and experiments were using different plants done in quadruplicate. The MDA concentration unit was μmol·g^{-1}DW.

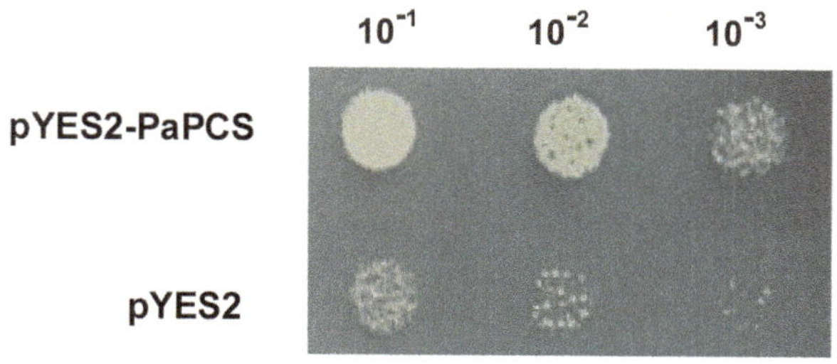

Figure 1. Functional testing of *PaPCS* in the Cd sensitive yeast strain YK44. Yeast cells harboring either an empty pYES2 vector or pYES2-PaPCS were cultured and inoculated into 40 μL YPGAL medium in 10^{-1}, 10^{-2}, and 10^{-3} concentrations in the presence of 100 μM $CdCl_2$. After 48 h, the growth of transgenic cells heterologously expressing *PaPCS* was compared to control cells carrying an empty pYES2 vector. All experiments were done in quadruplicate.

Enzyme activity measurement

Top most expanded leaves of transgenic and wild-type tall fescue with or without Cd treatment were collected. Peroxidase (POD) and superoxide dismutase (SOD) activity were measured as described by Scebba et al. [32]. Fresh leaves were immediately frozen in liquid nitrogen and then ground in liquid nitrogen. The powder was suspended in 0.5 mL 0.1 M Tris pH 8.0, 1 mM PMSF, 1% (w/v) polyvinylpyrrolidone, 1% (w/v) sodium ascorbate and 1% (v/v) β-mercaptoethanol. Extracts were centrifuged twice at 26,000 g for 20 minutes (4°C), and the clarified supernatant was used for determination of enzyme activity.

Total protein of the crude extract was measured using the method of Bradford [33]. Different amounts (0.005, 0.010, 0.020 and 0.040 mL) of crude extract were added to a reaction mixture. The mixture contained 50 mM sodium phosphate buffer pH 7.8, 0.1 mM ethylenediaminetetraacetic acid, 13 mM methionine, 2 μM riboflavine and 75 μM 3-(4,5)-dimethylthiazol-2-yl-2,5-diphenyl-tetrazolium bromide (MTT). Exposing the mixture to cool white fluorescent light for 15 minutes started the reaction. The light was then switched off, the tubes were stirred and the blue color was measured at 560 nm. POD (E.C.1.11.1.7) activity was assayed in a reaction mixture. The mixture contained 10 mM potassium phosphate buffer pH 7.0, 10 mM H_2O_2 solution, 20 mM guaiacol and 0.01 mL of crude extract. Adding H_2O_2 and guaiacol solution at the same time started the reaction. The activity was determined by monitoring the increase of absorbance at 470 nm, as a result of guaiacol oxidation. Both SOD and POD enzyme activities were measured after a 0 h, 6 h, 12 h, 24 h, 48 h, 72 h and 120 h exposure to 150 μM $CdCl_2$. All treatments and experiments were using different plants done in quadruplicate. POD activity unit was $\Delta 460 \cdot min^{-1} \cdot g^{-1}$. SOD activity unit was $Unit \cdot mg^{-1}$.

The non-protein thiol, GSH and PCs quantification

Total glutathione (GSH) content was determined as described by Gupta et al. [34]. Frozen tillers (700 mg FW) was homogenized

Figure 2. Transformation of *F. arundinacea* with *PaGCS*. (A) Callus induction, (B) Agroinfection, (C) Kanamycin selection of transgenic calli, (D) Regenerated plants, (E) Selected plants maintain their greenness in the presence of kanamycin, (F) Putative transgenic plants grown in soil.

in 0.1 M sodium phosphate buffer (pH,8.0) and 25% HPO_3, centrifuged at 20,000 g for 20 minutes and GSH was fluorometrically determined in the supernatant after 15 minutes incubation at 25°C with o phthaldehyde (OPT) and the fluorescent intensity was recorded at 420 nm after excitation at 350 nm on a Hitachi Fluorescence Spectrophotometer (Model No. 650-60). The non-protein thiol (TNP-SH) content was determined as described by Tu et al. [35]. The PC content were measured by HPLC as described by Gupta et al., using PC2 and PC3 as PC standards [34]. TNP-SH, GSH and PC content were measured after a 0 h, 6 h, 12 h, 24 h, 48 h, 72 h and 120 h exposure to 150 μM $CdCl_2$. All treatments and experiments were using different plants done in quadruplicate. TNP-SH, GSH and PC content unit were μmol·g−1DW.

Statistical analysis

For each statistical test, a threshold of $P<0.05$ was applied to indicate statistical significance. The data were analyzed using the SAS statistical package (v5.1, SAS Institute Inc., Cary, NC, USA).

Results

Isolation and function of *PaPCS*

The consensus regions shared by the PCS sequences present in bread wheat, pea, onion, *Zinnia elegans* and tomato provided the target for the design of PCR primers. A full-length *P. australis* cDNA was obtained by RACE-PCR. The *PaPCS* (GenBank accession number JX826285) sequence length was 1497 bp, and comprised a single open reading frame encoding a 498 residue polypeptide of predicted molecular mass ~54.9 kDa. A phylogenetic analysis showed that its nucleotide sequence shared considerable similarity with its homologs from bread wheat (84.3%), bermudagrass (84.3%) and rice (70.9%). The closest related peptide sequence was the PCS from bermudagrass (Fig. S2). The presence of the *pYES2-PaPCS* construct in the Cd-sensitive yeast strain YK44 supported cell growth when the yeast was challenged by 100 μM $CdCl_2$ (Fig. 1).

Transformation of tall fescue with *PaPCS* or *PaGCS* or *PaPCS-PaGCS*

Granular embryogenic calli were used for agroinfection (Figs. 2A, S3A, S4A). Two weeks after agroinfection with either pROK/PaGCS, pROK/PaPCS or p3301/PG (Figs. 2B, S3B, S4B), kanamycin resistant calli having a yellow granulated structure were selected (Figs. 2C, S3C, S4C), shoots and roots were induced (Figs. 2D, S3D, S4D) and a second round of kanamycin selection was applied to remove escapes (Figs. 2E, S3E, S4E). In all, 95 putative transgenic *PaPCS* plants, 87 *PaGCS* plants and 15 *PaPCS/PaGCS* plants were regenerated (Figs. 2F, S3F, S4F). The PCR-based screening of the putative transgenics identified seven *PaPCS*, eleven *PaGCS* transgenic lines and three *PaPCS/PaGCS* transgenic lines. Of these, two per transgene were taken forward. A semi-quantitative RT-PCR analysis showed that transgene transcript appeared in transgenic lines (Fig. 3A,C). Profiling the genomic DNA of the six selected transformants by Southern hybridization showed that a single copy of *PaPCS* was integrated into each of the two *PaPCS* lines, a single copy of *PaGCS* into one of the *PaGCS* lines (and two copies into the other), and a single copy of *PaPCS-PaGCS* into each of the *PaPCS/PaGCS* lines (Fig. 3B,D).

Growth and Cd accumulation of WT and transgenic tall fescue plants

After five days exposure to 150 μM $CdCl_2$, the total biomass of each of the three types of transgenic lines was greater than that of the WT, with the double transformant being the most productive, suggesting a better Cd tolerance. However, the total biomass increase for *PaPCS* transgenic lines was not significant (Fig. 4). The Cd content of each of the three types of transgenic lines was significant greater than that of the WT in leaf and root, in the order of double transformants>*PaGCS* single transformants> *PaPCS* single transformants>WT (Fig. 5). The root was higher than that of the leaf in each case, and the double transformant accumulated more Cd than did either of the single transformants. The ratio between the leaf and root Cd content was higher for the transgenic material than for the WT. In WT, the Cd content in the root was 1.9 fold higher than that of in the leaf, this ratio was significant decreased in *PaPCS* transgenic lines (1.4 fold) and *PaGCS* transgenic lines (1.1–1.28 fold). In the double transfor-

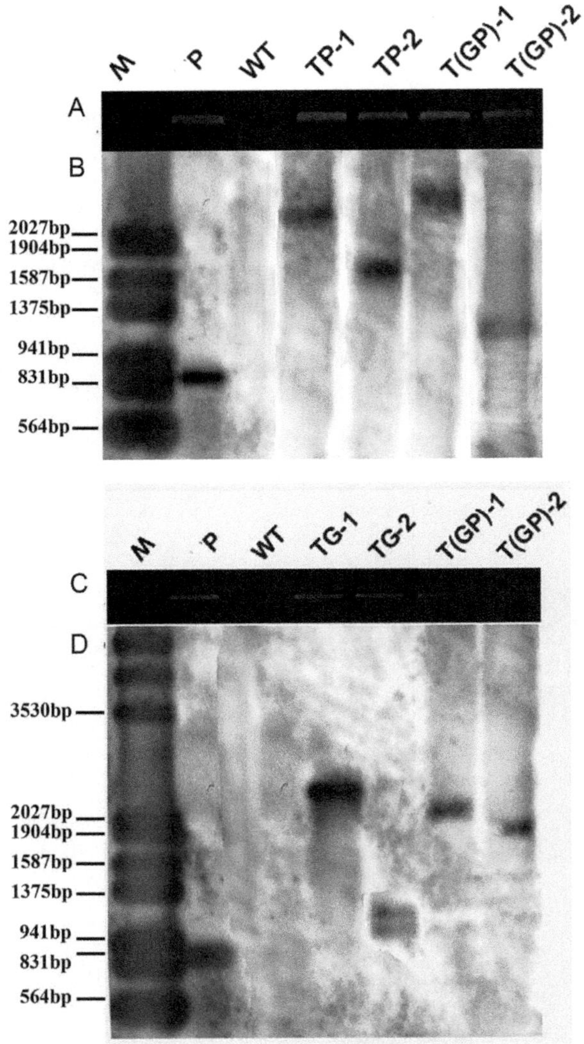

Figure 3. DNA fingerprinting to confirm the presence of *PaPCS* (A,B) and *PaGCS* (C,D). (A,C) PCR analysis, (B,D) Southern blot hybridization of *Eco*RI and *Sac*I digested genomic DNA. TP-1, TP-2: carriers of *PaPCS*; TG-1, TG-2: carriers of *PaGCS*; T(GP)-1, T(GP)-2: carriers of *PaGCS-PaPCS*; P: plasmid; WT: wild type *F. arundinacea*. M: size marker.

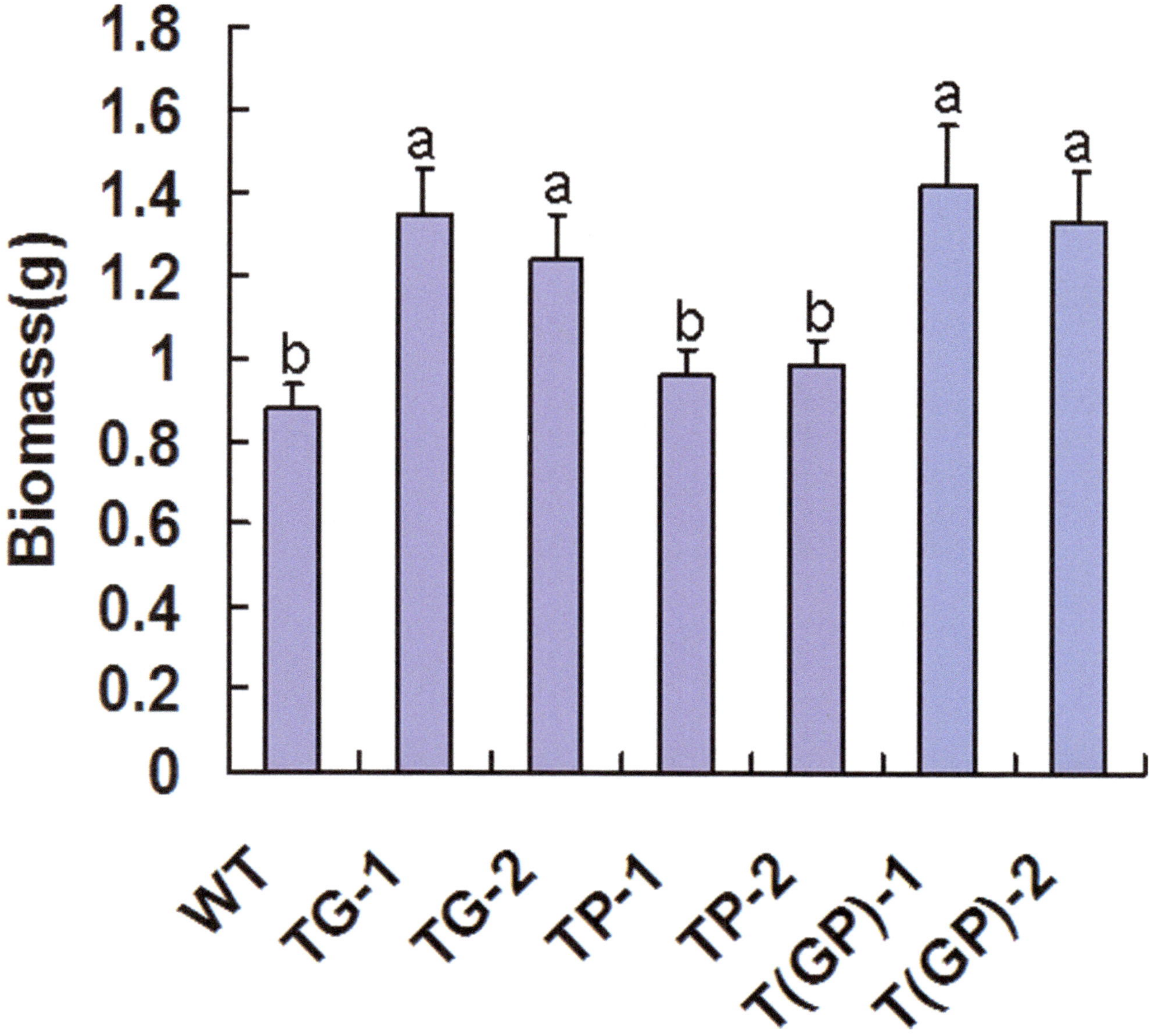

Figure 4. The total biomass accumulation after exposure to 150 μM $CdCl_2$ for five days. TP-1, TP-2: carriers of *PaPCS*; TG-1, TG-2: carriers of *PaGCS*; T(GP)-1, T(GP)-2: carriers of *PaGCS-PaPCS*; WT: wild type *F. arundinacea*. Vertical bars indicate the standard error of the mean (n = 4). Different letters indicate significant differences according to a Tukey test ($P<0.01$).

mant, the ratio was 1.0–1.2 fold, which was not significantly different with *PaGCS* transgenic lines (Tab. 1), suggesting these transgenic plant was particularly efficient in translocating Cd from the root to the leaf.

MDA content and enzyme activities under Cd stress

After 6 h of Cd stress, the MDA content of both the transgenic and WT plants began to rise, with the rate of increase being fastest in the WT plants. The rate of increase in MDA content was slowest in the two double transformants, followed first by the *PaGCS* single transformants and then by the *PaPCS* single transformants (Fig. 6A). Thus in the transgenic plants, lipid peroxidation was slowed, and this slowing was more effective in the presence of *PaGCS*.

After 6 h of Cd stress, the MDA content of both the transgenic and WT plants began to rise, with the rate of increase being fastest in the WT plants. The rate of increase in MDA content was slowest in the two double transformants and *PaGCS* single transformants, followed by the *PaPCS* single transformants (Fig. 6A).

Thus in the transgenic plants, lipid peroxidation was slowed, and this slowing was more effective in the presence of *PaGCS*. SOD and POD activity all peaked rather quickly in response to the Cd stress, the former by 6 h, and the latter by 12 h (Figs. 6B, C). At its peak, SOD activity was highest in the WT, followed by the *PaPCS* transgenics, the *PaGCS* transgenics and finally the double transformants (Fig. 6B). At its peak, POD activity had the same pattern as SOD activity (Fig. 6C). After peaking, both SOD and POD activity sharply decreased. Thus, after 24 h of Cd stress, both SOD and POD activity had dropped below their initial level. At this time, the ranking of enzyme activity was *PaGCS*/*PaPCS* transgenics>*PaGCS* transgenics>*PaPCS* transgenics>WT.

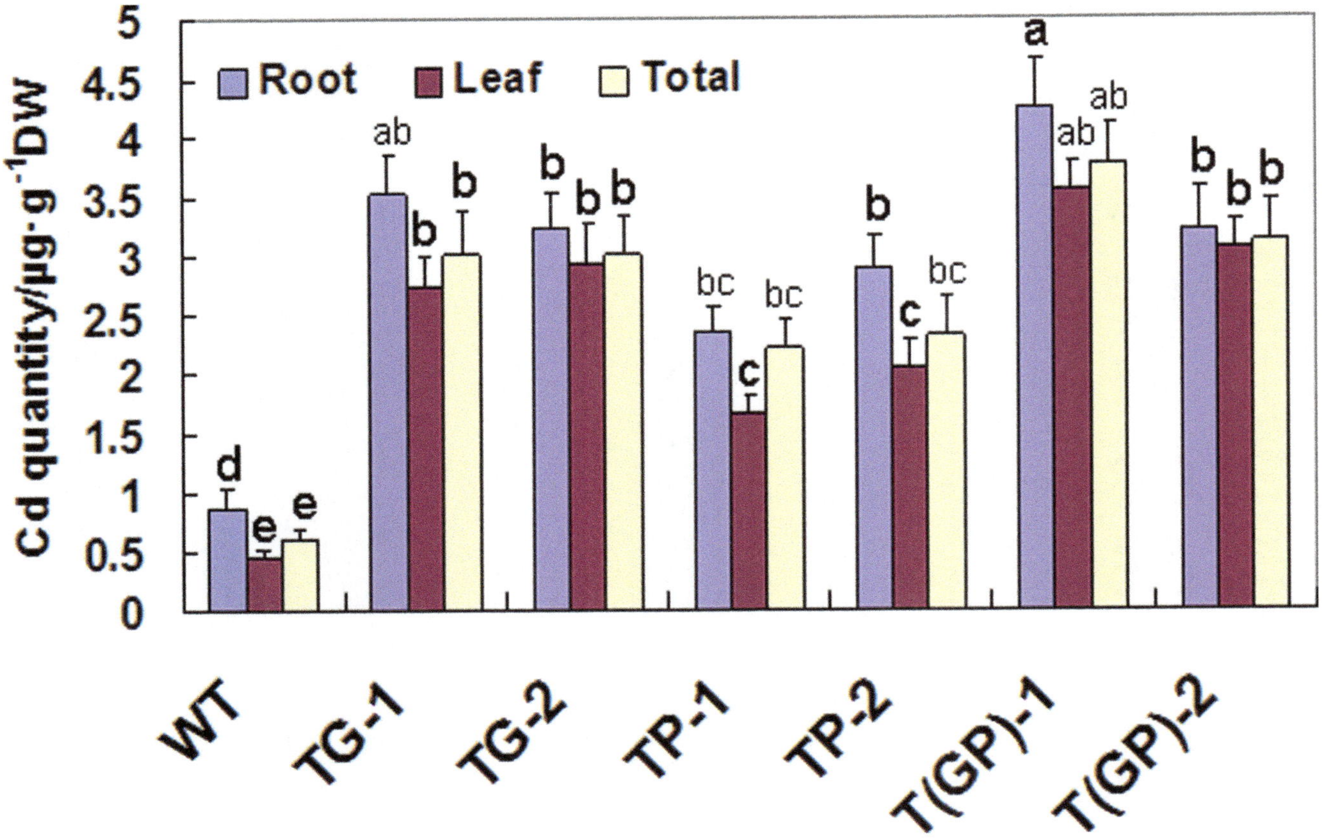

Figure 5. Cd^{2+} accumulation of transgenic lines and wild type after exposure to 150 μM $CdCl_2$ for five days. The Cd content of the roots, leaves and total plant was measured in four plants per line. TP-1, TP-2: carriers of *PaPCS*; TG-1, TG-2: carriers of *PaGCS*; T(GP)-1, T(GP)-2: carriers of *PaGCS-PaPCS*; WT: wild type *F. arundinacea*. Vertical bars indicate the standard error of the mean($n=4$). Different letters indicate significant differences according to a Tukey test (P<0.01).

GSH, PC and TNP-SH content of transgenic tall fescue under cadmium stress

Before exposure to Cd stress, the total GSH content of the three transgenic types was similar to that of WT plants, as well as the PC and TNP-SH content. As the Cd stress was applied, the total GSH content of WT and the transgenic plants did not significantly change (Fig. 7A), but both the PC and TNP-SH content increased (Figs. 7B, C, S5). Overall, with respect to PC, the double transformants out-performed each of the single transformants; the *PaGCS* transgenic was superior to the *PaPCS* transgenic, and both single transformants out-performed WT (Fig. 7C).

Table 1. Ratio of Cd root/leaf content.

plant lines	Ratio of Cd root/leaf content
WT	1.9312±0.165 a
TG-1	1.2878±0.103 c
TG-2	1.1059±0.111 c
TP-1	1.4341±0.121 b
TP-2	1.4058±0.134 b
T(GP)-1	1.1948±0.128 c
T(GP)-2	1.0522±0.110 c

Values are mean ± standard error (n = 4). For each variable, values followed by different letters differ significantly between the strains (Student's t-test, P< 0.05).

Discussion

PCS is the enzyme which catalyses the final step in the synthesis of phytochelatin, so it has been the major focus in the field of the genetic engineering of phytoremediation ever since it was first identified by Vatamaniuk et al. [36] and Clemens et al. [37]. Expression of PCS in both prokaryotes and yeast increases tolerance to heavy metals [38–40], as well as to various abiotic stresses in *Anabaena* [41]. In the present study, when the *PaPCS* gene was expressed in a strain of yeast sensitive to Cd, the tolerance of the cells to the presence of 100 μM $CdCl_2$ in the growth medium was noticeably improved. The introduction of *PCS* genes from various species into plants has also generally resulted in an improved ability to tolerate and/or accumulate heavy metals. Examples include the heterologous expression of *A. thaliana AtPCS1* in tobacco [42] and Indian mustard [43], bread wheat *TaPCS1* and *Ceratophyllum demersum CdPCS1* in tobacco [40], and that of *Nelumbo nucifera NnPCS* and *CdPCS1* in *A. thaliana* [44–46]. When a PCS deficient *A. thaliana* mutant was transformed with *AtPCS1* driven by a leaf-specific promoter, the resulting lines proved less sensitive to Cd toxicity [47]. However, although over-expression of *AtPCS1* helped raise the level of tolerance to arsenic [48], it also resulted in a higher sensitivity to Cd [49]. In the present study heterologous expression of *PaPCS* in tall fescue improved the plants' ability to accumulate biomass in

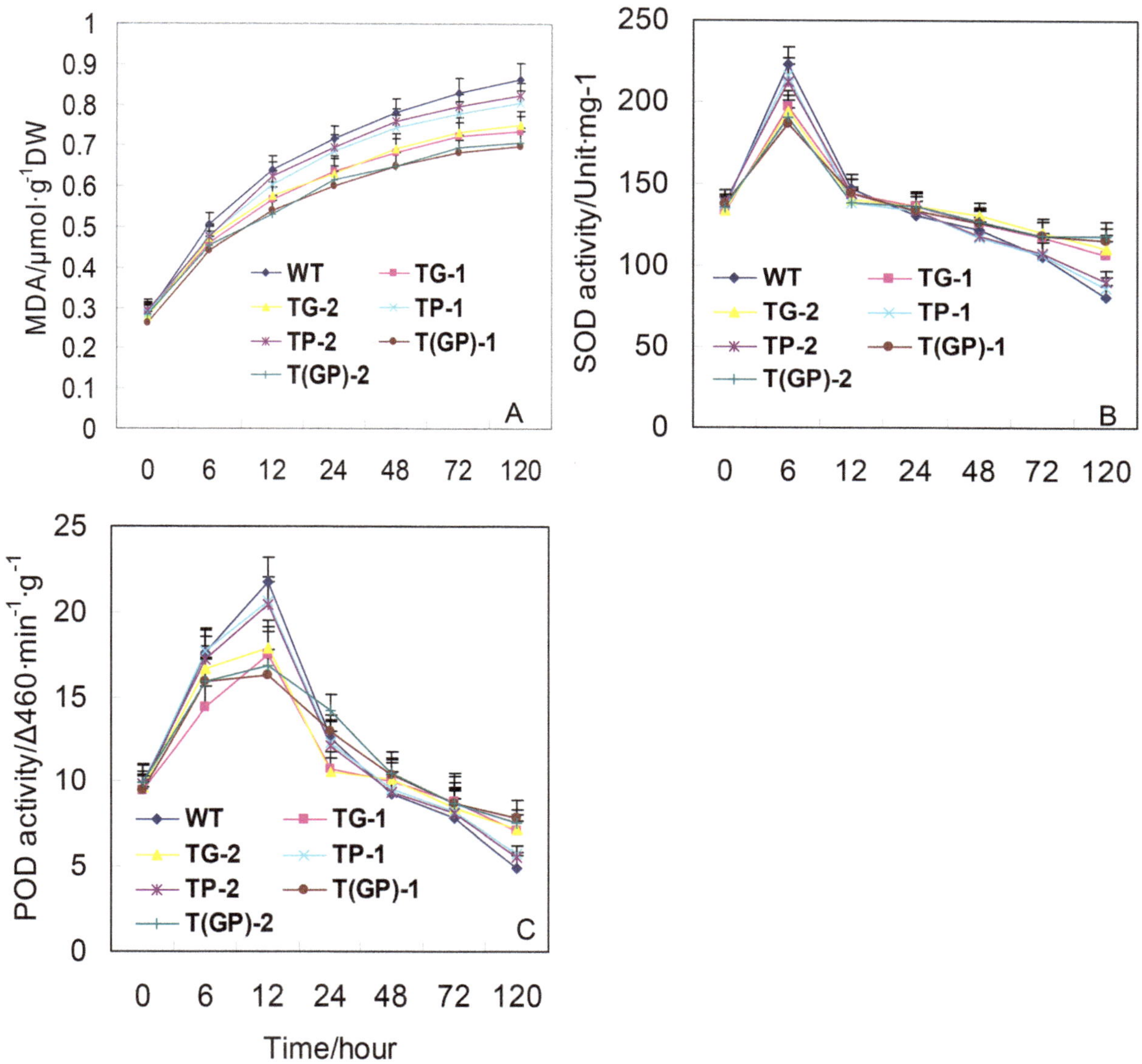

Figure 6. Short term response of plants to Cd exposure/After 0–120 h, the leaves were sampled for (A) MDA content, (B) SOD activity, (C) POD activity. TP-1, TP-2: carriers of *PaPCS*; TG-1, TG-2: carriers of *PaGCS*; T(GP)-1, T(GP)-2: carriers of *PaGCS-PaPCS*; WT: wild type *F. arundinacea*. Vertical bars indicate the standard error of the mean ($n = 4$).

the presence of 150 μM $CdCl_2$, which demonstrated the contribution that this gene makes to Cd tolerance.

However, the previous studies paid more attention on single overexpression of genes involved in chelation, and it indicated that overexpression of genes involved in PCs synthesis in plants results in either hypersensitivity or tolerance to Cd [50–51]. In our study, the line overexpressed single PaPCS or PaGCS, was tolerant to Cd stress (Fig. 4). It had higher content of PCs (Fig. 7) and accumulated higher amounts of this metal (Fig. 5) than WT. This result supported some of the above reports. In our study, the transgenic lines simultaneous overexpressed PaPCS and PaGCS gene had higher contents of Cd and PCs than WT and single-gene transgenic lines after Cd exposure (Fig. 5,7). This result is consistent with the previous study on the transgenic lines simultaneous overexpressed CsPCS and YCF1 which accumulated higher amounts of Cd than single-gene transformants [52]. This showed that multiple overexpression of genes involved in chelation can led to a higher increase on the tolerance and accumulation to heavy metal, it might be a way of acquiring more efficient hyperaccumulator.

The behaviour of MDA content and the patterns of SOD and POD activity confirmed this conclusion, since the leaf content of MDA in the non-transgenic plants remained substantially higher than in the *PaPCS* transgenic lines (Fig. 6A). The significance of MDA content lies in its being a diagnostic for the severity of oxidative stress. It is suggested that the level of oxidative damage in the transgenic lines was less severe than in the WT. Consistently, both SOD and POD activity were higher in the *PaPCS* transgenics than in the WT 24 h after the stress was imposed (Fig. 6B, 6C). Heavy metals in their ionic state are considered to be rather more toxic than when they are complexed with organic molecules, for instance the PCs. The *PaPCS* transgenic lines accumulated more Cd than the WT, but at the same time they were able to neutralize this additional Cd by enhancing the cellular level of PC.

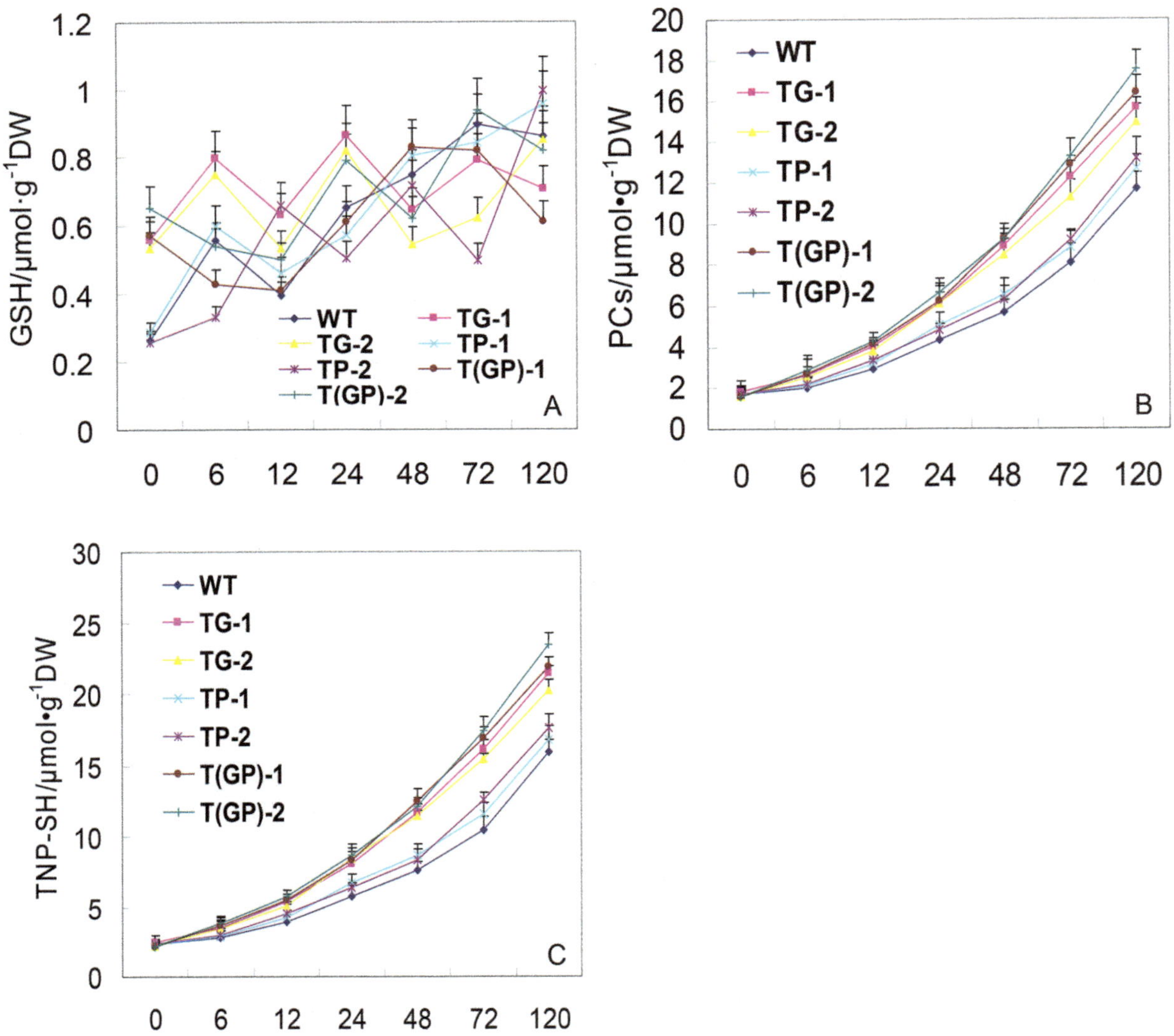

Figure 7. Short term response of plants to Cd exposure/After 0–120 h, the leaves were sampled for (A) Total glutathione (GSH) content, (B) Non-protein thiol (TNP-SH) content, (C) PC content. TP-1, TP-2: carriers of *PaPCS*; TG-1, TG-2: carriers of *PaGCS*; T(GP)-1, T(GP)-2: carriers of *PaGCS-PaPCS*; WT: wild type *F. arundinacea*. Vertical bars indicate the standard error of the mean ($n = 4$).

It has been noted that the tissue-specific expression of PC biosynthesis genes can result in the relocation of heavy metal mediated by the long distance xylem transport of γ-glutamylcysteine, and PCs [53,54]. However, little is known about the relationship between growth condition, PC content and redistribution of Cd of the transgenic lines with single or double PC biosynthesis genes. Here, we detected the variation of the redistribution of Cd of the WT and three transgenic lines, and found that the heterologous overexpression of *PaGCS* in tall fescue increased the ratio of Cd root/leaf content (Tab. 1), which is consistent with the previous study of *PaGCS* heterologous overexpression in *Agrostis palustris* [26]. Heterologous overexpression of *PaPCS* in tall fescue also increased the ratio of Cd root/leaf content (Table 1). This phenomenon has been observed in previous study of wheat PCS overexpressed in rice, in which the Cd accumulation were increased in shoot, but not changed in root [55]. However, other study did not show the redistribution of Cd in PCS transgenic plants, which suggested the redistribution might depend on the gene sources [52,53,56].

In previous study, the growth of *PCS* transgenic plants appeared to be inhibited under Cd stress, although when the plants were given exogenous GSH, this sensitivity disappeared. The interpretation of this behavior was that glutathione must be necessary for the operation of PCS. γ-GCS catalyzes the initial step of GSH synthesis, and it is generally understood that GSH synthesis is a rate limiting step for PCS activity. Heterologous expression of the *E. coli* γ-GCS gene *gshI* in poplar chloroplasts is known to enhance the accumulation of GSH [56]. When *PaGCS* and *PaPCS* were simultaneously introduced into tall fescue, the double transformant line out-performed both of the single transgene carriers, which were superior to the WT. This out-performance was measurable in terms of biomass accumulation, MDA content and antioxidant enzyme activity. The implication is that of the two genes *PaGCS* and *PaPCS*, the former has a stronger beneficial effect than the latter on heavy metal tolerance.

Supporting Information

Figure S1 The plant expression vectors (A) pROK/PaGCS, and (B) p3301/PG. RB and LB: right and left borders of the *Ti* plasmid; *Tnos*: nos terminator (260 bp).

Figure S2 Phylogeny of deduced phytochelatin synthase polypeptides.

Figure S3 Transformation of F. arundinacea *with PaPCS.* (A) Callus induction, (B) Agroinfection, (C) Kanamycin selection of transgenic calli, (D) Regenerated plants, (E) Selected plants maintain their greenness in the presence of kanamycin, (F) Putative transgenic plants grown in soil.

Figure S4 Transformation of *F. arundinacea* with *PaGCS-PaPCS.* (A) Callus induction, (B) Agroinfection, (C) Kanamycin selection of transgenic calli, (D) Regenerated plants, (E) Selected plants maintain their greenness in the presence of kanamycin, (F) Putative transgenic plants grown in soil.

Figure S5 PC2 and PC3 content after short term Cd exposure/After 0–120 h, the leaves were sampled for (A) PC2 and (B) PC3. TP-1, TP-2: carriers of *PaPCS*; TG-1, TG-2: carriers of *PaGCS*; T(GP)-1, T(GP)-2: carriers of *PaGCS-PaPCS*; WT: wild type *F. arundinacea.* Vertical bars indicate the standard error of the mean ($n = 4$).

Acknowledgments

We thank Prof. Dietrich Nies (Martin-Luther-Universität, Halle-Wittenberg) for providing yeast strain YK44, and the Department of Plant Pathology, Kansas State University, USA for access to their DNA sequencing facility. Thanks to Dr. Robert Koebner (UK), for linguistic advice.

Author Contributions

Conceived and designed the experiments: FX. Performed the experiments: CZ PW QL. Analyzed the data: FX CZ PW. Contributed reagents/materials/analysis tools: SL. Wrote the paper: CZ JX SL.

References

1. Ding TD, Ni WM, Zhang JY (2012) Research advances in heavy metals pollution ecology of diatom. Ying Yong Sheng Tai Xue Bao 23: 857–866.
2. Guo XL, Gu J, Chen ZX, Gao H, Qin QJ, et al. (2012) Effects of heavy metals pollution on soil microbial communities metabolism and soil enzyme activities in coal mining area of Tongchuan, Shaanxi Province of Northwest China. Ying Yong Sheng Tai Xue Bao 23: 798–806.
3. Ellis JK, Athersuch TJ, Thomas LD, Teichert F, Perez-Trujillo M, et al. (2012) Metabolic profiling detects early effects of environmental and lifestyle exposure to cadmium in a human population. BMC Med 10: 61.
4. Salt DE, Smith RD, Raskin I (1998) PHYTOREMEDIATION. Annu Rev Plant Physiol Plant Mol Biol 49: 643–668.
5. Suresh B, Ravishankar G (2004) Phytoremediation-A novel and promising approach for environmental clean-up. Crit Rev Biotechnol 24: 97–124.
6. Bonanno G, Lo Giudice R (2010) Heavy metal bioaccumulation by the organs of *Phragmites australis* (common reed) and their potential use as contamination indicators. Ecological Indicators 10: 639–645.
7. Olguín EJ, Sánchez-Galván G (2012) Heavy metal removal in phytofiltration and phycoremediation: the need to differentiate between bioadsorption and bioaccumulation. New Biotechnol 30: 3–8.
8. Bhargava A, Carmona FF, Bhargava M, Srivastava S (2012) Approaches for enhanced phytoextraction of heavy metals. J Environ Manage 105: 103–120.
9. Huang J, Xia H, Li Za, Xiong Y, Kong G (2009) Soil aluminium uptake and accumulation by *Paspalum notatum*. Waste Management & Research 27: 668–675.
10. Cassina L, Tassi E, Morelli E, Giorgetti L, Remorini D, et al. (2011) Exogenous cytokinin treatments of an Ni hyper-accumulator, Alyssum murale, grown in a serpentine soil: Implications for phytoextraction. Int J Phytoremediat 13: 90–101.
11. Seth C, Misra V, Chauhan L (2012) Accumulation, detoxification, and genotoxicity of heavy metals in indian mustard (*Brassica juncea* L.). Int J Phytoremediat 14: 1–13.
12. Bennett LE, Burkhead JL, Hale KL, Terry N, Pilon M, et al. (2003) Analysis of transgenic Indian mustard plants for phytoremediation of metal-contaminated mine tailings. J Environ Qual 32: 432–440.
13. Gendre D, Czernic P, Conéjéro G, Pianelli K, Briat JF, et al. (2007) TcYSL3, a member of the YSL gene family from the hyper-accumulator *Thlaspi caerulescens*, encodes a nicotianamine-Ni/Fe transporter. Plant J 49: 1–15.
14. Meyer C-L, Peisker D, Courbot M, Craciun AR, Cazalé A-C, et al. (2011) Isolation and characterization of *Arabidopsis halleri* and *Thlaspi caerulescens* phytochelatin synthases. Planta 234: 83–95.
15. Dhankher OP, Li Y, Rosen BP, Shi J, Salt D, et al. (2002) Engineering tolerance and hyperaccumulation of arsenic in plants by combining arsenate reductase and γ-glutamylcysteine synthetase expression. Nat Biotechnol 20: 1140–1145.
16. Gisbert C, Ros R, De Haro A, Walker DJ, Pilar Bernal M, et al. (2003) A plant genetically modified that accumulates Pb is especially promising for phytoremediation. Biochem Biophys Res Commun 303: 440–445.
17. Zitka O, Krystofova O, Sobrova P, Adam V, Zehnalek J, et al. (2011) Phytochelatin synthase activity as a marker of metal pollution. J Hazard Mater 192: 794–800.
18. Pal R, Rai J (2010) Phytochelatins: peptides involved in heavy metal detoxification. Appl Biochem Biotechnol 160: 945–963.
19. Zhang ZC, Chen BX, Qiu BS (2010) Phytochelatin synthesis plays a similar role in shoots of the cadmium hyperaccumulator *Sedum alfredii* as in non-resistant plants. Plant Cell Environ 33: 1248–1255.
20. Cobbett CS (2000) Phytochelatins and their roles in heavy metal detoxification. Plant Physiol 123: 825–832.
21. Hirata K, Tsuji N, Miyamoto K (2005) Biosynthetic regulation of phytochelatins, heavy metal-binding peptides. J Biosci Bioeng 100: 593–599.
22. Liu Z, Gu C, Chen F, Yang D, Wu K, et al. (2012) Heterologous expression of a *Nelumbo nucifera* phytochelatin synthase gene enhances cadmium tolerance in *Arabidopsis thaliana*. Appl Biochem Biotechnol 166: 722–734.
23. Brunetti P, Zanella L, Proia A, De Paolis A, Falasca G, et al. (2011) Cadmium tolerance and phytochelatin content of Arabidopsis seedlings over-expressing the phytochelatin synthase gene *AtPCS1*. J Exp Bot 62: 5509–5519.
24. Wawrzyński A, Kopera E, Wawrzyńska A, Kamińska J, Bal W, et al. (2006) Effects of simultaneous expression of heterologous genes involved in phytochelatin biosynthesis on thiol content and cadmium accumulation in tobacco plants. J Exp Bot 57: 2173–2182.
25. Wawrzynska A, Wawrzynski A, Gaganidze D, Kopera E, Piatek K, et al. (2005) Overexpression of genes involved in phytochelatin biosynthesis in *Escherichia coli*: effects on growth, cadmium accumulation and thiol level. Acta Biochim Pol 52: 109–116.
26. Zhao C, Qiao M, Yu Y, Xia G, Xiang F (2010) The effect of the heterologous expression of *Phragmites australis* γ-glutamylcysteine synthetase on the Cd^{2+} accumulation of *Agrostis palustris*. Plant Cell Environ 33: 877–887.
27. Ito H, Fukuda Y, Murata K, Kimura A (1983) Transformation of intact yeast cells treated with alkali cations. J Bacteriol 153: 163–168.
28. Murashige T, Skoog F (1962) A revised medium for rapid growth and bio assays with tobacco tissue cultures. Physiol Plant 15: 473–497.
29. Doyle JJ (1990) Isolation of plant DNA from fresh tissue. Focus 12: 13–15.
30. Vitória AP, Lea PJ, Azevedo RA (2001) Antioxidant enzymes responses to cadmium in radish tissues. Phytochemistry 57: 701–710.
31. Buege JA, Aust SD (1978) Microsomal lipid peroxidation. Method Enzymol 52: 302–310.
32. Scebba F, Arduini I, Ercoli L, Sebastiani L (2006) Cadmium effects on growth and antioxidant enzymes activities in *Miscanthus sinensis*. Biol Plantarum 50: 688–692.
33. Bradford MM (1976) A rapid and sensitive method for the quantitation of microgram quantities protein using the principle of protein-dye binding. Biochem 72: 248–254.
34. Gupta M, Tripathi R, Rai U, Chandra P (1998) Role of glutathione and phytochelatin in *Hydrilla verticillata* (l.f.) royle and *Valusneria spiraus* L. under mercury stress. Chemosphere 37: 785–800.
35. Tu S, Ma LQ, MacDonald GE, Bondada B (2004) Effects of arsenic species and phosphorus on arsenic absorption, arsenate reduction and thiol formation in excised parts of *Pteris vittata* L. Environ Exp Bot 51: 121–131.
36. Vatamaniuk OK, Mari S, Lu Y-P, Rea PA (1999) AtPCS1, a phytochelatin synthase from *Arabidopsis*: isolation and in vitro reconstitution. PNAS 96: 7110–7115.
37. Clemens S, Kim EJ, Neumann D, Schroeder JI (1999) Tolerance to toxic metals by a gene family of phytochelatin synthases from plants and yeast. EMBO J 18: 3325–3333.
38. Osaki Y, Shirabe T, Tamura S, Yoshimura E (2008) A functional putative phytochelatin synthase from the primitive red alga *Cyanidioschyzon merolae*. Biosci Biotechnol Biochem 72: 3306–3309.
39. Singh S, Lee W, DaSilva NA, Mulchandani A, Chen W (2008) Enhanced arsenic accumulation by engineered yeast cells expressing *Arabidopsis thaliana* phytochelatin synthase. Biotechnol Bioeng 99: 333–340.
40. Yong X, Chen Y, Liu W, Xu L, Zhou J, et al. (2014) Enhanced cadmium resistance and accumulation in *Pseudomonas putida* KT2440 expressing the phytochelatin synthase gene of *Schizosaccharomyces* pombe. Lett Appl Microbiol 58: 255–261.

41. Chaurasia N, Mishra Y, Rai LC (2008) Cloning expression and analysis of phytochelatin synthase (*pcs*) gene from *Anabaena* sp. PCC 7120 offering multiple stress tolerance in *Escherichia coli*. Biochem Biophy Res Commun 376: 225–230.
42. Pomponi M, Censi V, Di Girolamo V, De Paolis A, di Toppi LS, et al. (2006) Overexpression of *Arabidopsis* phytochelatin synthase in tobacco plants enhances Cd^{2+} tolerance and accumulation but not translocation to the shoot. Planta 223: 180–190.
43. Gasic K, Korban SS (2007) Transgenic Indian mustard (*Brassica juncea*) plants expressing an *Arabidopsis* phytochelatin synthase (AtPCS1) exhibit enhanced As and Cd tolerance. Plant Mol Biol 64: 361–369.
44. Shukla D, Kesari R, Mishra S, Dwivedi S, Tripathi RD, et al. (2012) Expression of phytochelatin synthase from aquatic macrophyte *Ceratophyllum demersum* L. enhances cadmium and arsenic accumulation in tobacco. Plant Cell Rep 31: 1687–1699.
45. Liu Z, Gu C, Chen F, Yang D, Wu K, et al. (2012) Heterologous expression of a *Nelumbo nucifera* phytochelatin synthase gene enhances cadmium tolerance in *Arabidopsis thaliana*. Appl Biochem Biotechnol 166: 722–734.
46. Shukla D, Kesari R, Tiwari M, Dwivedi S, Tripathi RD, et al. (2013) Expression of *Ceratophyllum demersum* phytochelatin synthase, CdPCS1, in *Escherichia coli* and *Arabidopsis* enhances heavy metal(loid)s accumulation. Protoplasma 250: 1263–1272.
47. Peterson AG, Oliver DJ (2006) Leaf-targeted phytochelatin synthase in *Arabidopsis thaliana*. Plant Physiol Biochem 44: 885–892.
48. Li Y, Dhankher OP, Carreira L, Lee D, Chen A, et al. (2004) Overexpression of phytochelatin synthase in *Arabidopsis* leads to enhanced arsenic tolerance and cadmium hypersensitivity. Plant Cell Physiol 45: 1787–1797.
49. Lee S, Moon JS, Ko T-S, Petros D, Goldsbrough PB, et al. (2003) Overexpression of *Arabidopsis* phytochelatin synthase paradoxically leads to hypersensitivity to cadmium stress. Plant Physiol 131: 656–663.
50. Shukla D, Tiwari M, Tripathi RD, Nath P, Trivedi PK (2013) Synthetic phytochelatins complement a phytochelatin-deficient *Arabidopsis* mutant and enhance the accumulation of heavy metal(loid)s. Biochem Biophys Res Commun 434: 664–669.
51. Wu G, Kang H, Zhang X, Shao H, Chu L, et al. (2010) A critical review on the bio-removal of hazardous heavy metals from contaminated soils: issues, progress, eco-environmental concerns and opportunities. J Hazard Mater 174: 1–8.
52. Guo J, Xu W, Ma M (2012) The assembly of metals chelation by thiols and vacuolar compartmentalization conferred increased tolerance to and accumulation of cadmium and arsenic in transgenic *Arabidopsis thaliana*. J Hazard Mater 199–200: 309–313.
53. Gong JM, Lee DA, Schroeder JI (2003) Long-distance root-to-shoot transport of phytochelatins and cadmium in *Arabidopsis*. PNAS 100: 10118–10123.
54. Li Y, Dankher O, Carreira L, Smith A, Meagher R (2006) The shoot-specific expression of γ-glutamylcysteine synthetase directs the long-distance transport of thiol-peptides to roots conferring tolerance to mercury and arsenic. Plant Physiol 141: 288–298.
55. Wang F, Wang Z, Zhu C (2012) Heteroexpression of the wheat phytochelatin synthase gene (*TaPCS1*) in rice enhances cadmium sensitivity. Acta Biochim Biophys Sin (Shanghai) 44: 886–893.
56. Noctor G, Strohm M, Jouanin L, Kunert K-J, Foyer CH, et al. (1996) Synthesis of glutathione in leaves of transgenic poplar overexpressing [gamma]-glutamylcysteine synthetase. Plant Physiol 112: 1071–1078.

Diverse Functions of mRNA Metabolism Factors in Stress Defense and Aging of *Caenorhabditis elegans*

Aris Rousakis[1,2]◑, Anna Vlanti[1]◑, Fivos Borbolis[1,3], Fani Roumelioti[1,3¤a], Marianna Kapetanou[1,4¤b], Popi Syntichaki[1]*

1 Biomedical Research Foundation of the Academy of Athens, Center of Basic Research II, Athens, Greece, **2** Faculty of Medicine, University of Athens, Athens, Greece, **3** Faculty of Biology, School of Science, University of Athens, Athens, Greece, **4** Department of Biology, School of Science and Engineering, University of Crete, Heraklio, Crete, Greece

Abstract

Processing bodies (PBs) and stress granules (SGs) are related, cytoplasmic RNA-protein complexes that contribute to post-transcriptional gene regulation in all eukaryotic cells. Both structures contain translationally repressed mRNAs and several proteins involved in silencing, stabilization or degradation of mRNAs, especially under environmental stress. Here, we monitored the dynamic formation of PBs and SGs, in somatic cells of adult worms, using fluorescently tagged protein markers of each complex. Both complexes were accumulated in response to various stress conditions, but distinct modes of SG formation were induced, depending on the insult. We also observed an age-dependent accumulation of PBs but not of SGs. We further showed that direct alterations in PB-related genes can influence aging and normal stress responses, beyond their developmental role. In addition, disruption of SG-related genes had diverse effects on development, fertility, lifespan and stress resistance of worms. Our work therefore underlines the important roles of mRNA metabolism factors in several vital cellular processes and provides insight into their diverse functions in a multicellular organism.

Editor: Dimitris L. Kontoyiannis, BSRC 'Alexander FLEMING', Greece

Funding: The work was supported by European Research Council under the European Union's Seventh Framework Program (FP/2007-2013)/ERC Grant Agreement n. [201975]. The funder had no role in study design, data collection and analysis, decision to publish, or preparation of the manuscript.

Competing Interests: The authors have declared that no competing interests exist.

* Email: synticha@bioacademy.gr

◑ These authors contributed equally to this work.

¤a Current address: Institute of Immunology, BSRC "Alexander Fleming", Athens, Greece
¤b Current address: Institute of Biology, Medicinal Chemistry and Biotechnology, National Hellenic Research Foundation, Athens, Greece

Introduction

A plethora of evidence in all eukaryotes has established that aging is a multifactorial process that can be modulated by several means, ranging from genes to lifestyle and pharmacological interventions. Genetic and genome-wide studies in many organisms shed light on the basic molecular mechanisms of aging and identified conserved signaling pathways as master modulators of lifespan [1]. Fundamental cellular functions that primarily alter metabolism or preserve homeostasis can influence the rate of aging and age-related disability/degeneration. Such cellular functions comprise the activation of stress response and repair mechanisms, the enhancement of catabolic processes and the reduction of anabolic processes. Protein synthesis is an anabolic process that, when reduced by gene mutations, drugs, hormonal or stress signals, increases lifespan in diverse species. The underlying mechanisms could involve energy conservation and/or reprogramming of gene expression that is essential for cell survival [2]. In accordance with this, several studies demonstrate that protein synthesis is tightly regulated by environmental stress in eukaryotes. Protein synthesis rates are mainly controlled at the level of translation initiation but post-transcriptional regulation of mRNAs, including processing, export, decay and localization, can also affect translation. Accordingly, the impact of post-transcriptional regulation of gene expression by RNA-binding proteins and cytoplasmic RNA granules is an emerging field in regulators of aging and age-related diseases [3–5].

Processing bodies (PBs) and stress granules (SGs) are major cytoplasmic RNA-protein complexes that regulate, in part, the abundance and translation of mRNAs in all eukaryotic cells. PBs contain non-translatable mRNAs and a variety of proteins involved in translational repression, 5′ to 3′ mRNA decay and RNA-induced silencing [6–8]. SGs function in mRNA storage following environmental stress and contain stalled 43S pre-initiation complexes and many RNA-binding proteins that regulate the stability, storage or translation of mRNAs [9]. Both PBs and SGs are highly dynamic structures that share specific components and can spatially interact to determine the fate of mRNAs in the cytoplasm, in equilibrium with translating polysomes [10–14]. However, the physiological role of these complexes has not yet been fully elucidated; the processes of translation inhibition and mRNA decay can occur normally in yeast, *Drosophila* and mammalian cells that are defective in PB or SG formation, suggesting that their accumulation is a consequence of rather than a perquisite for mRNA metabolism [15–19]. Furthermore, under some conditions in yeast, decapping and 5′ to 3′ decay occur while mRNAs are still associated with polyribosomes [20]. Nevertheless, the evolutionary conservation of both

RNA granules supports that they function to impart specificity in molecule interactions and efficiency in translation repression and mRNA degradation.

Both PBs and SGs participate in the transient translational silencing and global reprogramming of gene expression that occur in response to cellular stress [8]. They accumulate within minutes of stress and dissolve within a few hours of stress recovery. The aggregation of non-translatable mRNAs into PBs and SGs could either promote or inhibit mRNA decay and might be crucial for the adaptation or recovery of cells e.g. through the stabilization of specific transcripts. These include either stress-responsive transcripts with short half-lives [21] or energy-costly mRNAs, as those encoding the abundant cytoplasmic ribosomal proteins [22]. In addition, SGs can affect key signaling pathways involved in the global stress response, such the MAPK, TOR and HIF-1 pathways [23–26]. Studies in yeast and mammalian cells have shown that stress-induced granules can differ in composition and assembly rules, depending on the type of stress insult [11,27].

In *C. elegans* several types of RNA granules have been described during the processes of oogenesis and embryogenesis. P granules are a type of germ granule, required for germ cell development, with a primary role in post-transcriptional regulation of maternal mRNAs in the gonad [28–31]. Other RNA granules accumulate within the *C. elegans* gonad in response to developmental or environmental signals and differ in composition and function [32–34]. All these diverse RNA granules participate in proper mRNA control during early development or in aged/stressed oocytes and share some components with somatic PBs and SGs in other organisms[32–37]. Such components include the worm homologues of proteins implicated in mRNA decapping or translational repression, as the decapping enzyme DCAP-2 (Dcp2 in yeast and mammals) and its cofactor DCAP-1 (Dcp1), the translation regulator CGH-1 (Dhh1/RCK), the decapping activators PATR-1 (Pat1) and LSM-1 (Lsm1) or the 5′ to 3′ exonuclease XRN-1 (Xrn1). In addition, they harbor RNA-binding proteins, such as the worm homologues of human ataxin-2 (ATX-2), the poly(A) binding protein PABP (PAB-1) and TIA-1/TIAR proteins (TIAR-1, -2) that are SG components in mammalian cells. Although the structure, localization and function of these germline and early embryonic RNA granules have been thoroughly characterized, somatic PBs and SGs have not yet been well investigated in *C. elegans*. The developmental consequences of loss of several P granule components have been nicely described [34–36,38] but the effect of somatic PB and SG components on survival of the organism during stress conditions or aging are largely unknown.

Here, we present data on worm's stress survival and lifespan modulation by factors involved in mRNA metabolism. We first provide *in vivo* insights into the dynamic formation of somatic PBs and SGs in intact adult animals, in response to various environmental stresses. Moreover, an age-dependent accumulation of PBs, but not SGs, was observed in somatic cells of worms. Aggregation of PBs in young adults is also triggered in response to alterations in 5′ to 3′ mRNA degradation but not to reduced mRNA translation. Furthermore, we show that direct alterations in PB-related genes, by mutations or RNA interference (RNAi), can influence lifespan and stress resistance, beyond their developmental role. We reveal that stress conditions induce distinct modes of SG formation and deletion of SG-related genes has diverse effects on development, lifespan and stress response of worms. Our work provides genetic characteristics of proteins localized to PBs and SGs and offers new insights into the pattern and function of these RNA granules at the organismal level.

Materials and Methods

C. elegans strains and culture

Standard methods of culturing and handling worms were used [39]. Worms were raised on NGM plates seeded with *Escherichia coli* OP-50 or HT115 (DE3) for RNAi experiments, at the indicated temperature. Wild-type Bristol N2 and some mutant strains were provided by the *Caenorhabditis* Genetics Center (CGC, University of Minnesota), which is funded by NIH Office of Research Infrastacture Programs (P40 OD010440). Other mutant strains were provided by the Mitani Lab through the National Bio-Resource Project of the MEXT, Japan. Table S1 shows all strains used in this study. All single mutants were outcrossed 3–5 times with N2 and relevant mutations were tracked in F2 progeny by PCR (Table S2). Double mutants were made by crossing the relevant strains and PCR-based selection in F2 progeny. Transgenic animals were generated by microinjection of plasmid DNAs into the gonad of N2 young adults, using *rol-6(su1006)* as co-transformation marker [40]. Multiple lines were generated for each genotype and screened for the representative expression pattern. Transgenic mutants were generated by crossing N2 transgenic hermaphrodites with males of the desired mutant background.

Constructs

RNAi plasmids were constructed by inserting gene-specific PCR product, amplified from genomic DNA using the appropriate primers (Table S2), into the L4440 feeding vector (pPD129.36) [41]. For the double *dcap-1/-2(RNAi)* construct both gene fragments were cloned into a single L4440 vector [42]. See Table S3 for RNAi construct details. The presumptive promoter of *dcap-1* (~500-kb) was obtained by PCR (Table S2) and inserted into the Fire Lab vector pPD95.77 for the $P_{dcap\text{-}1}$::*gfp* transcriptional fusion. For the *dcap-1::gfp* transgene, the *gfp* sequence was inserted between the end of *dcap-1* coding region and its 3′ UTR in the plasmid FLAG::*dcap-1*, kindly provided by Dr. Min Han [43]. The *tiar-3::gfp* transgene was constructed by cloning the promoter (1969-bp) and coding region in pPD95.77. For *gfp::tiar-1* and *gfp::tiar-2* a modified version of pPD95.77 was used, allowing insertion of the promoter regions of *tiar-1* (1421-bp) or *tiar-2* (1614-bp) upstream of *gfp* with subsequent in-frame insertion of their corresponding coding regions followed by their own 3′ UTRs (489-bp for *tiar-1* and 680-bp for *tiar-2*). The intermediate constructs with the promoter only sequences fused to *gfp* of pPD95.77 were used as transcriptional reporters. The tagRFP fusions were constructed by replacing the GFP by tagRFP, obtaining the relevant sequence by PCR amplification from pHb9::tagRFP (a gift from Dr. Ivo Lieberam).

RNA interference

RNAi experiments were carried out by adding synchronized L4s or eggs to plates seeded with HT115(DE3) bacteria that express dsRNA for the indicated gene. HT115 bacteria transformed with the relevant RNAi vectors were grown at 37°C in LB medium with ampicillin (50 μg/ml) and tetracycline (10 μg/ml). On the following day, fresh cultures with ampicillin were induced with 1 mM isopropylb-D-thiogalactopyranoside (IPTG) and seeded on RNAi plates [44]. Bacteria carrying the empty vector (pL4440) and treated likewise were used as control cultures (Control(RNAi)).

Microscopy

The expression pattern of transgenic worms was monitored by mounting levamizole-treated animals on 3% agarose pads, on glass microscope slides. Animals (~20 animals per condition in at least three experiments) were imaged using a Leica TCS SP5 confocal imaging system. For imaging assays of transgenic worms at different ages, adults from age-synchronized cultures, raised at the indicated temperatures from hatching, were picked and monitored under the same microscopy settings. For heat-shock assays 1-day adults were shifted at 35°C for the indicated time and were immediately monitored. For recovery assays the heat-shocked animals were incubated for 0.5–3 h at 20°C before mounting. For oxidative stress 1-day adults were transferred on NGM plates seeded with UV-killed OP-50 bacteria containing 10–15 mM sodium arsenite and were visualized after 2–3 h. For osmotic stress 1-day adults were transferred on NGM plates containing 350 mM NaCl and were visualized after 0.5–1 h. All images were taken at 20× magnification, keeping the same microscopy settings for each strain per condition. Images shown from confocal are 2D maximal projections of z-stacks, or optical sections where is indicated, processed in Photoshop CS3.

The quantification of PBs formed under HS was performed as follows: three random squares of 50 μm×50 μm were drawn along a worm; the number of aggregates within each square was measured and their mean value, representing the number of granules per 2500 μm^2 per worm, was calculated. This procedure was repeated for a total of 20 worms from at least two independent experiments and the average of the mean values is presented in Table S4. The same method was applied to measure SGs visualized using the $P_{myo\text{-}3}$::GFP::TIAR-2 marker. In the case of age-induced PBs and SGs we measured aggregates formed in the area of the head (from the pharynx to the tip of the worm) to avoid the intestinal autofluorescence. The same method was used to measure TIAR-1 granules as well as SA-induced PBs. When using the GFP::TIAR-2 marker, we measured the number of granules formed in 100 μm length of excretory cell. Again a total of 15–20 worms were counted in each case from at least two independent experiments. All measurements were performed using the count tool in Photoshop CS3. Statistical analysis (unpaired *t*-test) was performed using GraphPad Prism version 5 (GraphPad Softwate, San Diego, California USA).

RNA isolation and quantitative reverse transcription PCR (qRT-PCR)

Total RNA was prepared from frozen worm pellets, of the indicated genetic backgrounds and developmental stages, using a NucleoSpin RNA XS kit (Macherey-Nagel) and measured by Quant-iT RNA Assay Kit (Invitrogen). Total RNA was reverse transcribed with iScriptTM cDNA Synthesis Kit (Biorad) and quantitative PCR was performed using the SsoFastTM EvaGreen Supermix (BioRad) in the MJ MiniOpticon system (BioRad). The relative amounts of mRNA were determined using the Comparative Ct method for quantification and gene expression data are presented as the fold change relative to control. qRT-PCR was performed in triplicates and each sample was independently normalized to endogenous reference *ama-1*. The mean ± the standard deviation (SD) of at least two independent experiments is presented. All statistical comparisons were performed by Student's *t*-test for unpaired samples in GraphPad Prism 5.The primer sequences for qRT-PCR are available upon request.

Western Blotting

About 1500–2000 synchronized worms of each strain, grown on OP50 plates at 25°C, were subjected to heat-shock (35°C for 3 h), in the first day of adulthood and were collected in M9 buffer, washed 2–3 to remove bacteria and frozen in ethanol dry ice. Control worms (no heat-shocked) of the same age were treated similarly. Likewise, ~1500 L4-synchronized worms of each strain at 25°C, were transferred on OP50 plates containing 50 μM 5-fluoro-2'-deoxyuridine (FUDR) to prevent progeny production and collected as either 1-day or 5-day adults in M9 buffer. After 2–3 washes to remove bacteria they were frozen in ethanol dry ice. In all cases, before loading onto SDS-PAGE, worm pellets were boiled in 2× SDS-sample buffer for 10 min. Worm lysates were resolved on a 10% SDS-PAGE, western blotted and analyzed with primary antibodies against to either GFP (1:10,000, BD Biosciences) or actin (1:5,000, Clone C4, Millipore). Secondary anti-rabbit and anti-mouse IgG antibodies (HRP) were used respectively for immunoblot signal detection with ECL (Thermo Scientific). Quantification of immunoblot signals was performed using ImageJ software. Ratio of DCAP-1::GFP to actin levels was measured in two independent experiments.

Lifespan assays

Lifespan analysis was conducted at 20°C or 25°C as described previously [45]. Briefly, eggs or mid-to-late L4 larvae of each strain (at least 70 animals per experiment) were transferred to NGM plates seeded with OP-50 or RNAi bacteria of interest and were moved to fresh plates every two days. Viability of the worms was daily scored and animals that failed to respond to stimulation by touch were referred as dead, whereas that bagged, ruptured or crawled off the plates were referred as censored in the analysis. For post-developmental assays, day 0 of adulthood was defined as the day that mid-to-late L4s were transferred to plates. Lifespan and statistical analysis were performed using GraphPad Prism version 5. Each population is compared with the appropriate control population and p-values were determined using the log-rank (Mantel-Cox) test.

Fertility assay

Worms of each genotype were grown at 20°C or 25°C and 5–15 L4 hermaphrodites were placed on individual NGM plates to lay eggs. Animals were transferred daily to fresh plates until egg-laying ceased and the hatched progeny were counted in each plate. The total number of progeny per worm (brood size) was counted and the average brood size (mean ± SD) of each strain was plotted. Unpaired *t*-test was used to calculate p-values in GraphPad Prism 5.

Stress resistance assays

For heat-shock assays, 1-day adult worms were shifted to 35°C for 6 h (8 h for *daf-2* mutants). After 16 h of recovery at 20°C the percentage of worms surviving was determined. For UV resistance assays 5-day adults were irradiated on plates without bacteria at 0.2 J/cm^2 and then were transferred to plates with food at 20°C. Three days later the percentage of worms surviving was determined. For oxidative stress 1-day adults of each strain were transferred on plates seeded with UV-killed OP-50 containing 5 mM sodium arsenite and the percentage of worms surviving was determined after 24 h (for transgenic worms) or 48 h (non-transgenic worms). For osmotic stress 1-day adults were transferred on NGM plates containing 400 mM NaCl and scored for survival after 24 h. The average (mean ± SD) of at least three independent experiments with ~100 individuals for each strain per experiment was plotted. Unpaired *t*-test was used to calculate p-values in GraphPad Prism 5.

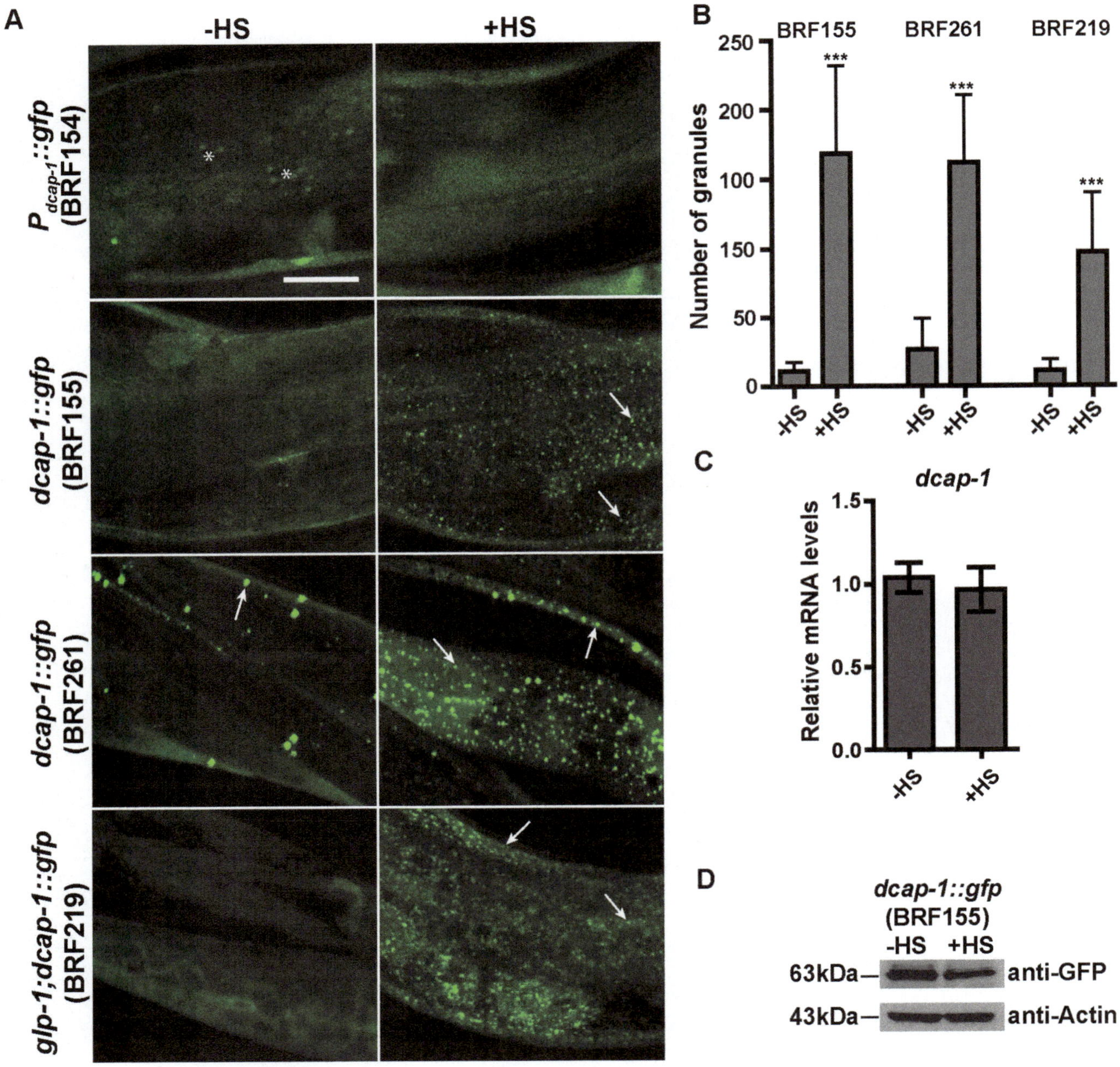

Figure 1. Aggregation of DCAP-1 protein in response to heat-shock. (A) Representative confocal images of somatic tissues in 1-day adult worms expressing the transcriptional fusion $P_{dcap\text{-}1}$::*gfp* (BRF154) or the translational fusion *dcap-1::gfp* in N2 (BRF155 and BRF261) and germline-deficient *glp-1(e2141)* worms (BRF219), normally grown at 25°C (-HS) or transiently subjected to heat-shock (+HS, 35°C for 3 h). Arrows point to DCAP-1::GFP granules in body wall muscles that are highly induced upon stress. Asterisks denote the intestinal autofluorescence. Scale bar: 25 μm. (B) Quantification of data in (A). Values on Y axis show the number of granules per 2500 μm^2 per worm (mean±SD, see Materials and Methods and Table S4). (C) Endogenous *dcap-1*(mRNA) levels in 1-day adults N2, grown at 25°C before (-HS) or after heat shock (+HS, 35°C for 3 h) were measured by quantitative RT-PCR and normalized to endogenous *ama-1(mRNA)* levels. Error bars represent the standard deviation of the means of five independent experiments (p-value = 0.3466, calculated by Student's *t*-test). (D) Western blot analysis of DCAP-1::GFP protein expression in 1-day adult BRF155 transgenic animals, before (-HS) or after heat shock (+HS), using anti-GFP or anti-Actin (as a loading control) antibodies. Band intensity of DCAP-1::GFP normalized to actin gives a value of 0.83, showing similar protein levels in the two conditions.

Results

Induction of PBs in *C. elegans* somatic cells by environmental stress

A well-described marker of PBs in all organisms is the decapping subunit Dcp1, which is encoded by the *dcap-1* (Y55F3AM.12.1) gene in worms (http://www.wormbase.org). DCAP-1 protein is a component of P granules in germline and PB-like structures in embryos [32,36–38]. We generated transgenic animals that carry either the transcriptional ($P_{dcap\text{-}1}$::*gfp*) or the translational ($P_{dcap\text{-}1}$::*dcap-1::gfp::3' UTR*$^{dcap\text{-}1}$ referred to as *dcap-1::gfp* for simplicity) fluorescent reporter gene to monitor the expression pattern in somatic cells of adult wild-type (N2) worms, under normal growth conditions (Fig. S1A). Worms expressing the $P_{dcap\text{-}1}$::*gfp* reporter (line BRF154, Fig. S1A) displayed a smooth and diffused fluorescent signal throughout the cytoplasm of most cells. Worms bearing the *dcap-1::gfp* fusion (BRF155, Fig. S1A) showed a similar ubiquitous expression pattern with occasional cytoplasmic puncta, varying among the worms. Although several protein components of PBs have an intrinsic capacity to aggregate [16,46–48], in mammalian cell lines PB components show a diffuse distribution in the cytoplasm and only few PBs are formed under normal conditions [49]. Also, high expression levels of fluorescent PB proteins in mammalian or yeast cells can further induce the aggregation of PBs and can alter the cellular stoichiometry of other PB components leading to aberrant structures [48–52]. Having established that the *dcap-1::gfp* fusion is functional (see text below and Fig. S5C), we showed by quantitative RT-PCR that the expression levels of *dcap-1* were not higher but underrepresented in BRF155 transgenic animals compared to the physiological levels of *dcap-1* in N2 (Fig. S1B). We interpreted that the transgene silencing effects resulted from the presence of vector backbone, as microinjection of the *dcap-1::gfp* reporter construct in linear, vector-free form [53] greatly improved its expression; a representative transgenic line (BRF261, Fig. S1A) had increased *dcap-1(mRNA)* levels (Fig. S1B) and more punctate pattern under normal conditions. Moreover, in a germline-deficient background (*glp-1*) the *dcap-1::gfp* transgene silencing was abolished (BRF219, Fig. S1A, B) confirming that such effects originate from the germline [54].

To validate that the DCAP-1::GFP-containing granules represent dynamic PBs in living animals we subjected all the above transgenic worms to heat-shock (+HS, Fig. 1A), a stress that generally enhances PB formation in other organisms, in a reversible manner [51,55]. Monitoring worms under confocal microscopy immediately after incubation at 35°C for 3 h, resulted in increased number and size of DCAP-1::GFP-containing cytoplasmic puncta in all transgenic lines (Fig. 1A, B). In contrast, there was no change in the expression pattern of worms carrying the transcriptional fusion transgene (BRF154, Fig. 1A), suggesting that the DCAP-1::GFP granules result from the accumulation of diffuse cytoplasmic protein rather than transcriptional or translational induction. This was verified by qRT-PCR for *dcap-1(mRNA)* levels and western blotting for DCAP-1::GFP protein levels, under normal or heat-shock conditions (Fig. 1C, D). The formation of DCAP-1::GFP aggregates in response to heat-shock was also reversible, with the majority of puncta dissolving within 3 h of recovery at 20°C (Fig. 2A, B). Similar enhanced formation of DCAP-1::GFP granules was observed in response to oxidative stress induced by sodium arsenite (BRF155, Fig. S2A). We further showed that RNAi-mediated knockdown of a core PB component, *cgh-1* (encoding a conserved DEAD-box RNA helicase) was sufficient to disrupt the formation of DCAP-1::GFP granules during heat (Fig. 2C, D) or oxidative stress (data not shown), in agreement with studies in yeast or human cells [8,56].

Another putative component of mammalian PBs is the mRNA-binding protein GW182 [57], which has two homologues in *C. elegans*, AIN-1 and AIN-2, both involved in miRNA-induced silencing function [58]. However, only AIN-1 interacts with components of the decapping complex and co-localizes with DCAP-1 in PB-like structures in *C. elegans*, under normal conditions [43]. Transgenic worms expressing the *ain-1::gfp* reporter have sporadic cytoplasmic puncta under normal conditions, which are further increased after heat-shock (Fig. S2B) without changes in *ain-1(mRNA)* levels (Fig. S2D), similar to *dcap-1::gfp*-expressing animals. We also tested whether the somatic PBs contain the translation initiation factor eIF4E, which localizes to PBs in unstressed mammalian cells but is found to both PBs and SGs upon exposure to stress [12,56]. In *C. elegans*, *ife-2* encodes the major somatic isoform of eIF4E [59]. Transgenic animals expressing the transcriptional fusion $P_{ife\text{-}2}$*::gfp* (BRF68, Fig. S2B) displayed a uniform fluorescence in all cells and did not show any aggregation under heat stress. In contrast, animals carrying the translational fusion *ife-2::gfp*, that drives expression of *ife-2* by its own promoter, had a similar diffused fluorescence pattern under normal growth conditions but rapidly formed numerous cytoplasmic granules in response to heat-shock (Fig. S2B) or oxidative stress (Fig. S2A). In co-localization experiments we monitored that several of heat-induced IFE-2::GFP granules also contain the PB marker DCAP-1::tagRFP (Fig. S2C). However, this partial co-localization could result from docking between SGs and PBs under stress, as it was shown to occur in yeast and mammalian cells [12,15,60]. The homogenous cytoplasmic fluorescence pattern of the strong *ife-2::gfp* reporter in unstressed animals argues that IFE-2 does not localize to constitutive PBs. Additionally, we found that *cgh-1(RNAi)*, which prevents stress-induced aggregation of PBs, had no significant effect on the formation and the relative number of IFE-2::GFP granules under heat-shock (Fig. S3A). This indicates that IFE-2 localizes mainly in SGs under stress conditions (see below Fig. S7A). We also found that deletion of *ife-2* did not affect the heat-induced accumulation of PBs (Fig. S3B). Thus, stress-induced PBs in the soma of intact adult worms resemble yeast and mammalian PBs in their pattern, formation and reversibility, albeit differ in localization of IFE-2/eIF4E factor.

Accumulation of somatic PBs with age

We next explored the pattern of PBs during the aging process, which is associated with marked alterations in protein synthesis and homeostasis [61,62]. Using the *dcap-1::gfp* reporter we demonstrated that the number and size of PBs gradually increased with age as, there was an increased punctate pattern in 5-day adults compared to 1-day adults, at 25°C (BRF155 and BRF261, Fig. 3A, B). This was not due to altered mRNA or protein levels with age, as verified by qRT-PCR for the endogenous *dcap-1(mRNA)* levels and western blotting for DCAP-1::GFP protein (Fig. 3C, D). The aggregation of PBs with age was also evident in worms grown at 20°C (Fig. S4A, B), albeit in later time-points compared to 25°C consistent with the slower aging process at lower temperatures. Similar age-dependent accumulation of PBs, was reported in worms expressing the DCAP-1::RFP fusion protein [63]. Cytoplasmic granules containing the AIN-1::GFP protein (MH2704, Fig. S2B) also accumulated in aged animals, without changes in *ain-1(mRNA)* levels (Fig. S2D), suggesting that aging can alter the profile of somatic PBs. Interestingly, we did not observe any age-dependent accumulation of the IFE-2::GFP fusion protein (BRF70, Fig. S2B), indicating that IFE-2/eIF4E factor is not part of the PB foci that accumulate with age. Taken together our observations led us to consider that the age-dependent formation of PBs in adult worms could not result from oxidative stress that takes place in aged cells and tissues [64]; this should have induced the assembly of SGs as well, something that we did not monitor by IFE-2 or other SG markers (see below).

We next considered the possibility that reduction of mRNA translation or degradation rates with age might cause or contribute to the accumulation of PBs in aged worms. Studies in yeast and mammalian cells have described the induction of PBs when either translation initiation or mRNA decay is inhibited, due to accumulation of non-translatable mRNAs into PBs [12,14]. Thus, we performed early in life RNAi of *ife-2* or other translation factors such as eIF2α (encoded by Y37E3.19), *ppp-1*/eIF2Bγ and *rsks-1*/S6K in N2 worms carrying the *dcap-1::gfp* transgene but we didn't observe any PB aggregation, in 1-day adults (Fig. 4A). We validated the efficiency of RNAi clones by qRT-PCR (Fig. 4C)

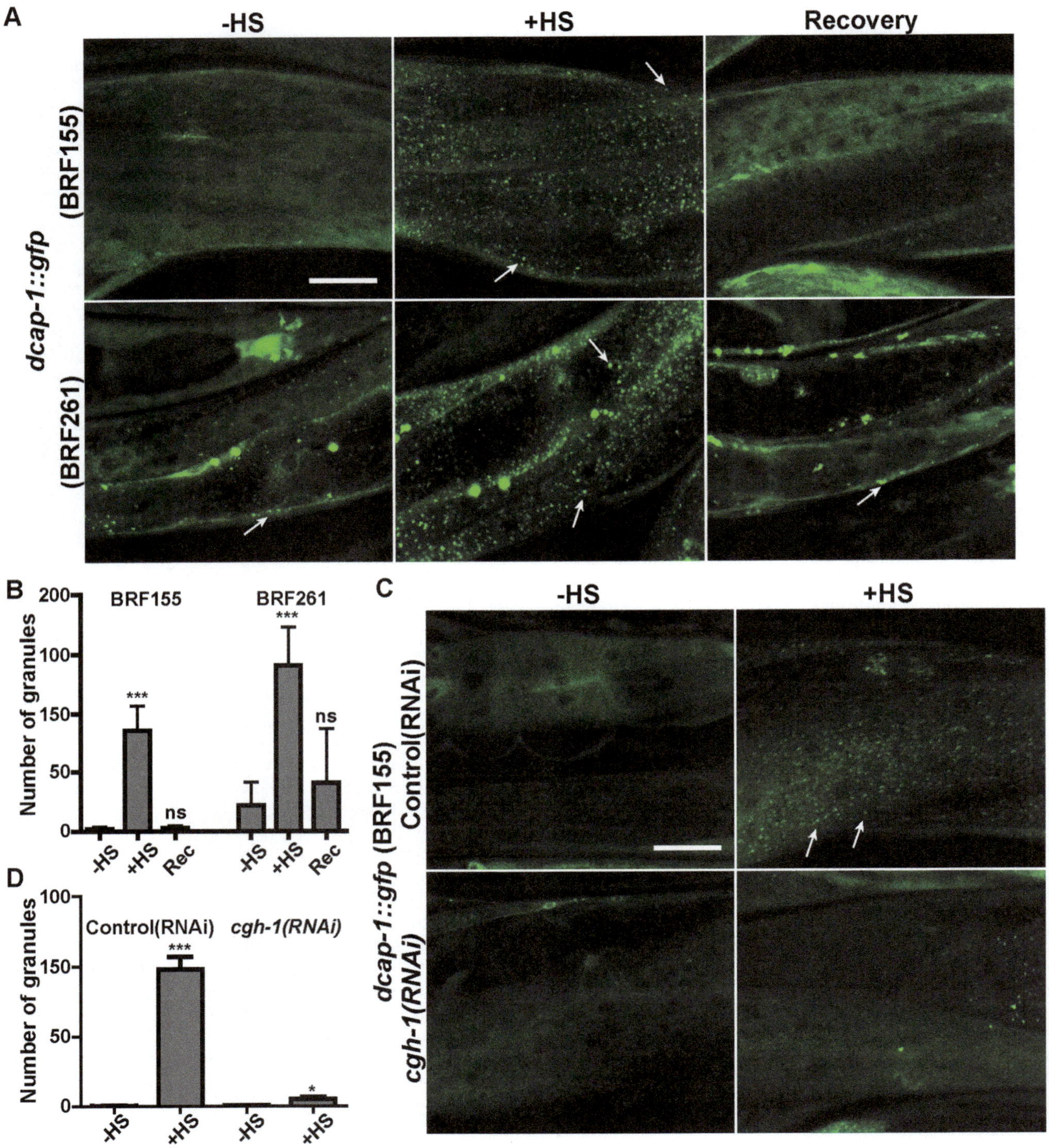

Figure 2. Accumulation of DCAP-1-containing granules under heat-shock is rapid, reversible and sensitive to *cgh-1(RNAi)*. (A) Representative confocal images of 1-day adult worms expressing *dcap-1::gfp* (BRF155 and BRF261 strains) normally grown at 25°C (-HS) or transiently subjected to heat-shock (+HS, 35°C for 3 h),following recovery for 3 h at 20°C (Recovery). (B) Quantification of data in (A). (C) Representative confocal images of 1-day adult worms expressing *dcap-1::gfp* (BRF155) fed from eggs with Control(RNAi) or *cgh-1(RNAi)* bacteria and grown at 25°C (-HS) or subjected to heat-shock at 35°C for 3 h (+HS). In all cases arrows point to DCAP-1::GFP granules in body wall muscles. (D). Quantification of data in (C). Values on Y axis show the number of granules per 2500 μm^2 per worm (mean±SD, see Materials and Methods and Table S4). Scale bar: 25 μm.

or by testing their effects in the treated adults and their progeny [45,65]. Consistent with *ife-2(RNAi)*, a strain carrying a null *ife-2(ok306)* mutation and exhibiting lower protein synthesis rates than N2 [45,65,66] did not form PBs in adult somatic tissues (*ife-2(ok306); dcap-1::gfp*, Fig. 4D). On the other hand, RNAi of *xrn-1*, the major cytoplasmic 5′ to 3′ exonuclease, resulted in the accumulation of DCAP-1::GFP granules under the same conditions (Fig. 4A, B). Thus, the dynamics of PBs appear to be more prone to regulation by alterations in mRNA degradation rather than in translation rates. We suggest that an age-dependent

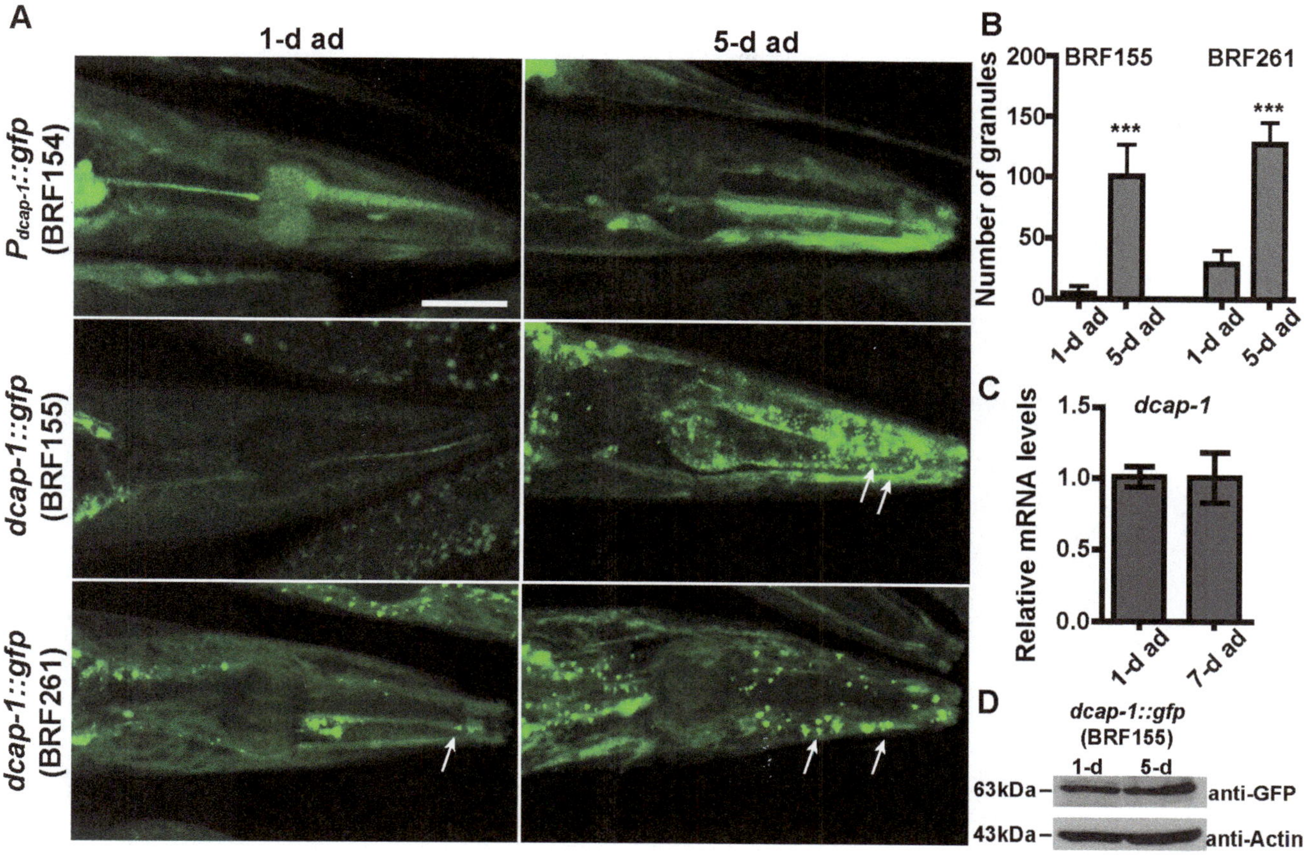

Figure 3. Accumulation of PBs with age. (A) Representative confocal images of 1-day and 5-day adults, grown at 25°C and expressing the transcriptional fusion P_{dcap-1}::*gfp* reporter (BRF154) or the translational fusions *dcap-1::gfp* (BRF155 and BRF261). Arrows point to age-induced DCAP-1::GFP granules. Scale bar: 25 μm. (B) Quantification of data in (A). Values on Y axis show the number of granules per head (mean±SD, see Materials and Methods and Table S4). (C) Endogenous *dcap-1*(mRNA) levels in 1-day and 7-day N2 adults grown at 25°C before (-HS) or after heat shock (+HS, 35°C for 3 h) were measured by quantitative RT-PCR and normalized to endogenous *ama-1(mRNA)* levels. Error bars represent the standard deviation of the means of six independent experiments (p-value = 0.9697, calculated by Student's *t*-test). (D) Western blot analysis of DCAP-1::GFP protein expression in 1-day and 5-day adult BRF155 transgenic animals, grown at 25°C, using anti-GFP or anti-β-Actin (as a loading control) antibodies. Band intensity of DCAP-1::GFP normalized to actin gives a value of 0.95, showing similar protein levels in the two conditions.

decline in mRNA decay process could be a causal factor in the onset of PBs aggregation in aged tissues but further experimental validation is needed. Interestingly, *xrn-1(RNAi)* did not cause aggregation of IFE-2::GFP, which remained cytoplasmic in worms that co-express DCAP-1::tagRFP and form distinct granules (Fig. 4E), showing that IFE-2 does not relocalize to PBs under impaired mRNA degradation conditions.

Disruption of PB-related genes affects development, lifespan and stress response of worms

We further set out to evaluate the importance of PB-related genes in the physiology of worms and whether disruption of their function could protect from or contribute to mortality. Worms bearing mutations in the decapping genes *dcap-1(tm3163)* and *dcap-2(ok2023)* showed a profound delay in development (Fig. S5A) and reduced brood size (Fig. S5B) compared to N2, at both growth temperatures (20 and 25°C). They also exhibited high levels (~50%) of matricidal death ("bagging" due to internal hatching of eggs) and uncoordinated locomotion. Both mutants had a significantly shorter lifespan than N2 (Fig. 5A and Table S5) and were more sensitive, as adults, in a range of environmental stressors (Fig. 5B). When we expressed the *dcap-1::gfp* reporter in *dcap-1(tm3163)* we rescued the growth and lifespan defects and restored resistance to stresses (Table S5 and Fig. S5C). We noticed that the mutant phenotypes of *dcap-1* were more severe at 25°C compared to 20°C due to a possible residual function of the mutated DCAP-1 protein; *dcap-1(tm3163)* harbors a 334-bp deletion that results in a truncated mRNA, generating a new amino-acid sequence at position 263 (E263) of the C′-terminus of wild-type protein (Fig. S5D). Since the C′ end of DCAP-1 contains a conserved trimerization domain, which in metazoan is required for incorporation of Dcp1 into PBs and mRNA decapping *in vivo* [67], it is possible that the allele *tm3163* doesn't completely remove gene function at 20°C, but at higher temperature (25°C) its phenotypes are greatly enhanced. Therefore, the lifespan of *dcap-1* worms is ~25% shorter than N2 at 20°C and ~45% shorter than N2 at 25°C, resembling the lifespan of the null *dcap-2* mutants (Table S5). Also, each gene disruption by RNAi from the time of egg hatching resulted in similarly decreased lifespan of N2 (Table S6).

We considered that the mortality of decapping mutants could emerge from their early developmental defects. To assess the effects of PB components on lifespan beyond development we performed RNAi post-developmentally, by initiating RNAi during late larval stages of N2. Although the effect of single decapping gene depletion was small we observed significantly decreased lifespan when *dcap-1* and *dcap-2* were simultaneously inactivated

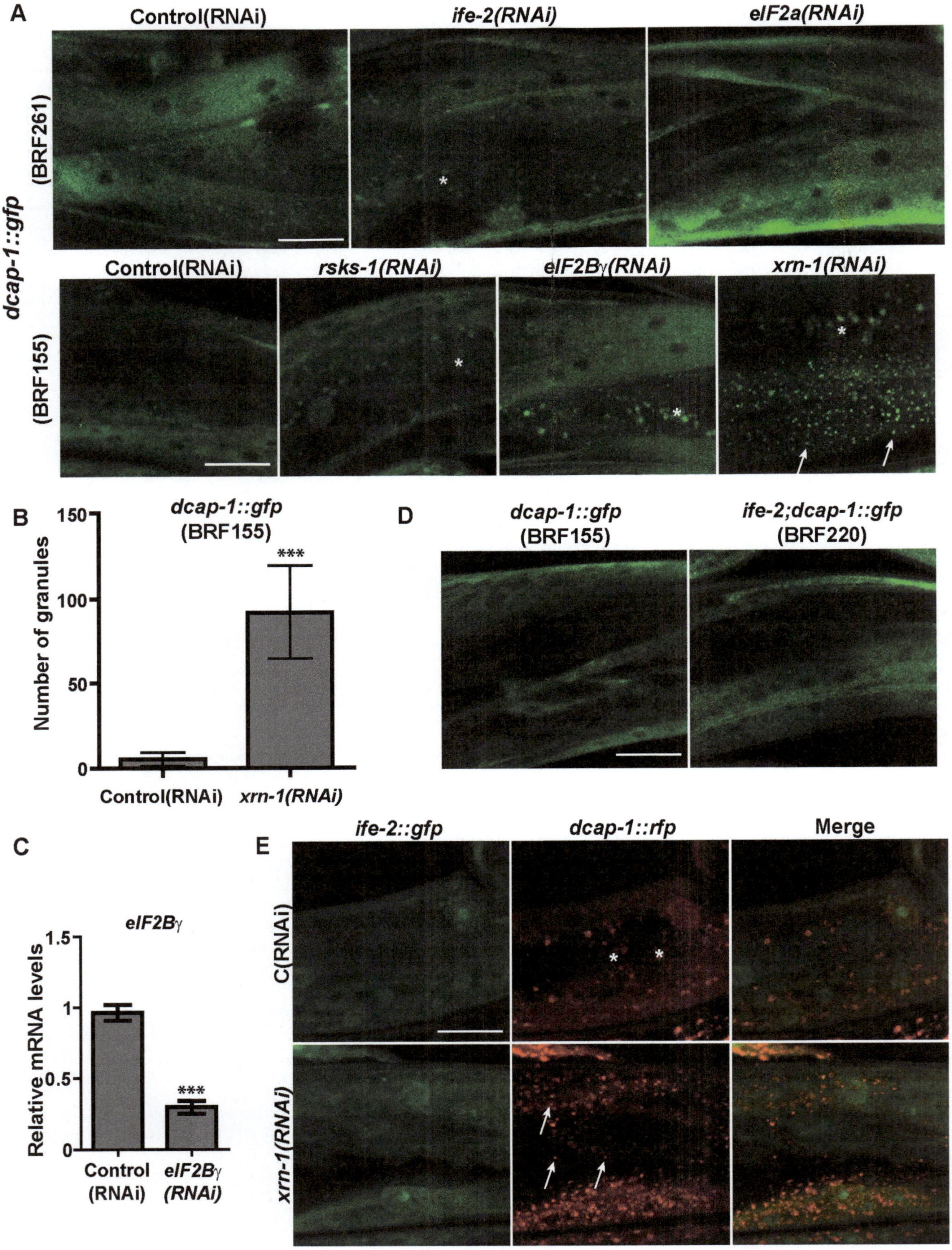

Figure 4. Disruption of mRNA degradation, but not of translation factors, triggers PBs formation that lack *ife-2*/eIF4E. (A) Representative confocal images of 1-day adults expressing *dcap-1::gfp* in N2 (BRF155 and BRF261) fed with *ife-2(RNAi)*, *eIF2a(RNAi)*, *rsks-1(RNAi)*,

eIF2Bγ(RNAi) or *xrn-1(RNAi)* bacteria. (B) Quantification of data in (A). Values on Y axis show the number of granules per 2500 μm^2 per worm (mean±SD, see Materials and Methods and Table S4). (C) Endogenous *eIF2Bγ*(mRNA) levels in 1-day adults BRF155 worms grown at 25°C and fed with *eIF2Bγ(RNAi)* from eggs, were measured by quantitative RT-PCR and normalized to *ama-1(mRNA)* levels. Error bars represent the standard deviation of the means of three independent experiments (***p-value<0.0001, in unpaired *t*-test). (D) Representative confocal images of 1-day adults expressing *dcap-1::gfp* in N2 (BRF155) or *ife-2(ok306)* (BRF220) background. (E) Representative confocal images of 1-day adults co-expressing *dcap-1::rfp* and *ife-2::gfp* (BRF313), and treated with *xrn-1(RNAi)* to monitor the lack of co-localization of the two proteins. In all cases, targeted RNAi was initiated from eggs at 25°C in parallel with Control(RNAi) to assess the localization of DCAP-1::GFP into PBs. Asterisks show the intestinal autofluorescence. Scale bar: 25 μm.

(dcap-1/2(RNAi) compared to Control(RNAi)-treated worms (Fig. 5C and Table S6). This is consistent with the model of a more efficient disruption of decapping activity when both decapping subunit genes are disrupted [36]. Similarly, post-developmental depletion of *xrn-1*, *patr-1* or *cgh-1* significantly shortened lifespan of N2 (Fig. 5C and Table S6). Given that *xrn-1(RNAi)* induced the accumulation of PBs we postulate that this accumulation is not sufficient to confer longevity, probably due to impaired degradation of resident mRNAs. Furthermore, disruption of *dcap-1* or *dcap-2* had an impact in the lifespan of several long-lived mutants. Post-developmental RNAi-treatment of worms lacking the insulin/IGF-1-like receptor *daf-2* gene [68,69] resulted in shortening of their extreme long life (Table S6). Similarly, RNAi of decapping genes during adulthood in *ife-2(ok306)* or *rsks-1(ok1255)* translation-defective worms [45], in dietary-restricted *eat-2(ad465)* animals [70] as well as in somatic tissues of the germline-deficient mutant *glp-1(e2141)* [71], reduced their longevity (Table S6). Likewise, introducing the *dcap-1(tm3163)* or *dcap-2(ok2023)* alleles in the genome of these long-lived mutants resulted in significantly reduced lifespan and stress resistance compared to controls (Fig. 5D, E for *daf-2* and Table S5), suggesting that the decapping mutant phenotypes were not rescued by any longevity pathway. On the other hand, overexpression of *dcap-1* in somatic tissues of *glp-1* mutants had a positive effect on their ability to cope with stress during adulthood, as it significantly enhanced survival under heat and oxidative stress (Fig. 5F). Taken together, our data indicate that genes directly related to PBs modulate adult lifespan, separately from their developmental role and alterations in functions of PB components increase the mortality rate and the sensitivity of adult worms to various stressors.

Accumulation of SGs in somatic cells upon stress but not with age

PBs are functionally related to SGs that have a crucial role in cellular response to environmental stress. SG components in human cells include TIA-1 (T-cell intracellular antigen-1) and TIAR (TIA-1-related) RNA-binding proteins that regulate mRNA stability, in addition to other cellular functions [72]. In *C. elegans* genome (http://www.wormbase.org), three genes, named *tiar-1* (C18A3.5), *tiar-2* (Y46G5A.13) and *tiar-3* (C07A4.1) encode homologues of human TIA-1/TIAR family proteins. The predicted proteins TIAR-1 and TIAR-2 contain three RNA-recognition motifs (RRMs) at their N′-termini and prion-related domains at their C′-termini, whereas TIAR-3 has only two RRMs and no prion-related domain (Fig. 6A). The latter domain in human TIA-1/TIAR proteins is essential for the assembly of SGs under stress [73]. We generated animals expressing transcriptional or translational *gfp*-reporters of *tiar* genes and through observation of several transgenic lines we concluded that: (a) the promoter of both *tiar-1* and *tiar-2* is expressed in most, if not all, cells (P_{tiar-1}*::gfp* and P_{tiar-2}*::gfp* in Fig. S6); (b) the *gfp::tiar-1* and *gfp::tiar-2* N′-terminal fusion genes, driven by their own promoters, display a diffused fluorescent signal dispersed in both nucleus and cytoplasm (Fig. S6); (c) *tiar-3::gfp*, a C′-terminal fusion driven by its own promoter, is exclusively nuclear, strongly expressed mainly in spermatheca, ventral nerve cord and some head and tail neurons (Fig. S6). We could not detect expression of GFP fusions in germline due to silencing of transgenes in this tissue [74] but previous studies have shown the expression of TIAR-1 [33,75] and TIAR-2 [32] in germ granules.

We next examined the pattern of the GFP fusion transgenes in adult worms subjected to heat or oxidative stress, conditions that elicit SG assembly in mammalian and yeast cells. In contrast to *tiar-3::gfp* protein, which constantly remained nuclear after each stress (data not shown), we observed the formation of numerous cytoplasmic aggregates containing *gfp::tiar-1* in response to heat-shock (HS) and sodium arsenite (SA) (Fig. 6B). In the case of *tiar-2*, the aggregates induced by heat-shock were more visible in the excretory cell of worms, but were also vigorously formed in muscles when we used a strong, muscle-specific promoter (P_{myo-3}*::gfp::tiar-2*) to enhance the fluorescent signal and avoid the intestinal autofluorescence at the confocal settings of *gfp::tiar-2* (Fig. 6B). Under oxidative stress we could monitor granules only in muscle-expressed *gfp::tiar-2* (Fig. 6B). The relative mRNA levels of each transgenic line are shown in figure 6C. The heat-shock induced granules, monitored by *gfp::tiar-1* or *gfp::tiar-2*, were not the result of increased transcription of *tiar-1* or *tiar-2*, respectively (Fig. 6D) and were not formed in worms expressing the transcriptional *gfp* fusions (data not shown). In accordance with the dynamic nature of SGs, the above heat-induced granules were formed rapidly (within 45 min) and could be dissociated within 2 h of recovery at 20°C (Fig. 7). Deletion of *tiar-1* did not affect the aggregation of *tiar-2*-containing granules and *vice versa* (data not shown). Moreover, under heat-shock conditions several RFP::TIAR-1 granules contain the IFE-2::GFP and partially colocalise with DCAP-1::GFP granules, while GFP::TIAR-2 also co-localizes partially with DCAP-1::RFP (Fig. S7A–C), again consistent with the spatial overlapping of PBs and SGs [12,15,60]. As opposed to PBs, RNAi-mediated depletion of *cgh-1* did not affect the formation of SGs upon heat-shock and *xrn-1(RNAi)* did not lead to accumulation of SGs under normal conditions of growth (data not shown), similar to our observations with IFE-2::GFP marker. Finally, our data using IFE-2 reporter suggest that aging cannot induce the formation of SGs, in sharp contrast to PBs. Indeed, we did not observe an age-dependent accumulation of TIAR proteins in SGs, comparing 6-day to 1-day adult worms at 25°C (Fig. 6B), which show similar expression levels of *tiar-1* or *tiar-2* (data not shown). These data strongly suggest that neither the mild stress accompanying aging nor the age-related alterations in mRNA degradation is sufficient to induce SG formation.

In the process of monitoring the fluorescent *tiar* granules, we noticed a tissue-specific response in their formation under heat-shock, with *gfp::tiar-1* aggregating mostly in head neurons, muscles, intestine, vulval and hypodermal cells, whereas *gfp::tiar-2* granules were visible mainly in excretory cell and less in intestine, muscle and hypodermal cells. Furthermore, by generating transgenic animals co-expressing *rfp::tiar-1* and *gfp::tiar-2* we observed the same tissue-specific aggregates in

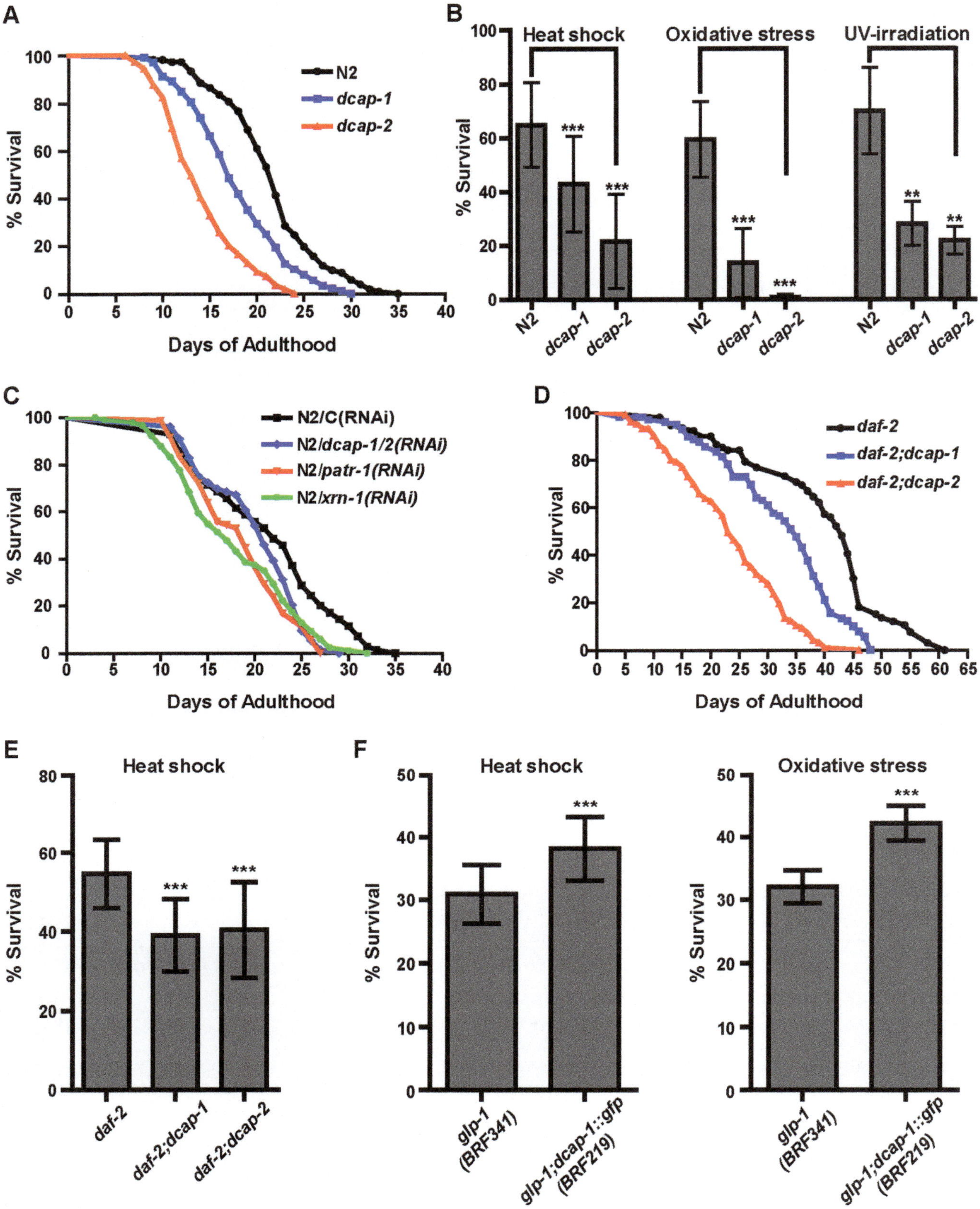

Figure 5. PB components are important for normal lifespan and stress response. (A) Survival curves of N2, *dcap-1* and *dcap-2* mutants at 20°C. (B) Survival of adults *dcap-1* and *dcap-2* mutants in heat-shock (1-day adults at 35°C for 6 h), oxidative stress (1-day adults in 5 mM sodium arsenite for 48 h) and UV-irradiation (5-day adults at 0.2 J/cm^2), compared to N2. (C) Survival curves of N2 worms treated post-developmentally with *dcap-1/-2(RNAi)*, *patr-1(RNAi)* or *xrn-1(RNAi)* at 20°C. (D) Lifespan of *daf-2*, *daf-2; dcap-1* and *daf-2; dcap*-2 mutants at 20°C and (E) survival of 1-day adults after heat-shock (35°C for 8 h). (F) Survival of 1-day adult *glp-1* germline-deficient mutants that overexpress *dcap-1::gfp* after heat-shock (at 35°C for 6 h) or oxidative stress (5 mM sodium arsenite for 48 h). In lifespan assays the p-values were determined using the log-rank test (see Tables

S5 and S6) and in stress assays the error bars show the SD in unpaired *t*-tests (see Materials and Methods). ** indicates very significant (p-value 0.001 to 0.01); *** indicates extremely significant (p<0.001).

response to heat-shock, with partial co-localization of the two fusion proteins only in muscles, hypodermal and intestinal cells (Fig. S7D). This implies some tissue-specific functions of each TIAR protein in cellular stress response. In support of this, only TIAR-2-containing granules were formed under osmotic stress (Fig. S8A). In addition, we observed a different requirement on the activity of GCN-2 kinase to initiate the formation of each type of SGs in response to oxidative stress. Having shown the conserved function of GCN-2 as an eIF2α kinase in worms [76] we tested whether SGs were formed in the *gcn-2(ok871)* loss-of-function mutant. While both TIAR-1 and TIAR-2 foci could assemble upon heat-shock in *gcn-2(ok871)*, only TIAR-2 aggregates were formed after treatment with sodium arsenite (Fig. S8B). Thus, TIAR-1 accumulation in response to oxidative stress requires phosphorylation of eIF2α by GCN-2 (and not by the other eIF2α kinase in worms, PEK-1, data not shown). Overall, we showed that the three TIAR proteins in *C. elegans* differ in cellular localization and only TIAR-1 and TIAR-2 participate in cytoplasmic SGs that have both overlapping and distinct patterns, with different assembly requirements depending on the stress stimulus.

Diverse function of SG-related genes in development, lifespan and stress response

Because of the differences in cellular responses of TIAR proteins we tested the impact of each *tiar* gene deletion on development, lifespan and stress survival. We used the mutant strains *tiar-1(tm361), tiar-2(tm2923)* and *tiar-3(ok144)* that harbor internal deletions in the coding region of *tiar* genes and accessed their phenotypes. Mutant *tiar-1* worms exhibited delayed developmental rate and reduced brood size compared to N2, while *tiar-2* and *tiar-3* mutants had milder (at 25°C) or no defect (at 20°C) on both development and fecundity (Fig. 8A). Also, *tiar-1* mutants displayed early behavioral decline, with premature onset of uncoordinated locomotion at day 6 of adulthood (Movies S1 and S2), which was not observed in *tiar-2* and *tiar-3* mutants. Animals lacking both *tiar-1* and *tiar-2* genes (*tiar-1; tiar-2*, Fig. 8A) displayed even slower development than the single mutants, with high matricide deaths at 25°C, suggesting additive effects.

We also assessed differences in the lifespan of *tiar* mutants; deletion of each *tiar* gene shortened normal lifespan, but the effect was more dramatic for *tiar-1* allele (Table S5). Since both the developmental (data not shown) and lifespan (Table S5) defects of *tiar-1* were rescued by the functional GFP::TIAR-1 fusion, these defects are linked to *tm361* allele and not to a secondary mutation. For *tiar-2* allele, there was a temperature-dependent effect on lifespan, as in development, with significantly shorter lifespan at 25°C (Table S5). The double mutant *tiar-1; tiar-2* had an even shorter lifespan than each single mutant at any temperature (Fig. 8B and Table S5). We obtained similar results on N2 lifespan by RNAi of *tiar* genes post-developmentally (Fig. 8B and Table S6). This suggests that the lifespan reduction was not solely due to the developmental abnormalities of each mutant. In addition, by introducing each mutation to the germline-deficient *glp-1(e2141)* worms we observed that loss of *tiar-1* and, to a greater extent, *tiar-2* from somatic cells significantly reduced the long lifespan of *glp-1* worms (Table S5). Surprisingly, deletion of *tiar-3* significantly improved the mean lifespan of *glp-1* worms, contrary to N2 (Table S5). Thus, the impact of *tiar* deletions on lifespan can be influenced by temperature or germline signaling.

Finally, we measured diverse responses of *tiar* mutants against various insults, which are, in some cases, related to their expression pattern. *tiar-1* deletion reduced the survival of worms in sodium arsenite-induced oxidative stress, or after UV-irradiation but increased survival under heat or osmotic stress (Fig. 8C). The impaired oxidative stress response of *tiar-1(tm361)* was rescued by the *gfp::tiar-1* transgene (Fig. 8D). In the imaging analysis we observed TIAR-1-positive granules upon heat and oxidative stress but not under osmotic stress (Fig. 6B and Fig. S8A). We assume that deletion of TIAR-1 affects only specific functions of SGs related to oxidative stress response and not to heat-shock, although it localizes to SGs formed under both conditions. Alternatively, since TIAR-1 is required to induce germ cell apoptosis under various stresses [75], the increased resistance of *tiar-1* mutants to heat-shock might be related to the systemic stress response that is observed in several DNA-damage checkpoint mutants affecting programmed cell death in germline [77,78]. In sharp contrast to *tiar-1* worms, *tiar-2* or *tiar-3* mutants had increased mortality under heat and osmotic stress but survived as N2 under oxidative or UV-irradiation stress (Fig. 8C). Interestingly, the expression of *tiar-2* driven by its own promoter was strong in the excretory system of worms which has a role in osmoregulation, analogous to the 'kidney' of higher animals [79]. TIAR-2 granules formed in oxidative stress, (BRF255 in Fig. 6B), were hardly visible while TIAR-1 granules under the same conditions, were readily formed in all tissues (Fig. 6B), suggesting a more crucial role of TIAR-1 under this stress. The double mutant *tiar-1; tiar-2* survived similar to the single *tiar-1*, at least under heat and oxidative stress (Fig. 8C). While further studies are required to address these issues, our data show that the three *tiar* genes exhibit diverse developmental, lifespan and stress response outputs at the organismal level that are possibly reflected by their specific functions, which are differentially regulated under various cellular contexts and stressors.

Discussion

Post-transcriptional regulation of eukaryotic gene expression is attained through several mRNA-specific control mechanisms, such as processing, export, turnover and translation. PBs and SGs are cytoplasmic aggregates of RNP complexes that mediate the subcellular localization, translation or decay of bulk mRNA in eukaryotic cells [6,8,11]. PBs consist of mRNAs associated with translation repressors or the 5' to 3' mRNA decay machinery, whereas SGs contain a pool of mRNAs stalled in the process of translation initiation in response to stress. Both are dynamic structures that transiently interact and share many components [12]. Hence, PBs and SGs are considered as hubs of mRNP trafficking that determine the fate of mRNAs but also as regulators of stress-responsive signaling pathways that determine the survival of cells under stress [80]. Although the function of these granules provides another level for fine-tuning gene expression to maintain cellular and tissue homeostasis, studies addressing their impact or pattern during stress or aging, at the level of intact organism, are limited. In *C. elegans*, related RNA-protein particles function in maternal mRNA metabolism and have been well-studied in germline development and embryogenesis [30–32,34,36].

Here, we documented the formation of PBs and SGs in somatic cells of *C. elegans* by using fluorescent reporters of known components of these granules. These include the decapping

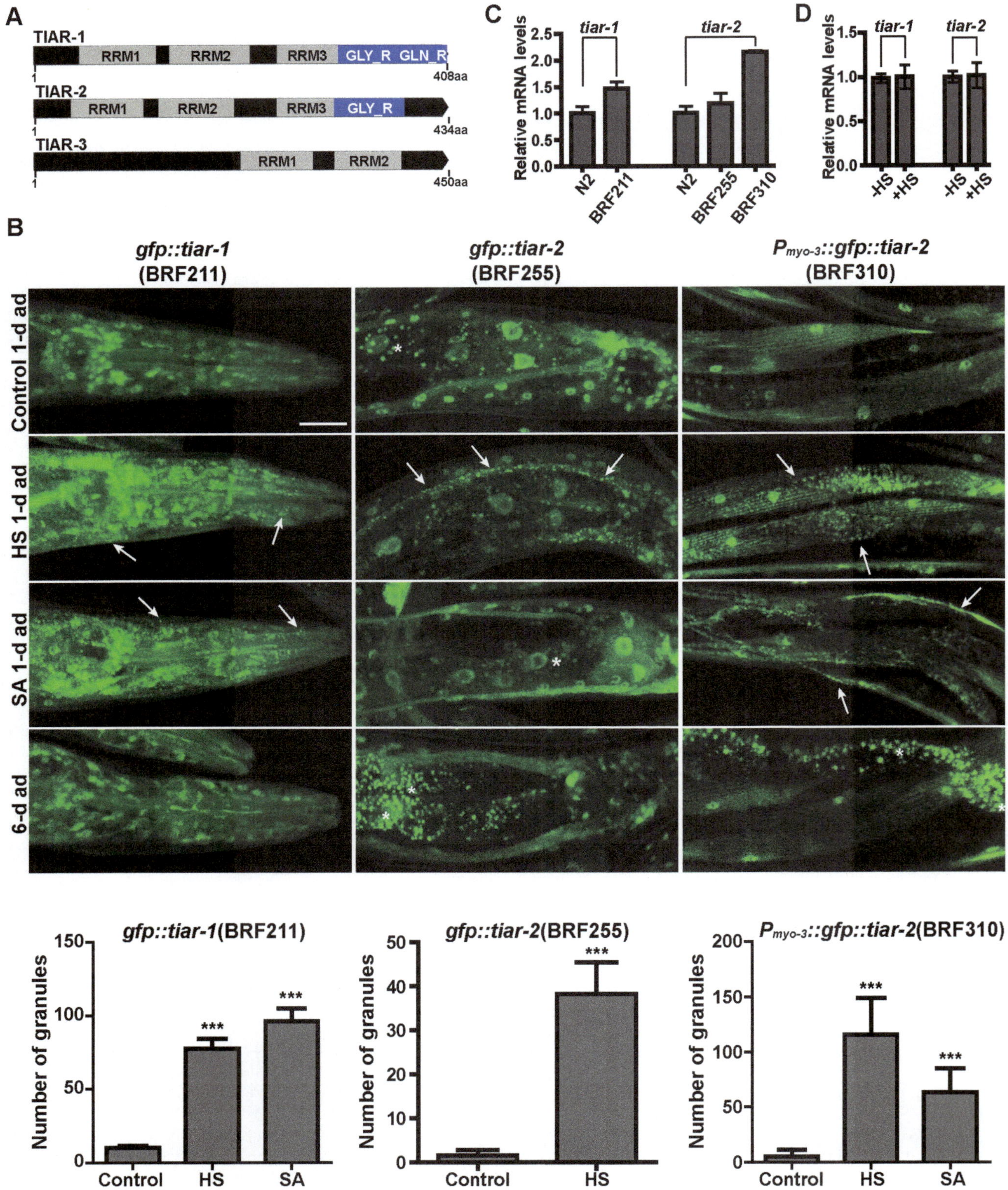

Figure 6. Aggregation of TIAR-1 and TIAR-2 proteins in response to stress but not to age. (A) The domain structure of TIAR proteins, designed using the Prosite MyDomains (http://prosite.expasy.org/mydomains/). The graphic was created using the Exon-Intron Graphic Maker (http://wormweb.org/exonintron). RRM: RNA-recognition motif, GLY_R: Glycine rich domain, GLN_R: Glutamine rich domain. (B) Representative confocal images of 1-day adults of the indicated transgenic strains grown at 25°C (Control 1-d ad), upon heat-shock (HS, 35°C for 3 h), oxidative stress (SA, 15 mM sodium arsenite for 3 h) or compared to 6-day adults under normal conditions. Arrows point to formed granules and quantification of the granule number is shown in the lower panel for each strain. Values on Y axis show the number of granules per head (BRF211), per 100 μM length of excretory cell (BRF255) or per 2500 μm^2 per worm (mean±SD, see Materials and Methods and Table S4). Asterisks show the intestinal auto-fluorescence. Scale bar: 25 μm. (C–D) Expression levels of *tiar-1* or *tiar-2* genes in (C) the indicated transgenic strains expressing *gfp::tiar-1* and *gfp::tiar-2* by their own promoter (BRF211 and BRF255, respectively) or the muscle specific P_{myo-3}*::gfp::tiar-2* (BRF310) and (D) N2 worms after heat

shock (35°C for 3 h). Quantification of each mRNA level, relative to *ama-1(mRNA)* levels, in 1-day adults and the mean ± SD of biological triplicates are shown (p>0.05 in Student's *t*-tests).

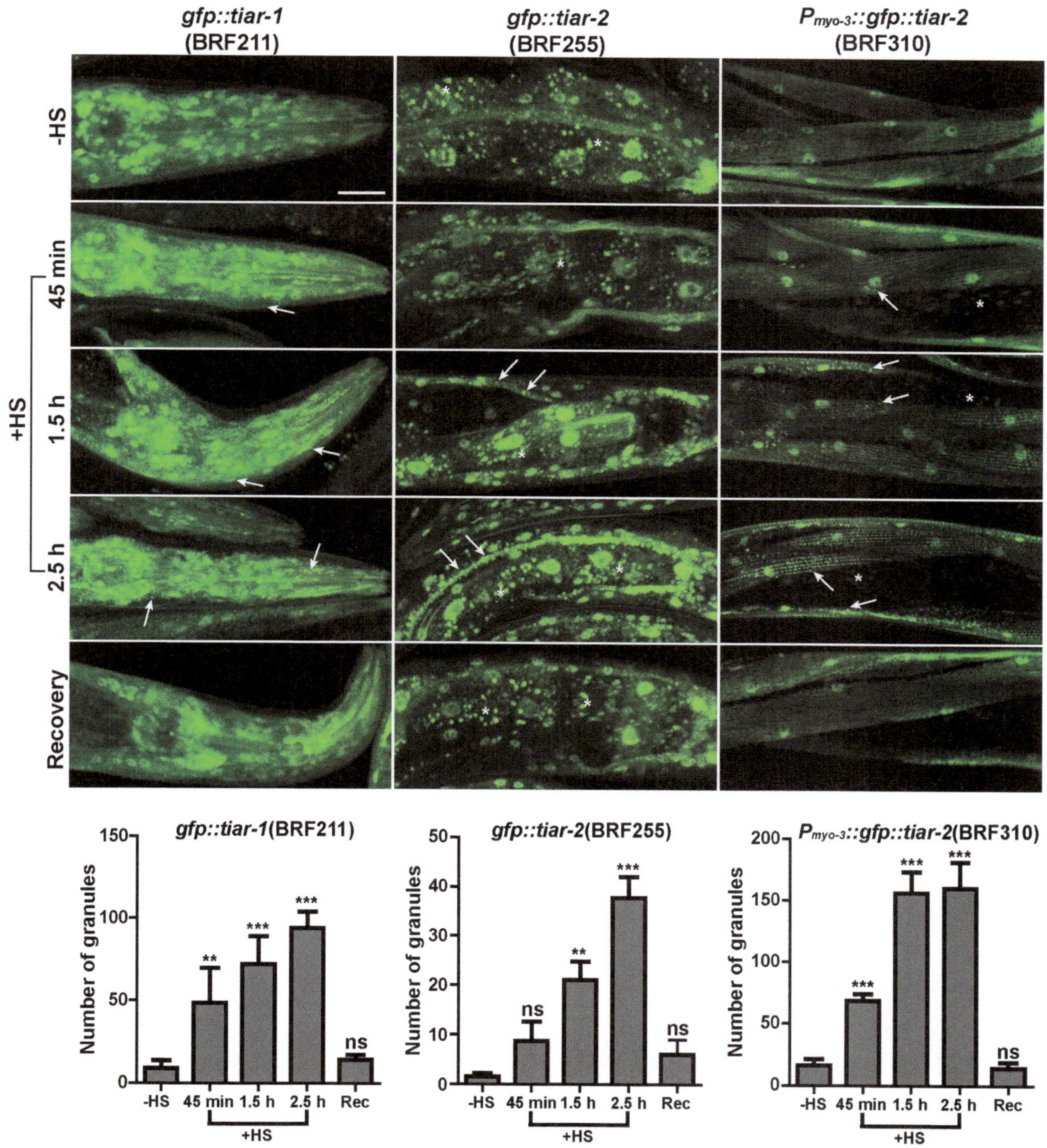

Figure 7. Rapid formation and dissociation of heat-induced SGs. Representative confocal images of 1-day adults expressing *gfp::tiar-1* and *gfp::tiar-2* by their own promoter (BRF211 and BRF255, respectively) or the muscle specific P_{myo-3}*::gfp::tiar-2* (BRF310) under normal growth conditions (-HS) and upon heat-shock (+HS, 35°C for 45 min to 2.5 h), following recovery at 20°C for 2 h. Arrows point to formed granules and quantification of the granule number is shown in the lower panel for each strain. Values on Y axis show the number of granules per head (BRF211), per 100 μM length of excretory cell (BRF255) or per 2500 μm^2 per worm (mean±SD, see Materials and Methods and Table S4). Asterisks show the intestinal auto-fluorescence. Scale bar: 25 μm.

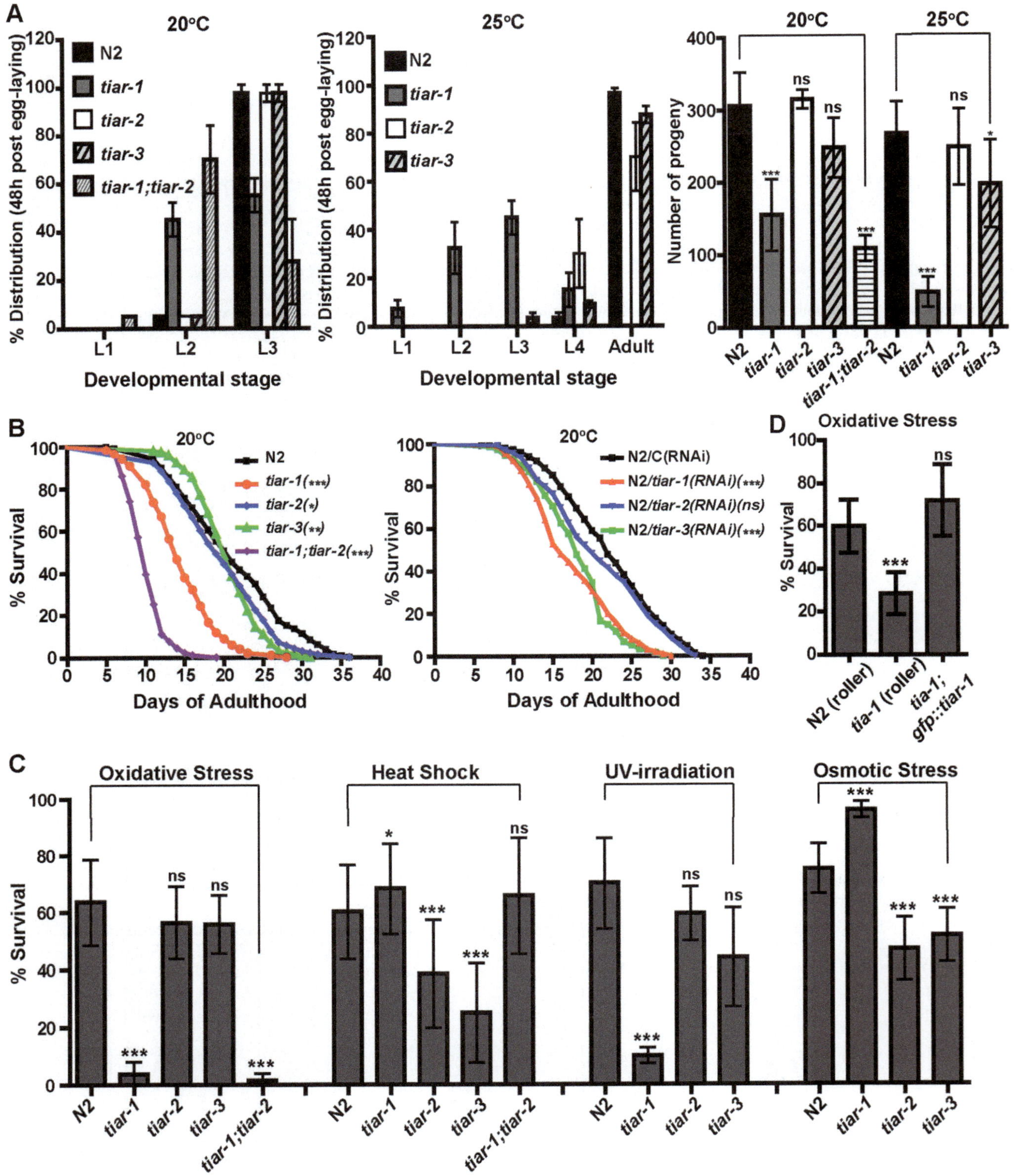

Figure 8. Effects of *tiar* genes deletion on development, fertility, lifespan and stress survival. (A) Developmental rate and brood size of *tiar-1*, *tiar-2*, *tiar-3* and *tiar-1;tiar-2* mutant worms, compared to N2. (B) Survival curves of the above mutants or of N2 worms treated post-developmentally with *tiar-1(RNAi)*, *tiar-2(RNAi)* or *tiar-3(RNAi)*. (C) Survival of adults of the above mutants in oxidative stress (1-day adults at 5 mM sodium arsenite for 48 h), heat-shock (1-day adults at 35°C for 6 h), UV-irradiation (5-day adults at 0.2 J/cm^2) or osmotic stress (1-day adults at 400 mM NaCl for 24 h, compared to N2. (D) Survival of *tiar*-1 mutants expressing the *gfp::tiar-1* rescuing transgene in oxidative stress (1-day adults at 5 mM sodium arsenite for 24 h) compared to N2 or *tiar-1* animals carrying only the *rol-6(su1006)* roller marker. In lifespan assays the p-values were determined using the log-rank test (see Tables S5 and S6) and in stress assays the error bars show the SD in unpaired *t*-tests (see Materials and Methods). ns indicates not significant ($p>0.05$); * indicates significant (p-value 0.01 to 0.05); ** indicates very significant (p-value 0.001 to 0.01); *** indicates extremely significant ($p<0.001$).

complex subunit DCAP-1/DCP1 and the miRNA-binding protein AIN-1/GW182, as PB markers and the homologues of human TIA-1/TIAR RNA-binding proteins, TIAR-1-to-3, as SG markers. Of the three worm TIAR proteins, only TIAR-1 and TIAR-2 contain a QN-rich C′-terminal domain and are distributed in both nucleus and cytoplasm. As shown by live imaging in adult transgenic worms there were no SGs and few PBs under normal conditions but there was robust aggregation of PB and SG reporters in the cytoplasm of most cells/tissues, in response to heat or oxidative stress. The dynamic nature of these aggregates is reminiscent of yeast and mammalian PBs and SGs; their formation is rapid and reversible, whereas the partial co-localization between the heat-induced DCAP-1 and TIAR-1 or TIAR-2 foci indicates that PBs and SGs could overlap or dock to each other in worms [12,15,49,51,60]. Additionally, we showed that the translation initiation factor IFE-2/eIF4E is not a component of constitutive PBs and localizes mainly to SGs in response to stress. The heat-induced accumulation of PBs, as monitored by DCAP-1::GFP, was affected by alterations in genes influencing PB assembly in other organisms; it was induced by RNAi of the 5′ to 3′ exonuclease *xrn-1* and prevented by depletion of the translation regulator *cgh-1*/RCK (Dhh1 in yeast). In *C. elegans* gonad, loss of *cgh-1*induces the formation of aberrant sheet-like structures through the relocalization of various PB components [34,35,81,82]. Interestingly, we did not observe such structures in somatic tissues of *cgh-1(RNAi)*-treated animals; instead PBs aggregation was prevented even upon stress. Possible reasons for this discrepancy could be related to different tissues or markers that we used since it has been suggested that the function of CGH-1 helicase depends on cellular and developmental context [34] and *cgh-1* loss induces relocalization of a subset of RNP factors (as CAR-1) into sheet-like structures but dissociates PATR-1 or PGL-1 [82,83]. Furthermore, the fact that knockdown of *cgh-1* prevents PB accumulation without affecting SG formation under stress, suggests that SGs can form independently of PBs. This supports evidence in yeast and mammalian cells, indicating that assembly of PBs and SGs is regulated by different signaling pathways [12,27,60,84].

We also investigated the impact of aging on the pattern of somatic PBs and SGs in *C. elegans*. We demonstrated that only PBs accumulated with age, showing an increase in their number and size. In sharp contrast, we did not observe SG formation in aged animals, despite that in aged oocytes both PB and SG components co-localize to large RNP structures [33]. Since IFE-2 and TIAR-1/-2 were localized in cytoplasmic aggregates induced by sodium arsenite, we reasoned that the accumulation of PBs with age does not result from oxidative stress in aged tissues. The accumulation of non-translatable mRNAs into PBs, due to reduction of either translation or degradation rates with age, provides another possible explanation for the increased PB aggregation during aging. However, knockdown of several translation factors or deletion of *ife-2* did not induce the formation of DCAP-1::GFP granules in adult worms, in contrast to the increased number and size of PBs caused by *xrn-1(RNAi)*. Similarly to the age-induced aggregates, the *xrn-1(RNAi)*-induced PBs did not contain SG markers (IFE-2 or TIAR-1/-2). Thus, we presume that the increased formation of PBs in aged tissues could be an outcome of age-related alterations in decay rates rather than translation but further experimentation is required to support this hypothesis. Alternatively, the accumulation of PBs in older ages could be just a consequence of the aging process, which is associated with a large increase in protein insolubility and aggregation of diverse proteins [85,86]. Among these are translation factors, ribosomal subunits and proteins involved in mitochondrial respiration, whose inhibition has been linked to longevity in many organisms [1]. Thus, we examined the effects of direct alterations in PB components on lifespan. In contrast to the above factors, post-developmental RNAi of genes related to PBs significantly shortened lifespan of N2 and several long-lived mutants. Moreover, increased expression of *dcap-1* in somatic cells of young adults improved their resistance in both heat and oxidative stress.

The maintenance of protein homeostasis, through regulation of both translation and protein degradation is considered a common longevity assurance mechanism [87]. Our findings support that mRNA metabolism factors contribute to such mechanisms and impaired function of PBs can limit lifespan. Interestingly, studies in *S. cerevisiae* have demonstrated that PBs are required for the long-term survival of stationary phase cells [88]. PBs functions are also vital for normal development and stress management. Mutant worms for the decapping genes exhibited severe defects in development, growth, fecundity, movement and impaired stress responses to a variety of environmental insults. A role in the protection of the nascent germline from stress has been described for P granules in worms [89]. Noteworthy, we noticed that *dcap-1* is highly expressed in the three pairs of coelomocytes, the scavenger cells that are considered as a primitive immune system in *C. elegans*. Moreover, we revealed diverse roles of worm TIAR proteins in the cellular stress response, as they displayed specific expression pattern and assembly requirements that can differ according to the stress stimulus. Consistent with such diverse roles, we measured differences in development, growth and lifespan of worms carrying mutations in any of the three *tiar* genes. The effects of *tiar* alleles on lifespan were influenced by temperature or germline signaling; loss of each *tiar* gene shortened normal lifespan, mainly at 25°C, whereas in the long-lived, germline-deficient mutant *glp-1* loss of *tiar-3* resulted in increased mean lifespan, in contrast to *tiar-1* or *tiar-2* deletion that reduced its longevity. The latter indicates a currently unknown role of the predicted RNA-binding protein TIAR-3 in germline-mediated longevity. We finally demonstrated diverse responses of *tiar* mutants against various stressors, suggesting different activities for these genes in worms. Such diverse cellular roles have been described for several RNA-binding proteins which regulate various mRNA subpopulations, often in a coordinated manner [90,91]. For example, several hundred mRNAs of various functions were identified as targets of the dsRNA binding protein STAU-1, the worm homologue of human and *Drosophila* Staufen proteins that are localized in neuronal RNPs and SGs under stress [92–94]. In conclusion, our work implicates factors related to PBs and SGs in the normal lifespan and stress responses, beyond their developmental roles in *C. elegans*. Our observations expand our knowledge on the formation, function and relationship between PBs and SGs in the somatic cells of worms and reveal their tissue- and stress-specific properties.

Supporting Information

Figure S1 Expression pattern and mRNA levels of *dcap-1* in the used transgenic lines. (A) Representative confocal images of 1-day adult worms expressing the transcriptional fusion $P_{dcap\text{-}1::gfp}$ (BRF154) or the translational fusion *dcap-1::gfp* in N2 (BRF155 and BRF261) or germline-deficient *glp-1(e2141)* worms (BRF219), normally grown at 25°C. m: muscles, n: neurons, sp: spermatheca, exc: excretory cell, v: vulva, i: intestine, cc: coeloemocytes. Scale bar: 50 µm. (B) Expression levels of *dcap-1* gene in N2, *glp-1(e2141)* or the indicated transgenic strains measured by quantitative RT-PCR and normalized to *ama-*

1(mRNA) levels. Error bars show the SD of the means of two independent experiments.

Figure S2 Monitoring of granule formation under oxidative stress, heat-shock or aging by using other reporter markers. (A) Representative confocal images of 1-day adults expressing the translational fusion *dcap-1::gfp* (BRF155 in Table S1) or *ife-2::gfp* (BRF70 in Table S1) under normal conditions (-SA) or after exposure to 10 mM sodium arsenite for 3 h (+SA) at 25°C. (B) Representative confocal images of 1-day adults expressing the translational fusion *ain-1::gfp* (MH2704 in Table S1), the transcriptional fusion of *ife-2* putative promoter to *gfp* (BRF68 in Table S1) or the translational fusion *ife-2::gfp* (BRF70 in Table S1) under normal conditions (-HS) or after heat-shock at 35°C for 3 h (+HS). The same transgenic animals are shown at the day 5 of adulthood, grown under normal conditions, at 25°C. (C) Representative confocal images of 1-day adults co-expressing *ife-2::gfp* and *dcap-1::rfp* (BRF313 in Table S1). Arrows point to induced granules. Scale bar: 25 μm. (D) Expression levels of *ain-1* gene in N2, as 1-day or 5-day adults grown under normal conditions (-HS) or as 1-day adults exposed to heat-shock at 35°C for 3 h (+HS), measured by quantitative RT-PCR and normalized to *ama-1(mRNA)* levels. Error bars show the SD of the means of two independent experiments.

Figure S3 Effects of *cgh-1(RNAi)* and *ife-2* deletion in heat-induced granule formation. (A) Representative confocal images of 1-day adults expressing the translational fusion *ife-2::gfp* (BRF70 in Table S1) fed from eggs either control Control(RNAi) or *cgh-1(RNAi)* bacteria, at 25°C, under normal conditions (-HS) or after heat-shock at 35°C for 3 h (+HS). (B) Representative confocal images of 1-day adults expressing the translational fusion *dcap-1::gfp* in N2 (BRF155 in Table S1) or in *ife-2(ok306)* mutant background (BRF220 in Table S1), grown under normal conditions (-HS) or after heat-shock at 35°C for 3 h (+HS). Arrows point to induced granules. Scale bar: 25 μm.

Figure S4 Accumulation of PBs with age at 20°C. (A) Representative confocal images of 1-day, 5-day, 10-day and 15-day adults grown at 20°C, expressing the translational fusion *dcap-1::gfp* (BRF155). Arrows point to DCAP-1::GFP granules. Scale bar: 25 μm. (B) Quantification of data presented in (A). Values on Y axis show the number of granules per head.

Figure S5 Decapping genes are important for development, growth, fertility and stress response. (A) Distribution of N2, *dcap-1* and *dcap-2* progeny in the indicated developmental stages (L1-L4 and adult) 48 h after egg-laying, at 20°C and 25°C. (B) Number of *dcap-1* and *dcap-2* progeny (brood size), compared to N2, at 20°C and 25°C. (C) Development and stress resistance of *dcap-1* mutants expressing the *dcap-1::gfp* rescuing transgene, compared to N2 and *dcap-1* worms carrying only the rol-6(su1006) roller marker, at 20°C. For stress assays see Materials and Methods. Error bars show the SD in unpaired t-tests. ns indicates not significant ($p>0.05$); ** indicates very significant (p-value 0.001 to 0.01); *** indicates extremely significant ($p<0.001$). (D) Reverse transcription (RT)-PCR of *dcap-1(mRNA)* in 1-day adult N2 and *dcap-1* mutants and alignment of the wild-type DCAP-1 protein and the one encoded by the *tm3163* allele using MultAlin (http://multalin.toulouse.inra.fr/multalin/multalin.html).

Figure S6 Expression pattern of *tiar* genes in the used transgenic lines. (A) Representative confocal images of 1-day adult worms expressing the transcriptional fusion P_{tiar-1}*::gfp* (BRF118 in Table S1), the translational fusion *gfp::tiar-1* (BRF211 in Table S1), the transcriptional fusion P_{tiar-2}*::gfp* (BRF238 in Table S1), the translational fusion *gfp::tiar-2* (BRF255 in Table S1) or the translational fusion *tiar-3::gfp* (BRF120 in Table S1), normally grown at 20°C. m: muscles, n: neurons, sp: spermatheca, exc: excretory cell, v: vulva, i: intestine, cc: coeloemocytes. Scale bar: 50 μm.

Figure S7 Spatial overlapping of PB and SG components in somatic cells of worms. Representative confocal images of 1-day adults, under normal conditions (-HS) or after heat-shock at 35°C for 3 h (+HS), co-expressing: (A) *rfp::tiar-1* and *ife-2::gfp* (BRF312 in Table S1), (B) *rfp::tiar-1* and *dcap-1::gfp* (BRF328 in Table S1), (C) *gfp::tiar-2* and *dcap-1::rfp* (BRF369 in Table S1), (D) *rfp::tiar-1* and *gfp::tiar-2* (BRF361 in Table S1). B and C are confocal optical sections. All fusions are driven by their own promoters. Arrows point to induced granules. Scale bar: 25 μm.

Figure S8 TIAR-1 and TIAR-2 granules show differences in their formation. Representative confocal images of 1-day adults expressing (A) *gfp::tiar-1* (BRF211 in Table S1), *gfp::tiar-2* (BRF255 in Table S1) or P_{myo-3}*::gfp::tiar-2* (BRF310 in Table S1) subjected to 350 mM NaCl and visualized after 0.5–1 h, (B) *gfp::tiar-1* or P_{myo-3}*::gfp::tiar-2* in N2 (BRF211 or BRF310, respectively in Table S1) and *gcn-2(ok871)* background (BRF294 or BRF339, respectively in Table S1), subjected to heat-shock (HS, 35°C for 3 h) or sodium arsenite (SA, 15 mM for 3 h). Arrows point to induced granules. Scale bar: 25 μm.

Table S1 List of the strains used in this study.

Table S2 Primers used in this study.

Table S3 RNAi plasmids used in this study.

Table S4 Quantification of number of granules.

Table S5 Lifespan assays in OP-50 plates.

Table S6 Lifespan assays in RNAi plates.

Movie S1 Motility of 6-day adults of N2 on OP-50 plates, at 25°C. Movie was taken in a Leica M205 FA fluorescence stereoscope with a Leica DFC340 FX camera.

Movie S2 Motility of 6-day adults of *tiar-1(tm361)* on OP-50 plates, at 25°C. Movie was taken in a Leica M205 FA fluorescence stereoscope with a Leica DFC340 FX camera.

Acknowledgments

We thank the BIU of BRFAA for using the Confocal system; Dr. Min Han and Dr. Ivo Lieberam for providing us with constructs; Michael Fasseas and Giota Poirazi for suggestions on this manuscript; Dimitris Stravopodis for insightful discussions. Some strains were provided by the Caenorhabditis Genetic Center, which is funded by the National Institutes for Health

National Center for Research Resources (Minneapolis). Other strains were provided by the Mitani Lab through the National Bio-Resource Project of the MEXT, Japan.

Author Contributions

Conceived and designed the experiments: PS AR AV FB. Performed the experiments: AR AV FB FR MK PS. Analyzed the data: PS AR AV FB. Contributed reagents/materials/analysis tools: AR AV FB FR MK. Wrote the paper: PS.

References

1. Kenyon CJ (2010) The genetics of ageing. Nature 464: 504–512.
2. Spriggs KA, Bushell M, Willis AE (2010) Translational regulation of gene expression during conditions of cell stress. Mol Cell 40: 228–237.
3. Renoux AJ, Todd PK (2012) Neurodegeneration the RNA way. Prog Neurobiol 97: 173–189.
4. Wang W (2012) Regulatory RNA-binding proteins in senescence. Ageing Res Rev 11: 485–490.
5. Westmark CJ, Malter JS (2012) The regulation of AbetaPP expression by RNA-binding proteins. Ageing Res Rev 11: 450–459.
6. Eulalio A, Behm-Ansmant I, Izaurralde E (2007) P bodies: at the crossroads of post-transcriptional pathways. Nat Rev Mol Cell Biol 8: 9–22.
7. Filipowicz W, Bhattacharyya SN, Sonenberg N (2008) Mechanisms of post-transcriptional regulation by microRNAs: are the answers in sight? Nat Rev Genet 9: 102–114.
8. Parker R, Sheth U (2007) P bodies and the control of mRNA translation and degradation. Mol Cell 25: 635–646.
9. Anderson P, Kedersha N (2008) Stress granules: the Tao of RNA triage. Trends Biochem Sci 33: 141–150.
10. Brengues M, Teixeira D, Parker R (2005) Movement of eukaryotic mRNAs between polysomes and cytoplasmic processing bodies. Science 310: 486–489.
11. Buchan JR, Parker R (2009) Eukaryotic stress granules: the ins and outs of translation. Mol Cell 36: 932–941.
12. Kedersha N, Stoecklin G, Ayodele M, Yacono P, Lykke-Andersen J, et al. (2005) Stress granules and processing bodies are dynamically linked sites of mRNP remodeling. J Cell Biol 169: 871–884.
13. Souquere S, Mollet S, Kress M, Dautry F, Pierron G, et al. (2009) Unravelling the ultrastructure of stress granules and associated P-bodies in human cells. J Cell Sci 122: 3619–3626.
14. Teixeira D, Sheth U, Valencia-Sanchez MA, Brengues M, Parker R (2005) Processing bodies require RNA for assembly and contain nontranslating mRNAs. Rna 11: 371–382.
15. Buchan JR, Muhlrad D, Parker R (2008) P bodies promote stress granule assembly in Saccharomyces cerevisiae. J Cell Biol 183: 441–455.
16. Decker CJ, Teixeira D, Parker R (2007) Edc3p and a glutamine/asparagine-rich domain of Lsm4p function in processing body assembly in Saccharomyces cerevisiae. J Cell Biol 179: 437–449.
17. Eulalio A, Behm-Ansmant I, Schweizer D, Izaurralde E (2007) P-body formation is a consequence, not the cause, of RNA-mediated gene silencing. Mol Cell Biol 27: 3970–3981.
18. Loschi M, Leishman CC, Berardone N, Boccaccio GL (2009) Dynein and kinesin regulate stress-granule and P-body dynamics. J Cell Sci 122: 3973–3982.
19. Stoecklin G, Mayo T, Anderson P (2006) ARE-mRNA degradation requires the 5'-3' decay pathway. EMBO Rep 7: 72–77.
20. Hu W, Sweet TJ, Chamnongpol S, Baker KE, Coller J (2009) Co-translational mRNA decay in Saccharomyces cerevisiae. Nature 461: 225–229.
21. Romero-Santacreu L, Moreno J, Perez-Ortin JE, Alepuz P (2009) Specific and global regulation of mRNA stability during osmotic stress in Saccharomyces cerevisiae. Rna 15: 1110–1120.
22. Arribere JA, Doudna JA, Gilbert WV (2011) Reconsidering movement of eukaryotic mRNAs between polysomes and P bodies. Mol Cell 44: 745–758.
23. Arimoto K, Fukuda H, Imajoh-Ohmi S, Saito H, Takekawa M (2008) Formation of stress granules inhibits apoptosis by suppressing stress-responsive MAPK pathways. Nat Cell Biol 10: 1324–1332.
24. Gottschald OR, Malec V, Krasteva G, Hasan D, Kamlah F, et al. (2010) TIAR and TIA-1 mRNA-binding proteins co-aggregate under conditions of rapid oxygen decline and extreme hypoxia and suppress the HIF-1alpha pathway. J Mol Cell Biol 2: 345–356.
25. Takahara T, Maeda T (2012) Transient sequestration of TORC1 into stress granules during heat stress. Mol Cell 47: 242–252.
26. Wippich F, Bodenmiller B, Trajkovska MG, Wanka S, Aebersold R, et al. (2013) Dual specificity kinase DYRK3 couples stress granule condensation/dissolution to mTORC1 signaling. Cell 152: 791–805.
27. Buchan JR, Yoon JH, Parker R (2011) Stress-specific composition, assembly and kinetics of stress granules in Saccharomyces cerevisiae. J Cell Sci 124: 228–239.
28. Pitt JN, Schisa JA, Priess JR (2000) P granules in the germ cells of Caenorhabditis elegans adults are associated with clusters of nuclear pores and contain RNA. Dev Biol 219: 315–333.
29. Scheckel C, Gaidatzis D, Wright JE, Ciosk R (2012) Genome-wide analysis of GLD-1-mediated mRNA regulation suggests a role in mRNA storage. PLoS Genet 8: e1002742.
30. Sheth U, Pitt J, Dennis S, Priess JR (2010) Perinuclear P granules are the principal sites of mRNA export in adult C. elegans germ cells. Development 137: 1305–1314.
31. Updike D, Strome S (2010) P granule assembly and function in Caenorhabditis elegans germ cells. J Androl 31: 53–60.
32. Gallo CM, Munro E, Rasoloson D, Merritt C, Seydoux G (2008) Processing bodies and germ granules are distinct RNA granules that interact in C. elegans embryos. Dev Biol 323: 76–87.
33. Jud MC, Czerwinski MJ, Wood MP, Young RA, Gallo CM, et al. (2008) Large P body-like RNPs form in C. elegans oocytes in response to arrested ovulation, heat shock, osmotic stress, and anoxia and are regulated by the major sperm protein pathway. Dev Biol 318: 38–51.
34. Noble SL, Allen BL, Goh LK, Nordick K, Evans TC (2008) Maternal mRNAs are regulated by diverse P body-related mRNP granules during early Caenorhabditis elegans development. J Cell Biol 182: 559–572.
35. Boag PR, Atalay A, Robida S, Reinke V, Blackwell TK (2008) Protection of specific maternal messenger RNAs by the P body protein CGH-1 (Dhh1/RCK) during Caenorhabditis elegans oogenesis. J Cell Biol 182: 543–557.
36. Lall S, Piano F, Davis RE (2005) Caenorhabditis elegans decapping proteins: localization and functional analysis of Dcp1, Dcp2, and DcpS during embryogenesis. Mol Biol Cell 16: 5880–5890.
37. Navarro RE, Shim EY, Kohara Y, Singson A, Blackwell TK (2001) cgh-1, a conserved predicted RNA helicase required for gametogenesis and protection from physiological germline apoptosis in C. elegans. Development 128: 3221–3232.
38. Squirrell JM, Eggers ZT, Luedke N, Saari B, Grimson A, et al. (2006) CAR-1, a protein that localizes with the mRNA decapping component DCAP-1, is required for cytokinesis and ER organization in Caenorhabditis elegans embryos. Mol Biol Cell 17: 336–344.
39. Brenner S (1974) The genetics of Caenorhabditis elegans. Genetics 77: 71–94.
40. Mello C, Fire A (1995) DNA transformation. Methods Cell Biol 48: 451–482.
41. Timmons L, Fire A (1998) Specific interference by ingested dsRNA. Nature 395: 854.
42. Min K, Kang J, Lee J (2010) A modified feeding RNAi method for simultaneous knock-down of more than one gene in Caenorhabditis elegans. Biotechniques 48: 229–232.
43. Ding L, Spencer A, Morita K, Han M (2005) The developmental timing regulator AIN-1 interacts with miRISCs and may target the argonaute protein ALG-1 to cytoplasmic P bodies in C. elegans. Mol Cell 19: 437–447.
44. Kamath RS, Martinez-Campos M, Zipperlen P, Fraser AG, Ahringer J (2000) Effectiveness of specific RNA-mediated interference through ingested double-stranded RNA in Caenorhabditis elegans. Genome Biol 2: RESEARCH0002.
45. Syntichaki P, Troulinaki K, Tavernarakis N (2007) eIF4E function in somatic cells modulates ageing in Caenorhabditis elegans. Nature 445: 922–926.
46. Ozgur S, Chekulaeva M, Stoecklin G (2010) Human Pat1b connects deadenylation with mRNA decapping and controls the assembly of processing bodies. Mol Cell Biol 30: 4308–4323.
47. Yang Z, Jakymiw A, Wood MR, Eystathioy T, Rubin RL, et al. (2004) GW182 is critical for the stability of GW bodies expressed during the cell cycle and cell proliferation. J Cell Sci 117: 5567–5578.
48. Yu JH, Yang WH, Gulick T, Bloch KD, Bloch DB (2005) Ge-1 is a central component of the mammalian cytoplasmic mRNA processing body. Rna 11: 1795–1802.
49. Aizer A, Brody Y, Ler LW, Sonenberg N, Singer RH, et al. (2008) The dynamics of mammalian P body transport, assembly, and disassembly in vivo. Mol Biol Cell 19: 4154–4166.
50. Fenger-Gron M, Fillman C, Norrild B, Lykke-Andersen J (2005) Multiple processing body factors and the ARE binding protein TTP activate mRNA decapping. Mol Cell 20: 905–915.
51. Kedersha N, Tisdale S, Hickman T, Anderson P (2008) Real-time and quantitative imaging of mammalian stress granules and processing bodies. Methods Enzymol 448: 521–552.
52. Swisher KD, Parker R (2010) Localization to, and effects of Pbp1, Pbp4, Lsm12, Dhh1, and Pab1 on stress granules in Saccharomyces cerevisiae. PLoS One 5: e10006.
53. Etchberger JF, Hobert O (2008) Vector-free DNA constructs improve transgene expression in C. elegans. Nat Methods 5: 3.
54. Dernburg AF, Zalevsky J, Colaiacovo MP, Villeneuve AM (2000) Transgene-mediated cosuppression in the C. elegans germ line. Genes Dev 14: 1578–1583.
55. Teixeira D, Parker R (2007) Analysis of P-body assembly in Saccharomyces cerevisiae. Mol Biol Cell 18: 2274–2287.
56. Andrei MA, Ingelfinger D, Heintzmann R, Achsel T, Rivera-Pomar R, et al. (2005) A role for eIF4E and eIF4E-transporter in targeting mRNPs to mammalian processing bodies. Rna 11: 717–727.
57. Eystathioy T, Jakymiw A, Chan EK, Seraphin B, Cougot N, et al. (2003) The GW182 protein colocalizes with mRNA degradation associated proteins hDcp1 and hLSm4 in cytoplasmic GW bodies. Rna 9: 1171–1173.

58. Zhang L, Ding L, Cheung TH, Dong MQ, Chen J, et al. (2007) Systematic identification of C. elegans miRISC proteins, miRNAs, and mRNA targets by their interactions with GW182 proteins AIN-1 and AIN-2. Mol Cell 28: 598–613.
59. Keiper BD, Lamphear BJ, Deshpande AM, Jankowska-Anyszka M, Aamodt EJ, et al. (2000) Functional characterization of five eIF4E isoforms in Caenorhabditis elegans. J Biol Chem 275: 10590–10596.
60. Grousl T, Ivanov P, Frydlova I, Vasicova P, Janda F, et al. (2009) Robust heat shock induces eIF2alpha-phosphorylation-independent assembly of stress granules containing eIF3 and 40S ribosomal subunits in budding yeast, Saccharomyces cerevisiae. J Cell Sci 122: 2078–2088.
61. Koga H, Kaushik S, Cuervo AM (2011) Protein homeostasis and aging: The importance of exquisite quality control. Ageing Res Rev 10: 205–215.
62. Syntichaki P, Tavernarakis N (2006) Signaling pathways regulating protein synthesis during ageing. Exp Gerontol 41: 1020–1025.
63. Sun Y, Yang P, Zhang Y, Bao X, Li J, et al. (2011) A genome-wide RNAi screen identifies genes regulating the formation of P bodies in C. elegans and their functions in NMD and RNAi. Protein Cell 2: 918–939.
64. Squier TC (2001) Oxidative stress and protein aggregation during biological aging. Exp Gerontol 36: 1539–1550.
65. Pan KZ, Palter JE, Rogers AN, Olsen A, Chen D, et al. (2007) Inhibition of mRNA translation extends lifespan in Caenorhabditis elegans. Aging Cell 6: 111–119.
66. Hansen M, Taubert S, Crawford D, Libina N, Lee SJ, et al. (2007) Lifespan extension by conditions that inhibit translation in Caenorhabditis elegans. Aging Cell 6: 95–110.
67. Tritschler F, Braun JE, Motz C, Igreja C, Haas G, et al. (2009) DCP1 forms asymmetric trimers to assemble into active mRNA decapping complexes in metazoa. Proc Natl Acad Sci U S A 106: 21591–21596.
68. Kenyon C, Chang J, Gensch E, Rudner A, Tabtiang R (1993) A C. elegans mutant that lives twice as long as wild type. Nature 366: 461–464.
69. Kimura KD, Tissenbaum HA, Liu Y, Ruvkun G (1997) daf-2, an insulin receptor-like gene that regulates longevity and diapause in Caenorhabditis elegans. Science 277: 942–946.
70. Lakowski B, Hekimi S (1998) The genetics of caloric restriction in Caenorhabditis elegans. Proc Natl Acad Sci U S A 95: 13091–13096.
71. Arantes-Oliveira N, Apfeld J, Dillin A, Kenyon C (2002) Regulation of life-span by germ-line stem cells in Caenorhabditis elegans. Science 295: 502–505.
72. Kedersha NL, Gupta M, Li W, Miller I, Anderson P (1999) RNA-binding proteins TIA-1 and TIAR link the phosphorylation of eIF-2 alpha to the assembly of mammalian stress granules. J Cell Biol 147: 1431–1442.
73. Gilks N, Kedersha N, Ayodele M, Shen L, Stoecklin G, et al. (2004) Stress granule assembly is mediated by prion-like aggregation of TIA-1. Mol Biol Cell 15: 5383–5398.
74. Kelly WG, Xu S, Montgomery MK, Fire A (1997) Distinct requirements for somatic and germline expression of a generally expressed Caernorhabditis elegans gene. Genetics 146: 227–238.
75. Silva-Garcia C, Navarro R (2013) The C. elegans TIA-1/TIAR homolog TIAR-1 is required to induce germ cell apoptosis. Genesis 51: 690–707.
76. Rousakis A, Vlassis A, Vlanti A, Patera S, Thireos G, et al. (2013) The general control nonderepressible-2 kinase mediates stress response and longevity induced by target of rapamycin inactivation in Caenorhabditis elegans. Aging Cell 12: 742–751.
77. Ermolaeva MA, Segref A, Dakhovnik A, Ou HL, Schneider JI, et al. (2013) DNA damage in germ cells induces an innate immune response that triggers systemic stress resistance. Nature 501: 416–420.
78. Judy ME, Nakamura A, Huang A, Grant H, McCurdy H, et al. (2013) A shift to organismal stress resistance in programmed cell death mutants. PLoS Genet 9: e1003714.
79. Nelson FK, Albert PS, Riddle DL (1983) Fine structure of the Caenorhabditis elegans secretory-excretory system. J Ultrastruct Res 82: 156–171.
80. Stoecklin G, Kedersha N (2013) Relationship of GW/P-bodies with stress granules. Adv Exp Med Biol 768: 197–211.
81. Audhya A, Hyndman F, McLeod IX, Maddox AS, Yates JR, 3rd, et al. (2005) A complex containing the Sm protein CAR-1 and the RNA helicase CGH-1 is required for embryonic cytokinesis in Caenorhabditis elegans. J Cell Biol 171: 267–279.
82. Hubstenberger A, Noble SL, Cameron C, Evans TC (2013) Translation repressors, an RNA helicase, and developmental cues control RNP phase transitions during early development. Dev Cell 27: 161–173.
83. Updike DL, Strome S (2009) A genomewide RNAi screen for genes that affect the stability, distribution and function of P granules in Caenorhabditis elegans. Genetics 183: 1397–1419.
84. Mollet S, Cougot N, Wilczynska A, Dautry F, Kress M, et al. (2008) Translationally repressed mRNA transiently cycles through stress granules during stress. Mol Biol Cell 19: 4469–4479.
85. David DC, Ollikainen N, Trinidad JC, Cary MP, Burlingame AL, et al. (2010) Widespread protein aggregation as an inherent part of aging in C. elegans. PLoS Biol 8: e1000450.
86. Reis-Rodrigues P, Czerwieniec G, Peters TW, Evani US, Alavez S, et al. (2012) Proteomic analysis of age-dependent changes in protein solubility identifies genes that modulate lifespan. Aging Cell 11: 120–127.
87. Ben-Zvi A, Miller EA, Morimoto RI (2009) Collapse of proteostasis represents an early molecular event in Caenorhabditis elegans aging. Proc Natl Acad Sci U S A 106: 14914–14919.
88. Ramachandran V, Shah KH, Herman PK (2011) The cAMP-dependent protein kinase signaling pathway is a key regulator of P body foci formation. Mol Cell 43: 973–981.
89. Gallo CM, Wang JT, Motegi F, Seydoux G (2010) Cytoplasmic partitioning of P granule components is not required to specify the germline in C. elegans. Science 330: 1685–1689.
90. Gerber AP, Herschlag D, Brown PO (2004) Extensive association of functionally and cytotopically related mRNAs with Puf family RNA-binding proteins in yeast. PLoS Biol 2: E79.
91. Keene JD (2003) Posttranscriptional generation of macromolecular complexes. Mol Cell 12: 1347–1349.
92. LeGendre JB, Campbell ZT, Kroll-Conner P, Anderson P, Kimble J, et al. (2013) RNA targets and specificity of Staufen, a double-stranded RNA-binding protein in Caenorhabditis elegans. J Biol Chem 288: 2532–2545.
93. Micklem DR, Adams J, Grunert S, St Johnston D (2000) Distinct roles of two conserved Staufen domains in oskar mRNA localization and translation. Embo J 19: 1366–1377.
94. Thomas MG, Martinez Tosar LJ, Desbats MA, Leishman CC, Boccaccio GL (2009) Mammalian Staufen 1 is recruited to stress granules and impairs their assembly. J Cell Sci 122: 563–573.

The Tumor-Educated-Macrophage Increase of Malignancy of Human Pancreatic Cancer Is Prevented by Zoledronic Acid

Yukihiko Hiroshima[1,2,3], Ali Maawy[2], Mohamed K. Hassanein[1,2], Rhiana Menen[1], Masashi Momiyama[3], Takashi Murakami[3], Shinji Miwa[1,2], Mako Yamamoto[1,2], Fuminari Uehara[1,2], Shuya Yano[1,2], Ryutaro Mori[3], Ryusei Matsuyama[3], Takashi Chishima[3], Kuniya Tanaka[3], Yasushi Ichikawa[3], Michael Bouvet[2], Itaru Endo[3], Robert M. Hoffman[1,2]*

1 AntiCancer, Inc., San Diego, California, United States of America, **2** Department of Surgery, University of California San Diego, San Diego, California, United States of America, **3** Yokohama City University Graduate School of Medicine, Yokohama, Japan

Abstract

We previously defined macrophages harvested from the peritoneal cavity of nude mice with subcutaneous human pancreatic tumors as "tumor-educated-macrophages" (Edu) and macrophages harvested from mice without tumors as "naïve-macrophages" (Naïve), and demonstrated that Edu-macrophages promoted tumor growth and metastasis. In this study, Edu- and Naïve-macrophages were compared for their ability to enhance pancreatic cancer malignancy at the cellular level in vitro and in vivo. The inhibitory efficacy of Zoledronic acid (ZA) on Edu-macrophage-enhanced metastasis was also determined. XPA1 human pancreatic cancer cells in Gelfoam co-cultured with Edu-macrophages proliferated to a greater extent compared to XPA1 cells cultured with Naïve-macrophages ($P = 0.014$). XPA1 cells exposed to conditioned medium harvested from Edu culture significantly increased proliferation ($P = 0.016$) and had more migration stimulation capability ($P < 0.001$) compared to cultured cancer cells treated with the conditioned medium from Naïve. The mitotic index of the XPA1 cells, expressing GFP in the nucleus and RFP in the cytoplasm, significantly increased in vivo in the presence of Edu- compared to Naïve-macrophages ($P = 0.001$). Zoledronic acid (ZA) killed both Edu and Naïve in vitro. Edu promoted tumor growth and metastasis in an orthotopic mouse model of the XPA1 human pancreatic cancer cell line. ZA reduced primary tumor growth ($P = 0.006$) and prevented metastasis ($P = 0.025$) promoted by Edu-macrophages. These results indicate that ZA inhibits enhanced primary tumor growth and metastasis of human pancreatic cancer induced by Edu-macrophages.

Editor: Keping Xie, The University of Texas MD Anderson Cancer Center, United States of America

Funding: This study was supported in part by National Cancer Institute grant CA132971 and JSPS KAKENHI Grant Numbers 26830081 to Y.H., 26462070 to I.E. and 24592009 to K.T. The funders had no role in study design, data collection and analysis, decision to publish, or preparation of the manuscript.

Competing Interests: Yukihiko Hiroshima, Shinji Miwa, Mako Yamamoto, Shuya Yano are affiliates of AntiCancer Inc. Mohamed K. Hassanein, Rhiana Menen, Masashi Momiyama, Fuminari Uehara and Takashi Chishima were former affiliates of AntiCancer Inc. Robert M. Hoffman is a non-salaried affiliate of AntiCancer Inc. AntiCancer Inc. markets animal models of cancer. There are no other competing interests. There are no patents, products in development or marketed products to declare.

* Email: all@anticancer.com

Introduction

The tumor microenvironment (TME) plays a major role in determining tumor behavior. We have previously shown that stromal cells are necessary for metastasis to occur [1]. Another indication of the role of the TME in influencing tumor behavior is orthotopic implantation in mouse models of intact tumor tissue, including the entire tumor stroma, which results in a much higher metastatic rate compared to orthotopic injection of cancer cells alone, where metastasis is rare [2,3].

Tumor-associated macrophages have been shown to correlate with poor prognosis in several studies [4,5]. Macrophages may promote tumor progression by chronic inflammation, matrix remodeling, promotion of tumor-cell invasion, intravasation, angiogenesis, and seeding at distant sites [6]. Vascular cell adhesion molecule-1 (VCAM-1) promotes lung metastasis in breast cancer by tethering cancer cells to lung metastasis-associated macrophages [7,8].

Our laboratory previously compared primary and metastatic tumor growth following the addition of either tumor-naïve-macrophages (Naïve) or macrophages previously exposed to pancreatic cancer in a mouse model, which were termed tumor-educated-macrophages (Edu) [8].

Our previous results suggest that macrophages influence tumors and tumors influence macrophages, and that Edu-macrophages promote tumor progression [8].

In the present study, Edu- and Naïve-macrophages were compared for their ability to promote malignancy at the cellular level in vitro and in vivo. The efficacy of Zoledronic acid (ZA) to inhibit Edu-macrophage-enhanced malignancy was also determined.

Materials and Methods

Cell line and culture conditions

The XPA1 human pancreatic cancer cell line was used in this study which was a kind gift from Dr. Anirban Maitra at Johns Hopkins University [6–16]. The XPA1 cell line was transformed to stably express GFP in the nucleus and RFP in the cytoplasm [9,17,18]. Cells were maintained in RPMI 1640 medium supplemented with 10% fetal bovine serum (FBS) and 2 mM glutamine (Gibco-BRL, Life Technologies Inc., Grand Island, NY). All media were supplemented with penicillin and streptomycin (Gibco-BRL). Cells were cultured at 37°C with 5% CO_2 [19].

Animals

NOD/SCID mice and athymic nude mice (*nu/nu*), 4–6 weeks, (AntiCancer Inc., San Diego, CA) were used. Transgenic nude C57/B6-GFP mice (AntiCancer, Inc., San Diego, CA). expressing green fluorescent protein (GFP) under the control of the chicken β-actin promoter and cytomegalovirus enhancer were also used [20–23]. Mice were kept in a barrier facility under HEPA filtration. Mice were fed with autoclaved laboratory rodent diet. All surgical procedures and imaging were performed with the animals anesthetized by intramuscular injection of 0.02 ml of a solution of 50% ketamine, 38% xylazine, and 12% acepromazine maleate. All animal studies were conducted with an AntiCancer Institutional Animal Care and Use Committee (IACUC)-protocol specifically approved for this study and in accordance with the principals and procedures outlined in the National Institute of Health Guide for the Care and Use of Animals under Assurance Number A3873-1.

Subcutaneous tumor cell implantation

Human XPA1 dual-color pancreatic cancer cells were harvested by trypsinization and washed twice with serum-free medium. Cells (2×10^6 in 100 µl serum-free media) were injected subcutaneously, within 30 min of harvesting, over the flanks in transgenic nude C57/B6-GFP mice or nontransgenic *nu/nu* mice between 4 and 6 weeks of age. Subcutaneous tumors were allowed to grow for 2–4 weeks until large enough for further experiments or subsequent orthotopic implantation.

Real-time imaging of the interactions between host macrophages and cancer cells in live mice

After the mice were anesthetized as described above, an arc-shaped incision was made in the abdominal skin, and then subcutaneous connective tissue was separated to free the skin flap without injuring the epigastric cranialis artery and vein [24]. The skin-flap was spread and fixed on a flat stand. XPA1-GFP-RFP cells (1×10^6 in 100 µl medium) were sprinkled over the surface of the skin–flap of mice (Fig. 1A) [25]. Twenty-four hours later, the inside surface of the skin-flap was directly observed with the FV1000 confocal microscope (Olympus, Tokyo, Japan) with excitation from semiconductor lasers at 473 nm for GFP and 559 nm for RFP excitation. Fluorescence images were obtained using the 20×/1.0 XLUMPLFLN objective [26]. The number of phagocytosed and mitotic cancer cells were counted, and the mean number was calculated from five visual fields at 20× magnification.

Macrophage harvest

Transgenic nude C57/B6-GFP mice with the dual-color XPA1-GFP-RFP subcutaneous tumors were the source for Edu-macrophages. Transgenic nude C57/B6-GFP mice without tumors were the source for Naïve-macrophages. After the mice were anesthetized as described above, 6 ml RPMI was injected into the intraperitoneal space of each mouse using a 10 ml syringe and the mice were gently agitated for 5 minutes. The syringe was then reinserted and all RPMI medium was removed and placed into a plastic petri dish. The dishes were then incubated at 37°C for 4 hours. RPMI medium was then removed and each plate was washed 10–15 times with 10 ml PBS, or until all red blood cells, fibroblasts, and other cellular debris were removed. Plates were visualized under an IX71 fluorescence microscope (Olympus, Tokyo, Japan) in order to confirm the presence of GFP-expressing macrophages. Macrophages were scraped from the petri dish, using a rubber spatula, collected with a wash of 10 ml PBS, and centrifuged at 850 rpm for 10 minutes. PBS was then removed and macrophages were resuspended in RPMI.

Proliferation assay using 3D culture

Naïve- or Edu-macrophages were co-cultured with dual-color XPA1-GFP-RFP cancer cells, but they were separated from each other by Gelfoam (Pharmacia & Upjohn Company, MI, USA) (Fig. 2A). Forty-eight hours later, images were obtained from the surface to 200 µm depth, every 1 µm, with the FV1000 confocal microscope. The images were stacked along the Z axis, and the mean number of cancer cells was calculated from three visual fields at 20× magnification.

MTS proliferation assay

XPA1-GFP-RFP cells in the exponential growth phase were trypsinized to yield a cell suspension and seeded onto 96-well plates (1×10^4 cells/well) in triplicate. Conditioned medium was harvested from each type of macrophages cultured in RPMI 1640 with 10% FBS for 72 h. The culture medium was removed from the wells with XPAI-GFP-RFP cells, and the conditioned medium was added after the cells were allowed to adhere for 24 h. After the plates were incubated at 37°C in 5% CO_2, the MTS assay was performed at various time points (0–72 h) using the Cell Titer 96 assay (Promega, Madison, WI, USA) according to the manufacturer's instructions.

Wound healing assay

Cells were grown in 24-well plates in 500 µl medium per well until confluence was reached. A wound was made by scratching the cells with a 10 µl pipette tip in PBS, followed by replacement with conditioned medium which was harvested from each macrophage type cultured in RPMI 1640 with 10% FBS for 72 h. The control group received the same volume of fresh culture medium without serum. The wounded monolayer was photomicrographed with the IX71 fluorescence microscope at different time points (0–72 h) after being scratched. Cell migration was assessed by measuring gap sizes in multiple fields using Image J (National Institute of Mental Health, Bethesda, MD).

Cytotoxicity assay

Naïve- or Edu-macrophages were seeded onto 6-well plates (2×10^5 cells/well). Zoledronic acid (ZA) (Novartis Pharmaceuticals Corporation, NJ, USA) was added to each well after the cells were allowed to adhere for 24 h. The final concentrations of ZA were as follows: 0, 10, 50 and 100 µM. After 48 h, the ZA-containing medium was removed and replaced with growth medium for 24 h.

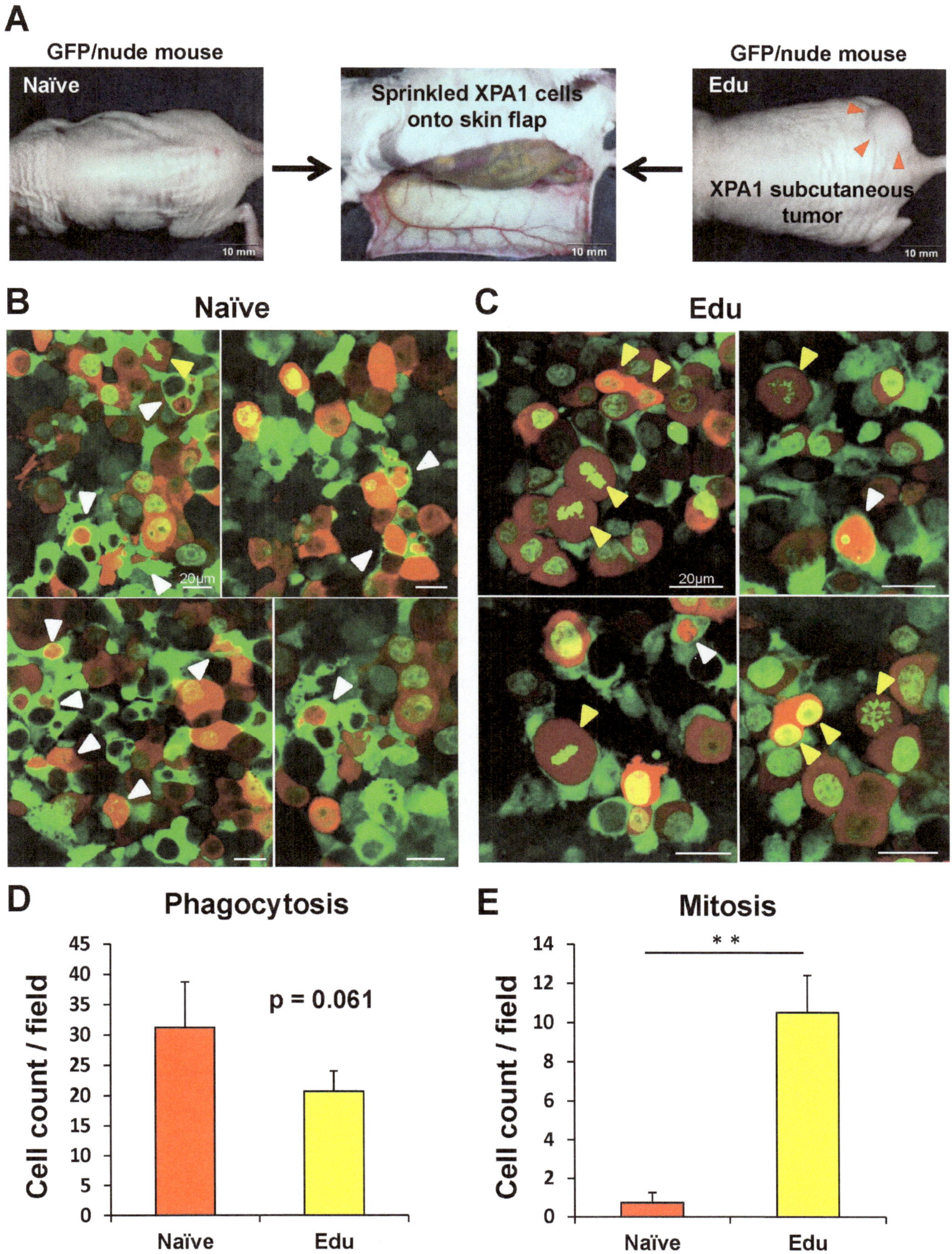
A
GFP/nude mouse
Naïve
10 mm
Sprinkled XPA1 cells onto skin flap
10 mm
GFP/nude mouse
Edu
XPA1 subcutaneous tumor
10 mm
B
Naïve
20μm
C
Edu
20μm
D
Phagocytosis
Cell count / field
p = 0.061
Naïve
Edu
E
Mitosis
Cell count / field
* *
Naïve
Edu

Figure 1. Imaging of the interactions between host macrophages and cancer cells in live mice. (**A**) Scheme of imaging the interactions between host macrophages and cancer cells in live mice. GFP nude mice with dual-color XPA1 subcutaneous tumors were the source of Edu-macrophages. GFP nude mice without tumors were the source of Naïve-macrophage. A skin-flap was spread and fixed on a flat stand. XPA1-GFP-RFP cells (1×10^6 in 100 μl medium) were sprinkled over the surface of the skin–flap of mice. Twenty-four hours later, the inside surface of the skin-flap was directly observed with the FV1000 confocal microscope (Olympus). Scale bars: 10 mm. GFP host macrophages and dual-color XPA1 cancer cells were observed in Naïve-macrophage mice (**B**) and Edu-macrophage mice (**C**). Phagocytosed cancer cells (white arrowheads) and mitotic cancer cells (yellow arrowheads) were detected in both groups. The mean values of phagocytosized and mitotic cancer cells were calculated from four fields with a 20× magnification objective (**D and E**). Mitosis significantly increased in cancer cells in mice with Edu-macrophages compared to cancer cells in mice with Naïve-macrophages ($P=0.001$). More phagocytosis of cancer cells tended to be detected in mice with Naïve-macrophages, ($P=0.061$). Scale bars: 20 μm.

After each well was washed 5 times with 3 ml PBS, the number of GFP-expressing macrophages was counted with the IX71 fluorescence microscope. The mean macrophage number was calculated from five visual fields at 10× magnification. All measurements were carried out in triplicate.

Orthotopic tumor implantation

A small 6- to 10-mm transverse incision was made on the left flank of the mouse through the skin and peritoneum. The tail of the pancreas was exposed through this incision and a single tumor fragment (3-mm^3) from a XPA1 dual-color subcutaneous tumor was sutured to the tail of the pancreas using 8-0 nylon surgical sutures (Ethilon; Ethicon Inc., NJ, USA). Upon completion, the tail of the pancreas was returned to the abdomen, and the incision was closed in one layer using 6-0 nylon surgical sutures (Ethilon) [2,27–29].

ZA treatment of the XPA1 orthotopic nude mouse model

Nude mice were orthotopically implanted with XPA1-GFP-RFP cells as described above. The mice were treated in the following groups: (1) saline (vehicle/control), (2) Naïve- macrophages (3) Edu-macrophages, (4) Edu-macrophages + ZA. Macrophages (1×10^6 cells in 200 μl PBS) were injected ip weekly as previously described [8]. ZA was subcutaneously injected daily from day 21 after tumor implantation for 4 weeks. Each treatment arm involved 8 tumor-bearing mice. No significant effects on body weight, morbidity, or severe toxicity were observed in any treatment arm. Animals underwent laparotomy at 7 weeks, and both primary tumors and metastases were imaged using the OV100 variable magnification Small Animal Imaging System (Olympus, Tokyo, Japan) [24] and weighed and harvested for analysis.

Data processing and statistical analysis

PASW Statistics 18.0 (SPSS, Inc) was used for all statistical analyses. The Student's t-test was used to compare continuous variables between two groups. Analysis of variance models were used to compare multiple groups. A p value of 0.05 was considered statistically significant for all comparisons.

Results and Discussion

Edu-macrophages stimulate cancer-cell mitosis in vivo

Dual-color XPA1-GFP-RFP cells (1×10^6) were sprinkled over the surface of a mouse skin-flap. Twenty-four hours later, the inside surface of the skin-flap was directly observed with the FV1000 confocal microscope (Olympus) (Fig. 1A). GFP-expressing host macrophages and dual-color XPA1 cancer cells were observed in Naïve-macrophage mice (Fig. 1B) and Edu-macrophage mice (Fig. 1C). Cancer-cell mitosis was significantly greater in cancer cells in Edu-macrophage mice compared to cancer cells in Naïve-macrophage mice ($P=0.001$) (Fig. 1E). Phagocytosis of cancer cells tended to be greater in Naïve-macrophage mice compared to Edu-macrophage mice ($P=0.061$) (Fig. 1D).

Edu-macrophages stimulate proliferation of cancer cells *in vitro*

Naïve- or Edu-macrophage were cultured with dual-color XPA1 cancer cells separated from each other by Gelfoam (Fig. 2A). Forty-eight hours later, confocal microscopy images were obtained from the surface to 200 μm depth every 1 μm and were stacked along the Z axis (Fig. 2B). Edu-macrophage-treated cancer cells proliferated to a greater extent compared to Naïve-macrophage-treated cancer cells ($P=0.014$) (Fig. 2C). Conditioned medium from Edu-macrophages stimulated the proliferation of cancer cells compared to control untreated cancer cells (24 h; $P=0.001$, 48 h; $P=0.003$, 72 h; $P=0.012$, respectively). In contrast, there was no significant difference between conditioned medium from Naïve-macrophages and fresh medium on proliferation of XPA1 cells (Fig. 2D).

Edu-macrophage-conditioned-medium-treated cancer cells covered a significantly larger wound area than untreated control cells at every time point (24 h; $P=0.008$, 48 h; $P=0.002$, 72 h; $P<0.001$, respectively) (Fig. 2E and 2F).

ZA inhibits XPA1 pancreatic tumor progression induced by Edu-macrophages

Edu-macrophages promoted primary tumor growth compared to control and Naïve- macrophage (control; $P=0.026$, Naïve; $P=0.03$). Edu-macrophages also promoted metastasis compared to control and Naïve-macrophages (control; $P=0.012$, Naïve; $P=0.015$) (Fig. 3C–E). ZA has been shown to kill macrophages and prevent metastasis [30,31]. The number of both Naïve- and Edu-macrophages were significantly reduced by ZA at every dose in vitro (Fig. 3A and 3B). In an orthotopic mouse model of pancreatic cancer, ZA significantly suppressed Edu-macrophage, stimulated tumor growth, and prevented metastasis compared to the mice treated with only Edu-macrophages (primary tumor; $P=0.006$, metastasis; $P=0.025$) (Fig. 3C–E).

In our previous study, GFP-expressing macrophages from GFP transgenic nude mice with a subcutaneous BxPC3-RFP human pancreatic tumor were used as a source of Edu-macrophages and compared to Naïve-macrophages from the transgenic GFP nude mice without tumors. When the Edu- or Naïve-macrophages were then implanted into nude mice with BxPC-3-RFP growing orthotopically, the Edu-macrophages stimulated tumor growth and metastasis to a greater degree than Naïve-macrophages [8].

In another previous study, human peripheral-blood mononuclear cells were exposed to conditioned medium from BxPC-3 human pancreatic cancer cells, in normal or high glucose levels. The pancreatic cancer cells educated the macrophages to be more invasive in vitro, which was further enhanced by hyperglycaemia [32].

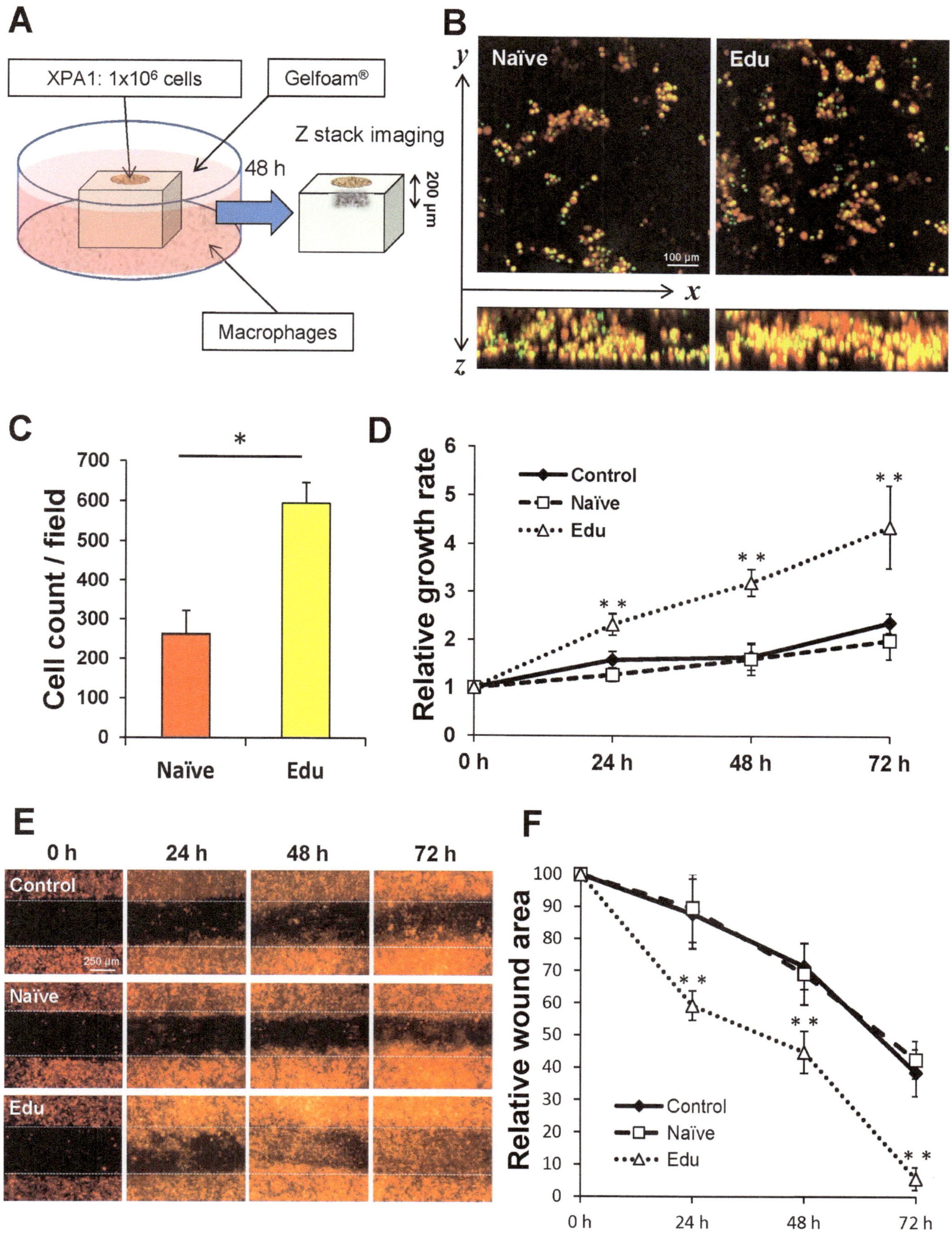

Figure 2. Conditioned medium from Edu-macrophages promotes pancreatic cancer cell proliferation and migration. (**A**) A schema of the separated co-culture system using Gelfoam. Naïve- or Edu-macrophages were co-cultured with dual-color XPA1 GFP-RFP cancer cells, but they were separated from each other by Gelfoam. Forty-eight hours later, images were obtained from the surface to 200 µm depth, every 1 µm, and

were stacked along the Z axis. (**B**) Representative Z-stack image using a 20× magnification objective. (**C**) The mean number of cancer cells was calculated from three fields of Z-stack images with the 20× magnification objective. The number of cancer cells treated with Edu-macrophages increased compared to the number of cancer cells treated with Naïve-macrophages ($P=0.014$). (**D**) MTS proliferation assay. Conditioned medium was harvested from each type of macrophage cultured in RPMI without serum for 72 h. The culture medium from the cancer cells was removed from the wells and the conditioned medium was added. Viable cancer cell numbers are indicated with the MTS assay at various time points. The line graph indicates the relative proliferation rate of each group. The proliferation rate of Edu-macrophage-condition-medium-treated cancer cells was significantly higher than control or Naïve-macrophage-conditioned-medium-treated cancer cells. Black diamonds: Naïve; white square: Edu, white triangle: control group. * $P<0.05$, ** $P<0.01$ (vs. control group). (**E**) Wounded areas were measured for 72 hours after scratching the monolayer of cancer cells and treatment with the conditioned medium from each type of macrophage. (**F**) Edu-macrophage-conditioned medium-treated cells covered a significantly larger wound area than other groups at every time point. Charts are plots of wounded areas of each group measured with ImageJ of three random areas in 3 days. Data from three repeat experiments are presented as mean ± SD (n = 5). * $P<0.05$, ** $P<0.01$ (vs. control group).

In another previous study, ZA reduced the macrophage-induced invasiveness of breast cancer cells. ZA affected macrophages but not the cancer cells [33].

Another previous study showed that both peritoneal and breast-tumor associated macrophages rapidly took up ZA in vivo, further suggesting macrophages are a target of the anti-tumor efficacy of ZA [34].

In another previous study, ZA directly caused apoptosis on pancreatic carcinoma cells and inhibited their invasiveness and as well as making the pancreatic cancer cells more vulnerable to T-cell attack [35].

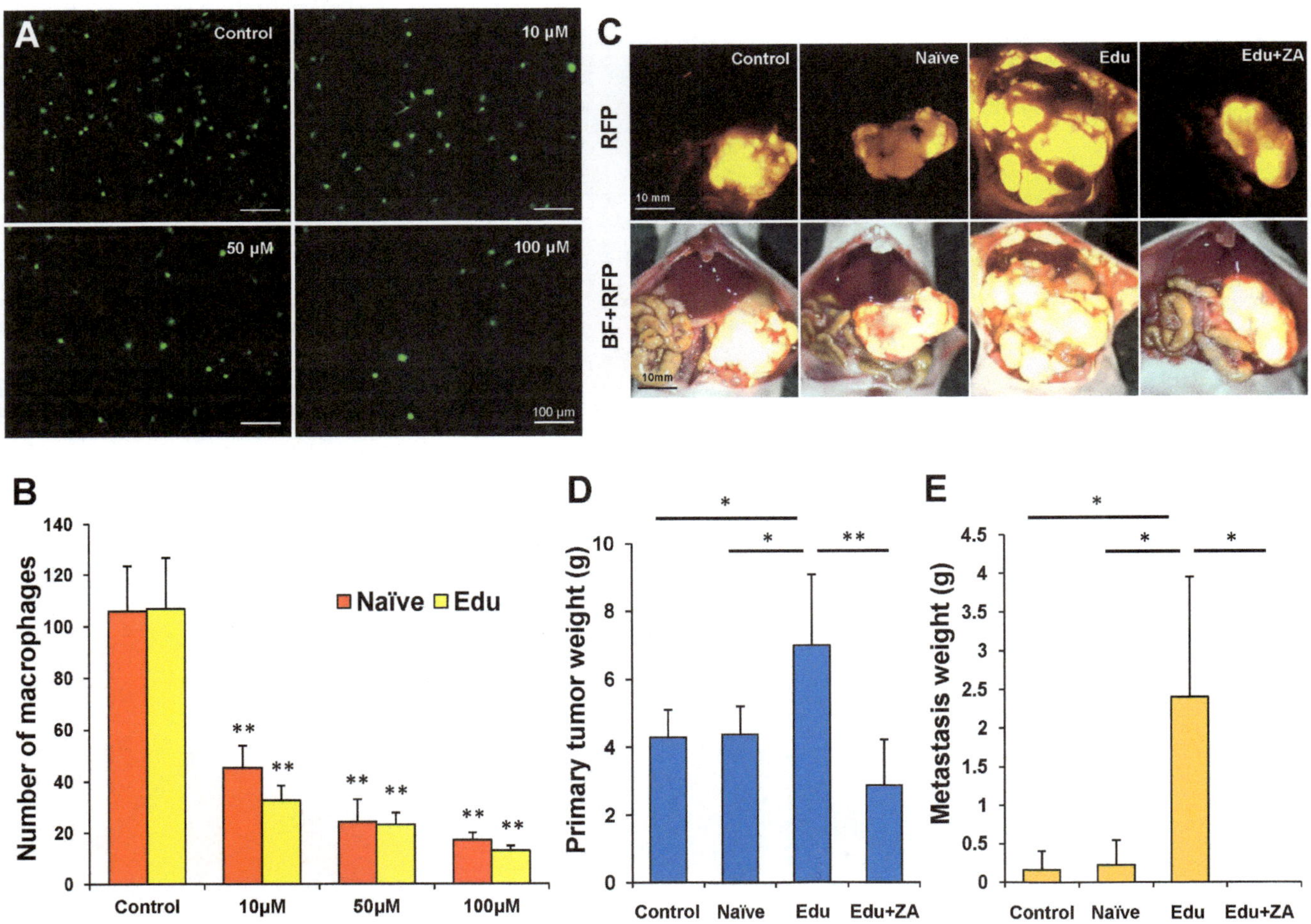

Figure 3. ZA inhibits Edu-macrophages and tumor progression induced by Edu macrophages in the orthotopic XPAI pancreatic cancer mouse model. (**A**) Representative fluorescence images of Edu-macrophages after ZA treatment. Scale bars: 10 μm. (**B**) Bar graphs of the number of Naïve- or Edu-macrophages after ZA treatment in vitro. The numbers of both Naïve- and Edu-macrophages treated with ZA were significantly reduced at every dose. ZA killed both Naïve and Edu in a dose-dependent manner. ** $P<0.01$ (vs. control group). (**C**) Intravital imaging of XPA1-RFP tumor-bearing mice at the termination of the experiment. Scale bars: 10 mm. (**D**) The primary tumor weight of Edu-macrophage-treated mice were significantly increased compared to control or Naïve-macrophage-treated mice (control: $P=0.026$; Naïve: $P=0.03$, respectively). The primary tumor weight of Edu-macrophage + ZA-treated mice were significantly decreased compared to Edu-macrophage-treated mice ($P=0.006$). * $P<0.05$, ** $P<0.01$. (**E**) The metastasis weight of Edu-macrophage-treated mice were significantly increased compared to control or Naïve-macrophage-treated mice (control; $P=0.012$, Naïve; $P=0.015$, respectively). No metastasis was detected in Edu + ZA-treated mice. There was a significant difference between Edu and Edu + ZA-treated mice ($P=0.025$). * $P<0.05$.

The present study used subcellular imaging techniques we have previously developed [36] to directly image that Edu-macrophages have reduced phagocytosis ability and can directly simulate mitosis of the cancer cells, as visualized in real time. Unlike previous studies, evaluating the efficacy of ZA, the present report used an orthotopic metastatic mouse model to demonstrate the effect of ZA on the stimulation by Edu-macrophages on metastasis as well as on the primary tumor. In the present study, the macrophages were isolated from the peritoneal cavity from tumor-bearing or control mice, far from the tumor and administered systemically. The results indicated that tumors had distant affect on macrophages and vice versa and that ZA could systemically inhibit such effects.

ZA reduced Edu-stimulated primary tumor growth and metastasis to approximately control values suggesting the major ZA effect was on the Edu-macrophages, although a relatively minor direct inhibitory effect on the tumor itself by ZA cannot be ruled out as described in a previous study [35]. ZA has been previously reported to have greater macrophage killing efficacy than other bisphosphonates and was therefore chosen for the present study [37].

The present study indicated that Edu-macrophages promote malignancy. This was shown by the stimulatory effect of Edu-macrophages on cancer-cell mitosis and subsequent proliferation as well as on metastasis. As we stated previously [8], tumors can affect macrophages which in turn can affect tumors. Tumor progression induced by Edu-macrophages was inhibited by ZA, which inhibited Edu-macrophage-enhanced primary tumor growth and prevented metastasis. The mechanisms by which Edu-macrophages stimulate cancer-cell proliferation and progression will be the subject of a future study.

Acknowledgments

Dedication. This paper is dedicated to the memory of A. R. Moossa, M.D.

Author Contributions

Conceived and designed the experiments: YH RMH. Performed the experiments: YH AM. Analyzed the data: YH AM MKH RM MM TM SM MY FU SY RM RM TC KT YI MB IE RMH. Contributed reagents/materials/analysis tools: RMH. Contributed to the writing of the manuscript: YH RMH.

References

1. Bouvet M, Tsuji K, Yang M, Jiang P, Moossa AR, et al. (2006) *In vivo* color-coded imaging of the interaction of colon cancer cells and splenocytes in the formation of liver metastases. Cancer Res 66: 11293–11297.
2. Hoffman RM. (1999) Orthotopic metastatic mouse models for anticancer drug discovery and evaluation: a bridge to the clinic. Investigational New Drugs 17: 343–359.
3. Fu X, Besterman J., Monosov A, Hoffman RM. (1991) Models of human metastatic colon cancer in nude mice orthotopically constructed by using histologically intact patient specimens. Proc Natl Acad Sci USA 88: 9345–9349.
4. Kurahara H, Shinchi H, Mataki Y, Maemura K, Noma H, et al. (2011) Significance of M2-polarized tumor-associated macrophage in pancreatic cancer. J Surg Res 167: e211–9.
5. Rogers TL, Holen I. (2011) Tumour macrophages as potential targets of bisphosphonates. J Transl Med 9: 177.
6. Lee CH, Espinosa I, Vrijaldenhoven S, Subramanian S, Montgomery KD, et al. (2008) Prognostic significance of macrophage infiltration in leiomyosarcomas. Clin. Cancer Res 14: 1423–1430.
7. Zeisberger SM, Odermatt B, Marty C, Zehnder-Fjallman AH, Ballmer-Hofer K, et al. (2006) Clodronate-liposome-mediated depletion of tumour-associated macrophages: a new and highly effective antiangiogenic therapy approach. Br J Cancer 95: 272–281.
8. Menen RS, Hassanein MK, Momiyama M, Suetsugu A, Moossa AR, et al. (2012) Tumor-educated macrophages promote tumor growth and peritoneal metastasis in an orthotopic nude mouse model of human pancreatic cancer. In Vivo 26: 565–569.
9. Amoh Y, Nagakura C, Maitra A, Moossa AR, Katsuoka K, et al. (2006) Dual-color imaging of nascent angiogenesis and its inhibition in liver metastases of pancreatic cancer. Anticancer Research 26: 3237–3242.
10. McElroy M, Kaushal S, Tran Cao HS, Moossa AR, Talamini MA, et al. (2009) Upregulation of thrombospondin-1 and angiogenesis in an aggressive human pancreatic cancer cell line selected for high metastasis. Mol Cancer Therap 8: 1779–1786.
11. Tran Cao HS, Reynoso J, Yang M, Kimura H, Kaushal S, et al. (2009) Development of the transgenic cyan fluorescent protein (CFP)-expressing nude mouse for "Technicolor" cancer imaging. J. Cell. Biochem 107: 328–334.
12. Snyder CS, Kaushal S, Kono Y, Tran Cao HS, Hoffman RM, et al. (2009) Complementarity of ultrasound and fluorescence imaging in an orthotopic mouse model of pancreatic cancer. BMC Cancer 9: 106.
13. Nagakura C, Hayashi K, Zhao M, Yamauchi K, Yamamoto N, et al. (2009) Efficacy of a genetically-modified *Salmonella typhimurium* in an orthotopic human pancreatic cancer in nude mice. Anticancer Res 29: 1873–1878.
14. Murphy EA, Shields DJ, Stoletov K, Dneprovskaia E, McElroy M, et al. (2010) Disruption of angiogenesis and tumor growth with an orally active drug that stabilizes the inactive state of PDGFRβ/B-RAF. Proc Natl Acad Sci USA 107: 4299–4304.
15. Hassanein MK, Suetsugu A, Saji S, Moriwaki H, Bouvet M, et al. (2011) Stem-like and non-stem human pancreatic cancer cells distinguished by morphology and metastatic behavior. J Cell Biochem 112: 3549–3554.
16. Hiroshima Y, Zhao M, Zhang Y, Maawy A, Hassanein MK, et al. (2013) Comparison of efficacy of *Salmonella typhimurium* A1-R and chemotherapy on stem-like and non-stem human pancreatic cancer cells. Cell Cycle 12: 2774–2780.
17. Tsuji K, Yang M, Jiang P, Maitra A, Kaushal S, et al. (2006) Common bile duct injection as a novel method for establishing RFP-expressing human pancreatic cancer in nude mice. J Pancreas 7: 193–199.
18. Jiang P, Yamauchi K, Yang M, Tsuji K, Xu M, et al. (2006) Tumor cells genetically labeled with GFP in the nucleus and RFP in the cytoplasm for imaging cellular dynamics. Cell Cycle 5: 1198–1201.
19. McElroy M, Kaushal S, Bouvet M, Hoffman RM. (2008) Color-coded imaging of splenocyte-pancreatic cancer cell interactions in the tumor microenvironment. Cell Cycle 7: 2916–2921.
20. Okabe M, Ikawa M, Kominami K, Nakanishi T, Nishimune Y. (1997) "Green mice" as a source of ubiquitous green cells. FEBS Lett 407: 313–319.
21. Yang M, Reynoso J, Jiang P, Li L, Moossa AR, et al. (2004) Transgenic nude mouse with ubiquitous green fluorescent protein expression as a host for human tumors. Cancer Res 64: 8651–8656.
22. Tran Cao HS, Kimura H, Kaushal S, Snyder CS, Reynoso J, et al. (2009) The cyan fluorescent protein (CFP) transgenic mouse as a model for imaging pancreatic exocrine cells. J Pancreas 10:152–156.
23. Yang M, Reynoso J, Bouvet M, Hoffman RM. (2009) A transgenic red fluorescent protein-expressing nude mouse for color-coded imaging of the tumor microenvironment. J Cell Biochem 106: 279–284.
24. Yamauchi K, Yang M, Jiang P, Xu M, Yamamoto N, et al. (2006) Development of real-time subcellular dynamic multicolor imaging of cancer-cell trafficking in live mice with a variable-magnification whole-mouse imaging system. Cancer Res 66: 4208–4214.
25. Yamauchi K, Tome Y, Yamamoto N, Hayashi K, Kimura H, et al. (2012) Color-coded real-time subcellular fluorescence imaging of the interaction between cancer and host cells in live mice. Anticancer Res 32: 39–43.
26. Uchugonova A, Hoffman RM, Weinigel M, Koenig K. (2011) Watching stem cells in the skin of living mice noninvasively. Cell Cycle 10: 2017–2020.
27. Fu X, Guadagni F, Hoffman RM. (1992) A metastatic nude-mouse model of human pancreatic cancer constructed orthotopically from histologically intact patient specimens. Proc Natl Acad Sci USA 89: 5645–5649.
28. Furukawa T, Kubota T, Watanabe M, Kitajima M, Hoffman RM. (1993) A novel "patient-like" treatment model of human pancreatic cancer constructed using orthotopic transplantation of histologically intact human tumor tissue in nude mice. Cancer Res 53: 3070–3072.
29. Bouvet M, Wang J-W, Nardin SR, Nassirpour R, Yang M, et al. (2002) Real-time optical imaging of primary tumor growth and multiple metastatic events in a pancreatic cancer orthotopic model. Cancer Res 62: 1534–1540.
30. Zhang W, Zhu XD, Sun HC, Xiong YQ, Zhuang PY, et al. (2010) Depletion of tumor-associated macrophages enhances the effect of sorafenib in metastatic liver cancer models by antimetastatic and antiangiogenic effects. Clin. Cancer Res 16: 3420–3430.
31. Baay M, Brouwer A, Pauwels P, Peeters M, Lardon F. (2011) Tumor cells and tumor-associated macrophages: secreted proteins as potential targets for therapy. Clin Dev Immunol 2011: 565187.
32. Karnevi E, Andersson R, Rosendahl AH. (2014) Tumour-educated macrophages display a mixed polarisation and enhance pancreatic cancer cell invasion. Immunol Cell Biol 2014, doi: 10.1038/icb.2014.22. [Epub ahead of print].
33. Rietkötter E, Menck K, Bleckmann A, Farhat K, Schaffrinski M, et al. (2013) Zoledronic acid inhibits macrophage/microglia-assisted breast cancer cell invasion. Oncotarget 4:1449–1460.

34. Rogers TL, Wind N, Hughes R, Nutter F, Brown HK, et al. (2013) Macrophages as potential targets for zoledronic acid outside the skeleton-evidence from in vitro and in vivo models. Cell Oncol (Dordr) 36:505–514.
35. Märten A, Lilienfeld-Toal MV, Büchler MW, Schmidt J. (2007) Zoledronic acid has direct antiproliferative and antimetastatic effect on pancreatic carcinoma cells and acts as an antigen for delta2 gamma/delta T cells. J Immunother 30:370–377.
36. Hoffman RM, Yang M. (2006) Subcellular imaging in the live mouse. Nature Protocols 1:775–782.
37. Rogers TL, Holen I. (2011) Tumour macrophages as potential targets of bisphosphonates. J Transl Med 9:177.

The Minor Wall-Networks between Monolignols and Interlinked-Phenolics Predominantly Affect Biomass Enzymatic Digestibility in *Miscanthus*

Zhengru Li[1]ↄ, Chunqiao Zhao[1]ↄ, Yi Zha[2], Can Wan[2], Shengli Si[2], Fei Liu[1], Rui Zhang[2], Fengcheng Li[1], Bin Yu[2], Zili Yi[3], Ning Xu[1,4], Liangcai Peng[1,2], Qing Li[5]*

1 National Key Laboratory of Crop Genetic Improvement, Biomass and Bioenergy Research Centre, College of Plant Science and Technology, Huazhong Agricultural University, Wuhan, China, **2** National Key Laboratory of Crop Genetic Improvement, Biomass and Bioenergy Research Centre, College of Life Science and Technology, Huazhong Agricultural University, Wuhan, China, **3** Department of Biotechnology, Hunan Agricultural University, Changsha, China, **4** Jiangsu Key Laboratory for Biomass-based Energy and Enzyme Technology, Huaiyin Normal University, Huaian, China, **5** National Key Laboratory of Crop Genetic Improvement, Biomass and Bioenergy Research Centre, College of Science, Huazhong Agricultural University, Wuhan, China

Abstract

Plant lignin is one of the major wall components that greatly contribute to biomass recalcitrance for biofuel production. In this study, total 79 representative *Miscanthus* germplasms were determined with wide biomass digestibility and diverse monolignol composition. Integrative analyses indicated that three major monolignols (S, G, H) and S/G ratio could account for lignin negative influence on biomass digestibility upon NaOH and H_2SO_4 pretreatments. Notably, the biomass enzymatic digestions were predominately affected by the non-KOH-extractable lignin and interlinked-phenolics, other than the KOH-extractable ones that cover 80% of total lignin. Furthermore, a positive correlation was found between the monolignols and phenolics at $p<0.05$ level in the non-KOH-extractable only, suggesting their tight association to form the minor wall-networks against cellulases accessibility. The results indicated that the non-KOH-extractable lignin-complex should be the target either for cost-effective biomass pretreatments or for relatively simply genetic modification of plant cell walls in *Miscanthus*.

Editor: Olga A. Zabotina, Iowa State University, United States of America

Funding: This work was supported in part by grants from 111 Project (B08032), the Open Project of Jiangsu Key Laboratory for Biomass-based Energy and Enzyme Technology (JSBEET1204), the 973 Specific Pre-project (2010CB134401) and the Fundamental Research Funds for the Central Universities (No2013PY061). The funders had no role in study design, data collection and analysis, decision to publish, or preparation of the manuscript.

Competing Interests: The authors have declared that no competing interests exist.

* Email: liqing@mail.hzau.edu.cn

ↄ These authors contributed equally to this work.

Introduction

Lignocellulose is the most abundant and sustainable biomass on the earth for biofuels and other chemical products [1–3]. The current biomass process mainly involves three major steps, namely, physical and chemical pretreatment for plant cell wall destruction, enzymatic hydrolysis for saccharification, as well as yeast fermentation for ethanol production [4]. However, biofuel production remains extremely costly due to biomass recalcitrance, and could result in secondary environmental pollution [5]. Generally, recalcitrance is determined by the wall polymer components and their various interactions [6–9]. For recalcitrance reduction, the genetic modification of plant cell walls is considered a promising solution in bioenergy crops and it has become essential in understanding the effects of wall polymer on biomass digestibility [10–12].

Plant cell walls are mainly composed of cellulose, hemicelluloses, and lignin with relatively small amounts of pectin and proteins [13,14]. In biomass hydrolysis, cellulose crystallinity is reportedly a negative parameter [15,16], whereas hemicelluloses are a positive and dominant factor [12]. Lignin is an amorphous polymer with phenylpropane units, which mostly comprise three monomers, namely, p-hydroxyphenyl (H), guaiacyl (G), and syringyl (S) [17]. These monomers are linked by ether-, ester- and C-C bonds that are irregularly repeated [18,19]. Lignin has been associated with biomass recalcitrance, and the biomass saccharification rate is strongly affected by the level and the monomer composition of lignin [16,20]. However, the phenolic acid-based interconnection among lignin and wall polysaccharides has not yet been fully understood [21]. Despite that the lignin level and the S/G ratio show the negative effects on biomass saccharification in transgenic switchgrass [22,23], the higher S/G ratio in natural populus was recently reported to exhibit a positive effect on sugar release while the negative effect of lignin level is less pronounced [20].

Miscanthus is a C4 perennial grass with an extremely high biomass yield, and is currently considered as the leading candidate for biofuel feedstocks. With its origin in East Asia, we collected over 1400 natural *Miscanthus* accessions with a rich and stable germplasm resource [12,24]. In this study, we initially selected 79 representative *Miscanthus* natural accessions with a diverse cell wall composition and structure [24,25]. An integrative analysis was performed among lignin level/composition, interlinked-phenolics and biomass digestibility of the selected *Miscanthus* samples. The mechanism was then interpreted about the lignin negative effect on biomass enzymatic saccharification under NaOH and H_2SO_4 pretreatments at three concentrations.

Results and Discussion

Diversity of lignin level and monomer composition in *Miscanthus*

Diverse lignin contents and monomer compositions were determined in 79 representative *Miscanthus* germplasm accessions (Fig. 1). The KOH-extractable and non-KOH-extractable lignin levels of *Miscanthus* accessions ranged from 831.47 μmol/g to 1582.40 μmol/g and from 46.35 μmol/g to 618.25 μmol/g, respectively (Table S1; Fig. 1A). In general, *Miscanthus* displayed a diverse lignin monomer composition. For example, an average of 48.63% G was much higher than the other two monomers, H at 24.16% and S at 27.21%. Notably, despite the non-KOH-extractable residue covered only 20% of total lignin (Table S1), it also showed a high variation in each monomer, i.e., H, 6.00 μmol/g to 119.18 μmol/g; G, 25.27 μmol/g to 284.39 μmol/g; and S, 15.08 μmol/g to 228.81 μmol/g (Fig. 1B). The diversity of lignin levels and monolignol compositions, particularly in the non-KOH-extractable form, suggested that lignin may play a distinct role in biomass enzymatic digestibility of the selected *Miscanthus* samples.

Variation of interlinked-phenolics in *Miscanthus*

Lignin interacts with other wall polymers through the interlinked-phenolics [26], and thus, we detected seven major ester- and ether-linked phenolic compounds in the *Miscanthus* (Fig. 2). The selected 79 *Miscanthus* samples exhibited the KOH-extractable phenolics that ranged from 22.47 μmol/g to 204.35 μmol/g, whereas the non-KOH-extractable phenolics ranged from 4.85 μmol/g to 39.82 μmol/g (Fig. 2A). Likewise, the non-KOH-extractable biomass residue also contained only 20% of total interlinked-phenolics. Among the seven interlinked-phenolics that were examined, *Miscanthus* samples showed a relatively high variation in syringaldehyde (S-), vanillin (G-), acetosyringone (AS-), and ferulic acid (FA-) (Fig. 2B). Hence, the selected *Miscanthus* accessions indicated diversity in monolignol and interlinked phenolic composition.

Lignin effect on biomass saccharification

Considering the selected *Miscanthus* samples that exhibited diversity in the monolignol and interlinked phenolic compositions, their biomass enzymatic digestions were detected after various chemical pretreatments. The biomass digestibility (saccharification) was defined by Huang et al. [24] by accounting for the hexoses yield (hexoses/cellulose) released from the enzymatic hydrolysis of a crude cellulase mixture of lignocellulose upon pretreatment. After pretreated with NaOH (0.5%, 1%, and 4%) and H_2SO_4 (0.25%, 1%, and 4%), the selected *Miscanthus* accessions exhibited a high variation in biomass digestibility, as hexoses yield. As a result, a few *Miscanthus* accessions could be considered as the potential bioenergy crops.

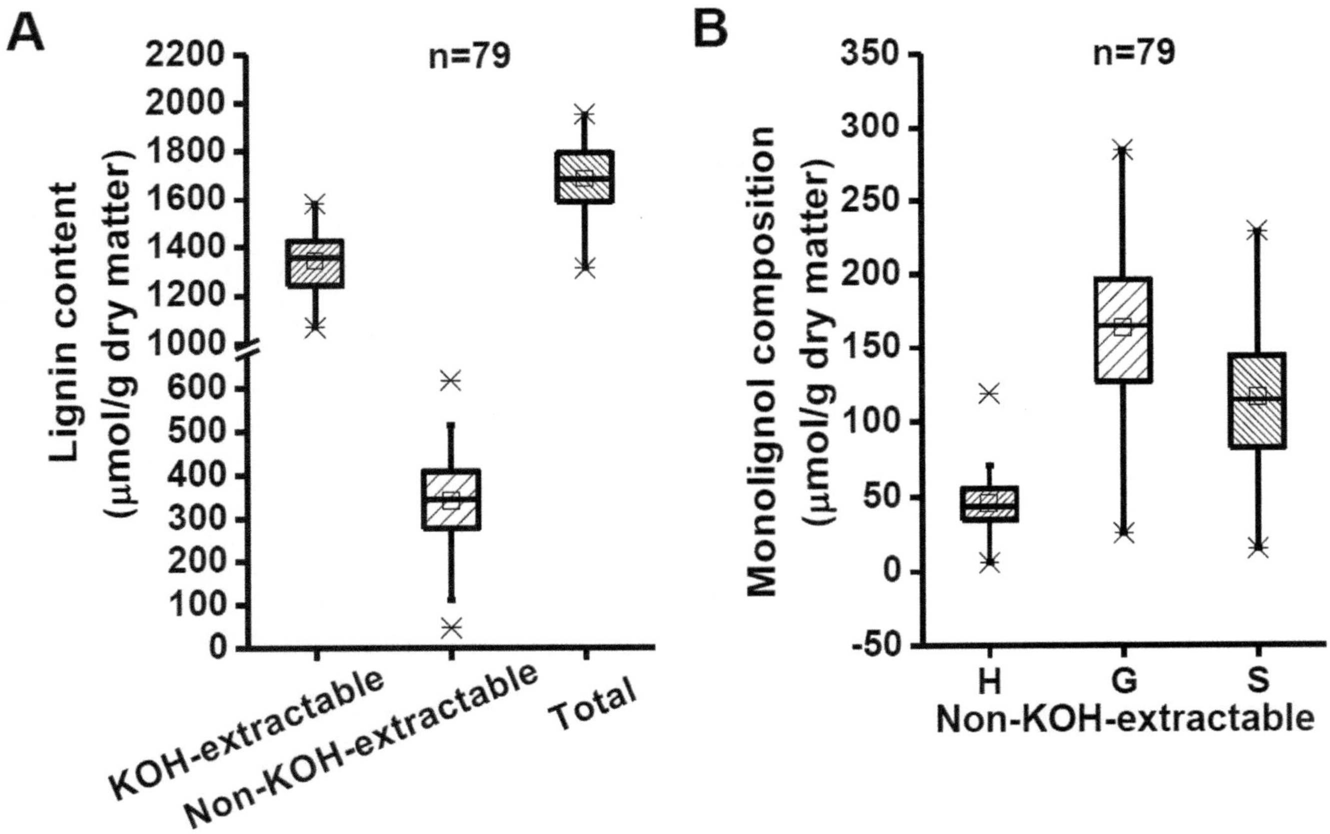

Figure 1. Variation of lignin content and monolignol composition in *Miscanthus* (n = 79). (A) Lignin content; (B) Monolignol composition.

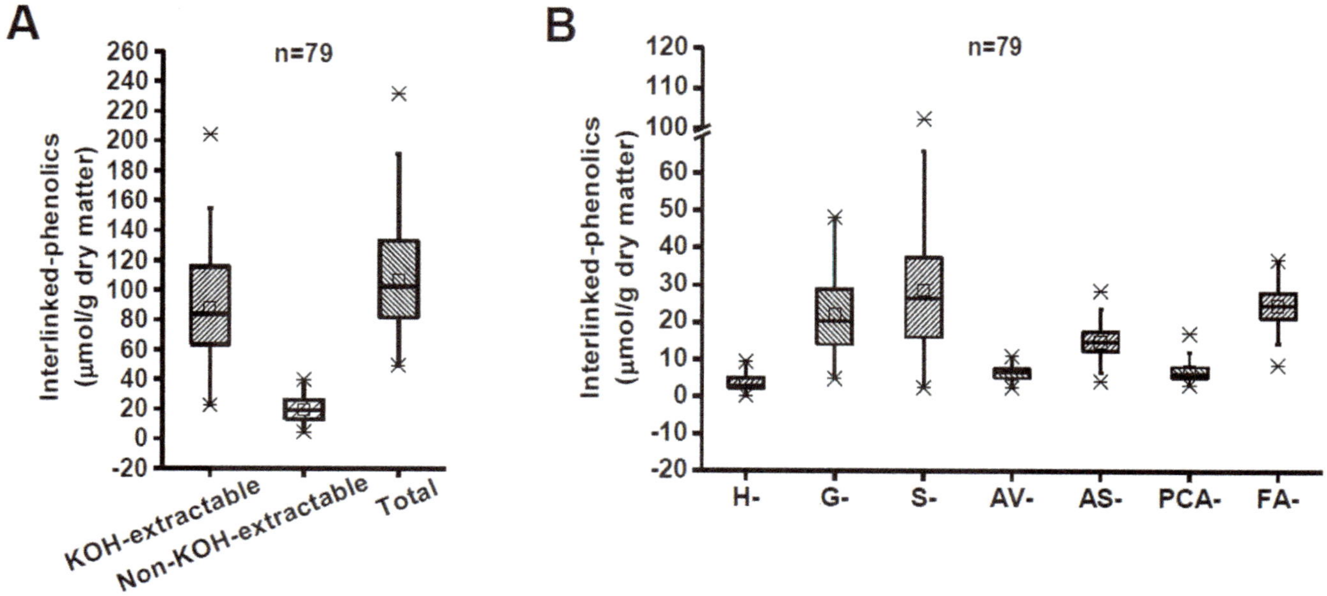

Figure 2. Variation of interlinked-phenolics in *Miscanthus* (n = 79). (A) Interlinked-phenolics in the KOH-extractable and non-KOH-extractable residues; (B) Total interlinked-phenolic compositions. H-: *p*-Hydroxybenzaldehyde, G-: Vanillin, S-: Syringaldehyde, AV-: Acetovanillone, AS-: Acetosyringone, PCA-: *p*-Coumaric acid, FA-: Ferulic acid, SA-: Sinapic acid.

With regard to the selected *Miscanthus* samples, a correlation analysis was performed among lignin, interlinked-phenolics, and biomass saccharification after pretreatments with NaOH and H_2SO_4 at three concentrations (Figs. 3 and S1). Among the 79 *Miscanthus* accessions, the total lignin level and interlinked-phenolics negatively affected the hexoses yield after the pretreatment with various concentrations of NaOH (0.5%, 1%) and H_2SO_4 (0.25%, 1%) (Figs. 3 and S1A). Only lignin exhibited a negative effect after a 4% NaOH or 4% H_2SO_4 pretreatment (Fig. S1B). Despite the relatively low correlation (R^2) values, the correlation coefficients reached significant levels at $p<0.01$ or $p<0.05$ (n = 79). Hence, lignin could exhibit the negative influence on biomass digestibility in *Miscanthus*. Although previous study has reported that the lignin level negatively affect biomass digestibility in *Miscanthus* [12], the characterization of diverse *Miscanthus* germplasm accessions in the current study could account for the mechanism of the negative effects of lignin on biomass saccharification as described below.

Monolignol and interlinked-phenolics impact on biomass digestibility

To gain insights on the negative effects of lignin on biomass digestions, we conducted a correlation analysis between three major monolignols and interlinked-phenolics in the KOH-extractable and non-KOH-extractable biomass residues of *Miscanthus* samples (Figs. 4, S2 and S3). In the KOH-extractable residue of *Miscanthus* accessions, S and S/G ratio negatively affected the biomass digestibility at $p<0.01$ or 0.05 levels (n = 79); however, the total monomers (H+G+S) produced no significant effect (Figs. 4A, S2A, and S3A). By contrast, S, S/G and the total monomers in the non-KOH-extractable exhibited a significantly negative effect on the hexoses yields after pretreated with various concentrations of NaOH (0.5% and 1%) and H_2SO_4 (0.25%, 1%, and 4%) (Figs. 4A, S2B, and S3B). In addition, despite H monomer in the non-KOH-extractable residue also exhibited high negative coefficient values, the S/H and H/G ratios did not show any significant correlations ($p>0.05$). Thus, the results indicated that the S level or S/G ratio (other than S/H and H/G ratios) in the non-KOH-extractable, and not in the KOH-extractable residue, could account for the negative effect of lignin on biomass digestibility in *Miscanthus*. Notably, the seven major interlinked-phenolics in the non-KOH-extractable residue also exhibited a significantly negative effect ($p<0.01$ or $p<0.05$) on biomass enzymatic digestibility after pretreated with three concentrations of NaOH and H_2SO_4, except for FA- (Figs. 4B, S2D, and S3D). In addition, compared with the KOH-extractable residues, the non-KOH-extractable residues contain much more S-monolignol and interlinked-S and AS compounds with less Ara and FA, suggesting that the non-KOH-extractable residues should present the S and AS-rich wall polymer networks.

Nevertheless, the most interlinked-phenolics in the KOH-extractable residue were not correlated with biomass saccharification (Figs. 4B, S2C, and S3C). Hence, despite that the non-KOH-extractable biomass residue contained only 20% of the total lignin in *Miscanthus* samples (Table S1), their monolignols and interlinked-phenolics could predominately determine the negative effect of total lignin on biomass digestibility under various pretreatments in *Miscanthus* accessions. It was consistent with the previous report that the non-KOH-extractable biomass residue was significantly correlated with the lignocellulose residues released from various physical and chemical pretreatments in terms of their biomass enzymatic digestibility [25]. The results also confirmed that biomass enzymatic saccharification could be fundamentally determined by plant cell wall structures [25,27].

To our knowledge, it was first time to report about the non-KOH-extractable lignin distinct effect on biomass enzymatic saccharification in plants. This finding indicated that the non-KOH-extractable lignin should be the target either for genetic modification of plant cell walls or for chemical and physical pretreatments of biomass. Notably, although three monomers (S, G, H) of the non-KOH-extractable lignin were all negatively correlated with biomass digestibility, only S/G ratio, other than S/H or H/G, displayed a significant effect (Fig. 4A). This result suggested that reducing S proportion in the non-KOH-extractable

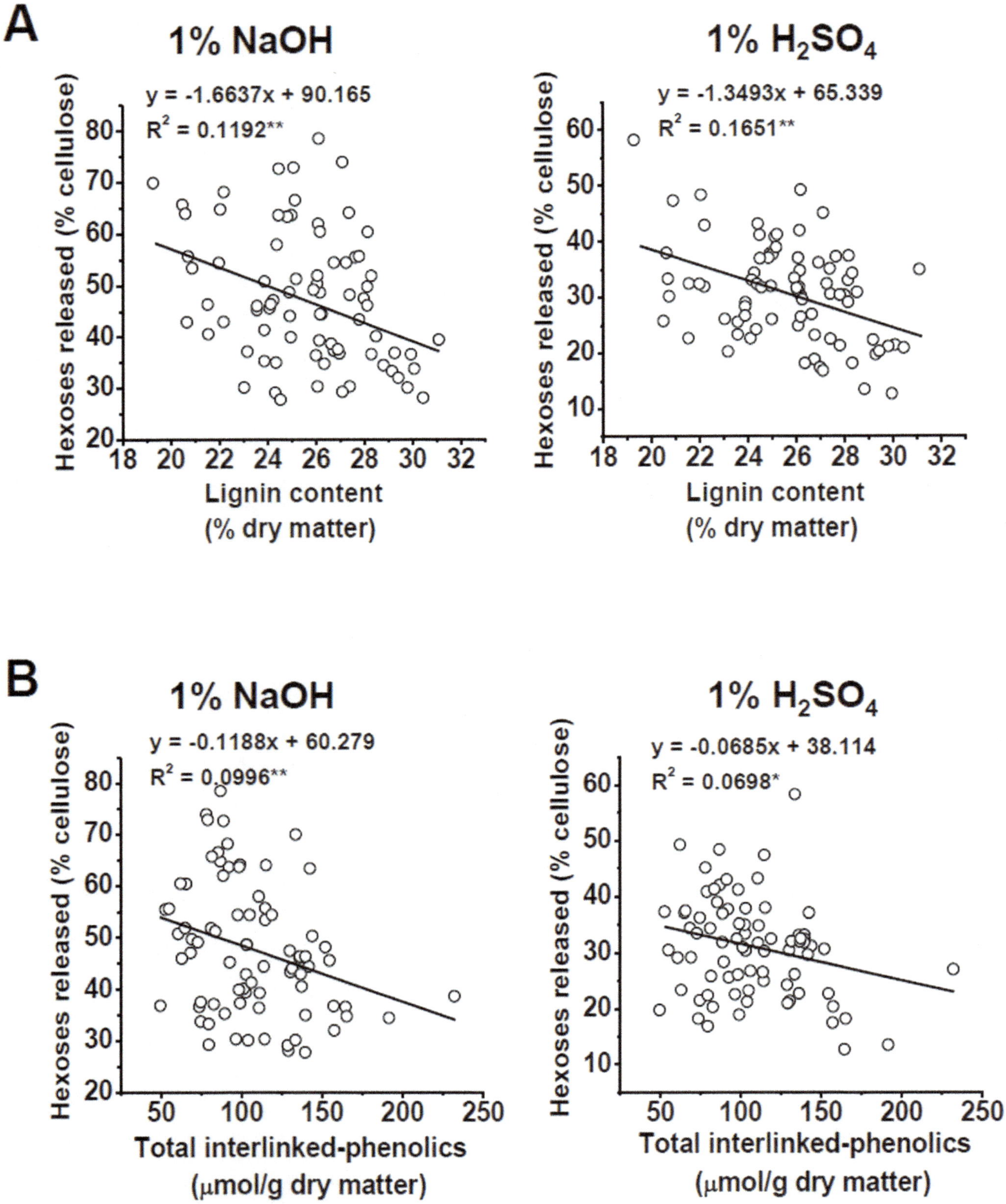

Figure 3. Correlation analysis among lignin, interlinked-phenolics and biomass saccharification in *Miscanthus*. (A) Correlation between lignin content and hexoses yield; (B) Correlation between total interlinked-phenolics and hexoses yield. * and ** Indicated the significant correlation coefficient values at $p<0.05$ and 0.01 (n = 79), respectively.

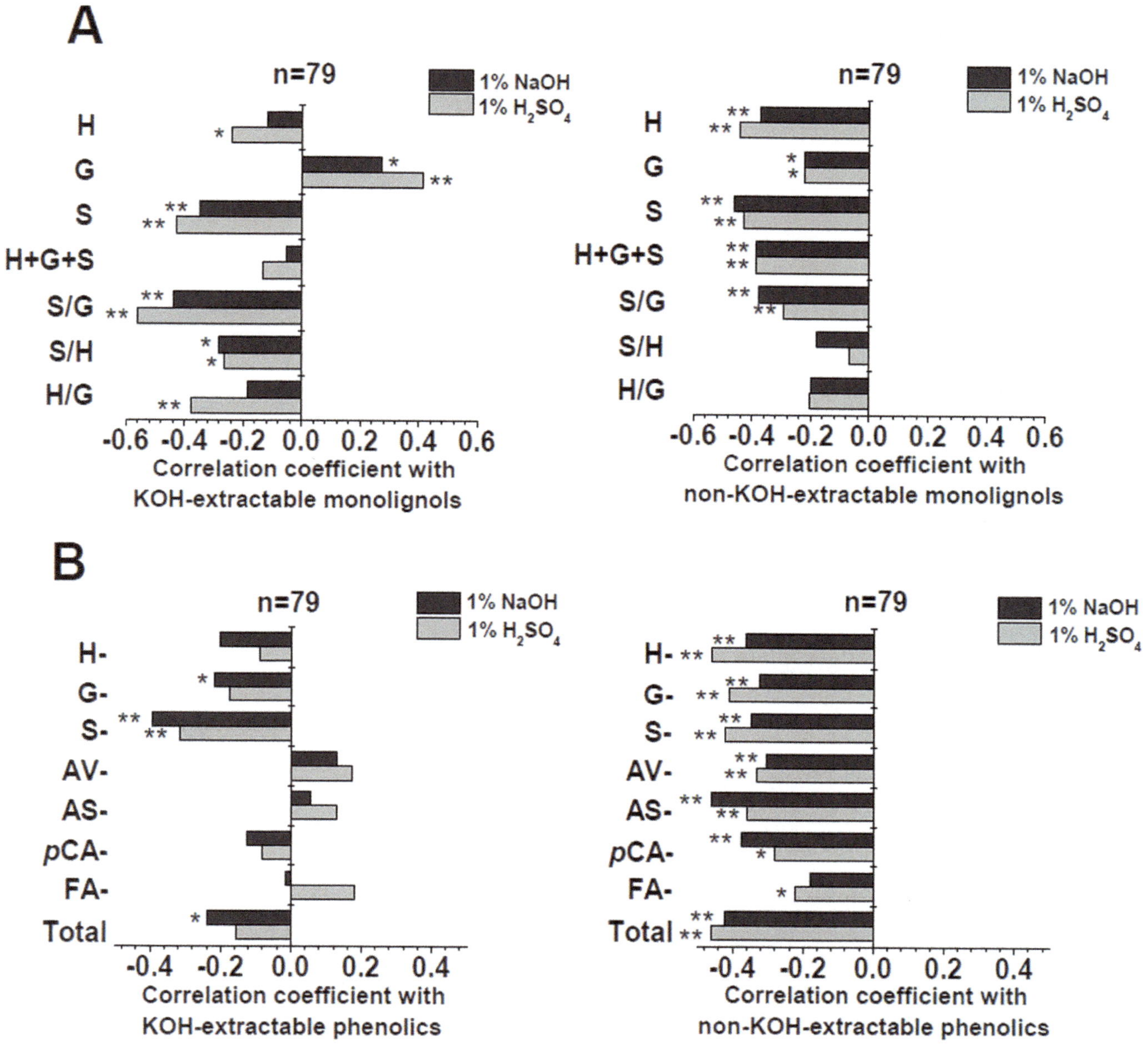

Figure 4. Correlation analysis among monolignols, seven interlinked-phenolics and biomass saccharification in *Miscanthus*. (A) The KOH-extractable and non-KOH-extractable monolignols; (B) The KOH-extractable and non-KOH-extractable interlinked-phenolics. * and ** Indicated the significant correlation coefficient values at $p<0.05$ and 0.01 (n = 79), respectively.

lignin should be the priority for lignin modification in *Miscanthus*. However, three monomers of lignin and their ratios have been determined with dual effects on biomass enzymatic hydrolysis in rice, wheat and sweet sorghum [16,28], suggesting that the specific lignin modification may depend on the plant species.

Mechanism on lignin negative effect on lignocellulose enzymatic hydrolysis

To further understand lignin negative impact on biomass digestibility, we performed a correlation analysis between three monolignols and interlinked-phenolics in the two types of biomass residues among total 79 *Miscanthus* accessions (Figs. 5 and S4). In general, no significant correlation was observed among the total monolignols and interlinked-phenolics in *Miscanthus* (Fig. 5A). However, *Miscanthus* accessions exhibited a positive correlation at $p<0.05$ level in the non-KOH- extractable, rather than in the KOH-extractable residue (Figs. 5B and 5C). Notably, seven major ester- and ether-interlinked-phenolics were positively correlated with the total lignin only in the non-KOH-extractable residues of *Miscanthus* at $p<0.01$ or 0.05 levels (Fig. S4B). Despite that the ester-interlinked-phenolics could be effectively extracted with alkali chemicals, they remained at minor levels in the non-KOH-extractable residues from 22.4% KOH extraction (data not shown), suggesting that the ester-interlinked-phenolics should also play a role in wall polymer network formation. Because the wall-networks formed by ester- and ether-interlinked-phenolics among lignin and other wall polymers may act as the barrier hindering the enzyme penetration into the cellulose surface [29], the significantly positive correlation between the non-KOH-extractable monolignols and interlinked-phenolics suggested their association to establish wall-networks against enzymatic accessibility. It also indicated that the pretreatments (0.5%, 1%, and 4% NaOH; 0.25%, 1%, and 4% H_2SO_4) used in this study could not remove the non-KOH-extractable lignin and interlinked-phenolics, due to the KOH extraction concentration at 22.4%. Nevertheless, while 8% NaOH or 8% H_2SO_4 was even used as one-step pretreatment, we found that the biomass enzymatic digestibility was not much increased in the *Miscanthus* samples examined in this study. It suggested that two-step pretreatments with alkali and acid or other integrated approaches may be attempted in order for removal of the non-KOH-extractable lignin and interlinked-phenolics.

In addition, despite the KOH-extractable residue contained almost 80% of total lignin (Fig. 1), the non-significant correlation between the KOH-extractable lignin and interlinked-phenolics suggested that most lignin could not well establish the linkages with the phenolics to form the complete wall-networks in *Miscanthus*. Hence, although the S/G ratio in the KOH-extractable residue displayed a significantly negative correlation with the lignocellulose enzymatic hydrolysis (Fig. 4A), it should not determine the lignin negative impact on biomass digestibility in *Miscanthus*. In other words, since the non-KOH-extractable biomass residues maintain 20% total hemicelluloses in *Miscanthus* [25], we assumed that the non-KOH-extractable hemicelluloses may strongly interact with lignin via interlinked phenolics. Hence, the minor non-KOH-extractable biomass residue should predominately affect lignocellulose enzymatic hydrolysis by its monolignols and interlinked-phenolics forming the tight wall-networks with hemicelluloses against cellulase accession in *Miscanthus*.

Potential cell wall modification for high biomass digestibility

Based on the previous analysis, one pair of representative samples were selected from the *Miscanthus* accessions (Figs. 6 and S5). The *Miscanthus* accession (Msi75) with a relatively higher biomass digestibility exhibited lower lignin level, interlinked phenolic composition, and S/G ratio compared with those of the paired sample (Mlu06) (Fig. 6). The results confirmed that lignin and interlinked-phenolics exhibited the negative effects on biomass digestibility in *Miscanthus*. Furthermore, *Miscanthus* accession (Msi75) exhibited an increased biomass enzymatic digestibility at a similar rate compared with the paired sample (Mlu06) after pretreatment with three concentrations of NaOH and H_2SO_4 (Fig. S5A), consistent with the previous findings about both hexoses yields (% cellulose) and total sugars yields in *Miscanthus* [25]. Using scanning electron microscopy, we also observed rougher biomass surface on the Msi75 sample compared with its paired sample (Mlu06) after NaOH and H_2SO_4 pretreatments and sequential enzymatic hydrolysis (Fig. S5B). The rough surface samples may enhance the cellulase compatibility and accessibility for high enzymatic activity as discussed by Xu et al (2012).

Miscanthus is a typical C4 plant with enormous biomass residues for biofuels. However, as plant biomass contains many different cell types with diverse wall components, it remains hard to identify desirable cell walls for high biomass digestibility [11]. Due to diversity of cell wall biological functions, any genetic modification of plant cell walls could consequently cause plant growth defect and mechanic strength reduction. In this work, however, we have identified several natural *Miscanthus* accessions with normal plant growth and high biomass digestibility from large collections of *Miscanthus* germplasms such as the Msi75 sample. This is because *Miscanthus* with self-incompatibility at flowering, can evolve into different species with a diverse germplasm that is suitable for plant growth under various environmental conditions [11,21,30]. Thus, the selected *Miscanthus* natural germplasm accessions should be directly used as desire bioenergy crops for biofuel purposes.

Notably, as the non-KOH-extractable residue only covers about 20% of total lignin, its genetic modification should cause relatively less defects on plant growth and development than that of the KOH-extractable one, suggesting that reducing of the non-KOH-extractable lignin complex could be the relatively simply way for genetic modification of plant cell walls in bioenergy crops. More importantly, as all *Miscanthus* accessions exhibited extremely lower interlinked-phenolics than that of lignin in the non-KOH-extractable residues (Figs. 1A and 2A), removal of the non-KOH-extractable interlinked-phenolics should be a cost-effective pretreatment approach for high biomass enzymatic digestibility.

On the other hand, replacing the non-KOH-extractable interlinked-phenolics with other soluble substances such as proteins, may be the way for satisfying with the needs of both plant normal growth under various environmental stresses, and high biomass enzymatic digestibility upon different pretreatments in *Miscanthus*.

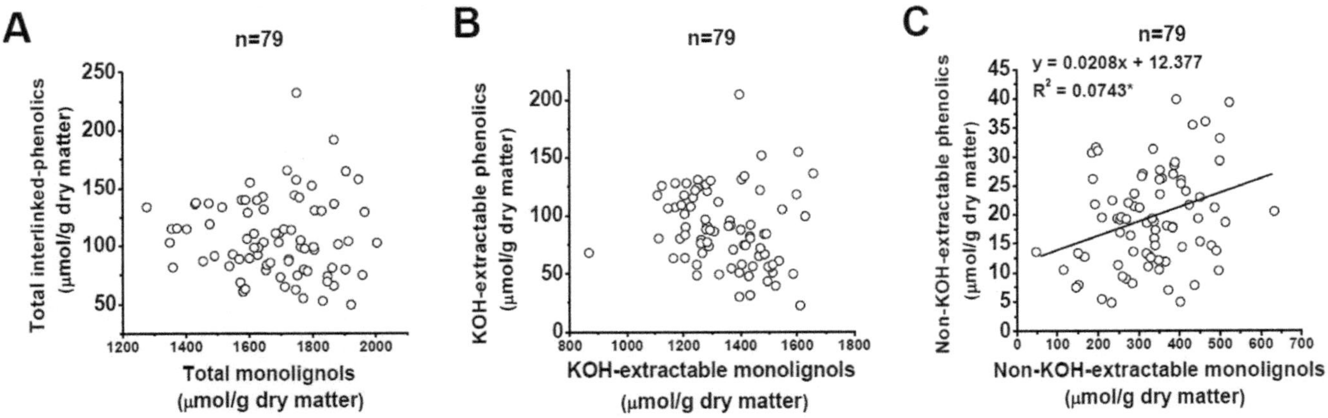

Figure 5. Correlation analysis between monolignols and interlinked-phenolics in *Miscanthus*. (A) Total monolignols and interlinked-phenolics; (B) The KOH-extractable monolignols and interlinked-phenolics; (C) The non-KOH-extractable monolignols and interlinked-phenolics. * Indicated the significant correlation coefficient value at $p<0.05$, (n = 79).

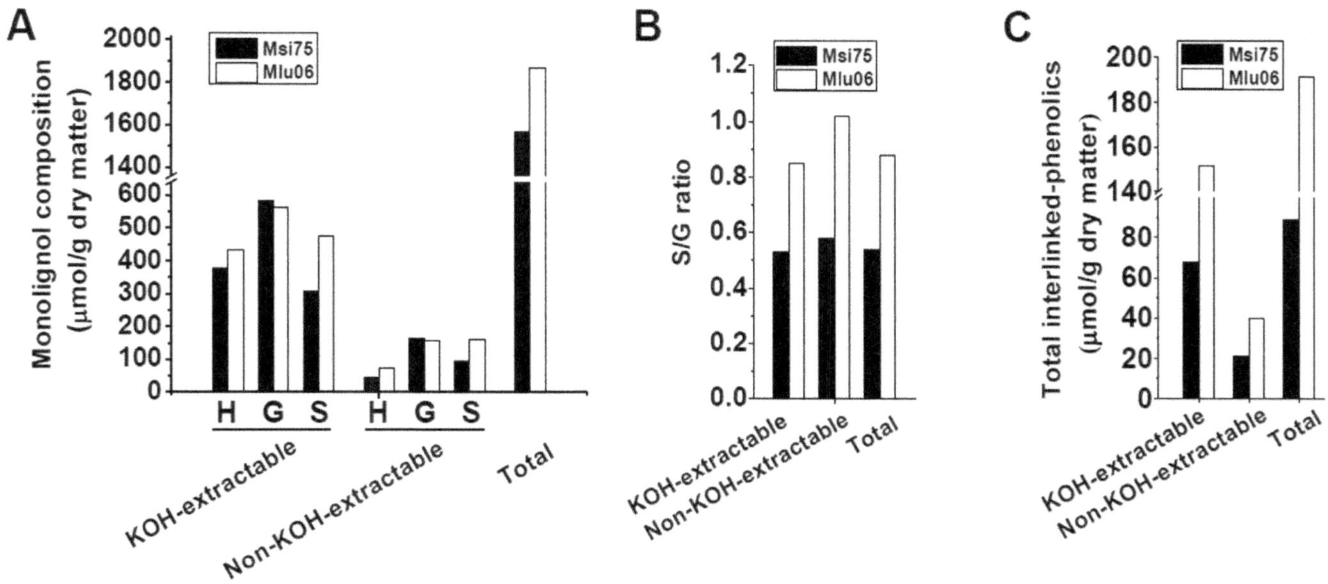

Figure 6. Analysis of monolignol and interlinked-phenolics compositions in the representative *Miscanthus* samples. (A) Monolignol composition; (B) S/G ratio; (C) Total interlinked-phenolics.

Conclusions

Total 79 representative *Miscanthus* samples have been determined with a variation of lignocellulose enzymatic digestibility and a diversity of three monolignols (S, G, H) and seven interlinked-phenolics compositions. Correlative analysis has indicated that either monolignols (S/G ratio) or interlinked-phenolics could negatively affect biomass enzymatic digestibility under NaOH and H_2SO_4 pretreatments. Integrative analysis has also suggested that the minor non-KOH-extractable monolignols and interlinked-phenolics should have a coordination to form the wall-network that could predominately determine lignin negative influence on lignocellulose saccharification. The findings could provide the potential approaches for simply genetic modification of plant cell walls in *Miscanthus*, as well as for cost-effective physical and chemical pretreatments toward high biomass digestibility.

Materials and Methods

Plant materials

Miscanthus samples were selected from wild *Miscanthus* germplasm resources collected nation-wide in China. The mature stem tissues of *Miscanthus* samples were supplied by the laboratory of Dr. Zili Yi at Hunan Agricultural University, Changsha, China[12,24]. The collected samples of mature stem tissues were inactivated at 105 °C for 10 min and dried at 50 °C until with constant weight. The dried tissues were ground through a 40 mesh screen, and stored in a dry container until use. All samples were collected from 5–10 individual mature stem tissues and the ground powders were mixed well prior to use for cell wall composition analysis, biomass pretreatments and enzymatic hydrolysis. No specific permission was required for the field study, which was carried out in the specific experimental field for energy plants. In addition, *Miscanthus* is not listed as the endangered or protected species in China.

Plant cell wall fractionation

The procedures of plant cell wall fractionation were described by Peng et al. [31] with minor modification by Li et al. [25]. The extracted crude cell wall residues were extracted with 0.5% (w/v) ammonium oxalate for 1 h in a boiling water bath. The remaining pellet was extracted with 4 M KOH containing 1.0 mg mL^{-1} sodium borohydride for 1 h at 25 °C, and the combined supernatant was neutralized, dialyzed and lyophilized as KOH-extractable fractionations. The remaining pellet contained the non-KOH-extractable fractionations. All samples were carried out in biological triplicate for wall fractionations.

Colorimetric assay for total hexoses and pentoses

UV/VIS Spectrometer (Shanghai MAPADA Instruments Co., Ltd. V-1100D) was applied for total hexoses and pentoses assay. The anthrone/H_2SO_4 method was used for total hexoses assay [32]. The orcinol/HCl method was applied for total pentoses assay [33].

Total lignin determination

Total lignin determination of crude cell wall residues or non-KOH-extractable residues was performed by two-step acid hydrolysis method according to Laboratory Analytical Procedure of the National Renewable Energy Laboratory. The details of the two-type of lignin assay were described by Xu et al. [12].

Three monolignols (S, G, H) determination by HPLC

The methods were described by Xu et al. [12] with minor modification. Standard chemicals: *p*-Hydroxybenzaldehyde(H), vanillin(G) and syringaldehyde (S) (Sinopharm Chemical Reagent Co., Ltd.) were used for three monolignol proportion determination by HPLC. Three monolignol levels were respectively calculated based on their percentages of total (%) detected by HPLC and total lignin levels determined by the two-step acid hydrolysis method described above. The lignin level of the KOH-extractable or the non-KOH-extractable residues was subjective to the sum up of three monolignols.

The crude cell wall residues or the non-KOH-extractable residues were extracted with benzene-ethanol (2:1, v/v) in a Soxhlet for 4 h, and the remaining pellet (0.05 g) was added with 5 mL 2 M NaOH and 0.5 mL nitrobenzene in the Teflon gasket with a stainless steel bomb. The bomb was sealed tightly and heated at 170 °C (oil bath) for 3.5 h and stirred at 20 rpm. The chromatographic internal standard (ethyl vanillin) was added to the oxidation mixture. The alkaline oxidation mixture was washed 3 times with 30 mL CH_2Cl_2/ethyl acetate mixture (1/1, v/v) to remove nitrobenzene and its reduction by-products. The alkaline solution was acidified to pH 3.0–4.0 with 6 M HCl, and extracted with CH_2CI_2/ethyl acetate (3×30 mL) to obtain the lignin oxidation products in the organic phase. The organic extracts were evaporated to dryness under reduced pressure 40 °C, and the oxidation products were dissolved in 10 mL chromatographic pure methanol. Hence, three monolignols of the KOH-extractable residue were calculated based on the subtraction between the crude cell wall residues and the non-KOH-extractable residues.

HPLC analysis: 20 µL solution was injected into HPLC (Waters 1525 HPLC) column Kromat Universil C18 (4.6 mm×250 mm, 5 µm) operating at 28 °C with CH_3OH:H_2O:HAc (25:74:1, v/v/v) carrier liquid (flow rate: 1.1 mL/min). Calibration curves of all analytes routinely yielded correlation coefficients 0.999 or better, and the detection of the compounds was carried out with a UV-detector at 280 nm.

Interlinked-phenolics determination by HPLC

The protocols were described by Xu et al. [12] with minor modification. Standard chemicals included *p*-Hydroxybenzaldehyde(H), vanillin(G) and syringaldehyde (S) (from Sinopharm Chemical Reagent Co., Ltd.), Acetovanillone (AV) and acetosyringone (AS) (from Biosharp Co., Ltd.), *trans-p*-CA (*p*CA) and *trans*-FA (FA) (purchased from Sigma-Aldrich Co. LLC.).

The crude cell wall residues (0.05 g) were added with 10.0 mL 4 M NaOH containing 1.0 mg/mL $NaHSO_3$ and stirred at 170 °C for 2 h in a 25 mL Teflon gasket sealed in a stainless steel bomb at 20 rpm. After transferred to a triangular flask, the lysate was acidified to pH 2.0 with 6 M HCl, the acidified solution was extracted with chloroform (3×10.0 mL), and the combined organic extracts were evaporated to dryness under reduced pressure at 40 °C. The extracts were re-dissolved with 2.0 mL elution phase, filtered by 0.22 µm membrane for total phenolics analysis.

Chemical pretreatment

The procedures were described by Huang et al. [24] and Zhang et al. [34] with minor modification. H_2SO_4 pretreatment: The biomass samples (0.5 g) were added with 10 mL H_2SO_4 at three concentrations (0.25%, 1%, 4%, v/v), respectively. The tube was sealed and heated at 121 °C for 20 min in autoclave (15 psi). The sample was shaken at 150 rpm for 2 h at 50 °C, and centrifuged at 3,000 *g* for 5 min. The remaining pellet was washed three times with 10 mL distilled water for enzymatic hydrolysis. The samples added with 10 mL distilled water were shaken for 2 h at 50 °C as the control, and all samples were carried out in biological triplicate.

NaOH pretreatment: The biomass sample (0.5 g) was treated with 10 mL NaOH at three concentrations (0.5%, 1%, 4%, w/v). The sample was shaken at 150 rpm for 2 h at 50 °C, and centrifuged at 3,000 *g* for 5 min. The pellet was washed three times with 10 mL distilled water for enzymatic hydrolysis. The samples with 10 mL distilled water were shaken for 2 h at 50 °C as the control, and all samples were carried out in biological triplicate.

Enzymatic hydrolysis

The methods were described by Huang et al. [24] and Li et al. [25] with minor modification. The remaining residues from pretreatments were washed 2 times with 10 mL distilled water, and once with 10 mL mixed-cellulases reaction buffer (0.2 M acetic acid-sodium acetate, pH 4.8). The residue sample was added with 10 mL (2 g/L) mixed-cellulases (containing β-glucanase≥6×10^4 U) and cellulase≥600 U and xylanase≥10×10^4 U from Imperial Jade Bio-technology Co., Ltd). During the enzymatic hydrolysis, the samples were shaken under 150 rpm at 50 °C for 48 h. After centrifugation at 3,000 *g* for 10 min, the supernatants were obtained for determining amounts of pentoses and hexoses released from enzymatic hydrolysis. The samples with 10 mL reaction buffer were shaken for 48 h at 50 °C as the control, and all samples were carried out in biological triplicate.

SEM observation

The observation was described by Li et al. [25]. The biomass powder residues were used for SEM observation after NaOH or H_2SO_4 pretreatment and enzymatic hydrolysis. The samples were washed with distill water, dried under air, and sputter-coated with gold in a JFC-1600 ion sputter (Mito City, Japan). The surface morphology of the treated samples was sputter-coated with gold and viewed by SEM (JSM-6390/LV, Hitachi, Tokyo, Japan).

Statistical analysis

Superior Performance Software Systems software package (SPSS 17.0, Inc., Chicago, IL) was applied for the related statistical analyses. Correlation analysis was performed using Spearman's rank correlation analysis at the two-sided 0.05 level of significance (*p <0.05, **p<0.01). The variation and regression analysis are developed using Origin 8.0 software (Microcal Software, Northampton, MA) for the best fit curve from the experimental data.

Supporting Information

Figure S1 Correlation of lignin and interlinked-phenolics with hexoses yield released from enzymatic hydrolysis. (A) Pretreated with 0.5% NaOH and 0.25% H_2SO_4; (B) Pretreated with 4% NaOH and 4% H_2SO_4 in *Miscanthus* accessions (n = 79). * and ** Indicated the significant correlation coefficient values at p<0.05 and 0.01, respectively.

Figure S2 Correlation of monolignols and interlinked-phenolics with hexoses yield released from enzymatic hydrolysis under various pretreatments. (A) KOH-extractable monolignols; (B) Non-KOH-extractable monolignols; (C) KOH-extractable phenolics; (D) Non-KOH-extractable phenolics. * and ** Indicated the significant correlation coefficient values at p<0.05 and 0.01, respectively (n = 79).

Figure S3 Correlation of monolignols and interlinked-phenolics with hexoses yield released from enzymatic hydrolysis under various pretreatments. (A) KOH-extractable monolignols; (B) Non-KOH-extractable monolignols; (C) KOH-extractable phenolics; (D) Non-KOH-extractable phenolics. * and ** Indicated the significant correlation coefficient values at p<0.05 and 0.01, respectively (n = 79).

Figure S4 Correlation between seven interlinked-phenolics and lignin in *Miscanthus*. (A) KOH-extractable lignin; (B) Non-KOH-extractable lignin. * and ** Indicated the significant correlation coefficient values at p<0.05 and 0.01, respectively (n = 79).

Figure S5 Biomass enzymatic digestibility and scanning electron microscopic observation in representative *Miscanthus* accessions. (A) Hexoses yields (% cellulose) released from enzymatic hydrolysis after pretreatments of NaOH and H_2SO_4 at three concentrations as means±SD (n = 3); (B) SEM imagines of the biomass residues obtained from pretreatments of 1% NaOH and 1% H_2SO_4 and sequential enzyme hydrolysis, Allow indicated a coarse face.

Author Contributions

Conceived and designed the experiments: LP QL. Performed the experiments: ZL CZ. Analyzed the data: YZ QL. Contributed reagents/materials/analysis tools: SS CW F.Liu F.Li BY ZY NX. Contributed to the writing of the manuscript: LP. Completed the microscopic observation: RZ.

References

1. Ragauskas AJ, Williams CK, Davison BH, Britovsek G, Cairney J, et al. (2006) The path forward for biofuels and biomaterials. Science 311: 484–489.
2. Pauly M, Keegstra K (2008) Cell-wall carbohydrates and their modification as a resource for biofuels. The Plant Journal 54: 559–568.
3. Chen P, Peng L (2013) The diversity of lignocellulosic biomass resources and their evaluations for use as biofuels and chemicals. In: Sun JZ, Ding SY, Peterson JD, editors. Biological Conversion of Biomass for Fuels and Chemicals: Explorations from Natural Biomass Utilization Systems. Oxfordshire: Royal Society of Chemistry. pp.83–109.
4. Saha BC, Cotta MA (2007) Enzymatic hydrolysis and fermentation of lime pretreated wheat straw to ethanol. Journal of Chemical Technology & Biotechnology 82: 913–919.
5. Himmel ME, Ding SY, Johnson DK, Adney WS, Nimlos MR, et al. (2007) Biomass recalcitrance: engineering plants and enzymes for biofuels production. Science 315: 804–807.
6. Himmel ME, Ruth MF, Wyman CE (1999) Cellulase for commodity products from cellulosic biomass. Curr Opin Biotechnol 10: 358–364.
7. Sun Y, Cheng J (2002) Hydrolysis of lignocellulosic materials for ethanol production: a review. Bioresour Technol 83: 1–11.
8. Cosgrove DJ (2005) Growth of the plant cell wall. Nat Rev Mol Cell Biol 6: 850–861.
9. Somerville C, Youngs H, Taylor C, Davis SC, Long SP (2010) Feedstocks for lignocellulosic biofuels. Science 329: 790–792.
10. Chen F, Dixon RA (2007) Lignin modification improves fermentable sugar yields for biofuel production. Nat Biotechnol 25: 759–761.
11. Xie G, Peng L (2011) Genetic engineering of energy crops: a strategy for biofuel production in China. J Integr Plant Biol 53: 143–150.
12. Xu N, Zhang W, Ren S, Liu F, Zhao C, et al. (2012) Hemicelluloses negatively affect lignocellulose crystallinity for high biomass digestibility under NaOH and H_2SO_4 pretreatments in *Miscanthus*. Biotechnol Biofuels 5: 58.
13. Keegstra K (2010) Plant cell walls. Plant Physiol 154: 483–486.
14. Somerville C, Bauer S, Brininstool G, Facette M, Hamann T, et al. (2004) Toward a Systems Approach to Understanding Plant Cell Walls. Science 306: 2206–2211.
15. Chang V, Holtzapple M (2000) Fundamental factors affecting biomass enzymatic reactivity. Applied Biochemistry and Biotechnology 84–86: 5–37.
16. Wu Z, Zhang M, Wang L, Tu Y, Zhang J, et al. (2013) Biomass digestibility is predominantly affected by three factors of wall polymer features distinctive in wheat accessions and rice mutants. Biotechnol Biofuels 6: 183.
17. Sun H, Li Y, Feng S, Zou W, Guo K, et al. (2013) Analysis of five rice 4-coumarate:coenzyme A ligase enzyme activity and stress response for potential roles in lignin and flavonoid biosynthesis in rice. Biochem Biophys Res Commun 430: 1151–1156.
18. Ralph J, Lundquist K, Brunow G, Lu F, Kim H, et al. (2004) Lignins: Natural polymers from oxidative coupling of 4-hydroxyphenyl- propanoids. Phytochemistry Reviews 3: 29–60.
19. Iiyama K, Lam T, Stone BA (1994) Covalent Cross-Links in the Cell Wall. Plant Physiol 104: 315–320.
20. Studer MH, DeMartini JD, Davis MF, Sykes RW, Davison B, et al. (2011) Lignin content in natural *Populus* variants affects sugar release. Proc Natl Acad Sci U S A 108: 6300–6305.
21. Lygin AV, Upton J, Dohleman FG, Juvik J, Zabotina OA, et al. (2011) Composition of cell wall phenolics and polysaccharides of the potential bioenergy crop -*Miscanthus*. GCB Bioenergy 3: 333–345.
22. Fu C, Mielenz JR, Xiao X, Ge Y, Hamilton CY, et al. (2011) Genetic manipulation of lignin reduces recalcitrance and improves ethanol production from switchgrass. Proc Natl Acad Sci U S A 108: 3803–3808.
23. Xu B, Escamilla-Trevino LL, Sathitsuksanoh N, Shen Z, Shen H, et al. (2011) Silencing of 4-coumarate:coenzyme A ligase in switchgrass leads to reduced lignin content and improved fermentable sugar yields for biofuel production. New Phytol 192: 611–625.
24. Huang J, Xia T, Li A, Yu B, Li Q, et al. (2012) A rapid and consistent near infrared spectroscopic assay for biomass enzymatic digestibility upon various physical and chemical pretreatments in *Miscanthus*. Bioresour Technol 121: 274–281.
25. Li F, Ren S, Zhang W, Xu Z, Xie G, et al. (2013) Arabinose substitution degree in xylan positively affects lignocellulose enzymatic digestibility after various NaOH/H_2SO_4 pretreatments in *Miscanthus*. Bioresour Technol 130: 629–637.
26. Grabber JH (2005) How Do Lignin Composition, Structure, and Cross-Linking Affect Degradability? A Review of Cell Wall Model Studies This paper was originally presented at the Lignin and Forage Digestibility Symposium, 2003 CSSA Annual Meeting, Denver, CO. Crop Sci 45: 820–831.
27. Jung H-J, Bernardo R (2012) Comparison of Cell Wall Polysaccharide Hydrolysis by a Dilute Acid/Enzymatic Saccharification Process and Rumen Microorganisms. BioEnergy Research 5: 319–329.
28. Li M, Feng S, Wu L, Li Y, Fan C, et al. (2014) Sugar-rich sweet sorghum is distinctively affected by wall polymer features for biomass digestibility and ethanol fermentation in bagasse. Bioresour Technol 167C: 14–23.
29. Achyuthan K, Achyuthan A, Adams P, Dirk S, Harper J, et al. (2010) Supramolecular self-assembled chaos: polyphenolic lignin's barrier to cost-effective lignocellulosic biofuels. Molecules 15: 8641–8688.
30. Sang TAO, Zhu W (2011) China's bioenergy potential. GCB Bioenergy 3: 79–90.
31. Peng L, Hocart CH, Redmond JW, Williamson RE (2000) Fractionation of carbohydrates in *Arabidopsis* root cell walls shows that three radial swelling loci are specifically involved in cellulose production. Planta 211: 406–414.
32. Fry SC (1988) The growing plant cell wall: chemical and metabolic analysis. London: Blackburn Press.
33. Dische Z (1962) Color reactions of carbohydrates. In: Whistler RL, Wolfrom ML, editors. Methods in Carbohydrate Chemistry. New York: Academic Press. pp.477–512.
34. Zhang W, Yi Z, Huang J, Li F, Hao B, et al. (2013) Three lignocellulose features that distinctively affect biomass enzymatic digestibility under NaOH and H_2SO_4 pretreatments in *Miscanthus*. Bioresour Technol 130: 30–37.

20

Expression of Wheat High Molecular Weight Glutenin Subunit 1Bx Is Affected by Large Insertions and Deletions Located in the Upstream Flanking Sequences

Yuke Geng[1,2]¶, Binshuang Pang[3]¶, Chenyang Hao[1], Saijun Tang[2], Xueyong Zhang[1]*, Tian Li[1]*

1 Key Laboratory of Crop Gene Resources and Germplasm Enhancement, Ministry of Agriculture/Institute of Crop Science, Chinese Academy of Agricultural Sciences, Beijing, China, **2** College of Biological sciences, China Agricultural University, Beijing, China, **3** Beijing Engineering and Technique Research Center of Hybrid Wheat, Beijing Academy of Agricultural and Forestry Sciences, Beijing, China

Abstract

To better understand the transcriptional regulation of high molecular weight glutenin subunit (HMW-GS) expression, we isolated four *Glu-1Bx* promoters from six wheat cultivars exhibiting diverse protein expression levels. The activities of the diverse *Glu-1Bx* promoters were tested and compared with β-glucuronidase (*GUS*) reporter fusions. Although all the full-length *Glu-1Bx* promoters showed endosperm-specific activities, the strongest GUS activity was observed with the *1Bx7*^{OE} promoter in both transient expression assays and stable transgenic rice lines. A 43 bp insertion in the *1Bx7*^{OE} promoter, which is absent in the *1Bx7* promoter, led to enhanced expression. Analysis of promoter deletion constructs confirmed that a 185 bp MITE (miniature inverted-repeat transposable element) in the *1Bx14* promoter had a weak positive effect on *Glu-1Bx* expression, and a 54 bp deletion in the *1Bx13* promoter reduced endosperm-specific activity. To investigate the effect of the 43 bp insertion in the *1Bx7*^{OE} promoter, a functional marker was developed to screen 505 Chinese varieties and 160 European varieties, and only 1Bx7-type varieties harboring the 43 bp insertion in their promoters showed similar overexpression patterns. Hence, the *1Bx7*^{OE} promoter should be important tool in crop genetic engineering as well as in molecular assisted breeding.

Editor: Leandro Peña, Instituto Valenciano De Investigaciones Agrarias, Spain

Funding: National Transgenic Research Project (ZX08002-004) and National Natural Science Foundation of China (30671293). The funders had no role in study design, data collection and analysis, decision to publish, or preparation of the manuscript.

Competing Interests: The authors have declared that no competing interests exist.

* Email: zhangxueyong@caas.cn (XZ); litian@caas.cn (TL)

¶ These authors contributed equally to this work.

Introduction

Hexaploid wheat (*Triticum aestivum* L.) is one of the most important human food sources. Its complex genetic background leads to great diversity in nutritional and processing qualities among cultivars. High molecular weight glutenin subunits (HMW-GSs) are the main grain storage proteins in the endosperms of wheat and related species [1,2]. Although HWM-GSs grain storage proteins account for only about 12% of the total protein [3], they play a key role in wheat gluten as the skeletal network that to a large extent determines its structure and formation [4]. The compositions and quantities of allelic variation in HMW-GS genes substantially affect the taste and appearance of dough products, such as Chinese noodles and European bread [5]. Therefore, improvement of flour quality based on superior HWM-GS alleles is necessary to meet changing consumer demands.

Both qualitative and quantitative effects of HMW-GS subunits are important for flour quality [6,7]. In the process of breeding, high dough strength is used as a predictor of good-quality bread wheat; and overexpression of Glu-1Bx7 by way of allele *1Bx7*^{OE} makes an important contribution to high dough strength in some cultivars [8,9]. Expression of HMW-GS is regulated by three major factors, which are at the genomic level (gene duplication), transcriptional level and translational level [10–13]. Transcriptional regulation driven by *Glu-1* 5′-upstream flanking regions might provide strategies for improving grain quality in wheat breeding programs [14]. A number of crucial *cis*-acting elements from HMW-GS promoters of various wheat cultivars have been investigated and characterized, as these could affect tissue specificity or expression activity, including conservative endosperm-specific motifs, such as the GCN4 motif [15], the prolamin box [16], AACA/TA motif [17], RY repeat motif [18], and Skn-1 [19,20], each of which is capable of exerting temporal expression [21,22]. However, the basis of transcriptional regulation of divergence caused by large insertion and deletion (InDel) alterations in HMW-GS promoter regions is still not clear. As reported earlier, a tandem 54 bp duplication, known as the "cereal-box" located at −400 bp in the *1Bx* promoter may enhance endosperm-specific expression [1], suggesting this duplicated region might be a key region for control of gene expression [23,24]. There is a 185 bp MITE insertion in the promoters of *1Bx14* and *1Bx20*, but functional verification indicated that this insertion had little effect on gene expression [25–27]. A 43 bp insertion found at −1000 bp in the *1Bx7^OE* promoter was significantly associated with the overexpression phenotype. It was speculated that the overexpression was brought about by gene duplication mediated by the insertion of a retroelement [13], and there was no further study concerning the 43 bp InDel effect on protein expression. Therefore, more experimental data are needed to clarify the effect of InDels in HMW-GS promoters.

Highly active endosperm-specific promoters serve as an important genetic resource for high-quality and high-yield wheat breeding. Use of seed storage protein gene promoters is an attractive strategy for obtaining target gene products exclusively from crop kernels. A number of seed-specific promoters from barley, rice, maize and other species have been investigated functionally [15,21,28]. Transgenic crops with favorable gene stacking require different tissue-specific promoters from various cereals, as this is helpful to reduce homology-based transcriptional gene silencing [29,30]. HMW-GS promoters from wheat, although containing endosperm-specific motifs, may not be spatially controlled in the same way as in their original genetic backgrounds due to subtle differences in respective regulation systems [31]. Hence, further research of key motifs from tissue-specific promoters would boost applicability in genetic engineering.

Among hexaploid wheat HMW-GSs, Glu-1Bx often shows the highest level of expression [32]. We therefore set out to analyse *1Bx* promoter sequence characteristics to uncover the transcriptional regulation mechanism. Based on diverse protein expression levels in six wheat cultivars, we isolated four *Glu-1Bx* promoters in approximately 2.2 kb of length and further validated their functions. By comparison with these upstream sequences, several large InDels such as a 43 bp InDel, a 54 bp duplication and a 185 bp MITE resulted in major divergences among the four promoters, including the *1Bx7* promoter (*Pro-1Bx7*), *1Bx7^OE* promoter (*Pro-1Bx7^OE*), *1Bx13* promoter (*Pro-1Bx13*) and *1Bx14* promoter (*Pro-1Bx14*). The promoter sequence variation was shown to be an important factor causing differential expression in transient expression systems and in transgenic rice plant assays. Notably, *Pro-1Bx7^OE* is a highly active endosperm-specific promoter that can be made available for crop improvement by transgenic methods. Moreover, we developed a new specific molecular marker in terms of the 43 bp insertion residing in the *1Bx7^OE* promoter, with which we screened 505 Chinese and 160 European cultivars [33]. We found that this functional marker is significantly associated with *1Bx7* overexpression. Our results further showed that transcriptional regulation might be responsible for *1Bx* expression diversity to a larger extent than initially expected.

Materials and Methods

Plant materials

Hexaploid wheat (*Triticum aestivum* L.) cultivars (cv.) Yanzhan 1, Atlas 66, Jimai 20, Xiaoyan 54, Yunmai 33 and Chinese Spring were grown in the field. Endosperm of Xiaoyan 54 was prepared for transient expression assays, and rice (*Oryza sativa* L. ssp *japonica*) cv. Kita-ake was used to produce stable transformants. Materials used for molecular marker screening included 505 Chinese and 160 European cultivars [33].

SDS-PAGE and quantification of HMW-GSs

Protein fractions were extracted from single wheat kernels using a previously reported HMW-GS extraction protocol [34]. Identical amounts of protein extracted from seeds of different varieties were separated by SDS-PAGE and visualized by Coomassie Blue staining as described by Zhang et al. [35]. Densitometric analyses of 1Bx subunits were carried out by Quantity One software (Bio-rad, USA). The value of the optical density multiplication area was used to quantify HMW-GS expression.

Promoter isolation and cis-element prediction

DNA extraction was performed as previously described [36]. Using a pair of specific primer sets, 1Bx2258F/R (Table S1), four full-length *1Bx* promoters, ~2.3 kb in size, from six wheat varieties were isolated, gel purified and sequenced. Putative regulatory elements within the *1Bx* promoter were predicted using the Plant Cis-acting Regulatory DNA elements (PLACE) database [37] combined with a previous report [38].

Construction of promoter-GUS chimeric genes and subsequent transformation

Several full-length and truncated *1Bx* promoters were obtained by PCR amplification with primers introducing DNA restriction enzyme sites for convenient subcloning (Table S1), and cloned into a modified vector PAHC25 [39] for transient expression, and then subcloned into binary vector pCAMBIA1391z, containing the reporter gene *GUS* under the control of different *1Bx* promoters. In transient expression experiments, immature embryos harvested at 12–14 days post anthesis (DPA) were used for bombardment as described by Ortiz et al. [40]. Different *1Bx* promoter-*GUS* constructs were tested in transient expression assays as described previously [41]. For stable transformation, the binary vector constructs were first introduced into *Agrobacterium tumefaciens* strain EHA105, and then rice transformation was carried out as described by Cho et al. [42]. Transgenic rice plants were selected on medium containing 50 mg L^{-1} of hygromycin, and positive lines were grown in the field for further analysis.

PCR and Southern blot analyses of transformed rice plants

Genomic DNA was isolated from leaf tissues of transformed rice plants as previously described [36]. PCR analysis for molecular identification of transgenic rice plants was performed using a set of specific primers for the *1Bx* promoter and *GUS* gene (listed in Table S1).

For Southern blot analysis, genomic DNA (10 µg) from different transgenic lines were digested with *Bam*HI or *Hind*III (New England Biolabs, USA). Digested DNA was separated by electrophoresis in 0.8% (w/v) agarose gels, and then transferred to Hybond-N^+ membranes (Amersham Biosciences, USA) and hybridized with a *GUS* gene fragment labeled with [α-^{32}P]dCTP as described previously [36].

Histochemical GUS assay

Wheat endosperms undergoing transient expression and different tissues of T_3 transgenic rice were used for histochemical GUS assays. GUS staining was performed as described by Kosugi et al. [43]. Images of stained samples were captured using an MZ16 High-tech Stereomicroscope (Leica, Germany). Stained GUS spots were counted, and for statistical comparison, data of each sample was expressed as the mean number of blue spots per endosperm.

Quantification of expression of the *GUS* gene under control of *1Bx* promoters in transgenic rice plants

Total RNA was extracted from kernels of 3 independent positive T_3 transgenic rice lines at 10–18 DPA. Transcriptional levels of *GUS* in all stable transgenic lines were quantified by quantitative real-time PCR (qRT-PCR) with a 7300 Real-time PCR system (Applied Biosystems, USA) using Power SYBR Green PCR Master Mix (Applied Biosystems, USA). Details of primer pairs used for qRT-PCR are given in Table S1. The specificity of the primer sets was assured by confirmation that the resulting products appeared as single peaks in real-time melting temperature curves and as single fragments after separation by agarose gel electrophoresis. To confirm adequate amplification PCR efficiency was assessed using a sample dilution series as templates [44]. Amplification plots and predicted threshold cycle values were obtained from three independent biological replicates with SDS software version 2.1 (Applied Biosystems, USA). *GUS* gene expression levels were presented as fold-changes calculated using the comparative threshold cycle (CT) method as described [45] with rice *GAPDH* used as the internal control.

Results

Identification of HMW-GSs by SDS-PAGE

HMW-GSs of six wheat cultivars were separated by SDS-PAGE (Figure 1A). Their subunit compositions varied from each other (Table S2). Evidently, expression levels of Glu-1A and Glu-1D are generally lower than Glu-1B. There were four allelic variants of *1Bx* among the six cultivars, namely *1Bx7*^{OE}, *1Bx14*, *1Bx13* and *1Bx7*. The protein level of 1Bx7^{OE} in Yunmai 33 was much higher than that of 1Bx in other cultivars, about 2.2-fold that of 1Bx13 and 1Bx7 and 1.8-fold that of 1Bx14 (Figure 1B).

Comparative analysis of upstream sequences of *Glu-1Bx*

Four types of 5' flanking sequences of *1Bx* alleles were isolated with the specific primer pair, 1Bx2258F/R (Table S1 and Figure S1). They were 2,294, 2,253, 2,185 and 2,433 bp in length for *Pro-1Bx7*^{OE}, *Pro-1Bx7*, *Pro-1Bx13*, and *Pro-1Bx14*, respectively. The four 5' proximal flanking regions contained five common motifs, including DOF recognition sites, bZIP recognition sites, MYB recognition sites, VP1 recognition sites and basal promoter elements (Figure 2 and Table S3), which are conserved in promoters of genes that encode most seed storage proteins [38]. In addition to several single-base substitutions or small deletions, the presence of sequence insertions or deletions (InDels) constituted the main differences among the four entire promoter regions (Figure S2). By comparison with the *1Bx7* promoter, the *1Bx13* promoter has a 54 bp deletion at −400 upstream from the start codon (54 bp duplication position), the *1Bx14* promoter contains a 185 bp MITE insertion at −874, consistent with previous reports [26,46], and the *1Bx7*^{OE} promoter possesses a 43 bp insertion at −1047, which is always associated with the overexpression phenotype [12].

The transient expression results for different *1Bx* promoters

We compared the expression efficiencies of the full-length promoters of *1Bx7*, *1Bx13* and *1Bx7*^{OE} by means of transient expression assays in wheat endosperms (Figure S3). and *GUS* driven by the *1Bx7*^{OE} promoter exhibited much higher activity than when driven by the *1Bx7* and *1Bx13* promoters (Figure 3A). Since the 43 bp InDel represents the difference between the *1Bx7* and *1Bx7*^{OE} promoter sequences, we speculated that the 43 bp insertion enhanced the endosperm-specific expression. In addition, the *1Bx13* promoter activity was lower than that of the *1Bx7* promoter, which lacks the 54 bp duplication present in the *1Bx13* promoter. We further investigated the effect of the 185 bp MITE on *1Bx14* expression. Two truncated *1Bx14* promoters were fused to *GUS*, and transient expression results showed that *GUS* expression driven by a 1,192 bp *Pro-1Bx14-D2* was slightly higher than that driven by the 873 bp *Pro-1Bx14-D1* (Figure 3B), confirming that the MITE might positively but weakly affected transcription of *1Bx14* [26].

Histochemical and quantitative assays in transgenic rice

Chimeras were constructed using different *1Bx* promoters fused to *GUS* and then transformed into rice. Through GUS staining assays in the T_3 generation of stable transgenic plants, we detected *GUS* activity driven by the three full-length *1Bx* promoters only in the seeds and not in stems or leaves collected at 15 DPA. These results indicated that the promoters are endosperm-specific (Figure 4A). In contrast, GUS staining was observed in all tissues of transgenic rice carrying the *Ubiquitin* promoter-*GUS* construct (Figure 4A). Therefore, the full-length *1Bx* promoters contained necessary *cis*-elements that specify endosperm-specific regulation in both wheat and rice. Consistent with the transient expression results, the full-length *1Bx7*^{OE} promoter with the 43 bp insertion exhibited much higher *GUS* activity than either the full-length *1Bx7* or *1Bx13* promoters (Figure 4A). Southern blot analysis confirmed that the transgenic rice lines had single copies of the *GUS* gene (Figure S4); therefore the comparative results of promoter activities were convincing.

We also investigated the activities of different truncated *1Bx* promoters in transgenic rice. Except for the *Pro-1Bx13-D* promoter, other truncated promoters, including *Pro-1Bx14-D1/D2* and *Pro-1Bx7-D*, retained endosperm-specific expression

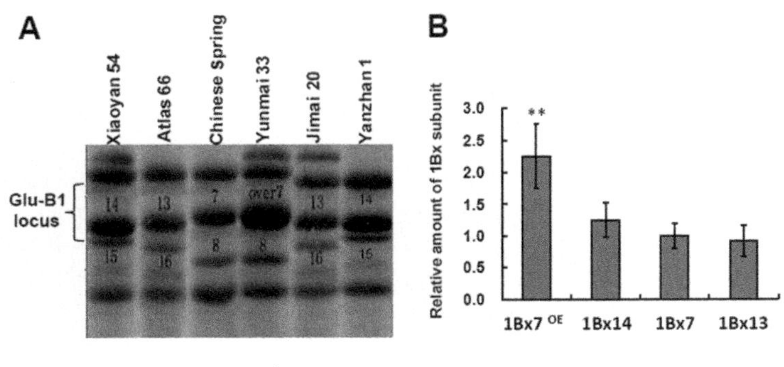

Figure 1. HMW-GSs from six common wheat cultivars. (A) SDS-PAGE profiles of HMW-GSs. 1Bx7 (Chinese Spring), 1Bx13 (Atlas 66 and Jimai 20), 1Bx14 (Xiaoyan 54 and Yanzhan 1) and 1Bx7^{OE} (Yunmai 33). (B) Relative amounts of four 1Bx subunits estimated by densitometric analysis.

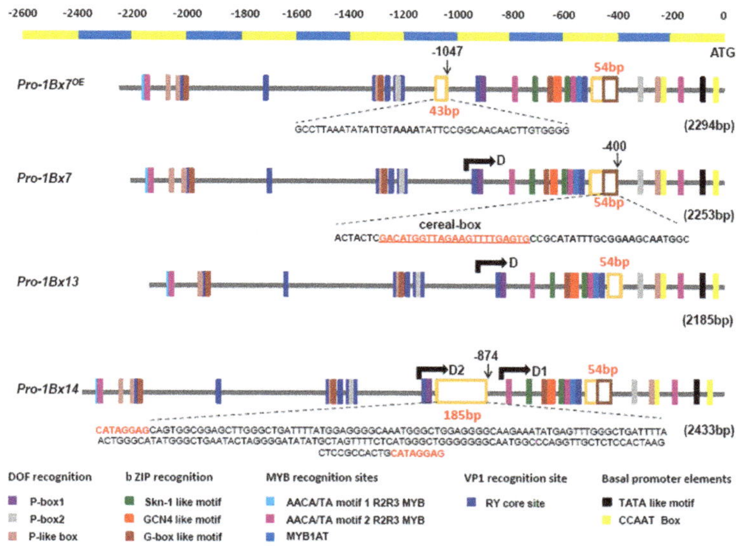

Figure 2. Schematic structure of four *1Bx* promoters. Regulatory elements are indicated by colored rectangles. InDels are labeled with hollow boxes and core sequences are listed in detail under the sketch map. The red underlined sequence shows the "cereal box" in the 54 bp duplication. The 8 bp bases in red at both ends indicate the target site duplication (TSD) of the MITE. Positions of upstream primers used for obtaining truncated promoters are indicated with black arrows.

activity (Figure 4C). Absence of the 54 bp duplication led to loss of endosperm-specific expression controlled by the *Pro-1Bx13-D* promoter, thus indicating that the fragment removed from the *1Bx13* promoter (−943~−2162) contained necessary *cis*-elements that restricted expression to the endosperm. Like transient expression results, *GUS* expression directed by *Pro-1Bx14-D2* harboring the 185 bp MITE was higher than that directed by *Pro-1Bx14-D1*.

To confirm the results of GUS staining in seeds of transgenic rice, we applied qRT-PCR to determine expression levels of *GUS*. *GUS* expression detected in seeds during 10–18 DPA showed that expression driven by *1Bx* promoters increased rapidly from 10 DPA to 14 DPA, and reached a peak level at 14–16 DPA (Figure 4B and 4D). As the control, *GUS* expression driven by the *Ubiquitin* promoter maintained a relatively constant level through 10 to 18 DPA (Figure 4B). The results of *GUS* expression at 14–16 DPA were highly consistent with those of GUS activities based on histochemical staining (Figure 4A and 4C), confirming that the protein expression pattern was similar to the gene expression pattern at the mRNA level.

Phylogenetic analysis of HMW-GS promoters

To address the question of whether these InDels are present in other *HMW-GS* promoters, we analyzed promoters from 14 different wheat *HMW-GS* genes [1,26,47] and identified the regulatory motifs related to endosperm-specific expression in the promoter regions about 1,200 bp upstream of the initiation codon. The numbers of regulatory motifs obviously differed among the

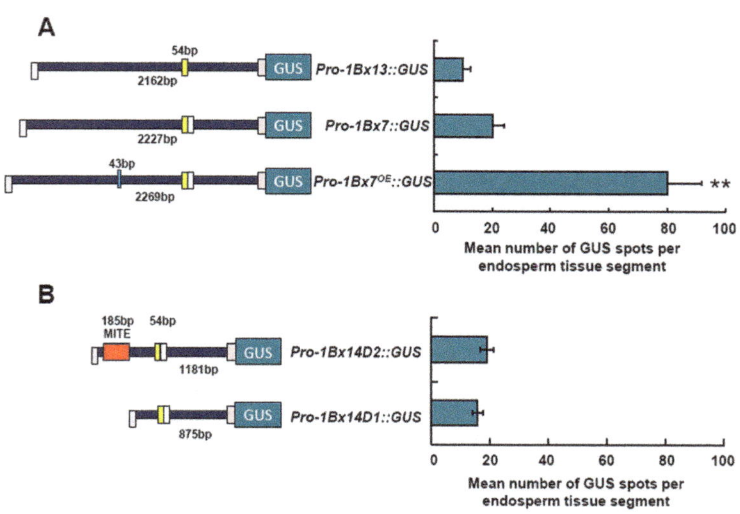

Figure 3. Schematic representation of constructs used for transient expression assays in wheat endosperms and *GUS* activities driven by full-length and truncated *1Bx* promoters. (A) Effects of full-length *1Bx7*, *1Bx13* and *1Bx7*^{OE} promoters on transient expression. (B) Effect of the 185 bp MITE from the *1Bx14* promoter on transient expression. **P<0.01 (student's *t*-test) indicates significant difference from others.

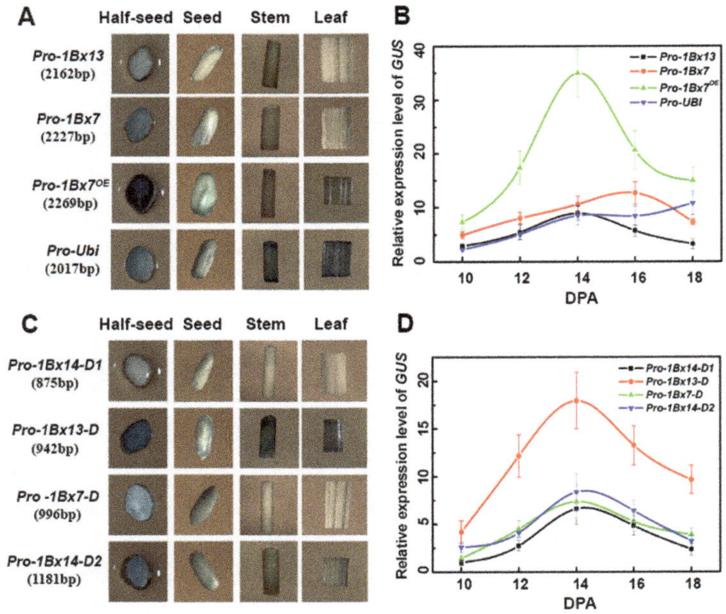

Figure 4. GUS staining of various tissues and quantitative analysis in developing endosperm of transgenic rice. (A) and (C) Histochemical analysis of *GUS* driven by different promoters in transgenic rice tissues collected at 15 DPA; (B) and (D) Relative expression levels of *GUS* in seeds from different transgenic lines during 10 to 18 DPA based on qRT-PCR. Rice *GAPDH* was used as the internal control. Values are shown as means ± s.d (standard deviation) of three independent experiments and three biological replicates. Colored lines at the top right corner represent different transgenic plants.

different HMW-GS promoter sequences (Figure 5). In phylogenetic analysis, all *Glu-1Bx* promoters clustered together in one branch. The 185 bp MITE insertion was present in both the *Glu-1Bx14* and *Glu-1Bx20* promoters. The 54 bp tandem duplication was absent in non-*Glu-1Bx* promoters, but present in all *Glu-1Bx* promoters except the *Glu-1Bx13* promoter. Therefore, InDels contributed to the diversity in HMW-GS promoters, which is an important means for evolution of HMW-GS genes. These large fragment InDels can be used as a potential resource for creating new alleles.

Development of a molecular marker for the 43 bp insertion and its distribution in natural populations

Based on the 43 bp insertion sequence in the *1Bx7*^{OE} promoter, we developed a new molecular marker that differed from those previously reported [34]. Our marker can precisely identify the insertion in HWM-GS promoters among common wheat varieties. PCR amplification resulted in two kinds of bands that distinguish promoters with the 43 bp insertion (a 476 bp fragment) from those without (a 433 bp fragment) (Figure 6A). Among 505 Chinese and 160 European accessions surveyed, we found 3 Chinese and 11 European varieties with the 43 bp insertion when we used this

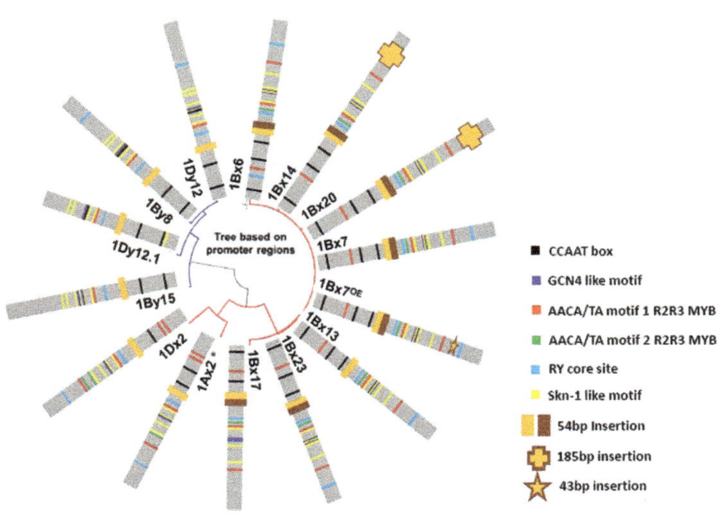

Figure 5. Phylogenetic analysis of 14 *HMW-GS* promoters. Neighbor-Joining Tree of partial length (-1 ~ -1200 bp upstream of the start codon) sequences of *Glu-1* from *Triticum aestivum* L. and other wheat-related grass species. This work was done with the MEGA program (Version 5.2). InDels and conserved cis-elements are shown by markers in different colors.

marker. HMW-GS profiles of accessions containing the 476 bp marker were later obtained by SDS-PAGE (Figure 6B). The presence of particular 1Bx alleles was determined by densitometric analysis; 10 accessions (3 from China and 7 from Europe) exhibited a 1Bx7 overexpression phenotype relative to Chinese Spring used as a control (Figure 6B and Table S4). We also identified some types of *1Bx6* and *1Bx14* with the 43 bp insertion (Figure 6B and Table S4), and confirmed its presence by DNA sequencing. However, these two types of 1Bx did not show overexpression at the protein level. Therefore, the 43 bp insertion in promoters preferentially enhanced 1Bx7 expression although no obvious differences were found among the upstream regions of *1Bx6*, *1Bx14* and *1Bx7*^{OE}.

Discussion

HMW-GSs represent a set of important seed-storage proteins, and both their composition and quantity significantly affect wheat flour quality [7]. Since gene transcriptional regulation is the dominant means of control in production of proteins [48], we isolated four promoter sequences of *Glu-1Bx* and investigated their effects on gene expression. Although the reporter gene driven by all the *1Bx* promoters exhibited an endosperm-specific expression pattern, the *1Bx7*^{OE} promoter from cv. Yunmai 33 produced a markedly stronger activity than other promoters. Previous studies showed that gene duplication was the cause of

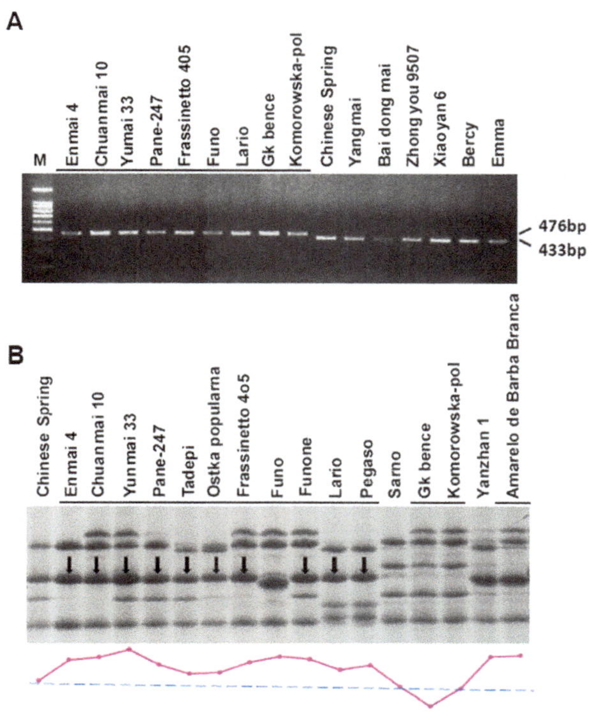

Figure 6. Electrophoretic separation of PCR products from *1Bx* promoters with or without the 43 bp insertion, and SDS-PAGE profiles of HMW-GS of wheat cultivars with the insertion. (A) PCR assays for *1Bx* promoters on a 2% agarose gel. M: 100 bp DNA Ladder. Underlined accessions possess the 43 bp insertion. (B) SDS-PAGE assay of HMW-GS from different accessions containing the 43 bp insertion. Underlined accessions possess the 43 bp insertion. Down black arrows indicate *1Bx7* with 43 bp insertion in accessions from China and Europe. Chinese Spring, Yanzhan 1 and Samo were used as controls. The purple curve represents the relative amounts from different 1Bx subunits.

1Bx7 overexpression in some wheat cultivars [49]. The connection between the strong activity of the *1Bx7*^{OE} promoter and high protein level produced by the 1Bx7^{OE} subunit clearly indicated that transcription regulation is also a factor in 1Bx7 overexpression. We therefore concluded that multiple factors, including gene duplication and transcriptional regulation determine the expression of *1Bx7*. This work revealed a complex regulatory network of HMW-GS expression in wheat.

The 43 bp insertion at −1047 bp is closely associated with high expression of 1Bx7^{OE}

Large InDels in promoter regions often result in higher rates of transcriptional divergence [50]. In this study, we identified a 43 bp InDel, a 54 bp duplication and a 185 bp MITE in different *Glu-1Bx* promoters, and they accounted for the main differences among *1Bx* promoter sequences. These InDels affected the expression levels of the genes. The presence of the 43 bp insertion at position −1047 upstream of the start codon was shown to be closely associated with high expression levels of the 1Bx7^{OE} subunit [12,13]. We verified that the 43 bp insertion can serve as a strong enhancer to improve expression of the gene by comparing the transcriptional activities between full-length *1Bx7* and *1Bx7*^{OE} promoters (Figure 3A; Figure 4A and 4B). Since there are no known *cis*-elements in the 43 bp insertion, this insertion may facilitate evolutionary tuning of gene expression by affecting local chromatin structure and nucleosome positioning [50].

The 185 bp MITE insertion at −874 bp of *1Bx14* might slightly affect transcription

The 185 bp MITE insertion located at −874 bp in the *1Bx14* promoter may be a remnant of an earlier transposition of a large element or of small, highly repeated elements [51]. In the present study, the *1Bx14* promoter with or without the 185 bp MITE, did not produce a significantly different activity in the transient system or in stable transgenic rice assays, suggesting that it might only slightly affect the transcriptional regulation (Figure 3B; Figure 4C and 4D). The 185 bp MITE exists in both hexaploid and tetraploid wheat, and may be linked to the polyploidization event affecting the constitutions and activities of the genomes of grass species [26].

A 54 bp cereal-box motif is necessary for endosperm-specific expression

The tandem 54 bp duplication at position −400 contains the "cereal-box" implicated in seed-specific expression [1]. Our data demonstrated that the *1Bx13* full-length promoter harboring one 54 bp deletion retains endosperm-specific activity, but a *1Bx13* promoter truncated at −942 bp lacks endosperm-specificity accompanied by increased activity (Figure 4C). We speculate that the 54 bp deletion might complement essential *cis*-elements in the region −940 to −2000 bp of the *1Bx13* promoter to effectively control gene endosperm-specific expression. Without the aid of the *cis*-elements in the region, only one 54 bp cereal-box motif may not be enough to restrict gene expression to endosperm. Based on phylogenetic analysis of HWM-GS promoters, only the *1Bx13* promoter and non-*1Bx* promoters contain a 54 bp deletion (Figure 5). This tandem 54 bp duplication must have occurred before hexaploidization because it is also present in tetraploid wheat. Flanking-sequence divergence was also noted from extensive DNA sequencing analysis of a-gliadin genes [52]. The basis of HMW-GS evolution is repeated sequence events that lead to new alleles [53].

A simple PCR marker was developed to target high expression of 1Bx7 and 1Bx7^{OE}

Since previous 43 bp InDel marker covers a region of 1.2~1.3 kb that also contains other InDels such as the 185 bp MITE and 54 bp duplication [34], a new specific marker based on the 43 bp insertion was developed and used effectively in two independent wheat populations combined with SDS-PAGE electrophoresis analysis to identify 1Bx7 overexpressing cultivars. Interestingly, the 43 bp insertion exists not only in the *1Bx7* promoter but also in other *1Bx* promoters such as those of *1Bx14* and *1Bx6* (Figure 6B). Despite harboring the 43 bp insertion in the promoters, the 1Bx14 and 1Bx6 subunits produce no significant increases in protein compared to subunit alleles without the insertion (Figure 6B). The likely reason is that a co-regulatory factor linking the 43 bp insertion to expression efficiency is present in the 1Bx7 alleles or regulation at the translational level might strongly influence the divergence in expression between 1Bx7 and non-1Bx7 subunits.

Putative additive effects of gene duplication and transcriptional regulation on 1Bx7 expression

According to the literature, it is concluded that the 1Bx7 overexpression phenotype is mediated by an LTR retroelement resulting in gene duplication along with the polyploidization event [13]. In the present study, we confirmed that a 43 bp insertion situated in the *1Bx7*^{OE} promoter is capable of strengthening transcriptional activity markedly through transient expression and transgenic rice assays. By using molecular markers which can be used to indicate *1Bx7* gene duplication [13], we found that only the cultivar Yunmai 33 has both the 43 bp InDel and two 1Bx7 copies (gene duplication), while other 9 cultivars with the 43 bp InDel have only one 1Bx7 copy (Table S4). Although the 1Bx7 subunit of Yunmai 33 is the most abundant in this study (Figure 6B), other cultivars with the 43 bp InDel demonstrate higher 1Bx7 expression than the control, especially Chinese cultivars Enmai 4 and Chuanmai 10 (Figure 6B). So it can be inferred that both gene duplication and transcriptional regulation can lead to 1Bx7 overexpression, and their effects on 1Bx7 expression can be accumulated.

Endosperm is the storage tissue for starch and protein in cereal crops, which are the major sources of carbohydrates and proteins for humans. Improved yield and quality of crops by genetic modification has huge potential, and some significant achievements have already been accomplished [54]. Because continuous high expression of foreign genes in all tissues may cause detrimental effects in host plants [55], identification and application of strong endosperm-specific promoters will attract interest from breeders and biologists. In the current work, we identified a highly active *1Bx7*^{OE} promoter that can enhance endosperm-specific gene expression at the transcriptional level, and it should be useful for wheat quality improvement by means of genetic transformation and molecular assisted breeding.

Supporting Information

Figure S1 PCR amplification of *1Bx* promoters by using 1Bx1007-F/R (A) and 1Bx2258-F/R (B) primer pairs. Lane 1–4: Chinese Spring (*Pro-1Bx7*); Yunmai 33 (*Pro-1Bx7*^{OE}); Yanzhan 1 (*Pro-1Bx14*); Atlas 66 (*Pro-1Bx13*); M is a DNA ladder. The PCR products were separated in 1.5% agarose gels.

Figure S2 Alignment of four *1Bx* promoters (*Pro-1Bx7*, *Pro-1Bx7*^{OE}, *Pro-1Bx13* and *Pro-1Bx14*).

Figure S3 Representative transient expression results of GUS driven by *Pro-1Bx* in wheat endosperms. (A) *Pro-1Bx13*; (B) *Pro-1Bx7*.

Figure S4 Southern blot analysis of transgenic rice lines with full-length *1Bx* promoters (A) or truncated *1Bx* promoters (B). Genomic DNA was digested by *Bam*HI and detected by GUS gene probes.

Table S1 Primers used in this study.

Table S2 HMW-GS compositions of six wheat accessions.

Table S3 Details of 12 known endosperm-specific cis-elements in *1Bx* promoters.

Table S4 Fourteen wheat cultivars harboring the 43 bp insertion in the *1Bx* promoter were identified by marker screening.

Acknowledgments

We are grateful to Dr. Jianmin Wan and Xiuping Guo for help with rice transformation, and gratefully acknowledge help with English editing from Prof. Robert A McIntosh, University of Sydney.

Author Contributions

Conceived and designed the experiments: YG BP XZ. Performed the experiments: YG BP TL. Analyzed the data: CH TL. Contributed reagents/materials/analysis tools: CH ST. Contributed to the writing of the manuscript: YG XZ TL.

References

1. Anderson OD, Greene FC (1989) The characterization and comparative analysis of high-molecular-weight glutenin genes from genomes A and B of a hexaploid bread wheat. Theor Appl Genet 77: 689–700.
2. Reddy P, Apples R (1993) Analysis of a genomic DNA segment carrying the wheat high molecular weight (HMW) glutenin Bx17 subunit and its use as an RFLP marker. Theor Appl Genet 85: 616–624.
3. Halford NG, Forde J, Anderson OD, Greene FC, Shewry PR (1987) The nucleotide and deduced amino acid sequences of an HMW glutenin subunit gene from chromosome 1B of bread wheat (*Triticum aestivum* L.) and comparison with those of genes from chromosomes 1A and 1D. Theor Appl Genet 75: 117–126.
4. Shewry PR, Tatham AS (1990) The prolamin storage proteins of cereal seeds: structure and evolution. Biochem J 267: 1–12.
5. Liu JJ, He ZH, Zhao ZD, Pena RJ, Rajaram S (2003) Wheat quality traits and quality parameters of cooked dry white Chinese noodles. Euphytica 131: 147–154.
6. Barro F, Rooke L, Békés F, Gras P, Tatham AS, et al. (1997) Transformation of wheat with high-molecular-weight subunit genes results in improved functional properties. Nat Biotechnol 15: 1295–1299.
7. Wieser H (2000) Comparative investigations of gluten proteins from different wheat species. I. Qualitative and quantitative composition of gluten protein types. Eur Food Res Technol 211: 262–268.
8. Gupta RB, Paul JG, Cornish GB, Palmer GA, Bekes F, et al. (1994) Allelic variation at glutenin subunit and gliadin loci, *Glu-1*, *Glu-3* and *Gli-1*, of common wheats. I. Its additive and interaction effects on dough properties. J Cereal Sci 19: 9–17.

9. Cornish GB, Békés F, Allen HM, Martin DJ (2001) Flour proteins linked to quality traits in an Australian doubled haploid wheat population. Aust J Agr Res 52: 1339–1348.
10. Harberd NP, Flavell RB, Thompson RD (1987) Identification of a transposon-like insertion in a *Glu-1* allele of wheat. Mol Genet Genomics 209: 326–332.
11. Marchylo BA, Lukow OM, Kruger JE (1992) Quantitative variation in high molecular weight glutenin subunit 7 in some Canadian wheats. J Cereal Sci 15: 29–37.
12. Butow BJ, Gale KR, Ikea J, Juhász A, Bedo Z, et al. (2004) Dissemination of the highly expressed *Bx7* glutenin subunit (*Glu-B1al allele*) in wheat as revealed by novel PCR markers and RP-HPLC. Theor Appl Genet 109: 1525–1535.
13. Ragupathy R, Naeem HA, Reimer E, Lukow OM, Sapirstein HD, et al. (2008) Evolutionary origin of the segmental duplication encompassing the wheat *GLU-B1* locus encoding the overexpressed Bx7 ($Bx7^{OE}$) high molecular weight glutenin subunit. Theor Appl Genet 116: 283–296.
14. Thomas MS, Flavell RB (1990) Identification of an enhancer element for the endosperm-specific expression of high molecular weight glutenin. Plant Cell 2: 1171–1180.
15. Zheng Z, Kawagoe Y, Xiao S, Li Z, Okita T, et al. (1993) 5′ distal and proximal cis-acting regulator elements are required for developmental control of a rice seed storage protein glutelin gene. Plant J 4: 357–366.
16. Dong G, Ni Z, Yao Y, Nie X, Sun Q (2007) Wheat Dof transcription factor WPBF interacts with *TaQM* and activates transcription of an *alpha-gliadin* gene during wheat seed development. Plant Mol Biol 63: 73–84.
17. Takaiwa F, Yamanouchi U, Yoshihara T, Washida H, Tanabe F, et al. (1996) Characterization of common cis-regulatory elements responsible for the endosperm-specific expression of members of the rice glutelin multigene family. Plant Mol Biol 30: 1207–1221.
18. Fujiwara T, Beachy RN (1994) Tissue-specific and temporal regulation of a β-conglycinin gene: roles of the RY repeat and other cis-acting elements. Plant Mol Biol 24: 261–273.
19. Sha S, Sugiyama Y, Mitsukawa N, Masumura T, Tanaka K (1996) Cloning and sequencing of a rice gene encoding the 13-kDa prolamin polypeptide. Biosci Biotech Bioch 60: 335–337.
20. Wu CY, Suzuki A, Washida H, Takaiwa F (1998) The GCN4 motif in a rice glutelin gene is essential for endosperm-specific gene expression and is activated by *Opaque-2* in transgenic rice plants. Plant J 14: 673–683.
21. Müller M, Knudsen S (1993) The nitrogen response of a barley *C-hordein* promoter is controlled by positive and negative regulation of the GCN4 and endosperm box. Plant J 4: 343–355.
22. Onodera Y, Suzuki A, Wu CY, Washida H, Takaiwa F (2001) A rice functional transcriptional activator, RISBZ1, responsible for endosperm-specific expression of storage protein genes through GCN4 motif. J Biol Chem 276: 14139–14152.
23. Forde J, Malpica JM, Halford NG, Shewry PR, Anderson OD, et al. (1985) The nucleotide sequence of a HMW glutenin subunit gene located on chromosome 1A of wheat (*Triticum aestivum* L.). Nucleic Acids Res 13: 6817–6832.
24. Halford NG, Ford J, Shewry PR, Kreis M (1989) Functional analysis of the upstream regions of a silent and an expressed member of a family of wheat seed-protein genes in transgenic tobacco. Plant Sci 62: 207–216.
25. Anderson OD, Larka L, Christoffers MJ, McCue KF, Gustafson JP (2002) Comparison of orthologous and paralogous DNA flanking the wheat high molecular weight glutenin genes: sequence conservation and divergence, transposon distribution, and matrix-attachment regions. Genome 45: 367–380.
26. Li W, Wan Y, Liu Z, Liu K, Li B, et al. (2004) Molecular charaterization of HMW glutenin subunit allele 1Bx14: further insights into the evolution of Glu-B1-1 alleles in wheat and related species. Theor Appl Genet 109: 1093–1104.
27. Jiang QT, Ma J, Zhao S, Zhao QZ, Lan XJ, et al. (2012) Characterization of HMW-GSs and their gene inaction in tetraploid wheat. Genetica 40: 325–335.
28. Marks MD, Lindell JS, Larkins BA (1985) Quantitative analysis of the accumulation of zein mRNA during maize endosperm development. J Biol Chem 260: 16445–16450.
29. Butaye KMJ, Cammue BPA, Delaurè SL, De Bolle MFC (2005) Approaches to minimize variation of transgene expression in plants. Mol breeding 16: 79–91.
30. Qu LQ, Xing YP, Liu WX, Xu XP, Song YR (2008) Expression pattern and activity of six glutelin gene promoters in transgenic rice. J Exp Bot 59: 2417–2424.
31. Furtado A, Henry RJ, Takaiwa F (2008) Comparison of promoters in transgenic rice. Plant Biotechnol J 6: 679–693.
32. Galili G, Feldman M (1983) Genetic control of endosperm proteins in wheat. Theor appl genet 66: 77–86.
33. Su Z, Hao C, Wang L, Dong Y, Zhang X (2011) Identification and development of a functional marker of *TaGW2* associated with grain weight in bread wheat (*Triticum aestivum L.*). Theor Appl Genet 122: 211–223.
34. Radovanovic N, Cloutier S (2003) Gene-assisted selection for high molecular weight glutenin subunits in wheat doubled haploid breeding programs. Mol breeding 12: 51–59.
35. Zhang X, Huang C, Xu X, Hew CL (2002) Identification and localization of a prawn white spot syndrome virus gene that encodes an envelope protein. J Gen Virol 83: 1069–1074.
36. Sambrook J, Fritsch EF, Maniatis T (1989) Molecular Cloning: A Laboratory Manual. Cold Spring Harbor Laboratory Press, New York.
37. Higo K, Ugawa Y, Iwamoto M, Korenaga T (1999) Plant cis-acting regulatory DNA elements (PLACE) database. Nucleic Acids Res 27: 297–300.
38. Juhász A, Makai S, Sebestyén E, Tamás L, Balázs E (2011) Role of conserved non-coding regulatory elements in LMW glutenin gene expression. PloS ONE 6: e29501.
39. Christensen AH, Quail PH (1996) Ubiquitin promoter-based vectors for high-level expression of selectable and/or screenable marker genes in monocotyledonous plants. Transgenic Res 5: 213–218.
40. Ortiz JP, Ravizzini RA, Morata MM, Vallejos RH (1997) A rapid system for studying foreign gene expression in wheat (*Triticum aestivum* L.). Theor Appl Genet 38: 123–130.
41. Oñate L, Vicente-Carbajosa J, Lara P, Diaz I, Carbonero P (1999) Barley BLZ2: a seed-specific bZIP protein that interacts with BLZ1 in vivo and activates transcription from the GCN4-like motif of B-hordein promoters in barley endosperm. J Biol Chem 274: 9175–9182.
42. Cho HJ, Brotherton JE, Widholm JM (2004) Use of the tobacco feedback-insensitive anthranilate synthase gene ASA2 as a selectable marker for legume hairy root transformation. Plant Cell Res 23: 104–113.
43. Kosugi S, Arai Y, Nakajima K, Ohashi Y (1990) An improved assay for *β*-glucuronidase (GUS) in transformed cells: methanol almost suppresses a putative endogenous *GUS* activity. Plant Sci 70: 133–140.
44. Rasmussen R (2001) Quantification on the Light Cycler instrument. In: Meuer S, Wittwer C, Nakagawara K, eds, Rapid Cycle Real-Time PCR, Methods and Applications. Springer Press, Heidelberg. pp.21–34.
45. Pfaffl MW, Tichopad A, Prgomet C, Neuvians TP (2004) Determination of stable housekeeping genes, differentially regulated target genes and sample integrity: BestKeeper–Excel-based tool using pair-wise correlations. Biotechnol Lett 26: 509–515.
46. Yang ZJ, Li GR, Liu C, Feng J, Zhou JP, et al. (2006) Molecular characterization of a HMW glutenin subunit allele providing evidence for silencing of x-type gene on *Glu-B1*. Acta Genet Sin 33: 929–936.
47. Yan Y, Zheng J, Xiao Y, Yu J, Hu Y, et al. (2004) Identification and molecular characterization of a novel y-type Glu-Dt1 glutenin gene of *Aegilops tauschii*. Theor Appl Genet 108: 1349–1358.
48. Shaw LM, McIntyre CL, Gresshoff PM, Xue GP (2009) Members of the Dof transcription factor family in Triticum aestivum are associated with light-mediated gene regulation. Funct Integr Genomics 9: 485–498.
49. Cloutier S, Banks T, Nilmalgoda S (2005) Molecular understanding of wheat evolution at the *Glu-B1* locus. In: Proceedings of the international conference on plant genomics and biotechnology: challenges and opportunities, Raipur, India, p 40.
50. Vinces MD, Legendre M, Caldara M, Hagihara M, Verstrepen KJ (2009) Unstable tandem repeats in promoters confer transcriptional evolvability. Science 324: 1213–1216.
51. Wessler SR (1998) Transposable elements and the evolution of gene expression. Symposia of the Society for Experimental Biology 51: 115–122.
52. Anderson OD (1997) Applications of molecular biology in understanding and improving wheat quality. In: Steele JL, Chung OK, eds. *Proc Int Wheat Quality Conf*, Grain Industry Alliance, Manhattan, KS. pp. 205–211.
53. SanMiguel P, Tikhonov A, Jin YK, Motchoulskaia N, Zakharov D, et al. (1996) Nested retrotransposons in the intergenic regions of the maize genome. Science 274: 765–768.
54. Bajaj S, Mohanty A (2005) Recent advances in rice biotechnology towards genetically superior transgenic rice. Plant Biol 3: 275–307.
55. Cheon BY, Kim HJ, Oh KH, Bahn SC, Ahn JH, et al. (2004) Overexpression of human erythropoietin (EPO) affects plant morphologies: retarded vegetative growth in tobacco and male sterility in tobacco and Arabidopsis. Transgenic Res 13: 541–549.

Simultaneous Over-Expression of *PaSOD* and *RaAPX* in Transgenic *Arabidopsis thaliana* Confers Cold Stress Tolerance through Increase in Vascular Lignifications

Amrina Shafi[1,2], Vivek Dogra[1], Tejpal Gill[1¤], Paramvir Singh Ahuja[1,2], Yelam Sreenivasulu[1,2*]

1 Division of Biotechnology, CSIR-Institute of Himalayan Bioresource Technology, Palampur, Himachal Pradesh, India, **2** Academy of Scientific and Innovative Research (AcSIR), CSIR- Institute of Himalayan Bioresource Technology, Palampur, Himachal Pradesh, India

Abstract

Antioxidant enzymes play a significant role in eliminating toxic levels of reactive oxygen species (ROS), generated during stress from living cells. In the present study, two different antioxidant enzymes namely copper-zinc superoxide dismutase derived from *Potentilla astrisanguinea* (*PaSOD*) and ascorbate peroxidase (*RaAPX*) from *Rheum austral* both of which are high altitude cold niche area plants of Himalaya were cloned and simultaneously over-expressed in *Arabidopsis thaliana* to alleviate cold stress. It was found that the transgenic plants over-expressing both the genes were more tolerant to cold stress than either of the single gene expressing transgenic plants during growth and development. In both single (*PaSOD, RaAPX*) and double (*PaSOD + RaAPX*) transgenic plants higher levels of total antioxidant enzyme activities, chlorophyll content, total soluble sugars, proline content and lower levels of ROS, ion leakage were recorded when compared to the WT during cold stress (4°C), besides increase in yield. In the present study, Confocal and SEM analysis in conjunction with qPCR data on the expression pattern of lignin biosynthetic pathway genes revealed that the cold stress tolerance of the transgenic plants might be because of the peroxide induced up-regulation of lignin by antioxidant genes mediated triggering.

Editor: Girdhar K. Pandey, University of Delhi South Campus, India

Funding: This work was supported by the Council of Scientific and Industrial Research, New Delhi, India. A.S., T.G. and V.D. acknowledge Fellowship by the CSIR, New Delhi, India. The funders had no role in study design, data collection and analysis, decision to publish, or preparation of the manuscript.

Competing Interests: The authors have declared that no competing interests exist.

* Email: sree_yelam@yahoo.com

¤ Current address: National Institute of Arthritis and Musculoskeletal and Skin Diseases, National Institute of Health, Bldg 10 CRC, 1-5256, Bethesda, MD, United States of America

Introduction

Abiotic stresses such as cold, drought, metal and salt stress are the major factors adversely affecting development and productivity in plants. Plants growing in different climatic niches have their own protective mechanisms to cope with the adverse conditions. Metabolic activities in various cellular compartments lead to the production of reactive oxygen species (ROS) such as superoxide anion (O_2^-), hydrogen peroxide (H_2O_2), hydroxyl radical (OH·), and singlet oxygen (O_2) [1]. These ROS are highly toxic, and are taken care of by the antioxidant system in the living cells. During abiotic and biotic stress conditions, the production of ROS increases to lethal levels, however, there is an up regulation of both enzymatic and non-enzymatic antioxidants, which modify the ROS to non toxic forms. Oxidative stress response involving antioxidant enzymes includes superoxide dismutase (SOD), ascorbate peroxidase (APX), catalase (CAT), glutathione reductase (GR) and peroxidase (POD) [2]. SOD (EC1.15.1.1) is the first enzyme in the plant's antioxidative defense mechanism that converts superoxide anion radicals ($O_2^{\cdot-}$) to hydrogen peroxide (H_2O_2) and water, thereby imparting protection against the harmful effects of highly reactive superoxide radical ($O_2^{\cdot-}$). APX is an integral component of the glutathione-ascorbate cycle [3] which detoxifies hydrogen peroxide in the presence of ascorbate to produce dehydroascorbate and water [4]. Genetic manipulation of antioxidant enzymes is one of the effective measures to impart stress tolerance in plants. Activities of superoxide dismutase and catalases were increased in plants subjected to cold stress [5]. It is commonly known that lignin evolved together with the plants adaptation to a terrestrial life to provide them with the structural support needed for an erect growth habit. Secondarily thickened cell walls were predominantly deposited by the lignin polymers, and making them rigid. In addition to developmentally programmed deposition of lignin, its biosynthesis can also be induced upon various biotic and abiotic stress conditions. The core enzymes of the phenylpropanoid pathway namely PAL, C4H and 4CL, converts the phenylalanine to *p*-Coumaroyl CoA. Thereafter, that the pathway flows to three directions, each of which is responsible for the production of individual components of monomers of lignin polymer i.e. *p*-hydroxyphenyl (H), guaiacyl (G) and syringyl (S) units. The generalized lignin biosynthetic pathway is shown in Fig. S1 and important enzymes which regulate the lignin pathway are also highlighted.

High altitude plants habitats are very specific because of their extreme niche areas that have naturally selected the best fit traits for adaptation in extreme environments. Their genes/proteins are being used as molecular tools for engineering crop and other plants for better stress tolerance and adaptability against the present scenarios of climate change. In *Potentilla atrosanguinea* - a Himalayan high altitude (4200 m msl) alpine plant, induction of a new copper-zinc superoxide dismutase (SOD) isozyme was recorded with the cold stress [6]. This gene encodes an isozyme of copper-zinc superoxide dismutase (EU532614.1) (*PaSOD*) and shared a homology of 80% (Expect = 5e-83) with the cytosolic *AtSOD* (NM_100757.3) [6]. Earlier studies confirmed that the protein encoded by this gene was functional from sub-zero temperature to >50°C and also the protein retained activity after autoclaving (heating at 121°C, at a pressure of 1.1 kg per square cm for 20 min) [7,8]. Increased SOD activity has been positively correlated against different stress tolerance such as low/high temperature drought, salinity, high light intensity, and ozone [9,10]. Over-expression of *Potentilla atrosanguinea* copper zinc superoxide dismutase (*PaSOD*) has improved copper and salt stress tolerance in *Arabidopsis thaliana* [11,12]. Apparently, over-expression of the same SOD in potato, enhanced photosynthetic performance under drought stress [13]. In another study, over accumulation of lignin in vascular bundles was found to be the molecular mechanism which underlies the improved stress tolerance induced by the over-expression of *PaSOD* in *Arabidopsis thaliana* [14]. *Rheum australe* - another high altitude alpine plant, up regulation of ascorbate peroxidase was recorded against cold stress (unpublished data). This APX gene from Rheum (*RaAPX*)(DQ078123.2) was one of the cold tolerant genes, which shared a homology of 75% (Expect = 4e-162) with the peroxisomal *APX* gene of *Arabidopsis thaliana* (NM_119666.3). Hence it was thought that high altitude plants can be a good source of potential antioxident genes which can be successfully utilized for the reprogramming of stress tolerance in plants without disturbing native physiology, after functionally validated in the model plant i.e. *Arabidopsis*. In view of the above discussion, the present study illustrates the efficiency of over expressing two important antioxidant genes from high altitude plants namely superoxide dismutase from *Potentilla atrosanguinea* (*PaSOD*) and ascorbate peroxidase from *Rheum australe* (*RaAPX*) in *Arabidopsis thaliana*. It was established that simultaneous over expression of *PaSOD* and *RaAPX* was more effective in alleviating cold stress over independent over expression of either *PaSOD* or *RaAPX*.

Materials and Methods

Plant growth and cold stress treatment

Arabidopsis (ecotype *coloumbia*) plants were grown on soil mixture of vermiculite: peat moss: perlite (1:1:1) in the greenhouse under a 16 h light and 8 h dark cycle at 20±1°C. For stress treatment, 21d old seedlings of wild type, and hygromycin selected transgenic seedlings were transferred to cold room (4°C). Samples were collected 0 h, 6 h, 12 h, 24 h, 48 h (2d), 96 h (4d) and 192 h (8d) during cold treatment for the analyses of transcript levels, enzyme activity, proline and soluble sugars accumulation, electrolyte leakage, ROS, chlorophyll and biomass accumulation estimation etc,.

Plasmid construction and plant transformation

Full length cDNAs of Copper-Zinc Superoxide Dismutase (*PaSOD*) and Ascorbate Peroxidase (now onwards *RaAPX*) from high altitude plants *Potentilla atrosanguinea* (which grows at day-time air temperatures of 3–10°C in Lahaul and Spiti districts of Himachal Pradesh: altitude 4517 m; 32° 24′ 20″N; 077° 38′ 40″ E) and *Rheum australe* (Rohtang Pass in Himachal Pradesh: altitude 4000 m; 32° 22′ 19″ N; 077° 14′ 46″ E), respectively, from Western Himalaya, were cloned in *Arabidopsis thaliana* as described earlier by Gill et al. [14]. Briefly, coding nucleotide sequences of these genes were amplified using the gene specific primers with incorporated *Nco*1 and *Bgl*II restriction sites at 3′ end. PCR products were cloned into a cloning vector pGEMT easy (Promega) and then sub-cloned into binary plant vector pCAMBIA1302 under the Cauliflower mosaic 35 S promoter. The prepared plasmid construct was mobilized into *Arabidopsis* plants via *Agrobacterium* mediated vacuum infiltration method [15]. Seeds were collected and screened in Murashige and Skoog [16] medium supplemented with 20 μg ml^{-1} hygromycin. Homozygous (T3 generation) transgenic *PaSOD* and *RaAPX* plants were crossed to obtain dual transgenic plants. Integration and expression of these transgenes in the dual transgenics was assessed by PCR using gene specific primers with genomic DNA as a template and semi-quantitative PCR from the total RNA isolated from leaf samples.

Gene-specific semi-quantitative and real time PCR

Total RNA was isolated from 0 hrs and 2d cold treated transgenic and the wild type *Arabidopsis* plants using Total RNA extraction kit (Real Genomics). One microgram of total RNA was used for oligo (dT) primed first-strand cDNA synthesis in 20 μl reaction using of Superscript III Reverse transcriptase (Invitrogen). Transcripts of *PaSOD* and *RaAPX* were quantified with PCR using gene specific primers. Constitutively expressed *GAP-C* (Glyceraldehyde-3-phosphate dehydrogenase C subunit) was amplified simultaneously in 27 cycles to ensure equal amounts of cDNA used.

For real time expression analysis of lignin pathway genes during cold stress, primers were designed using Primer 3 software [17]. Details of the primers used in this study are given in Table S1. Gene expression was performed on a Stratagene Mx3000P system (Agilent Technologies, Germany) using 2× Brilliant III SYBR@ Green qPCR Master Mix (Agilent Technologies, Germany). All qPCRs were run in triplicates with a no-template control to check for contamination. PCR was conducted under the following conditions: 10 min at 95°C (enzyme activation), 40 cycles each of 30 s at 95°C, 30 s at 55°C and 72°C for 30 s and a final melting curve analysis was performed (55° to 95°C) to verify the specificity of amplicons. The raw threshold cycle (Ct) values were normalized against a housekeeping gene encoding Actin [18] to calculate both the difference in expression between control and respective treatment samples and tissue specific gene abundance using the Relative Expression Software Tool (REST; [19]). Expression values were transformed (log2) to generate expression profiles. This experiment was repeated three times for the reproducibility and for statistical significance calculation.

SOD and APX enzyme activity assay

Total enzyme activity of SOD and APX was estimated at different time points during cold stress i.e. 0–192 h. Total SOD activity was estimated as described earlier [14] while APX activity was determined according to Nakano and Asada [20]. Briefly, leaf samples (100 mg) were homogenized in a pre-cooled mortar in homogenizing buffer containing 2 mM EDTA, 1 mM DTT, 1 mM PMSF, 0.5% (v/v) Triton-X100 and 10% (w/v) PVPP in 50 mM phosphate buffer pH 7.8. For APX activity homogenizing buffer contained ascorbate in addition and the buffer pH was 7.0. The homogenate was transferred to 1.5 ml Eppendorf and centrifuged at 13,000 rpm for 20 min at 4°C. The supernatant was used to estimate total SOD and APX activities. The total SOD activity was measured by adding 5 μl enzyme extract to a reaction mixture (200 μl) containing 1.5 μm Riboflavin, 50 μm NBT, 10 mM DL-Methionine and 0.025% (v/v) Triton-X100 in

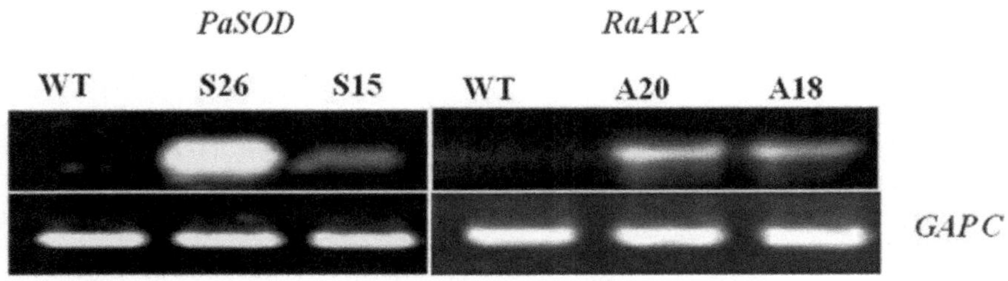

Figure 1. Confirmation of *PaSOD* and *RaAPX* gene expression in transgenic *Arabidopsis* plants at 20°C. The same plants were used for crossing to obtain the double transgenic plants. Expression of genes was (S26 and S15; A20 and A18) confirmed by estimating their transcripts from the leaf tissue by semi-quantitative RT-PCR. Constitutively expressed *GAP-C* was used as loading control.

50 mM phosphate buffer. One unit of enzyme activity was defined as the amount of enzyme required for 50% inhibition of NBT reduction at 25°C. Total protein content was estimated according to the dye binding method of Bradford [21] using BSA as standard. These experiments were repeated three times from three biological replicates.

In-situ ROS staining

In situ ROS staining was done in accordance with Beyer and Fridovich [22], on the basis of the principle of NBT (nitroblue tetrazolium) reduction to blue formazan by $O_2^{-\cdot}$ The intracellular concentration of ROS ($O_2^{-\cdot}$) was directly proportional to the development of intensity of blue color in the leaves. Briefly, leaf tissue was detached from the wild type and transgenic plants and vacuum infiltrated with 10 mM sodium azide (NaN_3) in 10 mM potassium phosphate buffer for 1 min. The infiltrated leaf tissue was incubated in 0.1% NBT (nitroblue tetrazolium) in 10 mM potassium phosphate buffer; pH 7.8 for 30 min. The stained leaf tissue was boiled in acetic acid:glycerol:ethanol (1:1:3) solution to remove other pigments and the stain content was visually documented under Carl-Zeiss Stereo DiscoveryV12 with Axiovision software. This experiment was repeated three times from three biological replicates.

Estimation of electrolyte leakage

Cold-induced cell membrane injuries in leaf tissues were quantified by estimating the electrolyte leakage from the leaf samples. Electrolyte leakage was measured using an electrical conductivity meter as described by Lutts et al. [23]. Briefly, leaves were excised and washed with de-ionized water. One gram of fresh leaves were cut into small pieces (about 1–2 cm) and then immersed in 20 mL of deionized water, incubated at 25°C. After 24 h, electrical conductivity (EC1) of the bathing solution was recorded. These samples were then autoclaved at 120°C for 20 min to completely kill the tissues. Samples were then cooled to 25°C and the final electrical conductivity (EC2) of all released electrolytes was measured. The electrolyte leakage (EL) was expressed following the formula $EL = EC1/EC2 \times 100$. This experiment was repeated three times from three biological replicates.

Estimation of total soluble sugars, proline and chlorophyll content

Total soluble sugar (TSS) was determined by anthrone method of Irigoyen et al. [24]. Free proline content was estimated using the acid ninhydrin method described by Bates et al. [25]. Fresh shoot and root samples (0.5 g) harvested at different time intervals were ground using mortar-pestle with 10 ml of 3% (w/v) sulfosalicylic acid aqueous solutions and the homogenate was filtered through Whatman no. 1 filter paper and taken for further analysis. Two ml

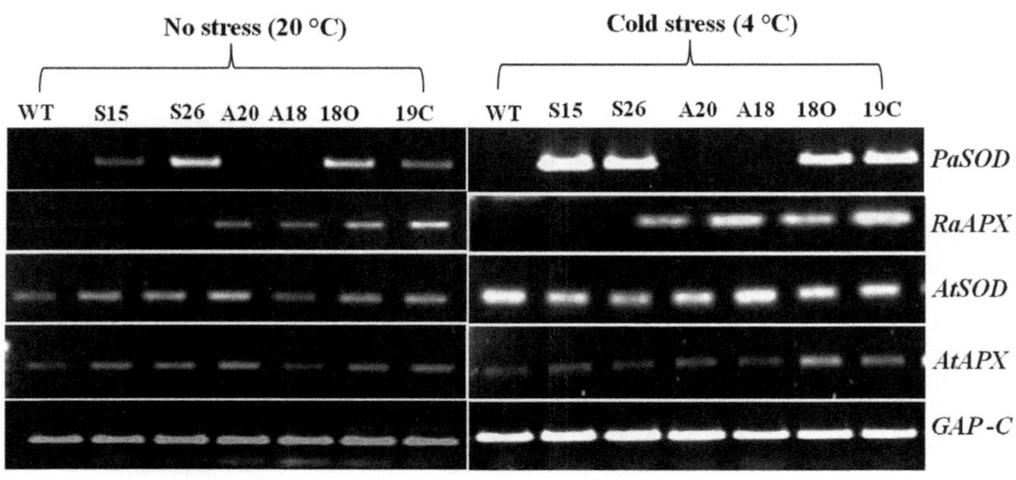

Figure 2. Assessment of native (*AtSOD, AtAPX*) and transgene (*PaSOD* and *RaAPX*) expression by estimating transcript levels under normal (20°C) and cold stress (4°C) conditions. Total RNA was isolated from the leaf samples of WT and transgenic plants, transcripts were amplified by using gene specific primers by semi-quantitative RT-PCR under normal and cold stress conditions. Constitutively expressed *GAP-C* was used as loading control.

Figure 3. Transgenic *Arabidopsis* plants showing tolerance to cold stress. Five weeks old seedlings were transferred to cold (4°C) and allowed to grow for 192 h and different stress tolerance parameters were estimated from the samples of these plants.

of filtered extract, combined with 2 ml acid ninhydrin (1.25 g ninhydrin warmed, in a mixture of 30 ml glacial acetic acid and 20 ml of 6 M phosphoric acid, until dissolved). The reaction mixture was incubated in a boiling water bath (100°C) for 1 h and the reaction was then terminated in an ice bath. Then 4 ml of toluene was added to the reaction mixture and the organic phase (chromophore containing toluene) was warmed to room temperature and its optical density was measured at 520 nm using toluene as blank in a UV-visible spectrophotometer. The amount of proline was determined from a standard curve.

Chlorophyll content was estimated by the method of Arnon [26]. The chlorophyll from the fresh leaves was extracted in 80% acetone and the absorbance was read spectrophotometer (Synergy HT Multi-Mode Microplate Reader, Bio-Tek, USA) at 663 and 645 nm. These experiments were repeated three times from three biological replicates.

Confocal and Scanning Electron Microscopy (SEM)

Confocal microscopy analysis was done as described earlier Gill et al. [14]. Stem of WT and transgenic lines were harvested and fixed in formalin, glacial acetic acid and 50% ethyl alcohol (FAA) (1:1:18) at room temperature. Samples were subsequently dehydrated in a tertiary butyl-alcohol series [27]; embedded in paraffin (melting point 58–60°C) and 8–10 mm sections were cut using a Finsee microtome. Sections were stained with 1% safranin in water and with 4% fast green in clove oil for 4 h and for 30 s, respectively. These were mounted in Canada balsam and examined using Confocal Laser Scanning Microscope (Zeiss LSM510 Meta Gmbh, Germany) equipped with a Zeiss Axiovert 100 M inverted microscope. Lignin auto fluorescence was collected by excitation/emission wave- lengths 488/505 nm.

For SEM analysis, stem cross-sections were fixed in two steps: first with a mixture of 2% paraformaldehyde and 2.5% glutaraldehyde in 0.1 mol/l cacodylate buffer, pH 7.4 for 1 h and then with 1% OsO_4 in 0.1 mol/l cacodylate buffer, pH 7.4 for 30 m. After critical point drying, the samples were sputter-coated with gold, and the coated samples were viewed with a Hitachi S-3400N field emission SEM using an accelerating voltage of 30 kV.

Statistical Analysis

All obtained data were analyzed statistically by Advance Linear/Non-linear models using General Linear Model (STATISTICA release 7, statsoft Wipro, Bangalore, Karnataka, India). Analysis of variance (ANOVA) followed by Duncan's multiple range tests was used for the analysis of the results. The means of each treatment (as mentioned in respective experiments) and their interactions were compared and found statistically significant ($P < 0.05$). The number of samples taken and their replicates are mentioned in respective experiments.

Results

Transgenic *Arabidopsis* over-expressing *PaSOD* in combination with *RaAPX* enhanced cold tolerance

Arabidopsis transgenic plants over expressing *PaSOD* under the control of CaMV35 S promoter were generated as reported earlier [12]. Similarly transgenic *Arabidopsis* plants over expressing ascorbate peroxidase gene from *Rheum* (*RaAPX*) under the control of CaMV35 S promoter were also developed. Presence of these transgenes was confirmed by PCR by using genomic DNA from these plants as template and their expression by the presence of its transcripts from leaf tissues by semi-quantitative RT-PCR (Fig. 1). On the basis of 3:1 segregation and with high enzyme activities, homozygous transgenic lines S26 (for *PaSOD*) and A20 (for *RaAPX*) were selected for further analysis as well as for crossings for raising the transgenics with both these genes. Transgenic *Arabidopsis* plants having both *PaSOD* and *RaAPX* genes were obtained by crossings of these individual gene

Table 1. Growth and biomass production under normal (20°C) and cold stress (4°C) conditions in WT and transgenic plants. ±mean SD values.

Plant type	No of leaves		Rosette diameter (cm)		Plant height (cm)		No of pods		Pod length (cm)		Total seeds	
	20°C	4°C	20°C	4°C	20°C	4°C	20°C	4°C	20°C	4°C	20°C	4°C
WT	18.6±1.2^{c}	19.4±1.2^{c}	3.9±0.4^{d}	3.3±0.3^{d}	26.5±1.07^{ef}	20.8±0.4 ^{g}	43.6±1.1^{de}	31±1.0 ^{g}	1.14±0.1^{fghi}	1.08±0.1^{fghi}	884±14.7^{d}	620±20.0 ^{g}
PaSOD (S15)	25.6±1.6^{ab}	25.4±0.6^{ab}	5.5±0.2^{bc}	6.6±0.2^{a}	27.3±0.94^{cde}	31.8±0.9^{ab}	39.2±0.3f	56.4±0.5^{c}	1.5±0.1^{bcd}	1.62±0.1^{abc}	784±7.4^{f}	1128±10.2^{c}
PaSOD (S26)	24.2±2.0^{b}	24±1.1^{b}	5.3±0.4^{c}	6.72±0.1^{a}	26.5±1.28^{ef}	31.4±1.1^{ab}	37.4±0.9^{f}	58.2±1.0^{c}	1.38±0.1^{cde}	1.68±0.1^{abc}	748±18.5^{f}	1164±21.3^{c}
RaAPX (A20)	17.4±0.6^{c}	27.8±.7^{ab}	5.4±0.2^{c}	6.9±0.2^{a}	23.1±1.12^{fg}	30.2±1.3^{bc}	43.2±1.0^{de}	68.4±0.9^{ab}	1.22±0.1^{fgh}	1.82±0.1^{ab}	864±20.4^{de}	1368±8.5^{ab}
RaAPX (A18)	18.2±0.3^{c}	29±1.0^{a}	5.5±0.2^{bc}	6.68±0.2^{a}	23.7±0.99^{efg}	30±1.4^{bcd}	41.6±0.6^{e}	67.2±0.3^{b}	1.3±0.0^{efg}	1.94±0.1^{a}	832±12^{e}	1344±7.4^{b}
PaSOD + *RaAPX* (18O)	20±1.3^{c}	28.4±0.5^{a}	5±0.4^{c}	6.9±0.5^{a}	25.8±0.73^{ef}	34±1.7^{a}	42.8±0.6^{de}	70.6±1.0^{a}	0.94±0.0^{hi}	1.78±0.1^{ab}	856±13.2^{de}	1412±21.5^{a}
PaSOD + *RaAPX* (19C)	20.4±1.2^{c}	27.2±0.9^{ab}	5.2±0.3^{c}	6.5±0.3^{ab}	26.6±0.68^{ef}	33.2±1.3^{ab}	43.6±0.6^{de}	69.4±0.4^{ab}	0.94±0.1^{hi}	1.96±0.04^{a}	872±12^{de}	1388±8.0^{ab}

Each value is a mean of three separate biological replicates, Mean ± Standard Deviation. Different parameters were analyzed using ANOVA to detect significant difference between means. Means were compared using Duncan's Multiple Range Test (DMRT) at $P<0.05$. Different combinations of alphabets abcdefghi denote the significantly different groups.

transgenic lines. Presence of transcripts for both genes in transgenic plants was confirmed by semi-quantitative RT-PCR (Fig. 2) and considered as transgenic line with both the genes. Two individual transgenic lines of each *PaSOD* (S15 and S26) and *RaAPX* (A18 and A20) were analysed and data was presented. Out of a number of transgenic lines having both the genes, the line 18O and 19C was selected for further analysis and data was presented.

Thus generated single and double transgenics *Arabidopsis* plants were found tolerant when exposed to cold (4°C) stress (Fig. 3). The morphological and vegetative growth of transgenic plants was more or less similar to the wild plants under normal growth conditions i.e. 20±1°C and 16 h/8 h light regime (Fig. S2) whereas, under cold stress the transgenic plants showed better growth (Fig. 3). Effects of cold stress were clearly observed in the growth of wild type plants, where leaves started yellowing which shows the symptoms of chlorosis with stunted growth but the transgenic plants showed better growth. *PaSOD* single transgenic plants (S26 and S15) and double transgenic plants (18O and 19C) were better than the *RaAPX* transgenic plants (A20 and A18). The same was confirmed by the accumulation of transcripts levels also (Fig. 2). *PaSOD* (S26, S15) and *RaAPX* (A20, A18) transgenic lines exhibited expression of *PaSOD* and *RaAPX*, respectively, whereas 18O and 19C exhibited expression of both genes under normal conditions. A low level of transcripts was observed for plants without stress at 20°C. Further, significant over accumulation of transcripts of transgenes were observed in single as well as in double transgenic plants under cold stress (Fig. 2).

PaSOD and *RaAPX* induced cold stress tolerance alters *Arabidopsis* root growth and biomass

Root growth and development pattern was also studied during cold stress. After 192 h of cold stress in the single PaSOD transgenics (S26, S15) and double transgenic lines (18O, 19C) showed increased root growth biomass (Fig. 4). But in A20 and A18 the APX transgenic Arabidopsis plants showed decreased root growth when compared to WT. In order to determine if this difference existed in young seedlings at the in vitro stage, root growth on normal medium without any stress was also measured. A marginal increase in root length was observed in between WT and transgenic plants under normal growth conditions (Fig. S3). Initially, WT and transgenic plants exhibited similar root growth under cold stress, but WT root growth rate was lower than that of transgenic, which had longer roots after a prolonged culture period (10 d). Transgenic plants efficiently tolerated the stress and exhibited better root growth and biomass production in cold stress conditions (Table 1). Transgenic *Arabidopsis* plants over expressing *RaAPX* perform more or less similar in certain aspects of vegetative growth like rosette diameter, number of leaves; reproductive growth in terms of number of pods, length and number of seeds per silique, with the transgenic over expressing both *PaSOD* and *RaAPX* genes under cold stress conditions (Table 1). But overall, the transgenic plants over expressing both the genes performed better under cold stress.

Reduced accumulation of ROS contents under cold stress

Accumulation of ROS may cause damage to many bio molecules of the cells. As the transgenic lines were more tolerant to cold stress than WT, an attempt was made to compare accumulation of ROS in their leaf samples at the end of the cold stress by staining with NBT. Histo-chemical staining of leaves from WT and transgenic plant with NBT revealed that it could also be stained ROS without cold stress also (Fig. 5). The figure showed that the transgenic lines had slightly lower ROS accumulation

Figure 4. Root growth of WT and transgenic plants under cold stress. A). Photograph showing root growth of transgenic plants after cold stress (192h). B). Graphical representation of root length of transgenic plants after cold stress (192 h). Values are the representation of Mean±SE of three biological replicates. Transgenic plants developed better root system and thus exhibited better stress tolerance.

relative to WT without stress. Cold resulted in significantly higher levels of ROS accumulation in WT leaves whereas the transgenic lines S26, S15, A20, A18, 18O and 19C low levels accumulation as evidenced by the lower intensity of the blue colour (Fig. 5). As expected, the double transgenic line 18O and 19C showed notable less blue colour than *PaSOD* (S26, S15) and *RaAPX* (A20, A18) transgenic plants. This implies that less ROS was accumulated in these transgenic lines.

Accumulation of *PaSOD* and *RaAPX* in transgenic *Arabidopsis* improved the cold stress tolerance

As the main objective of the present study was to improve the scavenging of ROS by the over expression of supplemental

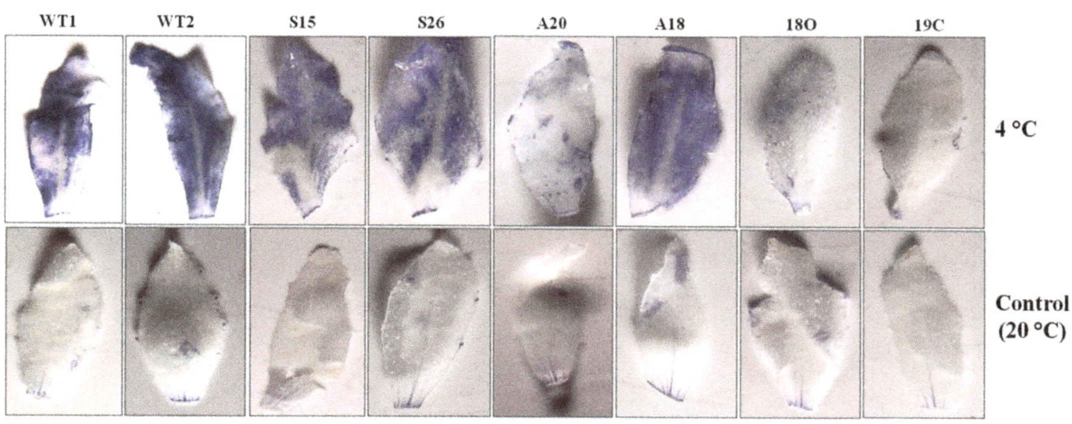

Figure 5. Histochemical staining assay of ROS accumulation with nitro blue tetrazolium (NBT) in the wild type and transgenics under cold stress. Leaves from potted plants that have been subjected to cold stress for 96 h were used for the accumulation of ROS.

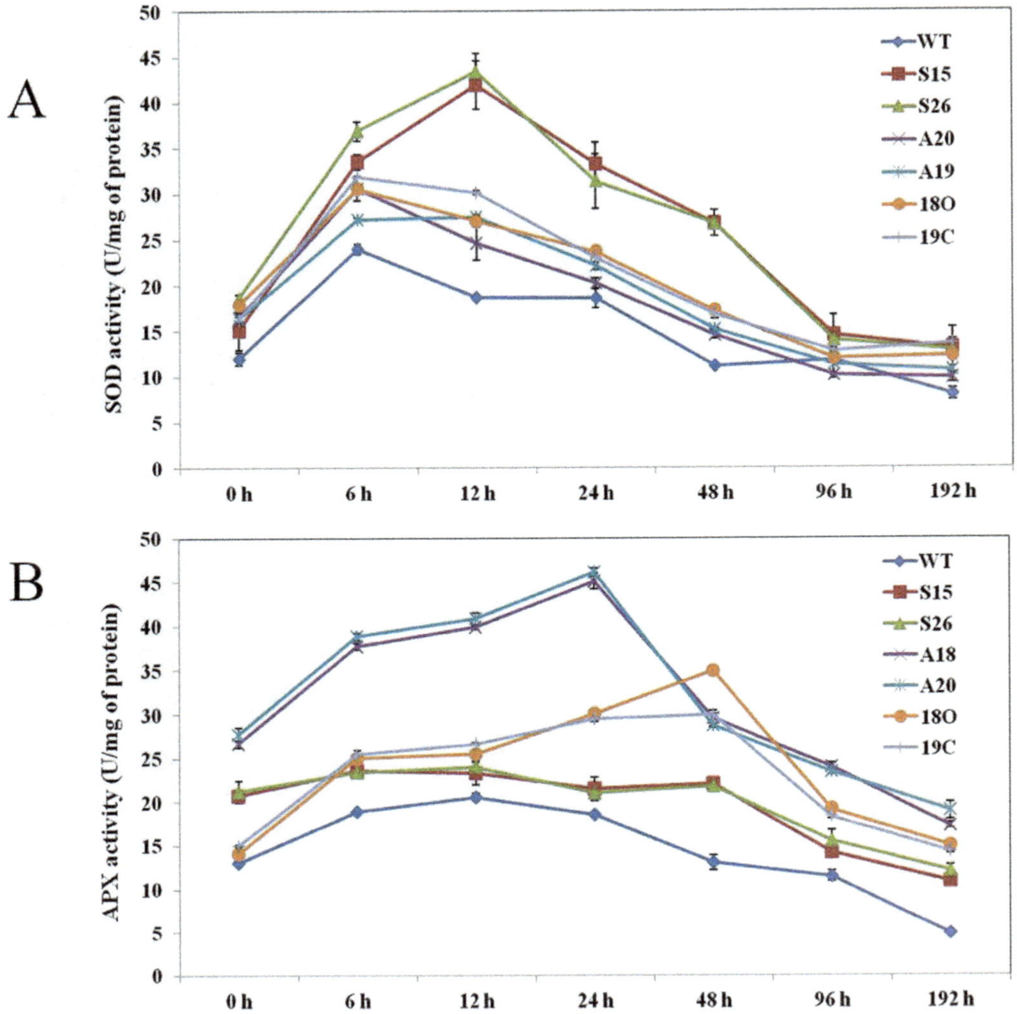

Figure 6. Estimation of total enzyme activities during cold stress both in WT and transgenics, A) superoxide dismutase and B) ascorbate peroxidase activity. Error bars represents ± SE of mean of three biological replicates.

antioxidant enzymes to the native ones, we checked the expression of the transgenes in the transgenics during cold stress. Superoxide dismutase and ascorbate peroxidase enzyme activities were estimated from the leaf samples of WT and transgenic plants at different time points during cold stress i.e. from 0–192 h under cold stress. As expected total SOD and APX activities were significantly higher in transgenic plants when compared to WT plants even under normal growth conditions, whereas 1.6, 1.2 and 1.3 folds increase in SOD levels was observed in SOD (S26, S15), APX (A20, A18) and double transgenic SOD and APX (18O, 19C) plants, respectively throughout 192 h of cold stress (Fig. 6). In the case of ascorbate peroxidase, 1.7-fold increase was recorded in transgenic plants A20 and A18 plants over WT plants under normal growth conditions and an increase of 2 fold was recorded throughout the cold stress. In double transgenic lines (18O and 19C) 1.3 folds higher activity was recorded. Abrupt elevation of these enzyme activities were recorded during the initial 12 h of cold in WT and all the transgenic lines and consequently a gradual decrease was recorded between 12 and 96 h, after which the minimal levels were maintained to sustain the growth and development of plants. Improved growth in terms of increase in biomass during cold stress in transgenic plants might be because of presence of the additional copy of antioxidant gene(s), implies increase in the activity of corresponding enzymes. Superoxide dismutase and ascorbate peroxidase enzyme activities were also estimated from the leaf samples of WT and transgenic plants at different time points without cold stress for comparisons. Slightly increased levels of SOD and APX activities were recorded in all the transgenic plants when compared to WT (Fig. S4).

Over expression of antioxidant enzymes protects cellular membranes from freezing injury

Stress accumulated ROS causes lipid peroxidation wherein the lipids in the cell membranes are damaged. Cold-induced cell membrane injuries in leaf tissues of transgenic lines were quantified by estimating the electrolyte leakage during *PaSOD* and *RaAPX* up regulation induced cold stress tolerance (Fig. 7). Chilling induced ion leakage was more in WT plants, when compared to transgenic plants. Whereas no difference was observed under normal conditions i.e. 20°C (Fig. S5). The relative electrolyte conductivity (REC) of transgenic plants was significantly lower than that of wild-type plants during cold stress. The results demonstrated that membrane damage was more in WT plants than that in the transgenic plants against cold stress. The ion leakage increased up to 48 h exposure to cold in *PaSOD*,

RaAPX and double transgenic lines, but after that there was a decline. After prolonged exposure to cold the ion leakage is more prominent in WT (Fig. 7), whereas transgenics were less affected by the cold. It seems that the over expressing antioxidant enzymes might be protecting the membrane damage.

Cold stress induces increase in aminoacid, soluble sugars and chlorophyll contents in transgenics

An increase in proline and simple sugars against stress is commonly known phenomena in *Arabidopsis* and other plants [28,29,30]. Under normal conditions, more or less similar amounts of proline content was recorded in all the plants, but higher levels of proline were found under stress in the transgenic lines when compared to corresponding WT plants (Fig. 8). Similar trend was observed in the case of soluble sugars under cold stress (Fig. 8). Overall, these findings suggested that the content of proline and soluble sugars, which are known osmo-protectants, accumulated more in transgenic plants under cold stress, thus providing better stress tolerance and better growth.

Chlorophyll content is susceptible to cold stress and usually continuously declines with cold stress [31]. In this study, chlorophyll content in the wild-type plants decreased at 4°C with the prolonged periods of cold stress. However, the transgenic plants have higher chlorophyll content than the WT plants under stress (Fig. 8). Thus cold tolerance of transgenic plants over expressing *PaSOD*, *RaAPX* and double transgenic plants showed increased chlorophyll accumulation.

Cold stress tolerance by simultaneous over-expression of *PaSOD* and *RaAPX* is because of the accumulation of lignin in the vascular bundles

Anatomical investigation of vascular changes during cold stress by using Confocal and scanning electron microscopy revealed that there was a dramatic alteration in lignin accumulation in transgenic plants as compared to the WT (Fig. 9A). Inter-fascicular lignification was evident under cold stress in single and double transgenics, whereas inflorescence stem of WT showed lignin deposition in the vascular bundles only (Fig. 9). Lignification of the inter-fascicular arcs was conspicuous along the entire stem and showed vascular bundles connected by lignified arcs in all transgenic lines, but more conspicuous in S26, S15 and 18O, 19C under stress condition only (Fig. 9). In addition, xylem elements of vascular bundles of the transgenic lines were larger than that of the WT plants.

The thickness of the walls of vascular bundles of the transgenic and WT were more or less the same (3μ) under normal conditions, whereas under cold stress the transgenic showed thicker walls. The thickness was more in double transgenics (18O and 19C) followed by *PaSOD* transgenics (S26 and S15) and *RaAPX* transgenics (A20 and A18) i.e. 12, 8 and 4 μ, respectively (Table 2).

PaSOD and *RaAPX* induced cold tolerance modulated lignin biosynthesis pathway in transgenic *Arabidopsis*

Higher vascular lignin thickenings were observed in transgenics, it was assumed that increased H_2O_2 levels might be responsible for these accumulations. Lignin is a polymer where different subunits of lignin (4-hydroxy cinnamyl aldehyde, coniferyl alcohol and sinapoyl alcohol) are produced via phenylpropanoid pathway and polymerized by peroxidase (*AtPRX9GE*) in to lignin. To understand the possible alterations in lignin biosynthesis pathway with the simultaneous over expression of *PaSOD and RaAPX* in transgenics, the expression profiles of different genes of this pathway namely *COMT*, *CAD*, *4CL*, *TRYT*, *PXR*, *C4 H*, *CcOMAT*, *C3 H*, *HCT*, *LAC*, *UGT*, *F5 H* and phenylalanine ammonia-lyase 1 (*PAL1*) were assessed by real time PCR. Generated expression profile data confirmed that in transgenic lines an up-regulation of the expression of all these genes tested was recorded under cold stress as compared to the WT. The heat map shows an overview of expression profiles of the phenylpropanoid pathway genes, and their interactions are shown in Fig. 10. Almost all the genes tested showed a basal level expression in

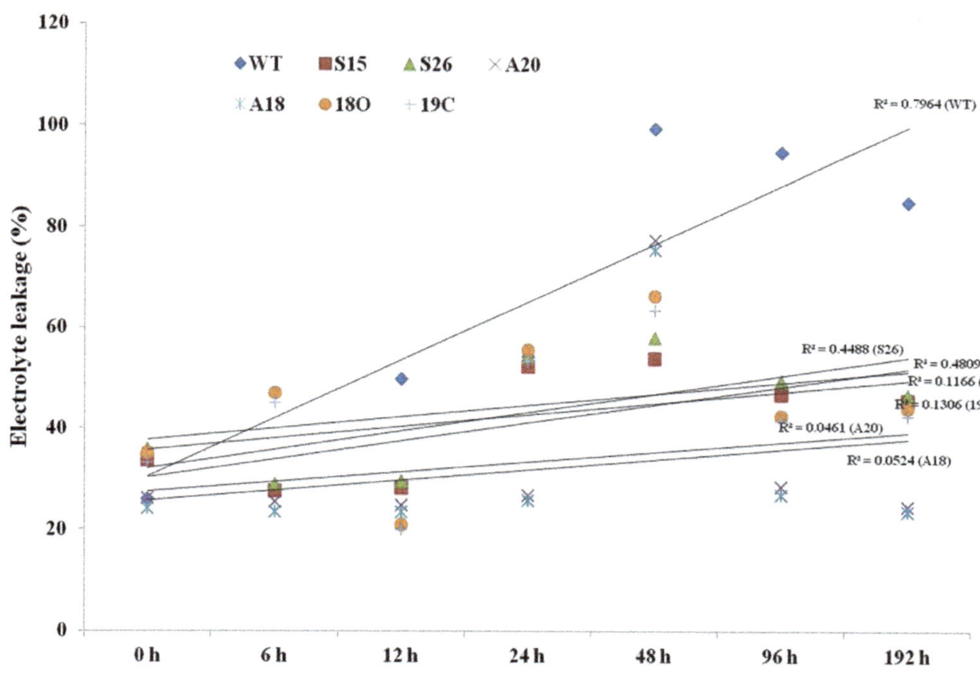

Figure 7. Change in relative electrolyte conductivity (electrolyte leakage) during cold stress in WT and transgenic plants. Error bars represents ± SE of mean of three biological replicates.

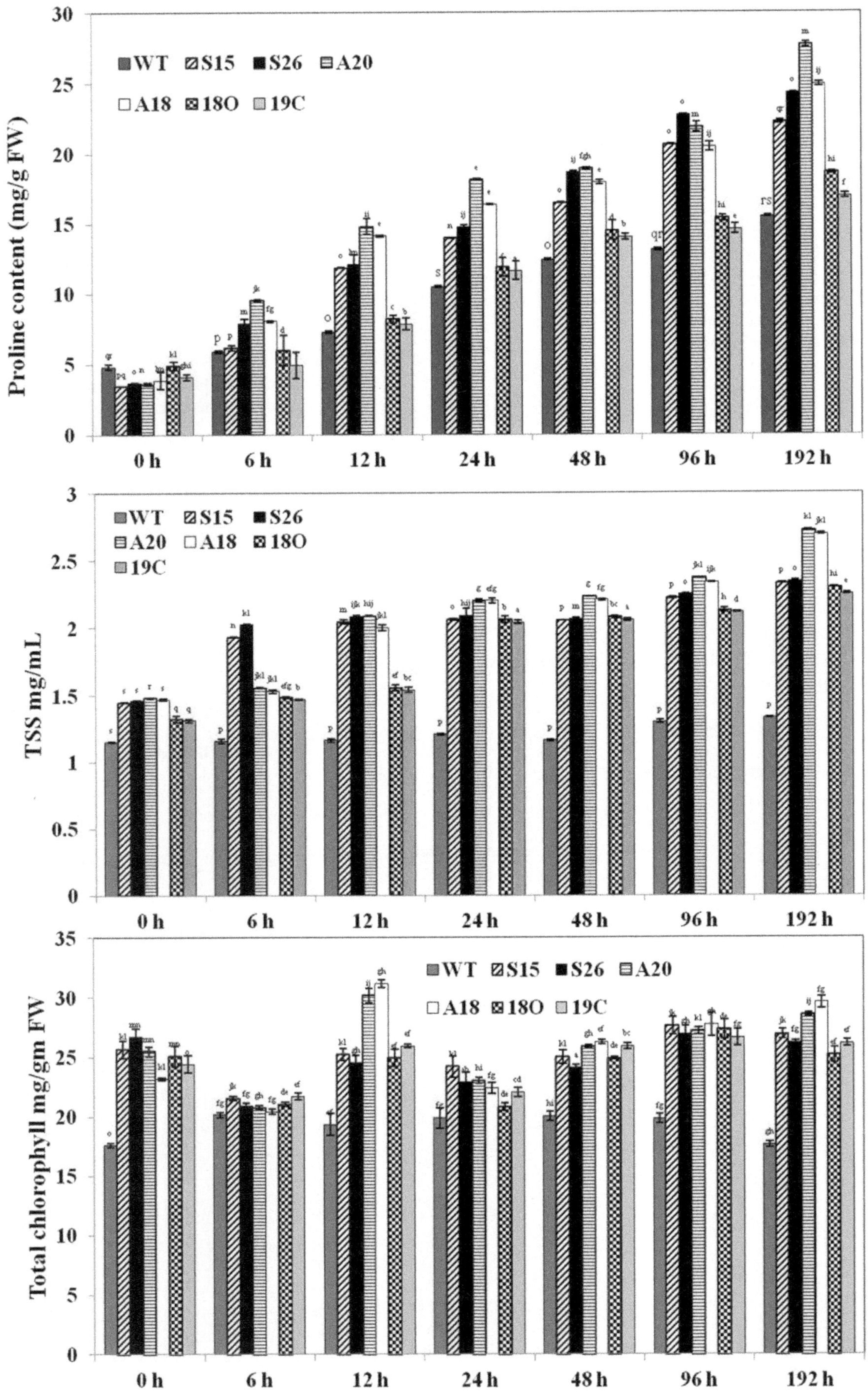
WT
S15
S26
A20
A18
18O
19C
Proline content (mg/g FW)
0 h
6 h
12 h
24 h
48 h
96 h
192 h
TSS mg/mL
Total chlorophyll mg/gm FW

Figure 8. Estimation of different stress tolerance parameters in transgenics during cold stress (at 4°C). (A) Estimation of Proline content (B) Estimation of soluble sugars and (C) estimation of chlorophyll content. In all the cases, data is the mean of three biological replicates. Duncan's multiple range tests was used to establish the statistical significance at $P<0.05$.

PaSOD (S15 and S26) and *RaAPX* (A18 and A20) transgenics, whereas, comparatively higher expression was recorded except *CAD, COMT, C3 H, PXR1* in double transgenics (18O and 19C) under normal conditions (20°C). An up regulation was observed in the expression of the core phenylpropanoid pathway genes, including *PAL*, cinnamate 4-hydroxylase (C4 H), and 4-coumarate CoA ligase (4CL) under cold stress. After that, the lignin pathway divides into three specific branches for the formation of H, G and S lignin monomers. Most of the identified genes involved in these three pathways were up-regulated during cold stress. *F5 H, LAC4, C3 H, PAL1, PXR1, CoMT1* were the commonly up regulated both in *PaSOD* and double transgenic plants. In *RaAPX* over-expressed lines i.e. A18 and A20 a basal level expression was recorded in *4CL2, CAD2, COMT1* even in cold stress. Overall, the results indicated that the activation of the phenylpropanoid pathway led to the synthesis of lignin in response to cold stress by the simultaneous over expression of *PaSOD* and *RaAPX*.

Discussion

As compared to other living organisms, plants need to protect themselves from different environmental, abiotic and biotic stresses. These stresses cause oxidative damage, ion toxicity and disruption of cellular homeostasis through the production of reactive oxygen species (ROS). ROS induces various complex biochemical, molecular, cellular, and physiological changes in plants including damage of DNA, proteins and membrane lipids [14,32]. The responses to abiotic stresses can generally be separated into two categories: those that regulate signal transduction and gene expression in response to the stress, and those that protect against environmental stress by redirection of metabolism. Resistance to cold and salt stress requires a number of changes in gene expression [33,34,35]. Plants have developed a wide range of enzymatic and non-enzymatic mechanisms to scavenge ROS. Among the enzymatic methods, superoxide dismutase (SOD) is one of the crucial enzymes in the plant's defense mechanism that converts superoxide anion radicals ($O_2^{-\cdot}$) to hydrogen peroxide (H_2O_2), there by imparting protection against the harmful effects of highly reactive $O_2^{-\cdot}$ [14,36]. This in turn with the help of other antioxidant enzymes such as APX and catalase converts them in to

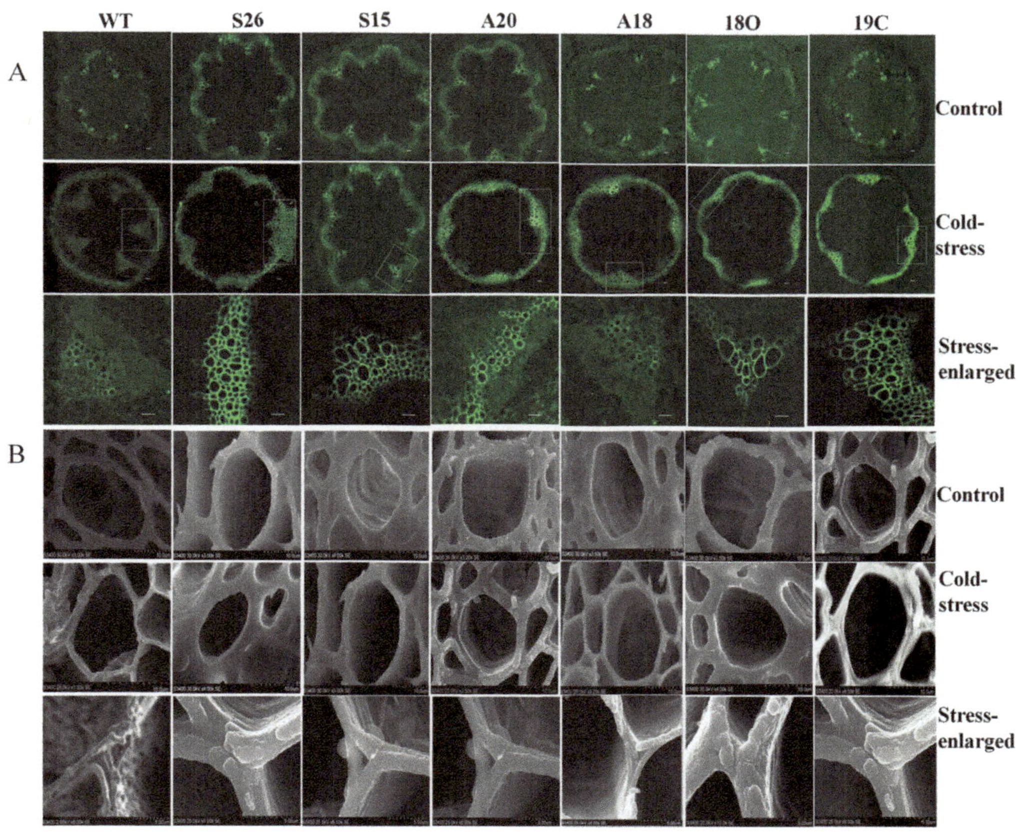

Figure 9. Pattern of lignin accumulation in transgenic *Arabidopsis* plants during cold stress. A) Transverse section of inflorescence stem of WT and transgenic lines (S26, S15, A20, A18, 18O and 19C) showing lignin accumulation pattern in WT and transgenic plants under cold stress. Close up view of the single vascular bundle of stress tissue is shown in the lower most panels (close view). Samples were collected after 192 h of the cold treatment. Lignin auto fluorescence was collected by excitation/emission wave lengths 488/505 nm by Confocal Laser Scanning Microscope. B) SEM analysis of vascular bundle walls showing thickness of lignin during cold stress.

Table 2. Details of the increase in lignifications of vascular bundle cell wall's width during cold stress in transgenic *Arabidopsis* over expressing *PaSOD, RaAPX* and both.

Plant type	Width of the vascular bundle wall (μ) ±SE	
	Control (20°C)	Cold stress (4°C)
WT	3±0.15	2±0.15
PaSOD (S26)	3±0.12	8±0.70
PaSOD (S15)	3±0.21	8±0.29
RaAPX (A20)	3±0.12	4±0.27
RaAPX (A18)	3±0.22	4±0.15
PaSOD + *RaAPX* (18O)	3±0.17	12±0.17
PaSOD + *RaAPX* (19C)	3±0.25	14±0.52

Each value is a mean of three separate biological replicates.

safer molecules like H_2O and O_2. The alpine climate is very harsh with low temperatures, the length of the daily photoperiod, and the quality of the incident light influences the plant growth in these areas. Because of having morphological, physiological and genetical adaptations to these stresses, the alpine vegetation can grow even in such harsh condition. So it is expected that these plants are known to have better enzymatic system whose activities are enhanced or are better adapted than their counterparts from plains. High altitude plants with efficient and active protective systems such as autoclavable superoxide dismutase from Potentilla [6], cell wall hydrolases with broad temperature range activity from Podophyllum [37], make these plants sustain under very low temperature conditions. Dissection of these adaptations in these plants has provided insights towards unraveling of the possible mechanisms of stress tolerance [14,11]. Cold stress was shown to enhance the levels of different ROS-scavenging enzymes [38,39] and accumulation of H_2O_2 in *Arabidopsis* callus cells was reported [40]. Scebba et al. [5] reported the induced expression of antioxidant enzymes SOD and CAT in wheat under cold stress. Over expression of *PaSOD* has improved salt and copper stress in *Arabidopsis thaliana* [11,12].

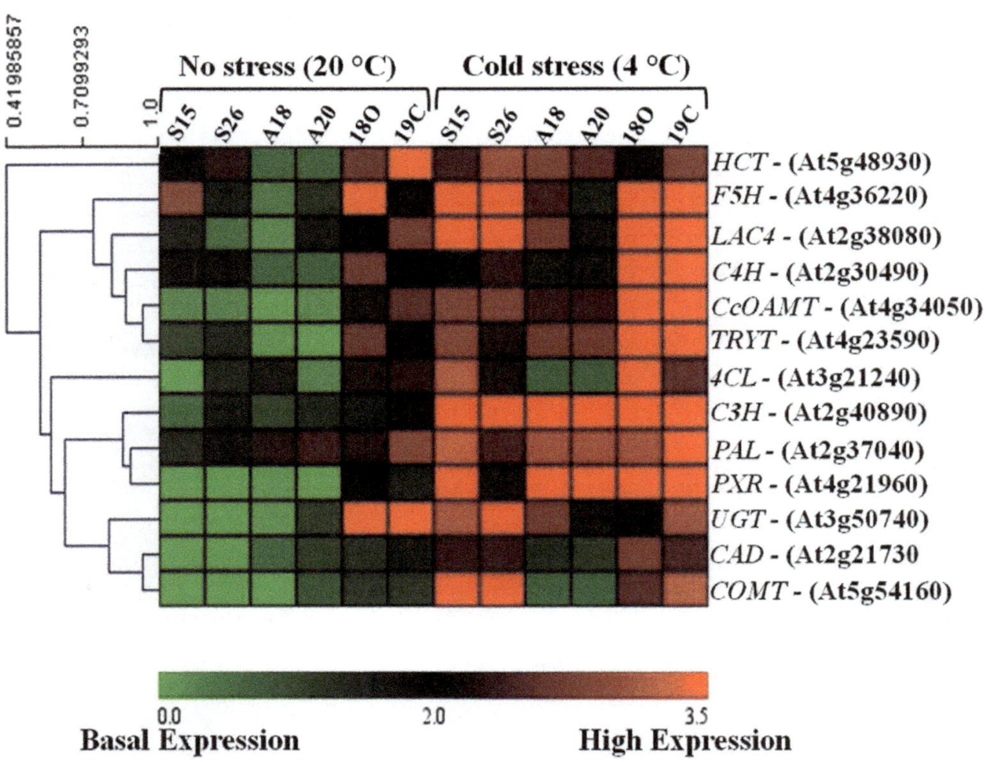

Figure 10. Heat Map of lignin biosynthesis pathway genes in transgenic *Arabidopsis* plants under cold stress (4°C). Phenylalanine ammonia lyase (*PAL1*), cinnamate 4-hydroxylase (*C4 H*), 4-coumarate:CoA ligase (*4CL-2*), ferulate 5-hydroxyase (*F5 H*), hydroxycinnamoyl-CoA transferase (*HCT*), *p*-coumarate 3-hydroxylase (*C3 H*), caffeoyl-CoA *O*-methyltransferase (*CCoAOMT1*), ferulate 5-hydroxylase (*F5 H*), caffeic acid *O*-methyltransferase (*COMT1*), cinnamyl alcohol dehydrogenase (*CAD2*), Laccase (*LAC4*), UDP-glucosyltransferase (*UGT 721 E1*) and Tyrosine ammonia lyase (*TRYT*) analysed in SOD, APX and double transgenic *Arabidopsis* plants.

Effects of low temperature stress includes reduced growth [41,42], wilting [43], chlorosis [44], impaired photosynthesis, increased electrolyte leakage across membranes [45] and impaired reproductive development which ultimately reduces the yield [46,47,48,49]. Reports were also available on successful manipulation of abiotic stresses with the simultaneous over expression of different antioxidant enzymes in Fescue plants, sweet potato and tobacco [50,51,52].

Present investigation dealt with the improvement of cold stress tolerance and studied the possible mechanism in *Arabidopsis* by the simultaneous over expression of antioxidant genes/enzymes *PaSOD* and *RaAPX* from the cold niche area plants. The present results showed that the WT plants were 'chilling sensitive' with stunted growth, reduced metabolism and with low biomass accumulation during cold stress. But the transgenic plants because of the manipulation of ROS detoxification mechanism by the over expression of antioxidant enzymes *PaSOD* and *RaAPX* performed better showing normal growth, more biomass, well developed root system, and higher yields in terms of more number of seeds per silique and thus have become 'chilling tolerant'. In the present results are in agreement with the work of Yuanyuan et al. [53], where he shows that increased soluble sugars play an important role during cold acclimation process, in transgenic *Arabidopsis* plants and also accumulation of soluble sugars and proline accumulation was recorded in response to cold stress. Soluble sugars exert their positive effects to protect plant cells from damage caused by cold stress through several ways, including serving as osmo-protectants, nutrient as well as interacting with the lipid bi-layer. These changes might also be responsible for maintaining the membrane stability and thus leading to reduced electrolyte leakage in the transgenic plants. Superoxide dismutase induced membrane damage tolerance was also reported by Kwon et al. [54] in transgenic tobacco plants. The photosynthetic performance of transgenic *Arabidopsis* plants was commendable with higher chlorophyll content throughout the cold stress whereas the WT plants were not able to sustain the lesser chlorophyll. Perhaps, this might be the reason in slower/stunted growth of WT during stress.

Lignins are the secondary metabolites, which maintain the structural integrity and provide mechanical strength to the stem and also impart hydrophobicity to vasculature allowing water and nutrient transportation [55]. The ROS catalysed by SOD and gets reduced into H_2O_2 which has a versatile role in the plant metabolic system. These molecules at low concentration acts as mediator of signalling pathways leading to stress acclimation at higher concentration it orchestrates the cellular damage and death [40]. Hydrogen peroxide acts as secondary messenger in stress signalling pathways due to its long half life and high permeability across membranes [56,57,58]. In addition to this the elevated levels of H_2O_2 have been shown to induce polymerization of lignin monomers (H-, S- and G- lignin) in to lignin [59].

Lignin induction has been correlated with cold, drought, or light stresses as well as mechanical injuries in a large number of different plant species, such as poplar, rice, pine, *Arabidopsis*, and soybean [60]. In our earlier study also, over expression of *PaSOD* induced lignin accumulation in the vascular bundles under NaCl stress which has been demonstrated to be the molecular basis for improved stress tolerance [14]. In the present study, anatomical investigation of vascular structures using Confocal and electron microscopy clearly showed that accumulation of lignin in the walls of the vascular strands of transgenic plants with cold stress. Transgenic plant expressing *PaSOD* alone and double transgenics were more lignified than *RaAPX* and WT plants. This high deposition of lignin in *PaSOD* transgenics, is probably due to the over-accumulation of H_2O_2 produced by scavenging of over produced superoxide ion ($O2^{-\cdot}$) under cold stress. Over-accumulation of H_2O_2 is toxic [40] and subsequent increased lignification may lead to choking of vascular system, thus cause damage and cell death in WT. In double transgenics, detoxification mechanism of increased levels of H_2O_2 can be achieved by the over expression of extra copy of downstream enzyme i.e. APX. This reduces the over-accumulated H_2O_2 levels and also probably distributes towards cell-signaling and lignifications and hence better adaptation to withstand the stress conditions. In the present study, the real time transcript analysis using qPCR also confirmed the up regulation of lignin biosynthetic pathway. A clear up regulation the core enzymes of the phenylpropanoid pathway namely PAL, C4 H and 4CL, a clear up regulation was observed in transgenics under cold stress. After that the pathway divides in to three directions, each is responsible for the production of individual components of monomers of lignin polymer. It is possible that the up regulated genes might have a significant role in antioxidant induced lignification during cold stress.

In conclusion, the simultaneous over expression of *PaSOD* and *RaAPX* from high altitude plants improved cold stress tolerance in *Arabidopsis*. This stress tolerance might be because of over expression of highly evolved and cold adapted genes, namely *PaSOD* and *RaAPX* as evidenced by the enzymes assays, which also detoxifies ROS as evidenced by the *in situ* ROS staining of leaves. Increased accumulation of TSS, proline in transgenic plants during cold stress, probably might be involved in protecting the membrane integrity as a result low electrolyte leakage was observed. Transgenic *Arabidopsis* plants had long roots with higher biomass without compromising yield, which is consistent with improved stress tolerance. No significant increase in all these parameters under normal growth condition (20°C) was also additional evidence to state that the cold stress tolerance is because of the over-expression of *PaSOD* and *RaAPX* only. This work clearly demonstrates that the modulation of endogenous ROS scavenging capacity against abiotic stresses can be successfully engineered by the simultaneous over expression of *PaSOD* and *RaAPX* in *Arabidopsis*. In addition, the results outlined the importance of the cytosolic antioxidant machinery in the cross-protection from multiple stresses in agriculturally important plants.

Supporting Information

Figure S1 Generalised scheme of lignin biosynthesis pathway. Each arrow shows a reaction in the pathway, and next to each arrow is the name of the enzyme that catalyzes the associated reaction, Abbreviations are listed as per pathway. CM, CHORISMATE MUTASE; PAT, PREPHENATE AMINOTRANSFERASE; AGT, AROGENATE DEHYDRATASE; PAL, PHENYLALANINE AMMONIA LYASE; C4H, CINNAMATE 4-HYDROXYLASE; 4CL, COUMARATE CoA LIGASE; CCR, CINNAMOYL-CoA REDUCTASE; CAD, CINNAMYL ALCOHOL DEHYDROGENASE; HCT, HYDROXYCINNAMOYL-CoA TRANSFERASE; C3H, *P*-COUMARATE 3-HYDROXYLASE; F5H, FERULATE 5-HYDROXYLASE; COMT, CAFFEIC ACID *O*-METHYLTRANSFERASE; UGT, UDP-GLUCOSYLTRANSFERASE; LAC, LACCASES; PXR, PEROXIDASES; CCoAOMT, CAFFEOYL-CoA 3-O-METHYLTRANSFERASE; H lignin, p-hydroxy phenyl lignin; G lignin, guaiacyl lignin monomers; S lignin, syringyl lignin monomers.

Figure S2 Growth of transgenic *Arabidopsis* plants under normal conditions (20°C) without cold stress. Five weeks old seedlings were transferred and allowed to grow at 20°C for 192 h and different stress tolerance parameters were estimated from the samples of these plants.

Figure S3 Root growth of WT and transgenic plants under normal conditions. A) Photograph showing root growth of transgenic plants without cold stress (192 h). B). Graphical representation of root length of transgenic plants without cold stress (192 h). Values are the representation of Mean±SE of three biological replicates. Transgenic plants developed better root system under normal growth conditions.

Figure S4 Estimation of total enzyme activities under normal growth conditions without cold stress both in WT and transgenics. A) superoxide dismutase and B) ascorbate peroxidase activity. Error bars represents ± SE of mean of three biological replicates.

Figure S5 Change in relative electrolyte conductivity (electrolyte leakage) in WT and transgenic plants under normal growth conditions. Error bars represents ± SE of mean of three biological replicates.

Table S1 Primer Sequence and PCR conditions for all the genes (including gene id and accession numbers) used for Semi-quantitative and Real Time expression analysis.

Acknowledgments

A.S., T.G. and V.D. acknowledge Fellowship by the CSIR, New Delhi, India. Authors thank Dr. Sanjay Kumar for providing cDNA of *RaAPX*. The manuscript represents CSIR-IHBT communication number 3576.

Author Contributions

Conceived and designed the experiments: PSA YS. Performed the experiments: AS VD TG. Analyzed the data: PSA YS. Contributed reagents/materials/analysis tools: AS VD TG PSA YS. Wrote the paper: PSA YS.

References

1. Scandalios JG (2005) Oxidative stress: molecular perception and transduction of signals triggering antioxidant gene defenses. Brazilian J. Med. Biol. Res. 38: 995–1014.
2. Hu D, Klann E, Theils E (2007) Superoxide dismutase and hippocampal function: age and isozymes matter. Antioxidant Redox. Signal. 9: 201–210.
3. Noctor G, Foyer CH (1998). Ascorbate and Glutathione: keeping active oxygen under control. Ann. Rev. Plant Physiol. Plant Mol. Biol. 49: 249–279.
4. Raven EL (2000) Peroxidase-catalyzed oxidation of ascorbate. Structural, spectroscopic and mechanistic correlations in ascorbate peroxidase. Subcell. Biochem. 35: 317–349.
5. Scebba F, Sebastiani L, Vitagliano C (1999) Protective enzymes against activated oxygen species in wheat (*Triticum aestivum* L.) seedlings: Responses to cold acclimation. J. Plant Physiol. 155: 762–768.
6. Sahoo R, Kumar S, Ahuja PS (2001) Induction of a new isozyme superoxide dismutase at low temperature in *Potentilla astrisanguinea* Lodd. variety *argyrophylla* (Wall.ex. Lehm) Griers. J. Plant Physiol.158: 1093–1097.
7. Yogavel M, Mishra PC, Gill J, Bhardwaj PK, Dutt S, et al. (2008) Structure of a superoxide dismutase and implications for copper-ion chelation. Acta Cryst. Biol. Cryst. D64: 892–901.
8. Kumar A, Dutt S, Bagler G, Ahuja PS, Kumar S (2012) Engineering a thermo-stable superoxide dismutase functional at sub-zero to >50°C, which also tolerates autoclaving. Sci. Rep. 2: 387: 1–8.
9. Wang Y, Ying Y, Chen J, Wang X (2004) Transgenic *Arabidopsis* overexpressing Mn-SOD enhanced salt-tolerance. Plant Sci. 167: 671–677.
10. Badawi GH, Kawano N, Yamauchi Y, Shimada E, Sasaki R, et al. (2004) Over-expression of ascorbate peroxidase in tobacco chloroplasts enhances the tolerance to salt stresses and water deficit. Plant Physiol. 121: 231–238.
11. Gill T, Kumar S, Ahuja PS, Sreenivasulu Y (2010b) Over-expression of *Potentilla* superoxide dismutase improves salt stress tolerance during germination and growth in *Arabidopsis thaliana*. J. Plant Genet. Transgen. 1: 1–10.
12. Gill T, Dogra V, Kumar S, Ahuja PS, Sreenivasulu Y (2012) Protein dynamics during seed germination under copper stress in *Arabidopsis* over-expressing *Potentilla* superoxide dismutase. J. Plant Res. 125: 165–172.
13. Pal AK, Acharya K, Vats SK, Kumar S, Ahuja PS (2013) Over-expression of *PaSOD* in transgenic potato enhances photosynthetic performance under drought. Biol. Planta. 57: 359–364.
14. Gill T, Sreenivasulu Y, Kumar S, Ahuja PS (2010a) Over-expression of superoxide dismutase exhibits lignifications of vascular structures in *Arabidopsis thaliana*. J. Plant Physiol. 167: 757–760.
15. Bechtold N, Ellis J, Pelletier GC (1993) *In planta* Agrobacterium mediated gene transfer by infiltration of adult *Arabidopsis* plants. C. R. Acad. Sci. Paris. 316: 1194–1199.
16. Murashige T, Skoog F (1962) A revised medium for rapid growth and bioassays with tobacco tissue cultures. Physiol. Plant. 15: 473–476.
17. Rozen S, Skaletsky H (2000) Primer3 on the WWW for general users and for biologist programmers. Methods Mol. Biol. 132: 365–386.
18. Shi LZ, Wang R, Huang G, Vogel P, Neale G, et al. (2011) HIF1α–dependent glycolytic pathway orchestrates a metabolic checkpoint for the differentiation of T_H17 and T_{reg} cells. J. Exp. Med. 208: 1367–1376.
19. Pfaffl MW, Horgan GW, Dempfle L (2002) Relative expression software tool (REST@) for group-wise comparison and statistical analysis of relative expression results in real-time PCR.Nucleic Acids Res. 30: e36.
20. Nakano Y, Asada K (1981) Hydrogen peroxide is scavenged by ascorbate-specific peroxidase in spinach chloroplasts. Plant Cell Physiol. 22: 867–880.
21. Bradford M (1976) A rapid and sensitive method for the quantitation of microgram quantities of protein utilizing the principles of protein dye binding. Analytical Biochem. 72: 248–254.
22. Beyer WF, Fridovich I (1987). Assaying for superoxide dismutase activity: some large consequences of minor changes in conditions. Anal. Biochem. 161: 559–566.
23. Lutts S, Kinet J, Bouharmont J (1996). NaCl induced senescence in leaves of rice (*Oryza sativa*) cultivars differing in salinity resistance. Annal. Bot. 78: 389–398.
24. Irigoyen JJ, Emerich DW, Sanchez-Diaz M (1992) Water stress induced changes in concentrations of proline and total soluble sugars in nodulated alfalfa (*Medicago sativa*) plants. Physiol. Plant. 84: 55–60.
25. Bates L, Waldren RP, Teare ID (1973) Rapid determination of free proline for water-stress studies. Plant and Soil. 39: 205–207.
26. Arnon DI (1949) Copper enzymes in isolated chloroplasts, polyphenoxidase in *Beta vulgaris*. Plant Physiol. 24: 1–15.
27. Jensen WA (1962) Botanical Histochemistry. p. 408. W. H. Freeman and Co., S. F. and London.
28. Koster KL, Lynch DV (1992) Solute accumulation and compartmentation during the cold acclimation of puma rye. Plant Physiol. 98: 108–113.
29. McKown R, Kuroki G, Warren G (1996) Cold responses of *Arabidopsis* mutants impaired in freezing tolerance. J. Exp. Bot. 47: 1919–1925.
30. Wanner LA, Junttila O (1999) Cold-induced freezing tolerance in *Arabidopsis*. Plant Physiol. 120: 391–399.
31. Gajewska E, Sklodowska M, Slaba M, Mazur J (2006) Effect of nickel on antioxidant enzyme activities, praline and chlorophyll contents in wheat shoots. Biol. Plant. 50: 653–659.
32. Munns R, Tester M (2008) Mechanisms of salinity tolerance. Ann. Rev. Plant Biol. 59: 651–681.
33. Fowler S, Thomashow MF (2002) *Arabidopsis* transcriptome profiling indicates that multiple regulatory pathways are activated during cold acclimation in addition to the CBF cold response pathway. Plant Cell 14: 1675–1690.
34. Zhu JK (2001) Plant salt tolerance. Trends Plant Sci. 6: 666–671.
35. Xiong L, Shumaker KS, Zhu JK (2002) Cell signaling during cold, drought and salt stresses. Plant Cell 14: 165–183.
36. Fridovich I (1975) Superoxide Dismutase. Ann. Rev. Biochem. 44: 147–159.
37. Dogra V, Ahuja PS, Sreenivasulu Y (2013) Change in protein content during seed germination of a high altitude plant *Podophyllum hexandrum* Royle. J. Proteom. 78: 26–38.
38. Prasad TK, Anderson MD, Martin BA, Stewart CR (1994) Evidence for chilling-induced oxidative stress in Maize seedlings and a regulatory role for hydrogen peroxide. Plant Cell 6: 65–74.
39. Sato Y, Murakami T, Funatsuki H, Matsuba S, Saruyama H, et al. (2001) Heat shock-mediated APX gene expression and protection against chilling injury in rice seedlings. J. Exp. Bot. 52: 145–151.
40. O'Kane D, Gill V, Boyd P, Burdon R (1996) Chilling, oxidative stress and antioxidant responses in *Arabidopsis thaliana* callus. Planta 198: 371–377.
41. Sowinski P, Rudzinska-Langwald A, Adamczyk J, Kubica I, Fronk J (2005) Recovery of maize seedling growth, development and photosynthetic efficiency after initial growth at low temperature. J. Plant Physiol. 162: 67–80.
42. Rymen B, Fiorani F, Kartal F, Vandepoele K, Inze D, et al. (2007) Cold nights impair leaf growth and cell cycle progression in maize through transcriptional changes of cell cycle genes. Plant Physiol. 143: 1429–1438.
43. Bagnall D, Wolfe JOE, King RW (1983) Chill-induced wilting and hydraulic recovery in mung bean plants. Plant Cell Environ. 6: 457–464.

44. Yoshida R, Kanno A, Kameya T (1996) Cool temperature-induced chlorosis in rice plants (II. Effects of cool temperature on the expression of plastid-encoded genes during shoot growth in darkness). Plant Physiol. 112: 585–590.
45. Nayyar H, Chander K, Kumar S, Bains T (2005a) Glycine betaine mitigates cold stress damage in Chickpea. Agro. Sustain. Deve. 25: 381–388.
46. Kaur G, Kumar S, Nayyar H, Upadhyaya HD (2008) Cold stress injury during the pod- filling phase in chickpea (*Cicer arietinum* L.): effects on quantitative and qualitative components of seeds. J. Agro. Crop Sci. 194: 457–464.
47. Ohnishi S, Miyoshi T, Shirai S (2010) Low temperature stress at different flower developmental stages affects pollen development, pollination, and pod set in soybean. Environ. Exp. Bot. 69: 56–62.
48. Kumar S, Malik J, Thakur P, Kaistha S, Sharma KD, et al. (2011) Growth and metabolic responses of contrasting chickpea (*Cicer arietinum* L.) genotypes to chilling stress at reproductive phase. Acta Physiol. Plant. 33: 779–787.
49. Suzuki K, Nagasuga K, Okada M (2008). The chilling injury induced by high root temperature in the leaves of rice seedlings. Plant Cell Physiol. 49: 433–442.
50. Lee HS, Ahsan N, Lee KW, Kim DH, Lee DG, et al. (2007) Simultaneous overexpression of both Cu-Zn superoxide dismutase and ascorbate peroxidase in transgenic tall fescue plants confers increased tolerance to a wide range of abiotic stresses. J. Plant Physiol. 164: 1626–1638.
51. Lu YY, Deng XP, Kwak SS (2010) Over expression of CuZn superoxide dismutase (CuZn SOD) and ascorbate peroxidase (APX) in transgenic sweet potato enhances tolerance and recovery from drought stress. African J. Biotech. 9: 8378–8391.
52. Faize M, Burgos L, Faize L, Piqueras A, Nicolas E, et al. (2011) Involvement of cytosolic ascorbate peroxidase and Cu/Zn-superoxide dismutase for improved tolerance against drought stress. J. Exp. Bot. 62: 2599–2613.
53. Yuanyuan M, Yali Z, Jiang L, Hongbo S (2009) Roles of plant soluble sugars and their responses to plant cold stress. African J. Biotech. 8: 2004–2010.
54. Kwon SY, Jeong YJ, Lee HS, Kim JS, Cho KY, et al. (2002) Enhanced tolerances of transgenic tobacco plants expressing both superoxide dismutase and Ascorbate peroxidase in chloroplasts against methyl viologen-mediated oxidative stress. Plant, Cell Environ. 25: 873–882.
55. Barriere Y, Riboulet C, Mechin V, Pichon M, Cardinal AJ, et al. (2007) Genetics and genomics of lignification in grass cell walls based on maize as a model system. G3-Genes Genome Genet. 1: 133–156.
56. Neill SJ, Desikan R, Clarke A, Hurst RD, Hancock JT (2002) Hydrogen peroxide and nitric oxide as signalling molecules in plants. J. Exp. Bot. 53: 1237–1247.
57. Huang C, Li J, Ke Q, Leonard SS, Jiang BH, et al. (2002) Ultraviolet-induced phosphorylation of p70(S6 K) at Thr(389) and Thr(421)/Ser(424) involves hydrogen peroxide and mammalian target of rapamycin but not Akt and atypical protein kinase C. Cancer Res. 62: 689–697.
58. Yang T, Poovaiah BW (2002) Hydrogen peroxide homeostasis: activation of plant catalase by calcium/calmodulin. Proc. Nat. Acad. Sci. USA 99: 4097–4102.
59. Denness L, McKenna JF, Segonzac C, Wormit A, Madhou P, et al. (2011) Cell wall damage-induced lignin biosynthesis is regulated by a reactive oxygen species- and jasmonic acid-dependent process in *Arabidopsis*. Plant Physiol. 156: 1364–1374.
60. Moura JCMS, Bonine CAV, de Oliveira Fernandes Viana J, Dornelas MC, Mazzafera P (2010) Abiotic and biotic stresses and changes in the lignin content and composition in plants. J. Integrat. Plant Biol. 52: 360–376.

PERMISSIONS

All chapters in this book were first published in PLOS ONE, by The Public Library of Science; hereby published with permission under the Creative Commons Attribution License or equivalent. Every chapter published in this book has been scrutinized by our experts. Their significance has been extensively debated. The topics covered herein carry significant findings which will fuel the growth of the discipline. They may even be implemented as practical applications or may be referred to as a beginning point for another development.

The contributors of this book come from diverse backgrounds, making this book a truly international effort. This book will bring forth new frontiers with its revolutionizing research information and detailed analysis of the nascent developments around the world.

We would like to thank all the contributing authors for lending their expertise to make the book truly unique. They have played a crucial role in the development of this book. Without their invaluable contributions this book wouldn't have been possible. They have made vital efforts to compile up to date information on the varied aspects of this subject to make this book a valuable addition to the collection of many professionals and students.

This book was conceptualized with the vision of imparting up-to-date information and advanced data in this field. To ensure the same, a matchless editorial board was set up. Every individual on the board went through rigorous rounds of assessment to prove their worth. After which they invested a large part of their time researching and compiling the most relevant data for our readers.

The editorial board has been involved in producing this book since its inception. They have spent rigorous hours researching and exploring the diverse topics which have resulted in the successful publishing of this book. They have passed on their knowledge of decades through this book. To expedite this challenging task, the publisher supported the team at every step. A small team of assistant editors was also appointed to further simplify the editing procedure and attain best results for the readers.

Apart from the editorial board, the designing team has also invested a significant amount of their time in understanding the subject and creating the most relevant covers. They scrutinized every image to scout for the most suitable representation of the subject and create an appropriate cover for the book.

The publishing team has been an ardent support to the editorial, designing and production team. Their endless efforts to recruit the best for this project, has resulted in the accomplishment of this book. They are a veteran in the field of academics and their pool of knowledge is as vast as their experience in printing. Their expertise and guidance has proved useful at every step. Their uncompromising quality standards have made this book an exceptional effort. Their encouragement from time to time has been an inspiration for everyone.

The publisher and the editorial board hope that this book will prove to be a valuable piece of knowledge for researchers, students, practitioners and scholars across the globe.

LIST OF CONTRIBUTORS

Germán Robert, Nacira Muñoz and Ramiro Lascano
Instituto de Fisiología y Recursos Genéticos Vegetales, Centro de Investigaciones Agropecuarias-INTA, Córdoba, Argentina
Cátedra de Fisiología Vegetal, Facultad de Ciencias Exactas Físicas y Naturales, Universidad Nacional de Córdoba, Córdoba, Argentina

Mariana Melchiorre
Instituto de Fisiología y Recursos Genéticos Vegetales, Centro de Investigaciones Agropecuarias-INTA, Córdoba, Argentina

Federico Sánchez
Departamento de Biología Molecular de Plantas, Instituto de Biotecnología, Universidad Nacional Autónoma de México, Cuernavaca, Morelos, México

Alejandra Droguett, Paola Krall, M. Eugenia Burgos, Graciela Valderrama, Leopoldo Ardiles and Sergio Mezzano
Division Nephrology, School of Medicine, Universidad Austral de Chile, Valdivia, Chile

Daniel Carpio
Hystopathology Division, School of Medicine, Universidad Austral de Chile, Valdivia, Chile

Bredford Kerr and Katherina Walz
Centro de Estudios Científicos, Valdivia, Chile

Sofia B. Rodriguez and Thomas S. Leyh
The Department of Microbiology & Immunology, The Albert Einstein College of Medicine, Bronx, New York, United States of America

Shunchao Sui, Jie Zhao, Ran Zhang, Chengdong Guo and Ning Li
State Key Laboratory of Agrobiotechnology, China Agricultural University, Beijing, PR China

Jianwu Wang
Beijing GenProtein Biotech Company Ltd., Beijing, PR China Wuxi
Kingenew Biotechnology Co., Ltd., Wuxi, Jiangsu Province, China

Tian Yu
Wuxi Kingenew Biotechnology Co., Ltd., Wuxi, Jiangsu Province, China

Yee Wen Choon, Mohd Saberi Mohamad, Safaai Deris, Chuii Khim Chong and Lian
En Chai
Artificial Intelligence and Bioinformatics Group, Faculty of Computing, Universiti Teknologi Malaysia, Johor, Malaysia

Rosli Md. Illias
Department of Bioprocess Engineering, Faculty of Chemical Engineering, Universiti Teknologi Malaysia, Skudai, Johor, Malaysia

Sigeru Omatu
Department of Electronics, Information and Communication Engineering, Osaka Institute of Technology, Osaka, Japan

Juan Manuel Corchado
Biomedical Research Institute of Salamanca/BISITE Research Group, University of Salamanca, Salamanca, Spain

Yuhui Cai, Xujun Chen and Qikai Xing
Key Laboratory of Plant Pathology, China Agricultural University, Beijing, China

Zongxiu Sun and Yawen Wu
China National Rice Research Institute, Hangzhou, China

Ke Xie, Jing Li and Caihong Du
Zhejiang University,Hangzhou, China

Zejian Guo
Key Laboratory of Plant Pathology, China Agricultural University, Beijing, China
Zhejiang University, Hangzhou, China

Nikenza Viceconte, Tomás McKenna and Maria Eriksson
Department of Biosciences and Nutrition, Center for Biosciences, Karolinska Institutet, Novum, Huddinge, Sweden

Rakesh Kumar Upadhyay
Plant Gene Expression Lab, CSIR-National Botanical Research Institute (Council of Scientific and Industrial Research), Lucknow, India
Sustainable Agricultural Systems Laboratory, USDA-ARS, Beltsville Agricultural Research Center, Beltsville, Maryland, United States of America

Department of Biology, Pennsylvania State University, Harrisburg, Pennsylvania, United States of America

Asmita Gupta, Pravendra Nath and Aniruddha P. Sane
Plant Gene Expression Lab, CSIR-National Botanical Research Institute (Council of Scientific and Industrial Research), Lucknow, India

Sanjay Ranjan, Ruchi Singh and Uday V. Pathre
Department of Plant Physiology, CSIR-National Botanical Research Institute (Council of Scientific and Industrial Research), Lucknow, India

Nikolay Vasilev and Christian Schmitz
Department Plant Biotechnology, Fraunhofer Institute for Molecular Biology and Applied Ecology IME, Aachen, Germany

Ulrike Grömping
Department II–Mathematics, Physics and Chemistry, Beuth University of Applied Sciences, Berlin, Germany

Rainer Fischer
Department Plant Biotechnology, Fraunhofer Institute for Molecular Biology and Applied Ecology IME, Aachen, Germany
Institute for Molecular Biotechnology, RWTH Aachen University, Aachen, Germany

Stefan Schillberg
Department Plant Biotechnology, Fraunhofer Institute for Molecular Biology and Applied Ecology IME, Aachen, Germany
Institute for Phytopathology and Applied Zoology, Justus-Liebig University Giessen, Giessen, Germany

Zhi-Yuan Zhang, Chaoyun Li, Caroline Zug and Hermann J. Schluesener
Division of Immunopathology of the Nervous System, Institute of Pathology and Neuropathology, University of Tuebingen, Tuebingen, Germany

Daisuke Tsugama
Asian Natural Environmental Science Center (ANESC), The University of Tokyo, Nishi-tokyo-shi, Tokyo, Japan
Laboratory of Plant Molecular Genetics, Graduate School of Agricultural and Life Sciences, The University of Tokyo, Bunkyo-ku, Tokyo, Japan

Shenkui Liu
Alkali Soil Natural Environmental Science Center (ASNESC), Northeast Forestry
University, Harbin, China

Tetsuo Takano
Asian Natural Environmental Science Center (ANESC), The University of Tokyo, Nishi tokyo-shi, Tokyo, Japan

Yanfang Ren
Joint Bioenergy Institute, Lawrence Berkeley National Laboratory, Berkeley, California, United States of America
College of Agriculture, Guizhou University, Guiyang, China

Sara Fasmer Hansen and Jane Lau
Joint Bioenergy Institute, Lawrence Berkeley National Laboratory, Berkeley, California, United States of America

Berit Ebert
Joint Bioenergy Institute, Lawrence Berkeley National Laboratory, Berkeley, California, United States of America
Department of Plant and Environmental Sciences, Faculty of Science, University of Copenhagen, Frederiksberg, Denmark

Henrik Vibe Scheller
Joint Bioenergy Institute, Lawrence Berkeley National Laboratory, Berkeley, California, United States of America
Department of Plant and Microbial Biology, University of California, Berkeley, California, United States of America

Navadagi B. Devanna, Joshitha Vijayan and Tilak R. Sharma
National Research Centre on Plant Biotechnology, Indian Agricultural Research Institute, New Delhi, India

Olaf Czarnecki, Wellington MucheroGerald A. Tuskan and Jin-Gui Chen
Biosciences Division, Oak Ridge National Laboratory, Oak Ridge, Tennessee, United States of America

Jun Yang
Biosciences Division, Oak Ridge National Laboratory, Oak Ridge, Tennessee, United States of America
National Key Laboratory of Plant Molecular Genetics, Institute of Plant Physiology and Ecology, Shanghai Institutes for Biological Sciences, Chinese Academy of Sciences, Shanghai, China

Xiaoping Wang
Biosciences Division, Oak Ridge National Laboratory, Oak Ridge, Tennessee, United States of America
Key Laboratory of Molecular Epigenetics of MOE, Institute of Genetics and Cytology, Northeast Normal University, Changchun, China

Shucai Wang
Key Laboratory of Molecular Epigenetics of MOE, Institute of Genetics and Cytology, Northeast Normal University, Changchun, China

Jingang Liang, Jun Ji, Fang Meng, Xiaobo Zheng and Zhengguang Zhang
Department of Plant Pathology, College of Plant Protection, Nanjing Agricultural University, and Key Laboratory of Integrated Management of Crop Diseases and Pests, Ministry of Education, Nanjing, China

Shi Sun and Cunxiang Wu
The National Key Facility for Crop Gene Resources and Genetic Improvement (NFCRI), MOA Key Laboratory of Soybean Biology (Beijing), Institute of Crop Science, The Chinese Academy of Agricultural Sciences, Beijing, China

Haiying Wu and Mingrong Zhang
Nanchong Academy of Agricultural Science, Nanchong, China

Cuizhu Zhao
The Key Laboratory of Plant Cell Engineering and Germplasm Innovation, School of Life Sciences, Shandong University, Jinan, Shandong, China
College of Agronomy, Northwest A&F University, Yangling, Shanxi, China

Jin Xu
Key Laboratory of Agricultural Water Resources, Center for Agricultural Resources Research, Institute of Genetics and Developmental Biology, Chinese Academy of Sciences, Shijiazhuang, China

Qiang Li, Shuo Li, Peng Wang and Fengning Xiang
The Key Laboratory of Plant Cell Engineering and Germplasm Innovation, School of Life Sciences, Shandong University, Jinan, Shandong, China

Aris Rousakis
Biomedical Research Foundation of the Academy of Athens, Center of Basic Research II, Athens, Greece
Faculty of Medicine, University of Athens, Athens, Greece

Anna Vlanti and Popi Syntichaki
Biomedical Research Foundation of the Academy of Athens, Center of Basic Research II, Athens, Greece

Fivos Borbolis and Fani Roumelioti
Biomedical Research Foundation of the Academy of Athens, Center of Basic Research II, Athens, Greece
Faculty of Biology, School of Science, University of Athens, Athens, Greece

Marianna Kapetanou
Biomedical Research Foundation of the Academy of Athens, Center of Basic Research II, Athens, Greece
Department of Biology, School of Science and Engineering, University of Crete, Heraklio, Crete, Greece

Yukihiko Hiroshima
AntiCancer, Inc., San Diego, California, United States of America
Department of Surgery, University of California San Diego, San Diego, California, United States of America
Yokohama City University Graduate School of Medicine, Yokohama, Japan

Ali Maawy and Michael Bouvet
Department of Surgery, University of California San Diego, San Diego, California, United States of America

Mohamed K. Hassanein, Shinji Miwa, Mako Yamamoto, Fuminari Uehara, Shuya Yano and Robert M. Hoffman
AntiCancer, Inc., San Diego, California, United States of America
Department of Surgery, University of California San Diego, San Diego, California, United States of America

Rhiana Menen
AntiCancer, Inc., San Diego, California, United States of America

Masashi Momiyama, Takashi Murakami, Itaru Endo, Ryutaro Mori, Ryusei Matsuyama, Takashi Chishima, Kuniya Tanaka and Yasushi Ichikawa
Yokohama City University Graduate School of Medicine, Yokohama, Japan

Zhengru Li, Chunqiao Zhao, Fei Liu and Fengcheng Li
National Key Laboratory of Crop Genetic Improvement, Biomass and Bioenergy Research Centre, College of Plant Science and Technology, Huazhong Agricultural University, Wuhan, China

Yi Zha, Can Wan, Shengli Si, Rui Zhang and Bin Yu
National Key Laboratory of Crop Genetic Improvement, Biomass and Bioenergy Research Centre, College of Life Science and Technology, Huazhong Agricultural University, Wuhan, China,

Zili Yi
Department of Biotechnology, Hunan Agricultural University, Changsha, China

Ning Xu
National Key Laboratory of Crop Genetic Improvement, Biomass and Bioenergy Research Centre, College of Plant Science and Technology, Huazhong Agricultural University, Wuhan, China
Jiangsu Key Laboratory for Biomass-based Energy and Enzyme Technology, Huaiyin Normal University, Huaian, China

Liangcai Peng
National Key Laboratory of Crop Genetic Improvement, Biomass and Bioenergy Research Centre, College of Plant Science and Technology, Huazhong Agricultural University, Wuhan, China
National Key Laboratory of Crop Genetic Improvement, Biomass and Bioenergy Research Centre, College of Life Science and Technology, Huazhong Agricultural University, Wuhan, China,

Qing Li
National Key Laboratory of Crop Genetic Improvement, Biomass and Bioenergy Research Centre, College of Science, Huazhong Agricultural University, Wuhan, China

Yuke Geng
Key Laboratory of Crop Gene Resources and Germplasm Enhancement, Ministry of Agriculture/ Institute of Crop Science, Chinese Academy of Agricultural Sciences,
Beijing, China
College of Biological sciences, China Agricultural University, Beijing, China

Binshuang Pang
Beijing Engineering and Technique Research Center of Hybrid Wheat, Beijing Academy of Agricultural and Forestry Sciences, Beijing, China

Saijun Tang
College of Biological sciences, China Agricultural University, Beijing, China

Xueyong Zhang, Tian Li and Chenyang Hao
Key Laboratory of Crop Gene Resources and Germplasm Enhancement, Ministry of Agriculture/ Institute of Crop Science, Chinese Academy of Agricultural Sciences,
Beijing, China

Amrina Shafi, Paramvir Singh Ahuja and Yelam Sreenivasulu
Division of Biotechnology, CSIR-Institute of Himalayan Bioresource Technology, Palampur, Himachal Pradesh, India
Academy of Scientific and Innovative Research (AcSIR), CSIR- Institute of Himalayan Bioresource Technology, Palampur, Himachal Pradesh, India

Vivek Dogra and Tejpal Gill
Division of Biotechnology, CSIR-Institute of Himalayan Bioresource Technology, Palampur, Himachal Pradesh, India

Index